geomorphology and natural hazards

geomorphology and natural hazards

Proceedings of the 25th Binghamton Symposium in Geomorphology, held september 24 – 25, 1994 at SUNY, Binghamton, USA

Edited by

M. Morisawa

State University of New York at Binghamton, Department of Geological Sciences and Environmental Studies, P.O. Box 6000, Binghamton, NY 13902-6000, USA

Reprinted from *Geomorphology*, Volume 10, Nos, 1–4

ELSEVIER

Amsterdam — Lausanne — New York — Oxford — Shannon — Tokyo 1994

ELSEVIER SCIENCE B.V.
Sara Burgerhartstraat 25
P.O. Box 211, 1000 AE Amsterdam, The Netherlands

ISBN 0-444-82012-4

This book is printed on acid-free paper.

Printed in The Netherlands

Dedication

Marie Morisawa (1919–1994)

As this volume went to press, Professor Marie Morisawa was killed in an automobile accident on June 10, 1994. She will be deeply missed by her friends and colleagues. But some of her contributions to geomorphology, as exemplified by this proceedings volume, will continue to help others strive for excellence far into the future. Marie will also be remembered for many other activities by those who knew her far better than I. In my view, though, two of Marie's most important contributions were her mentoring of women in the earth sciences, and her cooperation with Don Coates in establishment of the Binghamton Geomorphology Symposium.

Throughout her career, and especially in her later years, Marie served as an important role model for women seeking careers in the earth sciences, especially in geomorphology. She was rare in the earth sciences: a senior woman in a male-dominated discipline. She had risen to the top of her profession and gained international recognition for her educational and research accomplishments. In recognition of this she received the Outstanding Educator Award from the Association of Women Geoscientists in 1992. She recognized the importance and uniqueness of her position, and she actively encouraged and sought out women to be graduate students, to contribute to the symposia (as is clear from her remarks in the Preface to this volume), and to achieve prominent positions within the academic and scientific communities (such as governing boards of national organizations).

Marie will probably best be remembered, though, for establishing the Binghamton Geomorphology Symposium along with Don Coates. From rather humble beginnings in 1970, during her first year on the faculty at the State University of New York at Binghamton, she and Don (later assisted by a steering committee) established a program that has gained increasing international attention and visibility. After hosting (and often organizing) the conferences for 10 years, Marie and Don sent the series "on the road". Marie brought it back to Binghamton for the 15th meeting, and I convened the 21st symposium (with her invaluable guidance) for the 20-year anniversary in 1990. This 25th symposium could have stood as a culminating effort by Marie, but that is not how she viewed it. She was already involved with two colleagues in the planning process for the 28th symposium in conjunction with the International Geomorphology Congress in Italy in 1997.

This symposium meant a great deal to Marie because of how the series has helped the development of geomorphology. Her untimely death ends active participation, but her goals and values will be carried on by those she actively involved in so many different aspects of the profession. Marie developed an exciting program for this year's meeting, one which reflects the breadth of her own research in geomorphology and her commitment to applying geomorphic research to the public good by evaluating potential hazards. I am honored and challenged by the task of hosting this meeting in Marie's absence. The quality of the papers in this volume stands as a lasting testimony to her efforts at achieving excellence.

While we will all miss Marie, please share all of your fond memories of her with colleagues. She will certainly live on in this conference and would want us to continue her enthusiasm for geomorphology and the collegial interaction that occurs during the symposium. Marie's dedication to geomorphology shall long be remembered.

PETER L.K. KNUEPFER
Binghamton, New York

Preface

Twenty-five years ago when I came to the State University of New York at Binghamton, Don Coates and I decided it might be a good idea to have a conference where a group of geomorphologists could meet rather informally and discuss some topic of interest. Our goal was to have eminent researchers present their work and allow sufficient time for informal interaction and development of personal relationships among the geomorphic community. It seems we have succeeded.

After the tenth meeting in 1979, the symposium went "on the road" in response to an overwhelming sentiment from the participants to keep the event going and take the burden from Don and me. A steering committee was established to solicit and screen proposals and determine the hosts and topics of future meetings. Every now and again the symposium has returned "home", as it has this year for the 25th anniversary gathering.

The topic of the first symposium was Environmental Geomorphology and it seems we have gone full circle with this year's topic *Geomorphology and Natural Hazards*. Over the years the Binghamton Symposia have become internationally known and we are proud to say that ten of our twenty-two speakers are from abroad. In addition, we have gathered the largest number of women presenters, five. Disciplines of the speakers are geologists, geographers and engineers from academia, industry and government.

Humans live in a dynamic environment surrounded by natural hazards: earthquakes, hurricanes, landslides, tsunamis, floods and volcanic eruptions are the most spectacular. More unobtrusive, but just as disastrous over the long term is erosion of the land by streams and waves. The United Nations and the US National Academy of Sciences have declared the 90's as the decade of Natural Hazards and aptly so, as recent headlines have emphasized. During this decade it is hoped that we will come to a better understanding of the causes and effects of such natural hazards. It is to this end that we have organized this symposium.

Because of a growing population in hazardous areas, natural disasters claim an increasingly higher toll in destruction and loss of lives, both in the United States and world-wide. The United States, for example, have undergone a variety of hazards in 1993–94: the 1993 summer flooding of the Mississippi basin, the New Jersey, Oregon and California earthquakes, tornadoes throughout the south and midwest, landslides and debris flows in California, and winter ice-storms in the south.

Some of these hazards, such as hurricanes, tornadoes and floods, occur regularly and are well recognized. Others such as earthquakes and tsunamis are intermittent but just as ominous. Although we are accustomed to volcanic eruptions in Hawaii, those of Mt. St. Helens awakened the United States to the volcanic hazards in the Pacific Coastal states. Seventy million people in the United States face significant risk from earthquakes and associated hazards. The recent Northridge quake in California (January 17, 1994, mag. 6.8) resulted in property damage of over $30 billion. Coastal storms on Long Island, New York in 1992 caused over $300 million in damage. The Mississippi River flooding cost more than $20 billion economically. Landslides in the United States are common, costing $1–2 billion in economic losses yearly.

Recent tsunamis in Nicaragua, Indonesia and Japan have caused significant numbers of deaths and destruction. Volcanic eruptions (e.g. in Chile, Guatemala, Japan, the Philippines and New Guinea) and earthquakes (in e.g. India, Guam and Japan) are recent occurrences plaguing nations world-wide.

Attempts of humans to cope with these natural disasters have taken many forms. Early humans were

SSDI 0169-555X(94)00020-R

surrounded by spirits, some of whom were malevolent and brought on disaster. These humans coped with catastrophic events in two ways, appeasement and/or avoidance. One way was to appease the angry spirit who brought on the disaster and/or one could declare the disaster area ''taboo'' or banned from approach for the protection of the group.

Ancient Greeks and Romans envisaged volcanoes as the home of Vulcan whose forges sent up fiery blasts. Eruptions of Kilauea were brought on by the anger of Madame Pele, the volcano goddess. Often these gods/goddesses were appeased by human sacrifice. The death of one to save many.

The usual Euro-American attitude toward natural disasters has been one of controlling nature. A 1989 publication of the US Geological Survey on the environment was entitled ''Nature to be Commanded''. King Canute ordered the incoming tide to stop, to no avail. And our efforts to control natural disasters have been about as fruitless as were his. We have subjected the Mississippi (and other rivers) to numerous controls: levees, cutoffs, channelization and damming, yet in 1993 we had the greatest flood ever in that watershed. Now we are reexamining our attitude and turning more toward prediction and ''taboo'', i.e. avoidance by ordinance and planning.

The heavy toll in life and economy taken by these natural hazards can be reduced. Although we may seldom be able to prevent the hazard, the disastrous results may be predicted, avoided or mitigated. There is already much knowledge available on causes and effects of natural hazards that can do much to alleviate their disastrous results. An understanding and anticipation of the hazard and its effects will help us to cope with it in a better way.

Because many of these hazards are associated with surface features, it was thought that the latest research on geomorphic characteristics and processes associated with natural hazards would be a most pertinent theme for a symposium.

The presentations cover a wide gamut of types of disasters throughout the world, bringing before us the research and applications of studies in the United States and other countries. We open with a group of papers which give a basic background and philosophy of approaching an understanding of natural disasters. Gares et al. discuss paradigms for natural hazards research; Scheidegger presents a view of hazards as singularities in geomorphic systems, and Rosenfeld suggests a framework for a comprehensive, interdisciplinary approach to studying natural hazards.

We then have presentations of natural hazards in coastal areas. Coch gives us an understanding of hurricanes and their geomorphic effects; Lawrence presents the results of erosion along a developed shoreline; Dawson shows the geomorphic results of tsunami runup and backwash.

Moving inland, we have a discussion of the types of disasters occurring in mountainous regions of Japan as presented by Ohmori and Shimazu. Evans and Clague discuss catastrophic geomorphic processes brought about in mountainous areas by recent climatic changes. Schumm then chastises us on erroneous perceptions of fluvial hazards. After an overall view of flooding by Baker, we examine flooding in specific cases: in India by Kale et al. and in the Mississippi basin by Prestegaard et al. Boardman et al. discuss erosion of agricultural land in the United Kingdom and the European continent. Atwood then applies geomorphic studies to the flooding of closed lake basins using Great Salt Lake as an example.

The subject of landslides is presented in papers by Gerrard (in the Himalayas) and DeGraff (in the Sierra Nevada Mts., USA). Bull et al. are concerned with the dating of coseismic landslides; Keefer with earthquake induced mass wasting, slope processes and long term denudation rates. Olig et al. discuss earthquakes along the Oquirrh fault zone, Utah.

The disastrous effects of permafrost melting as a result of global warming is forecast for Eurasia by Demek. And finally, a practical approach using geomorphology and GIS in hazard research is presented by Meijia-Navarro et al.

We hope this symposium volume is helpful in gaining an understanding of natural hazards and their geomorphic relations. Such an understanding is needful for wise environmental planning in coping with disasters.

M. MORISAWA
Binghamton, NY, USA

GEOMORPHOLOGY AND NATURAL HAZARDS

Proceedings of the 25th Binghamton Symposium in Geomorphology, held September 24–25, 1994 at SUNY, Binghamton, N.Y., USA

Edited by

M. MORISAWA

State University of New York at Binghamton, Department of Geological Sciences and Environmental Studies, P.O. Box 6000, Binghamton, NY 13902-6000, USA

CONTENTS

ELSEVIER

Geomorphology 10 (1994) 1–18

GEOMORPHOLOGY

Geomorphology and natural hazards

Paul A. Gares[a], Douglas J. Sherman[b], Karl F. Nordstrom[c]

[a]Department of Geography, East Carolina University, Greenville, NC 27858, USA
[b]Department of Geography, University of Southern California, Los Angeles, CA, USA
[c]Institute of Marine and Coastal Sciences, Rutgers University, New Brunswick, NJ 08903, USA

Received March 7, 1994; revised March 30, 1994; accepted April 2, 1994

Abstract

Natural hazards research was initiated in the 1960's by Gilbert White and his students who promulgated a research paradigm that involved assessing risk from a natural event, identifying adjustments to cope with the hazard, determining people's perception of the event, defining the process by which people choose adjustments, and estimating the effects of public policy on the choice process. Studies of the physical system played an important role in early research, but criticisms of the paradigm resulted in a shift to a prominence of social science. Geomorphologists are working to fill gaps in knowledge of the physical aspects of individual hazards, but use of the information by social scientists will only occur if information is presented in a format that is useful to them. One format involves identifying the hazard according to seven physical parameters established by White and his colleagues: magnitude, frequency, duration, areal extent, speed of onset, spatial dispersion, and temporal spacing.

Geomorphic hazards are regarded as related to landscape changes that affect human systems. The processes that produce the changes are rarely geomorphic in nature, but are better regarded as atmospheric or hydrologic. An examination of geomorphic hazards in four fields — soil erosion, mass movement, coastal erosion and fluvial erosion — demonstrates that advances in those fields may be evaluated in terms of the seven parameters. Geomorphologists have contributed to hazard research by focusing on the dynamics of the landforms. The prediction of occurrence, the determination of spatial and temporal characteristics, the impact of physical characteristics on people's perception, and the impact of physical characteristics on adjustment formulation. Opportunities for geomorphologists to improve our understanding of geomorphic hazards include research into the characteristics of the events particularly with respect to predicting the occurrence, and increased evaluation of the impact of human activities on natural systems.

1. Introduction

The field of natural hazards research has a rich history in geography, appropriately so because it involves conflicts between physical processes and human systems. One area of conflict concerns landscape development processes that can also have a catastrophic impact on human systems. People are killed and property is damaged or destroyed by extreme geomorphologic events. An earthquake-induced debris flow from the flanks of Mt. Huarascan in Peru killed approximately 25,000 people in 1970 (Taype, 1979). Coastal erosion and flooding resulting from the "Ash Wednesday" northeaster of 1962 produced US$ 33.7 million in damages along the New Jersey and Delaware coastline (U.S. Army Corps of Engineers, 1962). Millions of dollars are spent to control, reduce, or eliminate the effects of similar geomorphologic events. These represent but a few of the ways that geomorphic events can impact or threaten human social systems.

Initially, studies of natural hazards focused on extreme events of nature and the way that they affect

SSDI 0169-555X(94)00013-H

human systems. An important goal of hazards research was to determine how careful planning could mitigate the effects of the hazardous events. The combination of both physical and human systems has afforded opportunities to scientists from many disciplines to undertake research in this field. Although both physical and social scientists recognize the importance of each other, each brings a bias for their area of expertise to the research table, and this tends to minimize the contribution of each group of scientists to the overall advancement of the field. It is worthwhile to review periodically the advancements made in fields of research to place the contributions of certain groups into the context of the whole field and to identify areas where additional contributions can be made by geomorphologists.

We have three purposes in writing this paper: (1) to provide an overview of the field of natural hazard research; (2) to review contributions made by geomorphologists to natural hazards research; and (3) to identify areas of promising research. Our primary contention is that, although geomorphologists have made tremendous contributions to understanding the processes that shape the earth's surface and have added to the large literature on extreme events of nature, we have not placed those studies into the broader natural hazards research paradigm. As a result, natural hazards research has developed a strong social science flavor, and the physical component has been reduced in stature despite its prominence in the original hazards paradigm. Geomorphologists can help to reinforce the importance of the physical component by effectively integrating their work into the natural hazards paradigm.

2. Natural hazards paradigms

Early research into natural hazards was pioneered by Gilbert White and several of his students. White's initial work resulted in the discovery that losses from flooding had increased between 1942 and 1956 despite an expenditure of over US$ 5 billion on flood control works such as dams, levees and channel improvements (White et al., 1958). White concluded that increasing numbers of people were settling in seemingly protected flood-prone areas, resulting in more damage when flooding occurred. These early observations led White to develop a research paradigm that involved 5 components (White, 1974a):

(1) estimate the extent of human occupancy in areas subject to natural hazards;

(2) determine the range of possible adjustments by social groups to those extreme events;

(3) examine how people perceive the extreme events and resultant hazards;

(4) examine the process of choosing damage-reducing adjustments;

(5) estimate the effects of varying public policy upon that choice process.

This paradigm involves a classification and categorization of hazards, of people affected by the hazards, and the range of responses (termed adjustments) available for the specific hazard. The original research largely consisted of case studies that examined communities' response to specific catastrophic events (e.g. White, 1974b). As the research paradigm developed over time, risk assessment became an important component of the analysis and involved predicting the vulnerability of people on the basis of probability of occurrence of an event (Whyte and Burton, 1980). Implicit in the hazards paradigm is the notion that individuals (or perhaps communities) make rational decisions regarding their adjustments to the problem. White (1960) suggests that individuals and communities use a theoretical analytical methodology to choose the adjustment(s) that best applies to their case. There is a further recognition that the choice adopted by the individual or community is a function of how that entity perceives the hazard (Kates, 1971; Mitchell, 1974). The choice could also be affected by public policies established by the government.

During the last 10–15 years, challenges were made to White's original paradigm. Hewitt (1983) argues that it is a mistake to focus research on the extreme event. He suggests that the extreme event is simply at one end of the range of activities that represent a person's life, and to treat the extreme event as separate from "ordinary life" results in a misunderstanding of the factors that affect hazard response. This response may be as much a function of the way society is organized as of an individual's assessment of available response options. Hewitt asserts that the "dominant view" (White's paradigm) focuses on the physical event by emphasizing monitoring, prediction, and post-disaster recovery that includes immediate relief efforts and long-term reconstruction. This view, Hewitt feels, restores the very infrastructure that is partly responsible for the human disasters associated with normal (in the long-term) geophysical events. Hewitt encourages the traditional hazards researchers to extend their analysis to understand how socio-political and economic aspects of society contribute to creating disasters.

Marxist theorists have extended Hewitt's (1983) view of natural hazards by advocating that natural hazards are caused by the economic system that forces people to occupy hazardous places that are beneficial to the welfare of the powerful but that are detrimental to the common people (Waddell, 1977, 1983; Watts, 1983). Marxists argue that by referring to natural disasters as "Acts of God", the blame is placed on nature, but in doing so one overlooks the role of economic and political structure in creating the disaster. This role is best exemplified in the Third World where increasing population, and lack of resources and capital are typically viewed as problems of underdevelopment. To Marxists, these problems are not symptoms of underdevelopment, but are effects of underdevelopment (Susman et al., 1983). In this view, underdevelopment leads to marginalization that involves the creation of a class of people who are forced to occupy undesirable locations because of their lack of capital or resources. Thus, marginals live on steep hillslopes, on floodplains, on mudflats of tidal estuaries, in drought-prone regions or in other locations that are hazard-prone. The problem of marginalization is applicable to developed countries (Warrick, 1983), although perhaps at a different level. A key to understanding marginalization is placing the existing system into a longer time frame, so the historical context provides background explanation for the problem, and solutions involve large-scale socio-political changes rather than adoption of adjustments suited to local constraints.

These theoretical assessments have led some traditional hazards researchers to develop new frameworks for hazards analysis. Mitchell et al. (1989) propose a contextual model of natural hazards, with four basic components: (1) physical processes; (2) human populations; (3) adjustments; and (4) net losses. Each component in this model interacts with others in a cause–effect relationship. Thus, physical processes affect human populations, but the populations in turn may affect the physical processes, and feedbacks occur between adjustments and the physical processes and between net losses and human populations. The model differs little from the basic White model except that each of the four components undergoes changes through time (Mitchell et al., 1989). As human populations grow, they may exacerbate the physical processes, while technology develops new adjustments to the hazard. The components of the model are modified by the timing of the events and by the social, economic, political, spatial and physical context of the hazard. The timing context is exemplified by a hurricane that arrives at a coastal resort on a summer holiday weekend rather than at a time when fewer people would be at the coast. The spatial context of a hazard is evident when a hurricane makes landfall on a heavily developed coastline rather than an unoccupied one. In advocating the contextual approach, Mitchell et al. (1989) incorporate some of the socio-political variables in their analysis, but they do not explain the vulnerability of the population in terms of the historical economic framework as the Marxists do. The focus on the context of the event conforms more to Hewitt's human ecology framework.

Palm (1990) goes further in incorporating socio-political variables in her analysis of earthquake hazards, but she does not embrace the Marxist point of view. Palm recognizes two levels of hazards analysis that she terms micro-level and macro-level. The former primarily involves analysis that follows the traditional hazards paradigm; the latter focuses on the roles of history and economics in explaining vulnerability. Palm recognizes that the micro-level yields management schemes that can alleviate vulnerability in the short-term, but she also believes that long-term solutions must come about as a result of social changes. She suggests that understanding the hazard must involve analysis at all levels simultaneously.

In a recent review of the natural hazards paradigm, Burton, Kates and White (1993) discuss developments in the field since their previous overview (Burton et al., 1978), and, like Palm, they recognize the complexity of the field. They discuss two new theoretical ideas — the lessening and catastrophe concepts. The "lessening" concept implies that societies are able to cope with recurrent hazards more successfully because of the development of improved adjustments to the hazard. The same event through time produces a diminishing amount of losses and damages, but this situation creates the potential for greater catastrophe during the extraordinary event. The protection afforded by the improved technology attracts additional inhabitants to the hazard zone, so that when the extraordinary event occurs, more people and structures are affected, creating a catastrophe.

This overview of natural hazards research frameworks shows the development of the theoretical side of the field from a basic paradigm to a complex one. The changes in the paradigm have been fueled by the criticisms of human ecologists who felt that the early paradigm was too focused on the physical event at the expense of the socio-economic factors. Contributions of Mitchell et al. (1989) and Palm (1990) acknowledge the role of the physical event while giving greater emphasis to the human side of the problem. Although the paradigm shift would seem to reduce the role of the physical scientist in hazard research, the question of the amount of risk associated with various physical phenomena still remains a key aspect. Of particular importance is the issue of increased risk due to human activities (Burton et al., 1993; Mitchell et al., 1989).

3. Hazards and physical science

Over the last thirty years, physical science has made great strides forward in explaining extreme natural events, but the role of physical science in natural hazards research remains fairly undefined. Although White and his colleagues recognized that "social systems of resource use cannot operate independently of atmospheric, hydrologic, geomorphic, and biotic systems" (White, 1974a), the physical system is only indirectly referred to in the five components of the natural hazards paradigm. However, understanding the physical system is crucial to fulfilling the goals of the paradigm, so it is impossible to accomplish the first goal (the estimation of the extent of human occupancy) if the spatial and temporal attributes of the hazard are not known. In order to fulfill the second goal (determining the range of adjustments), it is necessary to understand the physical characteristics of the extreme event. The physical characteristics of the hazard play smaller roles in the remaining three goals, but they are used to establish engineering design criteria for specific adjustments (e.g. U.S. Army Corps of Engineers, 1984), to partially explain people's perception of the hazard (e.g. Mitchell, 1974) and to establish public policy for a hazard (e.g. Platt, 1976).

Burton et al. (1978) emphasize the importance of the physical event to natural hazards research by describing seven physical parameters of an event that affect human response and a qualitative scale for each parameter: magnitude (high–low); frequency (frequent–rare); duration (long–short); areal extent (widespread–limited); speed of onset (slow–fast); spatial dispersion (diffuse–concentrated), and temporal spacing (regular–random). The nature of most of these parameters is obvious to most geomorphologists. Magnitude refers to the amount of energy involved in the event. Frequency implies how often an event of a given magnitude occurs in a given amount of time. Duration is the length of time the event persists. The areal extent refers to the physical space affected, for example the path of hurricane Andrew across south Florida and south Louisiana. Speed of onset is the amount of time between the first appearance of the event and its peak. Spatial dispersion applies to the space likely to be affected by all events of a specific hazard type. Thus, the spatial dispersion of landslide hazards is relatively concentrated, but it is more diffuse in the case of soil erosion. The temporal spacing of events can be relatively regular in the case of seasonal or cyclic hazards such as hurricanes and tornadoes, or random, as in the case of earthquakes.

Each hazard event can be categorized qualitatively for each variable. The combined evaluation allows the event to be placed along a continuum that represents the physical characteristics of the event in terms of a pervasive to intensive scale. Relatively frequent events of long duration, widespread areal extent, slow speed of onset, diffuse spatial dispersion and regular temporal spacing are at the pervasive end of the continuum (e.g. soil erosion, sea level rise, or drought). Intensive hazards in time and space, located at the opposite end of the continuum, include earthquakes, volcanic eruptions or tornadoes. Burton et al. (1993) recognize that hazards often assume different intensive–pervasive characteristics at different times or at different locations. A tropical cyclone might affect a limited section of coastline for a short period of time, as hurricane Bob did in New England in 1991, or a large inland region in addition to the landfall location, as tropical storm Agnes did in 1972.

Sherman and Nordstrom (1994) have argued that causal linkages also affect the event. This concept suggests that an individual extreme event may initiate another extreme event of a different type. Although we argue later that geomorphic hazards are associated principally with landform response, we recognize that there must always be a triggering mechanism (process) that initiates the hazard. Landslides, for example, are causally linked with earthquakes, or heavy precipitation. Soil erosion is causally linked with aeolian processes, surface hydrology, or rainsplash. The study of processes is, therefore, integral to understanding geomorphological hazards. Causal linkages can also involve more normal activities that contribute to the occurrence of an extreme event when a threshold is crossed. Thus, changing land-use on a watershed affects the magnitude and frequency characteristics of the discharge within the stream. Although the resulting hazard is related to the flow of water in the river, the factor that causes the hazard is land-use change. This concept of causal linkages between hazards components is consistent with Mitchell et al.'s (1989) notion of hazards in context.

An alternative way of evaluating hazards is to consider the "hazardousness of a place" (Hewitt and Burton, 1971; Palm and Hodgson, 1993) This concept focuses on a specific location rather than on a specific hazard, and evaluates the likelihood of all potential hazards for that location. Thus, the Los Angeles area could be evaluated for the potential of earthquakes, slope failures, coastal storm damage, or tsunami. The New Madrid area along the Mississippi river could be evaluated for the potential of floods, earthquakes, droughts, blizzards, tornadoes, amongst others.

Geomorphologists are familiar with many of the parameters used by hazards researchers to evaluate extreme events of nature, but few geomorphologists have adopted their methods of evaluation. Other than frequency/magnitude analyses, the variables that describe the physical nature of the event are rarely evaluated as distinct hazard characteristics, although they may be referred to indirectly in the investigation. The notion of hazardousness of a place is rarely considered by geomorphologists, perhaps due to a reluctance to venture beyond their perceived area of expertise. A notable exception is Cooke's (1984) *Geomorphological Hazards in Los Angeles* which focuses on slope and river channel processes in an area affected by a multitude of hazards. In this paper, we advocate that physical scientists adopt the hazards paradigm in order to promote comparability between studies.

4. Defining geomorphic hazards

Although many social scientists place the physical event into a black box and focus on the societal aspects of hazards, geomorphologists will wish to define their field precisely to emphasize their contribution. From a strict geomorphological point of view, geomorphic hazards must be regarded as the suite of threats to human resources arising from instability of the surface features of the earth. The threat arises from landform response to surficial processes, although the initiating processes may originate at great distances from the surface. Thus, this definition excludes earthquakes, per se, but not the landform response, e.g. slope failure, to earthquakes. Similarly, sea level rise is not a geomorphic hazard, but enhanced coastal erosion as a result of sea level rise is. According to this definition, processes such as wind, floods, and tsunami are not geomorphic hazards because they are not geomorphic processes until they change the landscape. They are better classified as atmospheric or hydrologic hazards, although many of these processes are associated with geomorphic events. It is difficult, for instance, to distinguish the effects of coastal erosion from coastal flooding, as the latter often produces the former.

This definition de-emphasizes the importance of the high-magnitude/low-frequency event in geomorphic hazards because surficial changes are ongoing. An examination of the types of hazardous events that have occurred over the last 5 years (Table 1) shows that the high-magnitude events, such as earthquakes or hurricanes, have occurred most frequently and have resulted in the largest number of casualties, but the true geomorphic hazards, such as landslides, occurred less frequently and had considerably fewer deaths. The high-magnitude event often produces spectacular change, but events of moderate frequency often do as much or more work cumulatively over the longer term (Wolman and Miller, 1960). Thus, the geomorphic component of hazards tends to be more at the pervasive end of the hazards continuum than hydrologic or seismic components. In addition to lower magnitude and higher frequency, geomorphic hazards would generally

Table 1
Frequency and number of deaths associated with selected catastrophic events worldwide, 1988–1992. (Source: New York Times Index)

Year		Avalanches	Landslides	Volcanoes	Earthquakes	Tornadoes	Floods	Hurricanes	Tsunami	Total
1988	Freq.	5	2	0	29	10	3	14	0	63
	Deaths	49	323	0	26,771	32	10,000 +	2958	0	40,133 +
1989	Freq.	0	3	3	25	13	3	11	0	58
	Deaths	0	286	0	1336	691	0	812	0	3125
1990	Freq.	4	1	4	30	11	6	11	0	67
	Deaths	46	21	0	37,720	53	158	351	0	38,349
1991	Freq.	5	1	5	24	6	0	9	0	50
	Deaths	28	64	141	526	55	0	143,629	0	144,443
1992	Freq.	2	2	5	26	10	0	6	2	53
	Deaths	217	330	1	3637	37	0	23	100 +	4345 +
Total	Freq.	16	9	17	134	50	12	51	2	291
	Deaths	340	1024	142	69,990	868	10,158 +	147,773	100 +	230,395 +

have slower speed of onset, longer duration, more widespread areal extent, more diffuse spatial dispersion and more regular temporal spacing. There are geomorphic hazards (e.g. slope failure) that do not conform to this characterization of hazards, but usually significant landform changes occur over the long term at slow rates.

In following sections, we review developments regarding four geomorphological hazards: soil erosion by water; mass movement; fluvial erosion; and coastal erosion. We hope that this overview provides insights to hazards research in other geomorphic specialties such as wind-blown sediment, lahars, glacial movement, periglacial effects such as frost heaving, or solutional effects such as sinkholes. Some of these other hazards have been recently reviewed (e.g. Sherman and Nordstrom, 1994), whereas others, such as glacial movement or periglacial effects, are limited in their effect on human populations, or have received little attention as hazards.

We have chosen to place geomorphology within the context of the original natural hazards paradigm (White, 1974a) for ease of discussion. We recognize the importance of socio-economic factors in creating hazardous conditions and point to this as an important area of future research. Ultimately, resolution of natural hazards problems must incorporate economics and society, and geomorphologists must be able to recognize how the natural systems are affected by human systems.

In accordance with the first two goals of the natural hazards paradigm (White, 1974a), we contend that the physical aspects of a hazard event can be evaluated in terms of 5 components: (1) the dynamics of the physical processes; (2) the prediction of the occurrence; (3) the determination of the spatial and temporal characteristics; (4) an understanding of the impact of physical characteristics on people's perception; and (5) knowledge about how the physical aspects can be used to formulate adjustments to the event.

4.1. Soil erosion by water

Unlike many natural disasters, there is no direct loss of life as a result of soil erosion, but it has a widespread distribution, high remediation costs, a potential for soil deterioration, and reduced food production. Evans (1990) shows that 36% of arable soils in England and Wales are at risk from moderate- to very high-erosion. An evaluation of costs associated with soil erosion in southern Ontario, Canada (Table 2) shows the degree to which the problem impinges on the human system. According to Wall and Dickinson (1978), total cost approached US$ 95 million in 1976; using the consumer price index, this cost is equivalent to about US$ 250 million in 1994 dollars. Comparable costs have been observed in other locations (Stocking, 1988). Pimental et al. (1976) demonstrate that the loss of 25 mm of topsoil from a layer 300 mm thick would cause a decline in corn yield of as much as 10 bushels ha^{-1}

Table 2
Costs associated with soil erosion in Southern Ontario, Canada, 1975–1976. Data are in millions of 1976 US dollars. (From Wall, G.J., and W.T. Dickinson, 1978, as presented in Boardman, 1988)

Activity	Cost/yr	% of total cost
Lake Erie Harbor sedimentation	7.7	8
Drainage system sedimentation	23.6	25
Lakes and reservoir sedimentation	4.3	5
Urban sedimentation	3.8	4
Water treatment costs	8.2	9
Cropland soil losses	14.3	15
Fertilizer losses	23.4	25
Agriculture energy savings	6.5	7
Fish and wildlife values	2.6	2
Total	94.4	100

yr^{-1}. Similar conclusions have been reached by others (e.g. Larson et al., 1983).

Soil erosion by water is minimal in natural environments (Bennett, 1939; Morgan, 1986) but is widespread on farmed land where human impact is great (Trimble and Lund, 1982; Higgitt, 1991; Dearing, 1992). The many components that contribute to soil erosion have been reviewed by Morgan (1986), Evans (1990), and Higgitt (1991) among others. Empirical studies demonstrate the complexity of the conditions that produce erosion; they emphasize the variability in the amount of soil erosion relative to individual rainfall events or to various morphologic characteristics of a specific site; and they suggest the difficulties that are associated with predicting soil erosion on agricultural land.

Soil erosion prediction is traditionally accomplished with the Universal Soil Loss Equation (USLE) (Wischmeier and Smith, 1978) that provides an annual estimate of soil erosion and disregards the effects of specific events. The empirical studies show that there are problems with certain USLE variables. Attempts have been made to improve on USLE by modifying input parameters to accommodate local conditions (Sinzot et al, 1989), combining USLE with GIS (Barnes, 1988) or digital terrain models (Flacke et al., 1990), or developing alternative models such as ANSWERS (Areal Nonpoint Source Water Environmental Response Simulation) (Beasley and Huggins, 1982) or GAMES (Guelph model for evaluating the effects of Agricultural Management systems on Erosion and Sedimentation) (Dickinson et al., 1985). Alternative approaches focus on energetics (De Ploey, 1990), or on process-based models, such as the Water Erosion Prediction Project (WEPP) (Foster, 1990).

Determining the extent of the soil erosion hazard represents a third component of soil erosion research. There are numerous approaches to this issue, but few actually represent the spatial distribution of erosion based on detailed empirical data. Soil erosion can be implied from rainfall intensity (Babu et al., 1978; Wischmeier and Smith, 1978), but other variables affect soil erosion, and this approach can only show potential for erosion. Actual erosion has been mapped on the basis of soil profile data (U.S. Dept. of Agriculture, 1957; Trimble, 1974), but the necessity of interpolating for areas between the location of the soil profile sample sites, and the inability to pinpoint especially critical areas limits the usefulness of these data in a hazards context. The accuracy of soil erosion assessments increases as the area evaluated diminishes in size because an increasing number of factors responsible for soil erosion can be incorporated into the analysis (Morgan, 1986). The broad erosivity indices and soil profile maps could be integrated with other factors to identify regions that are likely to have high risk of soil erosion. These critical regions could then be analyzed in greater detail (e.g. Morgan, 1986).

The perception of the hazard by the individual(s) affected involves the farmers whose fields are eroding. An important determinant of perception of risk is the degree to which the farmer physically experiences the hazard. In hazards terminology, soil erosion is an ongoing, widespread, long-term, slow and diffuse problem. The amount of soil erosion that occurs on an individual field on an annual basis is generally small in area (1–2% of total field area) and volume (0.5–1 m^3 ha^{-1}) (Colborne and Staines, 1985; Fullen and Reed, 1986; Evans, 1990). Boardman (1990) reports that erosion in agricultural fields occurs only once every 5 to 20 years. Because the annual erosion may only amount to a few millimeters per year, the annual decrease in productivity is negligible and the farmer is unlikely to recognize the existence of a problem on his farm although he may be familiar with the problem on a general level (Shelton et al., 1991). Thus, the importance of soil erosion as a hazard is not recognized by those directly affected because of its pervasive nature and because the actual costs of the hazard are spread across society (Table 1).

The perception of the hazard dictates the adjustment that is adopted, and the focus of conservation efforts is on agricultural fields (Morgan, 1985). The pervasive nature of the soil erosion hazard does not generally provoke the farmer to adopt significant adjustments to this hazard on his own. Soil erosion management occurs through direct governmental intervention, led in the United States by the Soil Conservation Service which has taken a proactive role since the dust bowl. Compliance with management policy has evolved from voluntary participation to coercion (Napier, 1990; Steiner, 1990). Efforts have also been made to demonstrate to farmers the economic benefits of adopting soil conservation practices, such as the use of winter cover crops (Corak et al, 1991). The adjustments available to the farmer include: mechanical measures that alter topography through such measures as terrace building or establishing shelterbelts, soil management which focuses on promoting dense vegetation growth through soil preparation; agronomic measures that involve using cover crops to reduce impact of rainfall; and farming practices that focus on soil tillage techniques and crop management (Morgan, 1986).

Accelerated soil erosion is a widespread problem (Higgitt, 1991), and although it is not recognized, the associated deterioration of the soil productivity makes this a geomorphological hazard. The amount of research into the problem of soil erosion is enormous, although the large majority comes from developed countries of the temperate zone (Higgitt, 1991). A wide range of management techniques designed to control soil erosion have been developed (Morgan, 1986) despite evidence that the adoption of adjustments does not occur voluntarily (Napier, 1990). Better models can improve prediction of soil erosion and they may suggest alternative erosion control techniques, but the adoption of the adjustments by farmers is unlikely unless they are mandated by law or unless farmers can be convinced that the actions will improve their economic situation. It is economic incentive, not geomorphologic rationale, that governs choice of adjustment to this hazard.

4.2. Mass movements

On a global scale, slope failures are responsible for a small percentage of all deaths and damages resulting from catastrophic events. In the period 1947–1980, only 29 of 1062 natural catastrophic events with over 100 deaths or over 100 persons injured were landslides (Shah, 1983), although the number of deaths is considerable at times (Table 3). Lower magnitude events are more frequent and more widespread. Numerous small slides in a confined geographic space over a limited time frame can have a significant impact, as in January, 1969 in the Santa Monica Mountains of southern California (Campbell, 1975) or in January, 1982 in the San Francisco bay area (Ellen and Wieczorek, 1988). Data compiled from many sources for specific events suggests that resulting damage costs are extremely high (Table 4). It is difficult to evaluate monetary damages associated with the myriad of small landslides or rock falls that come down on the many highways of the world. Schuster (1978) estimates that for the USA alone there were US$ 100 million in annual damages to the Federal Interstate Highway System during the 1970's. The impression from these spotty data is that the impact of slope failures is enormous and widespread.

In hazards terms, mass movements range from low to high magnitude, occur relatively frequently, are of short duration, have a moderately fast speed of onset, a limited areal extent and a concentrated spatial dispersion with random temporal spacing. The classification is complicated because mass movements occur in a variety of forms, involving different materials moving at different speeds (Varnes, 1978). Rockfalls have a fast speed of onset, whereas some mudflows can have a slow speed of onset. Individual debris slides can be limited in size, but several may occur over a short period of time in a small geographic area making their impact on human systems cumulatively more significant. For the most part, the slides that have caused the most deaths can be classified as debris slides (Alexander, 1989).

Processes of mass movement involve the interaction of shear strength and shear stress. These are difficult to measure in situ., so studies have sought to correlate slope failures with prevailing environmental conditions. Recent developments in understanding the conditions that produce slope failures are reviewed by Allison (1990, 1991 and 1993) and by Bovis (1993). Increased understanding of the physical processes offers hope of providing vulnerable populations with warnings of imminent failures. Environmental condi-

Table 3
Major landslide disasters (200 or more killed) since 1900. (Adapted from: Crozier, 1989; Alexander, 1989; and NY Times index)

Year	Location	Deaths	Type
1919	Mt. Kelut, Java	5000	loess failure
1920	Kansu, China	10,000	earthquake
1921	Alma Ata, Kazakhstan	500	debris slide
1938	Kobe, Japan	461	mudflow/slide
1941	Quebrada de Cojup, Peru	5000	glacial lake burst
1945	Kure, Japan	1154	precipitation
1949	Tienshan, Tadzhikstan	12,000	seismic slides
1958	Tokyo region, Japan	1100	precipitation
1962	Ranrachirca, Peru	4000+	avalanching
1963	Viaont Gorge, Italy	2117	groundwater and water wave
1966	Rio de Janeiro, Brazil	1000	precipitation
1967	Rio de Janeiro, Brazil	1700	Precipitation
1970	Mt. Huascaran, Peru	18,000	avalanche
1971	Chungar, Peru	400	slide and water wave
1972	Yungay, Peru	23,000	earthquake
1974	Mantaro River, Peru	450	precipitation
1976	Guatemala City, Guatemala	240	earthquake
1981	Mt Semeru, Java	500	precipitation
1982	Monrovia region, Liberia	200	mine waste collapse
1983	Yacitan and Cashipampa, Peru	300+	precipitation
1983	Mt Sale, China	277	Groundwater rise
1985	Nevado del Ruiz, Colombia	22,000	lahars
1988	Rio de Janerio, Brazil	200	debris flows
1989	Sharora, Tadzhikistan	1000	earthquake
1992	Gormec, Turkey	205	earthquake
1993	Ecuador	200+	precipitation

Table 4
Costs of landslide damages at selected locations in selected years (in millions of US dollars unadjusted for inflation). (Adapted from Lundgren, 1986)

Location	Year	Damage costs	Source
San Jose, California	1968–70	1.76	Nilsen and Brabb (1972)
San Francisco Bay Area	1968–69	25.18	Taylor and Brabb (1972)
Los Angeles, California	1969	6.3	Slosson and Krohn (1977)
Los Angeles, California	1978	60+	Slosson and Krohn (1978)
Seattle, Washington	1971–72	0.46	Tubbs (1975)
Allegheny Co. Pennsylvania	1974–76	6.71	Fleming and Taylor (1981)
Hamilton Co. Ohio	1974–76	17.44	Fleming and Taylor (1981)

tions that can be monitored relatively easily, such as rainfall, provide a quick measure of the temporal characteristics of the hazard. The determination of the spatial distribution of the hazard is more difficult because slope failure is the result of the interaction of several conditions, many of which are site specific.

An alternative to these process-based studies is to identify landslide potential through geomorphological surveys that rely on visual evidence. This involves delineation of areas that have a history of slope failure, identification of the geomorphologic characteristics of the slopes susceptible to failure, and analysis of past geologic and geomorphologic processes that have affected the area (Nilsen and Brabb, 1977; Cooke and Doornkamp, 1990). Geomorphological mapping of landslide hazards involves a variety of spatial analysis techniques, including mapping of landslides based on field observation (Kertez and Schweitzer, 1991) or on aerial photographs (Chandler and Cooper, 1988), use of GIS to record geomorphological characteristics of the area on a grid by grid basis (Carrara et al., 1991), or use of a descriptive check-list of variables involved in slope failure (Cooke and Doornkamp, 1990). These approaches make it possible to indicate on a regional basis where landslides are likely to occur, but it is difficult to pinpoint specific slopes that are candidates for failure. Precision can be accomplished with detailed engineering studies (Cooke and Doornkamp, 1990), but these are complex, time-consuming and costly, and their widespread usage in risk assessment is not feasible. An alternative is to install instruments to monitor the environmental conditions correlated with slope failure in landslide-prone areas identified by spatial analysis techniques and to use these to warn the population of the existence of those conditions (Keefer et al., 1987). The widespread applicability of these warning systems remains limited because prediction involves the establishment of sophisticated and expensive networks of instruments to measure soil moisture conditions and rainfall. They may be useful in areas where landslide hazard mapping reveals high risk, thus maximizing the application of technological resources.

Once landslide-prone areas are identified, adjustments to the hazard can be implemented. Adjustments include land-use controls on development of hazardous areas, evacuation, insurance and engineering approaches designed to reinforce hazardous slopes (Nordstrom and Renwick, 1984; Kockelman, 1986). Unfortunately, human occupation of steep slopes has proceeded as people strive for better views, as developers promote construction on unstable slopes, or as the poor are constrained by scarcity of land. Damages from landslides continue to mount worldwide (Alexander, 1989), with losses expected to approach US$ 1 billion per year in the United States at the turn of the century (U.S. Geological Survey, 1982). Managers of landslide hazards need to adopt different approaches in developed and developing countries. In developed countries, there seems to be a reverse marginalization where the most vulnerable locations are also highly desirable because of their scenic views. This situation warrants adjustments that either involve bearing the cost of the loss, or preventing occupation of the hazardous location. In less developed countries, Susman et al.'s (1983) concept of marginalization is at the root of the vulnerability, and relief from the risk can only come about through economic and social action. As in the case of soil erosion, risk cannot effectively be mitigated by geomorphologic solutions alone.

4.3. Coastal processes and erosion

Coastal hazards consist of four primary components: wind, waves, flooding and erosion. We concentrate on erosion, the most geomorphologic of these components. The other components are important in an assessment of erosion because of their causal linkages to erosion. Coastal erosion can be considered both as a short-term problem related to the occurrence of storms and as a long-term problem resulting from sediment starvation, sea-level rise, increased storminess or a combination of these. Coastal erosion is generally a low to moderate magnitude/moderate- to high-frequency event, of long duration, widespread areal extent, slow speed of onset, diffuse spatial dispersion, and regular temporal spacing. The generally slow process is punctuated by higher magnitude events associated with coastal storms. Erosion hazard from storms is more confined in terms of areal extent and of shorter duration, but because storms are meteorological events whose predictability in the short term is fairly accurate, the speed of onset of this hazard is relatively slow.

Coastal erosion is widespread. Kamphuis (1980) has estimated that 95% of the world's shorelines are eroding at varying rates. Increasing numbers of people are at risk worldwide as they move to the shore. The population of the United States coastal zone nearly doubled between 1940 and 1980, and developed land on U.S. Atlantic and Gulf coast barriers increased by 150% between 1945 and 1975 (Platt, 1987). In recent years, coastal storms have created much damage along the northeast coast of the United States (Table 5).

Shoreline changes are the result of wave and current processes. The basic theoretical knowledge of these processes is well-understood (Carter, 1988; Horikawa, 1988). The difficulty is translating that knowledge into an ability to predict the resulting sediment movement at the small scale, and beach response at the larger scale. Much is known about the hydrodynamic forces that operate on sediment particles, the initiation of particle motion, the bedforms that characterize hydrodynamic regimes, and sediment transport rates (Allen, 1988).

Table 5
Selected major coastal storms of the New York/New Jersey coastline since 1974. Damages estimates unadjusted for inflation. (Source: New York Times)

Date	Storm	Damages	Remarks
12/1/74	Northeaster	US$ 3.5 million	
8/11/76	Hurricane Belle	US$ 5 million	
2/9/78	Northeaster	US$ 20 million	
3/29/84	Northeaster	US$ 250 million	includes inland flooding
9/28/85	Hurricane Gloria	US$ 220 million	NY estimate only
12/12/92	Northeaster	US$ 234 million	NY estimate only
3/14/93	Northeaster	US$ 800 million	estimate for entire storm path

However, the accuracy of predictive sediment transport models for the surf zone is poor (Sherman and Bauer, 1993) and models of larger scale response to coastal processes (Wright and Short, 1984) tend to be static, in that they do not account for variations in wave and current regimes, in sediment delivery rates, in tidal levels, or in beach form.

Knowledge of wave and current processes and the resulting landform changes is only relevant to hazards research if it improves the ability to predict the occurrence of the catastrophic event. Research has established the connection between coastal hydrodynamics and shoreline erosion, but the models currently in use are fairly simplistic. Shoreline erosion is modeled in two-dimensional form (Dean, 1991) and is based on Bruun's (1962) rule that the equilibrium profile migrates laterally in response to water level changes. The SBEACH model (Larson et al., 1990) incorporates elements of the Bruun approach in that it is based on the concept of an equilibrium profile, but, it portrays beach response to storm conditions more realistically because it incorporates time-dependent sediment transport. Like the Bruun approach, it does not produce nearshore bathymetry that includes barred systems and thus has accuracy problems. Models based on the concept of an equilibrium profile are widely used in coastal engineering projects ranging from beach nourishment to dune erosion modeling (Hales et al., 1991; Vellinga, 1982), but the models have been criticized for their assumption that an equilibrium profile develops in response to storm conditions (Dubois, 1992; Pilkey et al., 1993).

In light of the problems of modeling beach response to storm conditions based on theoretical relationships, Hallermeier (1987) used empirical data to determine volumes of sediment eroded during storms of varying recurrence intervals. The data are then used to prescribe a dune volume necessary to provide protection against the 100 year storm (FEMA, 1988). This single value does not account for variations in the profile or for convergent wave orthogonals that might exacerbate erosion in certain locations, increasing their vulnerability. The issue of spatial variation in vulnerability is not addressed in any of the models and this is a crucial issue in protecting against the hazard. Spatial variability was examined in terms of the related hazard of shoreline flooding (Gares, 1990), This approach needs to be extended to the assessment of risk to coastal erosion.

The spatial aspects of vulnerability can also be approached by evaluating change in the position of the shoreline through time. Ideally, this change would be determined from survey data, but with certain exceptions (e.g. McCluskey and Stephenson, 1985) there are only a few sets of survey data that extend far enough into the past to document long-term change or that provide regional coverage. The alternative is to use data obtained from aerial photographs and from historical maps and charts. The difficulties of rectifying the various sources can be overcome, and shoreline changes can be mapped (e.g. Dolan et al., 1992). The difference between these approaches and the modeling attempts is that these do not account for changes in response to single storms; they represent shoreline change over specified time intervals. Human alterations of the shoreline through beach nourishment projects or construction of shore protection structures during these time intervals disguise the impact of specific erosional events.

There is a wide range of adjustments available to shoreline residents to deal with erosion. Traditional approaches involve the use of engineering structures (seawalls, groins, bulkheads), but these may exacerbate the destruction of the beach (Pilkey, 1981). Beach nourishment is now the preferred method of offsetting erosion losses, but these projects can be costly, and sources of sediment for use as beach fill can be difficult to find (Leonard et al., 1990). There have been several proposals (e.g. Gares et al., 1980; Godfrey, 1987) to protect and enhance the dune systems that line the shoreline with the objective of taking advantage of the natural protective functions of dunes. Flood insurance is widely used in the coastal zone because a majority of coastal communities joined the Federal Flood Insurance Program (Dawson, 1987). Initially, the program focused on coastal flooding, but in recent years FEMA has factored erosion into the regulations (Davison, 1993). An important component of the new regulations is the provision for removal of houses that are imminently threatened by erosion (Rogers, 1993), an action that conforms with the position of some that development is incompatible with shoreline processes (Pilkey, 1981).

4.4. Fluvial processes and erosion

Fluvial hazards involve both flooding and erosion, although only the erosional aspects can be considered as geomorphic hazards. The geomorphologic and hydrologic literature is replete with information concerning flooding problems, but there are few references to fluvial erosion problems. There is a general understanding of the relationship between river discharge and the hydraulic geometry of the river channel, of the principles of sediment transport, of drainage systems, and channel patterns (e.g. Knighton, 1984; Morisawa, 1985). Recent developments in fluvial geomorphology have been reviewed by Richards (1988) and Rhoads (1992). The basic conditions associated with bank erosion have been studied (Knighton, 1984; Richards, 1982) that contributes to a process-based predictive model of bank erosion, but there are a number of variables that control the erosion that must be effectively integrated to produce an effective model. Progress has been made in modeling the development and migration of meander curves (Furbish, 1991; Rhoads, 1992) that represent a major location of bank erosion. Much bank erosion is associated with the dynamics of the outer cut banks of meander curves, and this line of research provides promise in the effort to predict the occurrence of the hazard.

Table 6
Selected rates of bank erosion. (After Knighton, 1984)

River	Ave. rate of retreat ($m\ yr^{-1}$)	Period of measurement
Axe, GB	0.15–0.46	1974–76
Bollin-Dean, GB	0–0.9	1967–69
Cound, GB	0.64	1972–74
Crawfordsburn, N. Ireland	0–0.5	1966–68
Exe, GB	0.6–1.2	1974–76
Mississippi, USA	4.5	1945–62
Torrens, Australia	0.58	1960–63
Watts Branch, USA	0.5–0.6	1955–57
Wisloka, Poland	8–11	1970–72

Bank erosion is not a common problem along many rivers in developed countries because of the widespread use of channel stabilization measures (Brookes, 1985). Average annual erosion rates in unstabilized rivers in temperate climates can reach 10 m (Table 6). The rates may be higher in tropical zones; the Brahmaputra River migrated 200–400 m yr^{-1} in 1975–1981 (Haque and Hossain, 1988). There seems to be a general relationship between bank erosion and drainage area that can be expressed as a power function (Hooke, 1980). These data provide an estimate of the magnitude of the bank erosion problem, but they give little indication of the frequency of erosion. There is a direct relationship between bank erosion and the shear stress on the surface that is imparted by the velocity of the flowing water (Hooke, 1979), implying that there should be increased bank erosion with higher discharge. The analysis of at-a-station hydraulic geometry parameters, in particular channel width, show a direct relationship to discharge (Knighton, 1975), but the relationship between discharge and bank erosion has not received great attention. The lack of information about the frequency of bank erosion precludes identifying the critical hazards parameters, speed of onset, duration, and temporal spacing.

The identification of vulnerable locations is accomplished by examining the history of erosion along a particular river channel (e.g. Hooke, 1980; Haque and Hossain, 1988), resembling the approach used in coastal research. The problem with the historical approach is the assumption that the pattern operating in the past will continue in the future. An alternative approach would assume that the critical erosion will occur on the most vulnerable parts of rivers (e.g. the outside of meander bends) and classify all such locations as highly vulnerable. This information could be combined with the discharge/erosion relationship to refine the vulnerability assessment in terms of frequency/magnitude conditions.

The relationships between precipitation, discharge and bank erosion leads to some conclusions regarding the placement of fluvial geomorphic hazards on the Burton et al. (1978) continuum. The regular occurrence of rainfall in temperate regions, and the knowledge that even moderate amounts of precipitation increase discharge suggest that fluvial geomorphic hazards are fairly high frequency, moderate magnitude events of fairly short duration. They appear to be relatively regular in term of temporal spacing because of seasonal or even cyclical precipitation patterns. Knowledge about runoff/discharge relationships implies that the speed of onset is not particularly fast. The areal extent of these hazards is limited due to the confined nature of river channels, but the spatial dispersion is fairly diffuse because the hazard is tied to regional climatology and geology.

Adjustments to cope with river bank erosion are often subsumed within projects that deal with flooding. As in the case of adjustments for coastal erosion problems, engineering structures were the first choice until recently but these often exacerbate the erosion hazard by transferring it to another location (Brookes, 1985). There are numerous publications that review the engineering techniques used to cope with fluvial hazards (e.g. Chow, 1959; Jansen et al., 1979). Other approaches that control river flow involve channel manipulation and control of riparian vegetation, (Brookes, 1985). Requirements of the Federal Flood Insurance Program (Platt, 1976) were created ostensibly to control the occupation of floodplains although making insurance readily available to individuals appears to have had the undesirable effect of drawing people to this hazardous location (Burton et al., 1993). Other attempts to control fluvial hazards have involved creating greenways along the river or controlling development through wetland regulations (Kusler, 1985). Other than the Flood Insurance Program, there is no coordinated effort to manage floodplains in the USA. On the other hand, the extent of stabilized river systems has limited the hazard from bank erosion. This situation is not true in less developed countries, where engineering efforts have been limited and the poorer people who occupy the river banks are faced with little choice but to absorb any loss and to relocate.

5. Implications

In devising our definition of geomorphic hazards we adopted a traditional view of the field to avoid straying into other different, though adjunct, fields. We prefer a focus on the way the landscape and its changes affect human activities. Although the landscape changes are the product of physical processes which themselves are responsible for hazardous events, we believe that it is important to separate the disasters that are the product of landscape alterations from those that are produced by the agents of landscape alterations to avoid confusion and focus on issues that make geomorphology a unique discipline.

The adoption of this definition causes us to focus on erosion as the primary geomorphic hazard, although deposition may also be a hazard (e.g. Sherman and Nordstrom, in press). Defined this way, geomorphic hazards are ongoing rather than tied to individual catastrophic occurrences (although these contribute to progressive erosion). Our analysis of slope failure, soil, fluvial and coastal erosion, shows that these geomorphic hazards occur at the pervasive end of the hazards continuum (Table 7). They generally are associated with high-frequency/low-magnitude conditions, are widely distributed in space, are long-term situations, and have slow speeds of onset. A brief examination of literature on aeolian erosion and subsidence indicates that these basic hazard characteristics apply to other geomorphic hazards (Table 7). The physical characteristics of geomorphic hazards make them relatively inconsequential to the populations affected, and individuals may be less likely to adopt adjustments to geomorphic hazards than hydrologic or atmospheric hazards.

Geomorphologists have the opportunity to demonstrate the nature of geomorphic hazards and to argue for the adoption of management strategies that deal with the losses that will result, but we have not yet been proficient at this transfer of information. We tend to

Table 7
Qualitative assessment of hazard parameters for selected geomorphic hazards

Parameter	Hazard					
	Soil erosion	Mass movement	Coastal erosion	Fluvial erosion	Aeolian erosion	Subsidence
Frequency	moderate to high	moderate to high	moderate to high	moderate to high	moderate	low to moderate
Magnitude	low to moderate	low to high	low to moderate	low to moderate	low to moderate	low to high
Duration	moderate to long	short to moderate	moderate to long	moderate to long	moderate to long	short to moderate
Areal extent	widespread	moderate to limited	moderate to wide	moderate to limited	moderate to limited	moderate to limited
Speed of onset	slow	mod to rapid	slow	slow	slow	slow to rapid
Spatial dispersion	moderate to diffuse	moderate to concentrated	moderate	moderate	moderate to concentrated	moderate to concentrated
Temporal spacing	regular	random	regular	regular	regular	random

teach ourselves instead of teaching managers, planners and decision-makers who are the people most in need of the result of our hazards-oriented research. Of all of the general natural hazards text books, there are only a few that emphasize the natural side of the topic (e.g. Bryant, 1991; Alexander, 1993). Information about the physical aspects of natural hazards tends to be presented in environmental geology (e.g. Lundgren, 1986) or applied geomorphology (e.g. Cooke and Doornkamp, 1990) texts that are generally not used by teachers of planning courses. Most hazards texts focus on the social system response to, or anticipation of, natural hazards and disasters and give the physical aspects of hazards limited coverage (e.g. Foster, 1980; Petak and Atkisson, 1982).

There is a paucity of coursework aimed at educating non-scientists about the risks posed by the geomorphic environment. In a recent survey of 82 graduate schools of planning in North America, asking about curriculum content associated with natural hazards, only 3 courses were named that had the words "natural hazards" or "disaster" in the title (Havlick and Dorsey, 1994). One implication of the survey is that most planners (and we assume, other decision-makers in similar situations) rely on "on the job training" to develop expertise in risk assessment and disaster reduction. This provides us with a challenge - to seek ways of improving paths of communication with risk managers. This involves interacting with social scientists involved in hazards research, and publishing results in journals that are likely to be read by people involved in the field of natural hazards.

Geomorphologists must familiarize themselves with information produced by social scientists in order to make contributions in the policy arena. Management strategies cannot be based solely on geomorphic principles, but must be designed with political or economic constraints in mind so that their implementation will occur. A common problem is to propose a strategy based on sound geomorphic theory but with little empirical proof. This strategy is easily refuted in a public forum (Weinberg, 1985; Gares, 1989). We must also examine the human systems in a way that helps account for the influences of human activities on the physical landscape. Humans are an integral part of the landscape and geomorphologists must incorporate human actions into physical landscape models that are used to describe or understand the system (Phillips, 1991; Nordstrom, 1994). In order to do this, we can no longer view human alteration of the geomorphic system as an aberration, but as one of many variables that affect that system.

Socio-economic conditions affect hazards in other ways. The concept of marginalization (Susman et al., 1983) is *apropos* to geomorphic hazards because population pressures and inaccessibility to capital and natural resources push people to occupy marginal lands, such as steep slopes or tropical forests with thin soils. These marginal lands are susceptible to small alterations brought about by human presence that increase the vulnerability of the occupants. Landslides, soil erosion, coastal erosion and fluvial erosion have all resulted in loss of life or costly damages. A reverse marginalization may also occur where the vulnerable locations are desirable resort sites that attract wealthy people. One result of this occupation is the desire to protect one's investment once the hazard has finally been recognized, and this results in the implementation of increasingly complex adjustments, lessening the hazard but increasing the potential of a catastrophic event. The migration of people to these hazardous locations is facilitated by their wealth and the associated political power that demands governmental involvement in implementing adjustments. The increased occupancy of hazardous locations magnifies the difficulty of determining optimum management solutions because the presence of human structures and human activities affect the geomorphic processes that produce the hazard. This necessitates understanding the geomorphologic responses that result from the interaction of human and physical systems. Thus, despite the predominance of social science in hazards research, there is still an important role for geomorphologists.

References

Alexander, D., 1989. Urban landslides. Prog. Phys. Geogr., 13: 157–191.

Alexander, D., 1993. Natural Disasters. Chapman and Hall, New York. 632 pp.

Allen, J.R., 1988. Nearshore sediment transport. Geogr. Rev., 78: 148–168.

Allison, R.J., 1990. Slopes and slope processes. Prog. Phys. Geogr., 14: 373–388.

Allison, R.J., 1991. Slopes and slope processes. Prog. Phys. Geogr., 15: 423–437.

Allison, R.J., 1993. Slopes and slope processes. Prog. Phys. Geogr., 17: 92–101.

Babu, R., Tejwani, K.G., Agarwal, M.P. and Bhushan, L.S., 1978. Distribution of erosion index and iso-erodent map of India. Ind. J. Soil Conserv., 6: 1–12.

Barnes, K.B., 1988. Cartographic modeling of nonpoint pollutant surfaces for a coastal drainage area. Symp. Coastal Water Resources, Am. Wat. Res. Assoc., pp. 133–146.

Beasley, D.B. and Huggins, L.F., 1982. ANSWERS — User's Manual. Dept. of Agric. Eng., Purdue Univ., W. Lafayette, IN, 78 pp.

Bennett, H.H., 1939. Soil Conservation. McGraw-Hill, New York, 993 pp.

Boardman, J., 1988. Public policy and soil erosion in Britain. In: J.M. Hooke (Editor), Geomorphology in Environmental Planning. Wiley, Chichester, pp. 33–50.

Boardman, J., 1990. Soil erosion on the South Downs: A review. In: J. Boardman, I.D.L. Foster and J.A. Dearing (Editors), Soil Erosion on Agricultural Land. Wiley, Chichester, pp. 87–105.

Bovis, M.J., 1993. Hillslope geomorphology and geotechnique. Prog. Phys. Geogr., 17: 173–189.

Brookes, A., 1985. River channelization: traditional engineering methods, physical consequences and alternative practices. Prog. Phys. Geogr., 9: 44–73.

Bruun, P., 1962. Sea level rise as a cause of shore erosion. J. Waterways, Harb. Div., ASCE, 88: 117–130.

Bryant, E.A., 1991. Natural Hazards. Cambridge Univ. Press, Cambridge, 325 pp.

Burton, I., Kates, R.W. and White, G.F., 1978. The Environment as Hazard. 1st ed. Oxford Univ. Press, New York, 240 pp.

Burton, I., Kates, R.W. and White, G.F., 1993. The Environment as Hazard. 1st ed. The Guilford Press, New York, 290 pp.

Campbell, R.H., 1975. Soil slips, debris flow, and rainstorms in the Santa Monica Mountains and vicinity, southern California. U.S. Geol. Surv. Prof. Pap., 851: 51 pp.

Carrara, R., Cardinali, M., Detti, R., Guzetti, F., Pasqui, V. and Reichenbach, P., 1991. GIS techniques and statistical models in evaluating landslide hazard. Earth Surf. Proc. Landforms, 16: 427–445.

Carter, R.W.G., 1988. Coastal Environments: An Introduction to the Physical, Ecological, and Cultural Systems of Coastlines. Academic Press, London, 617 pp.

Chandler, J.H. and Cooper, M.A.R., 1988. Monitoring the development of landslides using archival photography and analytical photogrammetry. Land Min. Surv., 6: 576–584.

Chow, V.T., 1959. Open-Channel Hydraulics. McGraw-Hill, New York, 680 pp.

Colborne, G.J.N. and Staines, S.J., 1985. Soil Erosion in South Somerset. J. Agric. Sci. Cambridge, 104: 107–112.

Cooke, R.U., 1984. Geomorphological Hazards in Los Angeles. George Allen and Unwin, London, 206 pp.

Cooke, R.U. and Doornkamp, J.C., 1990. Geomorphology in Environmental Management. Oxford Univ. Press, New York, 410 pp.

Corak, S.J., Prye, W.W. and Smith, M.S., 1991. Legume mulch and nitrogen fertilizer effects on soil water and corn production. Soil Sci. Am. J., 55: 1395–1400.

Crozier, M.J., 1989. Landslides: Causes, Consequences, and Environment. Routledge Press, London, 252 pp.

Davison, A.T., 1993. The national flood insurance program and coastal hazards. Proc. Coastal Zone '93. Am. Soc. of Civil Engineers, New York, pp. 1377–1391.

Dawson, A.D., 1987. The NFIP and developed coastal barriers. In: R.H. Platt, S.G. Pelczarski and B.K.R. Burbank (Editors), Cities on the Beach. Univ. of Chicago Press, Chicago, pp. 245–259.

De Ploey, J., 1990. Modeling the erosional susceptibility of catchments in terms of energy. Catena, 17: 175–183.

Dean, R.G., 1991. Equilibrium beach profiles: characteristics and applications. J. Coast. Res., 7: 53–84.

Dearing, J.A., 1992. Sediment yields and sources in a Welsh upland lake — Catchment during the past 800 years. Earth Surf. Proc. Landforms, 17: 1–17.

Dickinson, W.T., Rudra, R.P. and Wall, G.J., 1985. Discrimination of soil erosion and fluvial sediment areas. Can J. Earth Sci., 22: 1112–1117.

Dolan, R., Fenster, M.S. and Holme, S.J., 1992. Spatial analysis of shoreline recession and accretion. J. Coast. Res., 8: 263–285.

Dubois, R.N., 1992. A re-evaluation of Bruun's rule and supporting evidence. J. Coast. Res., 8: 618–628.

Ellen, S.D. and Wieczorek, G.F. (Editors), 1988. Landslides, floods and marine effects of the storm of January 3–5, 1982 in the San Francisco Bay region. U.S. Geol. Surv. Prof. Pap., 1434: 310 pp.

Evans, R., 1990. Water erosion in British farmer' fields — Some causes, impacts, predictions. Prog. Phys. Geogr., 14: 199–291.

FEMA (Federal Emergency Management Agency), 1988. Guidelines and Specifications for Wave Elevation Determination and V Zone Mapping, Second Draft. Fed. Ins. Agency, Washington, DC, 113 pp.

Flacke, W., Auerswald, K. and Neufang, L., 1990. Combining a modified universal soil loss equation with a digital terrain model for computing high resolution maps of soil loss resulting from rain wash. Catena, 17: 383–397.

Fleming, R.W. and Taylor, F.A., 1980. Estimating the costs of landslide damage in the United States. U.S. Geol. Surv. Circ., 832: 21 pp.

Foster, G.R., 1990. Process-based modeling of soil erosion by water on agricultural land. In: J. Boardman, I.D.L. Foster and J.A. Dearing (Editors), Soil Erosion on Agricultural Land. Wiley, Chichester, pp. 429–445.

Foster, H.D., 1980. Disaster Planning — The Preservation of Life and Property. Springer, New York, 275 pp.

Fullen, M.A. and Reed, A.H., 1986. Rainfall, runoff and erosion on bare arable soils in east Shropshire, England. Earth Surf. Proc. Landforms, 11: 413–425.

Furbish, D.J., 1991. Spatial autoregressive structure in meander evolution. Geol. Soc. Am. Bull., 103: 1576–1589.

Gares, P.A., 1989. Geographers and policy-making: Lessons learned from the failure of the New Jersey dune management plan. Prof. Geogr., 41: 20–29.

Gares, P.A., 1990. Predicting flooding probability for beach/dune systems. Environ. Manage., 114: 115–123.

Gares, P.A., Nordstrom, K.F. and Psuty, N.P., 1980. Delineation and implementation of a dune management district. Proc. Coastal Zone '80. Am. Soc. Civ. Eng., New York, pp. 1269–1288.

Godfrey, P.J., 1987. A successful local program for preserving and maintaining dunes on a developed barrier island: Mantoloking NJ. In: R.H. Platt, S.G. Pelczarski and B.K.R. Burbank (Editors), Cities on the Beach. Univ. of Chicago Press, Chicago, pp. 163–170.

Hales, L.Z., Byrnes, M.R. and Dowd, M.W., 1991. Numerical modeling of storm-induced beach erosion, Folly Beach, South Carolina. Beach fill alternatives. Proc. Coastal Zone '91. Am. Soc. Civ. Eng., New York, pp. 495–509.

Hallermeier, R.J., 1987. Applying large replicas of shore erosion by storms. Proc. Coastal Sediments '87. Am. Soc. Civ. Eng., New York. pp. 1415–1429.

Haque, C.E. and Hossain, Ziarat M.D., 1988. Riverbank erosion in Bangladesh. Geogr. Rev., 78: 20–31.

Havlick, S.W. and Dorsey, B., 1994. To plan or not to plan: Opportunities for natural hazards education in graduate planning programs. Nat. Hazards Observer, 18: 4–5.

Hewitt, K., 1983. The idea of calamity in a technocratic age. In: K. Hewitt (Editor), Interpretations of Calamity: From the Viewpoint of Ecology. Allen and Unwin, London, pp. 3–32.

Hewitt, K. and Burton, I., 1971. The hazardousness of a place: A regional ecology of damaging events. Dept. of Geog. Res. Paper #6. Univ. of Toronto Press, Toronto, 154 pp.

Higgitt, D.L., 1991. Soil erosion and soil problems. Prog. Phys. Geogr., 15: 92–101.

Hooke, J.M., 1979. An analysis of the processes of river bank erosion. J. Hydrol., 42: 39–62.

Hooke, J.M., 1980. Magnitude and distribution of rates of river bank erosion. Earth Surf. Proc. Landforms, 5: 143–157.

Horikawa, K. (Editor), 1988. Nearshore Dynamics and Coastal Processes: Theory, Measurement and Predictive Models. Univ. of Tokyo Press, Tokyo, 500 pp.

Jansen, P., Bendegom, L., Berg, J., Vries, M. and Zanen, A., 1979. Principles of River Engineering — The Non-Tidal Alluvial River. Pitman, London. 509 pp.

Kamphuis, J.W., 1980. Coastal engineering. In: A.J. Bowen (Convenor), Basic Nearshore Processes — Short Course Lecture Notes, ACROSES. National Research Council, Ottawa, pp. 4.1–4.45.

Kates, R.W., 1971. Natural hazard in human ecological perspective: hypotheses and models. Econ. Geogr., 47: 438–451.

Keefer, D.K., Wilson, R.C., Mark, R.K., Brabb, E.E. et al., 1987. Real-time landslide warning during heavy rainfall. Science, 238: 921–925.

Kertez, A. and Schweitzer, F., 1991. Geomorphological mapping of landslides in Hungary with a case study of mapping Danubian bluffs. Catena, 18: 529–536.

Knighton, D., 1975. Variations in at-a-station hydraulic geometry. Am. J. Sci., 275: 186–218.

Knighton, D., 1984. Fluvial Forms and Processes. Edward Arnold, London, 218 pp.

Kockelman, W.J., 1986. Some techniques for reducing landslide hazards. Bull. Assoc. Eng. Geol., 23: 29–52.

Kusler, J.A., 1985. Roles along the rivers — regional programs meet national policy. Environment, 27: 18–20, 37–44.

Larson, M., Kraus, N.C. and Byrnes, M.R., 1990. SBEACH: Numerical Model for Simulating Storm-Induced Beach Change. Report #2. US Army Coast. Eng. Res. Cent. Tech Rep. #CERC-89-9, 262 pp.

Larson, W.E., Pierce, F.J. and Dowdy, R.H., 1983. The threat of soil erosion to long-term crop production. Science, 219: 458–465.

Leonard, L., Clayton, T. and Pilkey, O.H., 1990. An analysis of replenished beach design parameters on US east coast barrier islands. J. Coast. Res., 6: 15–36.

Lundgren, L., 1986. Environmental Geology. Prentice-Hall, Englewood Cliffs, NJ, 576 pp.

McCluskey, J.M., and Stephenson, R.A., 1985. Effects of dune stabilization in North Carolina. Proc. Coastal Zone '85. Am. Soc. Civ. Eng., New York, pp. 2171–2185.

Mitchell, J.K., 1974. Community response to coastal erosion. Dept. of Geog. Res. Paper #156. Univ. of Chicago Press, Chicago, 209 pp.

Mitchell, J.K., Devine, N. and Jagger, K., 1989. A contextual model of natural hazard. Geogr. Rev., 79: 391–409.

Morgan, R.P.C., 1985. Assessment of soil erosion risk in England and Wales. Soil Use Soil Manage., 1: 127–131.

Morgan, R.P.C., 1986. Soil Erosion and Conservation. Longman, Harlow, 298 pp.

Morisawa, M., 1985. Rivers — Form and Process. Longman, New York 222 pp.

Napier, T.L., 1990. The evolution of US soil conservation policy: from voluntary adoption to coercion. In: J. Boardman, I.D.L. Foster and J.A. Dearing (Editors), Soil Erosion on Agricultural Land. Wiley, Chichester, pp. 419–436.

New York Times Index, 1985–1992.

Nilsen, T.H. and Brabb, E.E., 1972. Preliminary Photointerpretation and Damage Maps of Landslide and other Surficial Deposits in Northeaster San Jose, Santa Clara Co., California. US Geol. Surv. Misc. Field Studies Map MF-361.

Nilsen, T.H. and Brabb, E.E., 1977. Slope stability studies in the San Francisco Bay area, California. In: D.B. Coates (Editor), Landslides. Geol. Soc. Am. Rev. Eng. Geol., 3: 235–244.

Nordstrom, K.F., 1994. Beaches and dunes of human-altered coasts. Prog. Phys. Geogr., in press.

Nordstrom, K.F. and Renwick, W.H., 1984. A coastal cliff management district for protection of eroding high relief coasts. Environ. Manage., 8: 197–202.

Palm, R.I., 1990. Natural Hazards: An Integrative Framework for Research and Planning. John Hopkins Press, Baltimore, 184 pp.

Palm, R.I. and Hodgson, M.E., 1993. Natural hazards in Puerto Rico. Geog. Rev., 83: 280–289.

Petak, W.J. and Atkisson, A.A., 1982. Natural Hazard Risk Assessment and Public Policy — Anticipating the Unexpected. Springer- Verlag, New York, 489 pp.

Phillips, J.D., 1991. The human role in earth surface systems: some theoretical considerations. Geogr. Anal., 23: 316–331.

Pimental, D., Terhune, E.C., Hudson, R., Rechereau, S., Samis, R., Smith, E.A., Denman, D., Reifscheider, D. and Shepard, M., 1976. Land degradation: effects on food and energy resources. Science, 194: 149–155.

Pilkey, O.H., 1981. Geologists, engineers, and a rising sea level. Northeast. Geol., 3/4: 150–158.

Pilkey, O.H., Young, R.S., Riggs, S.R., Smith, A.W., Wu, H. and Pilkey, W.D., 1993. The concept of shoreface profile of equilibrium: a critical review. J. Coast. Res., 9: 255–278.

Platt, R.H., 1976. The national flood insurance program: some midstream perspectives. Am. Inst. Plan. J., 42: 302–313.

Platt, R.H., 1987. Overview of developed barriers. In: R.H. Platt, S.G. Pelczarski and B.K.R. Burbank (Editors), Cities on the Beach. Univ. Chicago Press, Chicago, pp. 1–14.

Rhoads, B.L., 1992. Fluvial geomorphology. Prog. Phys. Geogr., 16: 456–477.

Richards, K., 1982. Rivers — Form and Process in Alluvial Channels. Methuen, London, 358 pp.

Richards, K., 1988. Fluvial geomorphology. Prog. Phys. Geogr. 12: 435–456.

Rogers, S.M., 1993. Relocating erosion-threatened buildings: a study of North Carolina housemoving. Proc. Coastal Zone '93. American Society of Civil Engineers, New York, pp. 1392–1405.

Shah, B.V., 1983. Is the environment becoming more hazardous? A global survey, 1947 to 1980. Disasters, 7: 202–209.

Shelton, D.P., Dickey, E.C., Jasa, P.J., Biere, D.A. and Krotz, S.S., 1991. A cooperative educational program to reduce soil erosion. J. Soil Water Conserv., 46: 169–172.

Sherman, D.J. and Bauer, B.O., 1993. Coastal geomorphology through the looking glass. Geomorphology, 7: 225–249.

Sherman, D.J. and Nordstrom, K.F., 1994. Hazards of wind blown sand and sand drift. J. Coast. Res., in press.

Schuster, R.L., 1978. Introduction. In: R.L. Schuster and R.J. Krizek (Editors), Landslides — Analysis and Control. Transportation Research Board, National Research Council Special Report 176. National Academy of Sciences, Washington, DC, pp. 1-10.

Sinzot, A., Bollinne, A., Laurant, A., Erpicum, M. and Pissart, A., 1989. A contribution to the development of an erosivity index adapted to the prediction of erosion in Belgium. Earth Surf. Proc., 14: 509–515.

Slosson, J.E. and Krohn, J.P., 1977. Effective building codes. Calif. Geol., 30: 138–139.

Slosson, J.E. and Krohn, J.P., 1979. AEG building code review — mudflow/debris flow damage, February 1978 storm, Los Angeles area. Calif. Geol., 32: 8–11

Steiner, F.R., 1990. Soil Conservation in the United States: Policy and Planning. Johns Hopkins Univ. Press, Baltimore, 288 pp.

Stocking, M., 1988. Socioeconomics of soil conservation in developing countries. J. Soil Water Conserv., 43: 381–385.

Susman, P., O'Keefe, P. and Wisner, B., 1983. Global disasters a radical interpretation. In: K. Hewitt (Editor). Interpretations of Calamity: From the Viewpoint of Ecology. Allen and Unwin, London, pp. 263– 283.

Taylor, F.A. and Brabb, E.E., 1972. Maps showing distribution and cost by counties of structurally damaging landslides in the San Francisco Bay Region, California, Winter of 1968–69. US Geol. Surv. Misc. Field Studies Map MF-327.

Taype, V., 1979. Los desastres naturales como problema del al defensa civil. Bol Soc. Geol. Peru, 61: 101–111.

Trimble, S.W., 1974. Man-Induced Soil Erosion on the Southern Piedmont, 1700–1970. Soil Cons. Soc. Am., Ankeny, IA, 180 pp.

Trimble, S.W. and Lund, S.W., 1982. Soil Conservation and the reduction of erosion and sedimentation in the Coon Creek Basin, Wisconsin. U.S. Geol. Surv. Prof. Pap., 1234: 35 pp.

Tubbs, D.W., 1975. Causes, mechanisms, and prediction of landsliding in Seattle. PhD. Dissertation, Univ. Washington, Seattle, 97 pp.

U.S. Army Corps of Engineers, 1962. Post flood report — Coastal storm of 6–7 March, 1962. U.S. Army Engineers District, Philadelphia, 45 pp.

U.S. Army Corps of Engineers, 1984. Shore Protection Manual. Coastal Engineering Research Center, Vicksburg, MS, 1242 pp.

U.S. Dept. of Agriculture, 1957. Yearbook of Agriculture — Soil. U.S. Dept. Agric., Washington, 748 pp.

U.S. Geological Survey, 1982. Goals and tasks of the landslide part of the ground failure hazards reduction program. U.S. Geol. Surv. Circ., 880: 49 pp.

Varnes, D.J., 1978. Slope movement types and processes. In: R.L. Schuster and R.J. Krizek (Editors), Landslides — Analysis and Control. Transportation Research Board, National Research Council Special Report 176. National Academy of Sciences, Washington, DC, pp. 12–33.

Vellinga, P., 1982. Beach and dune erosion during storm surges. Coast. Eng., 6: 361–387.

Waddell, E., 1977. The hazards of scientism. Human Ecol., 5: 69–76.

Waddell, E., 1983. Coping with frosts, governments and disaster experts: some reflection based on a New Guinea experience and a perusal of the relevant literature. In: K. Hewitt (Editor), Interpretations of Calamity: From the Viewpoint of Ecology. Allen and Unwin, London, pp. 33–43.

Wall, G.J. and Dickinson, W.T., 1978. Economic impact of soil erosion in Ontario. Notes Agric., 14: 10–12.

Watts, M., 1983. On the poverty of theory: natural hazards research in context. In: K. Hewitt (Editor), Interpretations of Calamity: From the Viewpoint of Ecology. Allen and Unwin, London, pp. 231–262.

Warrick, R.A., 1983. Drought in the US Great Plains: shifting social consequences? In: K. Hewitt (Editor), Interpretations of Calamity: From the Viewpoint of Ecology. Allen and Unwin, London, pp. 67–82.

Weinberg, A.M., 1985. Science and its limits: the regulator's dilemma. Iss. Sci. Technol., 2(1): 59–72.

White, G.F., 1960. The choice of use in resource management. Nat. Res. J., 1: 23–40.

White, G.F., 1974a. Natural hazards research: concepts, methods and policy implications. In: G.F. White (Editor), Natural Hazards: Local, National, Global. Oxford Univ. Press, New York, pp. 3–16.

White, G.F. (Editor), 1974b. Natural Hazards: Local, National, Global. Oxford Univ. Press, New York, 288 pp.

White, G.F., Calef, W.C., Hudson, J.W., Mayer, H.M., Sheaffer, J.R. and Volk, D.J., 1958. Changes in human occupance of flood plains in the United States. Dept. of Geogr. Res. Paper #57. Univ. Chicago Press, Chicago, 235 pp.

Whyte, A.V and Burton, I., 1980. Environmental risk assessment. SCOPE Report 15. Wiley, New York, 198 pp.

Wischmeier, W.H. and Smith, D.D., 1978. Predicting Rainfall Erosion Losses — A Guide to Conservation Planning. Handbook #537. U.S. Dept. of Agric., Washington, 58 pp.

Wolman, M.G. and Miller, J.P., 1960. Magnitude and frequency of forces in geomorphic processes. J. Geol., 68: 54–74.

Wright, L.D. and Short, A.D., 1984. Morphodynamic variability of surf zones and beaches: a synthesis. Mar. Geol., 56: 93–118.

ELSEVIER

Geomorphology 10 (1994) 19–25

GEOMORPHOLOGY

Hazards: singularities in geomorphic systems

A.E. Scheidegger

Section of Geophysics, Technical University, Gusshausstrasse 27-29-128/2, A-1040 Vienna, Austria

Received November 20, 1993; revised March 8, 1994; accepted April 1, 1994

Abstract

The development of landscapes is commonly interpreted as the outcome of the operation of a few simple phenomenological principles amongst which the Antagonism Principle is the most fundamental one. It states that there are two types of processes active in the formation of a landscape at the same time: the endogenic (tectonic) and the exogenic (climate-driven) processes. These two types of processes may more or less balance each other, in which case the actual aspect of a landscape corresponds to the instantaneous dynamic equilibrium between them. If one of the external parameters is changed, the equilibrium is supposed to be re-established by a corresponding continuous change of the other parameters (process–response theory). The dependences between various landscape parameters may become multivalued at junctions, cusps, etc. In that case, the system can jump from one branch of the process–response curve to another: a natural disaster occurs. However, modern views on the behavior of *complex systems* allow a somewhat different interpretation of the above phenomenological landscape interpretation, but also in the context of the Principle of Antagonism. However, a "stationary" landscape-state does now not correspond to a dynamic equilibrium, but to self-organized exogenic order at the edge of chaos. Under conditions of such self-organized criticality there are no natural space and time scales, so that fractal statistics are applicable; spatial and temporal correlations follow a power law. Extremely small accidental perturbations can cause the system to become unstable, leading to rapid changes in the landscape which are experienced as hazard-events. These follow a power law distribution for size versus frequency, provided the endogenic input occurs at a constant rate; they are therefore simply part and parcel of geomorphic systems.

1. Introduction and background: phenomenological landscape principles

The development of landscapes can be described phenomenologically as the outcome of the operation of a few simple fundamental principles (Scheidegger, 1987): (i) the Antagonism Principle, (ii) the Instability Principle with (iii) the Catena Principle as corollary, (iv) the Selection Principle and (v) the Tectonic Predesign Principle.

The Antagonism Principle is the most fundamental principle of landscape evolution. It refers to the fact that there are essentially *two* types of processes active in the formation of a landscape at the same time: the endogenic (tectonic) and the exogenic (climate-driven) processes. Of fundamental importance is the fact that the two processes involved in the geomorphic antagonism have a *fundamentally different stochastic nature*: Endogenic processes are essentially non-random, exogenic ones essentially random (Scheidegger, 1979). The reason for this is that endogenic processes are connected with plate tectonic movements which are autocorrelated over thousands of kilometers, whereas exogenic processes are connected with turbulence phenomena in liquids (water and air) which are essentially chaotic. Although the endogenic processes must be basically primary (without uplift there could be no degradation), the two processes are often assumed to more or less balance each other; thus the actual aspect of a landscape corresponds to the instantaneous

SSDI 0169-555X(94)00014-I

dynamic equilibrium between them. If one parameter is changed by, say, tectonic processes or human activity, the others adjust themselves accordingly to re-establish equilibrium (*Process–Response Theory* — cf. Hack, 1960; Carson and Kirkby, 1972 Strahler, 1980; Terjung, 1982 and many others).

This idea has to be modified by the observation that the "balance" is quite often unstable (Scheidegger, 1983). Apart from the fact that individual surface features tend to be impermanent, although their overall appearance may seem to be constant corresponding to the process–response theory, the dynamic equilibrium may actually be inherently unstable: any deviation from uniformity tends to grow; there is, so to speak, a positive feed-back between the size of the deviation that has already been reached and the rate by which this size increases. Although the dynamic equations of the system are unknown, the (presumably) nonlinear differential equations can be linearized for the purpose of stability analysis according to Taylor (1950), at least in the neighborhood of the deviation for a short time range. Then, for any parameter $x(t)$ (this may be a vector) describing a landscape feature, the corresponding linearized differential equation is:

$$dx(t)/dt = \lambda x \tag{1}$$

where λ is a parameter. Its solution is an exponential function of time; thus the growth-equation at the beginning has the form:

$$x(t) = C \exp(\lambda t) \tag{2}$$

where C is a constant of integration. If λ is positive, one has positive feedback and the condition is unstable, the deviation $x(t)$ will grow exponentially and the value of λ is a measure of the degree of the instability. The exponential function would increase infinitely so that it would lead to a catastrophic situation; however, the linearization holds true only for short time ranges so that an unstable state need not necessarily lead to a catastrophe: the growth process may eventually come to a stop when a *saturation stage* is reached.

The instability model is evidently correct for many purely exogenic geomorphic features: the longitudinal profile of stepped valleys, the terraced transverse profiles of valleys, sequences of pools and riffles in river beds, the meandering of rivers, the formation of glacial cirques and certain dissolution phenomena (such as "onion-skin" weathering). Some of the above features can also be described as a consequence of a *Catena Principle*, which is very closely related to the instability principle. In effect, such sequences are "chains" (Latin: *catenae*) of geomorphological elements that are repeated again and again; this is the consequence of instabilities that arise at certain points but reach a "saturation stage" so that the corresponding features are repeated in a sequence. One can express this as a separate principle, namely the *Catena principle* (Scheidegger, 1986). The concept of a *catena* was originally invented by Milne (1935, 1947) in soil research who noted that certain definite sequences (catenas) of soil types recur on a slope; it was then adopted generally for self-repeating morphological features, such as eluvial–colluvial–alluvial zones on slopes (Scheidegger, 1986), slow (wide) – fast (narrow) – slow (wide) moving stretches in rivers (Scheidegger, 1986), flat (slow-moving) – steep (fast moving) – flat (slow-moving) sections on chronic clay slides (Gerber and Scheidegger, 1984) and fine–coarse–fine grained mass flow sequences on scree slopes (Gerber and Scheidegger, 1974).

A further aspect of the exogenic processes is that they tend to occur in such a fashion that statically stable forms (with reference to the self-induced stresses) are preferentially "selected"; this fact is stated by the *Selection Principle* of Gerber (1969).

Finally, the remaining landscape principle of those mentioned at the beginning states that many landscape features are designed by deep-seated tectonic processes (*Tectonic Predesign Principle* — Twidale, 1971; Morisawa, 1975; Scheidegger and Ai, 1986). This principle is really part of the Antagonism Principle; namely that part which refers to the endogenic processes.

2. Complexity theory

Landscapes are very complex systems. It is therefore only natural that geomorpholgists have tried to apply general systems theory to them (e.g. Chorley, 1962). The term "system" was created by Von Bertalanffy (1950). To speak of a "system" one needs (i) a set of elements with some variable attribute of objects, (ii) a set of relationships between attributes and objects, and (iii) a set of relationships between those attributes of objects and the "environment", i.e., the region outside the system. *Macroscopically*, in geomorphology, the "objects" are landforms, the "attributes" are quantifiable properties of these landforms, and the "relationships" consist of exchanges of energy and mass between the landforms so that causal links are formed between the attributes. *Microscopically*, however, landscapes are *mechanical* systems, i.e., they consist of a great number of particles each of which has a position and a momentum. The state of such a system is represented by a point in phase space, i.e., in the multidimensional space which has one dimension for each coordinate and each momentum of each particle. The *evolution* of the state of the system is described by a *trajectory* in phase space.

First, landscapes were treated as closed systems; i.e., systems with no interactions with any "outside". One can then immediately draw an analogy with equilibrium thermodynamics and define analogs of temperature and entropy. The conclusion is that such landscapes tend to develop towards an equalization of all relief at an average level. In a closed system in equilibrium, the phase trajectories over time are equally dense throughout the phase space (ergodic hypothesis). The condition that the entropy tends towards its possible maximum, allows one to calculate the most probable values of relevant variables in particular systems such as meander trains (Langbein and Leopold, 1966).

More realistic is a view of landscapes as *open* systems, i.e., systems which have an interaction with an "environment" (in this, the distinction between "system proper" and "environment" is by arbitrary fiat). The landscape is then considered as a system through which mass and energy flow or "cascade" from an input to an output (Terjung, 1982). The analogy with equilibrium thermodynamics is no longer possible; the macroscopic variables behave in a complex way (hence their study has been called the "Science of Complexity"), the phase trajectories are no longer ergodic; in fact, the latter tend towards certain "attractors" in phase space. If the attractors are fixed points, one has a trend towards stable equilibrium, if they are limit-cyclical or toroidal, one has certain stable evolution patterns, if they are otherwise (strange attractors), the behavior is generally unstable. Such systems that are nonlinear, nonequilibrium, deterministic, and incorporate randomness so that they are sensitive to initial conditions, and have strange attractors, are defined as *chaotic* (definition after Çambel, 1993, p. 14). In such systems, the entropy need not tend towards a maximum, in fact, it may well decrease (in contradiction to the second law of thermodynamics — Prigogine and Stengers, 1984). Due to the form of the attractors some order can establish itself at the edge of chaos, at least for a certain length of time. Empirically, certain "laws" have been found to hold in self-ordered systems at the edge of chaos (Bak et al., 1988): The pertinent characteristic observables are spacially and temporally scale-invariant, i.e. they have been found to be fractal. In a fractal set of dimension D, there exists a power law of subsets: the number N of subsets of (linear) size D is proportional to L^{-D}. In terms of size (M)–frequency (N number of events per unit time) distributions, the power law mentioned above is represented by a "Gutenberg–Richter (e.g. 1949) type" law:

$$\log N(M) = a - bM \tag{3}$$

which is a "power law" for N. Such power laws have been found everywhere in self-ordered natural systems. Since there are positive feedbacks in such systems, the process–response theory is no longer tenable.

3. Interpretation of landscape principles in the light of complexity theory

The modern views on the behavior of complex systems sketched above allow a somewhat different interpretation of the phenomenological landscape principles than was common hitherto. The basis is evidently the Principle of Antagonism. The endogenic processes, and therewith also the Tectonic Predesign Principle, correspond to the input into a mechanical system originating from the "environment". At the other end of the spectrum are the exogenic processes based on atmospheric and aquatic turbulence whose input is essentially chaotic (see Section 1). The aspect of actual landscapes, then, does not correspond to a "dynamic equilibrium," but to self-organized order at the "edge of chaos" in an open, dissipative system. Orderly evolution is still subject to Taylor instabilities (positive feedback) for deviations from this order; these grow at the beginning exponentially as implied by Eq. (1); however, since the system is inherently nonlinear these deviations do not follow an unbridled growth equation of the the type of Eq. (1), but rather a "logistic equation'' like that of Verhulst (cf. the review by Çambel, 1993):

$$dx(t)/dt = rx(t)[1 - x(t)] \quad (4)$$

where x is a normalized deviation-variable and r a system parameter. Although the Verhulst equation is nonlinear, it can still be integrated in closed form; its solution is:

$$x(t) = 1/[1 + \exp(-rt)] \quad (5)$$

This equation represents a sigmatoid growth curve for the deviation $x(t)$ which tends towards a *saturation value* of 1. Examples of such a saturation occurring are represented for instance by the fact that meander loops cannot grow indefinitely in size, slopes cannot become infinitely high, pools and riffles in river beds cannot become arbitrarily large, etc. (Scheidegger, 1983).

Steady-state landscapes correspond to reasonably stationary conditions of self-organized order at the edge of criticality with a power law (Eq. 3) for spacial and temporal correlations (Bak et al., 1988). In the critical state, there are no natural length and time scales, so that fractal statistics are applicable (Turcotte, 1992). Computer simulations have shown that such a self-organization can indeed establish itself in stochastic situations solely under the assumption of micro-scale relationships between individual elements of the system. Thus, Werner and Hallet (1993) have shown that solely the assumption of preferential growth patterns of individual ice-needles in stone-free regions during the freezing cycle on (periglacial) slopes automatically produces self-organized stone stripes in a corresponding numerical model. The results closely agree with the stone stripes found on such slopes in the glacial region of Mauna Kea, Hawaii.

The basis of the process–response theory is that one *presupposes* small changes or perturbations in some of the external parameters describing the system to cause only small and continuous adjustments in the others so as to re-establish equilibrium: The pattern of the landscape development is described by a smooth, well-behaved curve representing the dependence of one parameter on the values of the others; a model that hardly corresponds to the manifest facts.

4. Hazards

4.1. General remarks

Hazards are represented by the possibility that a reasonably stable orbit in system–phase space may change abruptly. In this, both, endogenic and exogenic processes can be involved: if the normally continuous relations between the parameters describing geomorphic variables (process–response relations) contain singularities, and if such a singularity is encountered during the geomorphic evolution (a perfectly ''natural'' possibility), an event appearing to an observer as a ''catastrophe'' occurs.

First of all, it is possible that the primary input from the endogenic processes is not steady and continuous: The endogenic processes occur normally in a steady-state fashion so that they contribute to an orderly development of the Earth's surface; however, they, too, can become unstable and become chaotic which expresses itself to an observer in the form of earthquakes, volcanic eruptions, etc. However, our main concern in this paper is not with such ''primary'' instabilities in the endogenic input, but rather with the self-created instabilities occurring during the otherwise continuous and regular geomorphic evolution.

Second, it is well known that even basically regular process–response dependences may become multivalued at junctions, cusps etc. (cf. the ''catastrophe'' theory of Thom, 1972); in such cases, small random perturbations can cause the system to jump from one branch of the process–response curve to another: a natural disaster occurs.

Third, a general description of instability theory can be given in terms of attractors. The phase point describing a (temporarily) stable landscape system, as we have seen above, generally lies in a self-organized domain at the edge of chaos under the influence of a strange attractor: thus it should be clear that extremely small accidental perturbations can cause the system to become unstable leading to rapid changes in the landscape which are experienced as a disaster. Cases where minute changes in the initial conditions lead to a ''catastrophic'' long-term development occur under many ''ordinary'' conditions.

4.2. Slopes and landslides

As a first example, we use denudational slope-development for which a general differential equation has been developed and discussed at length regarding its applicability by the author (Scheidegger, 1961):

$$\partial y/\partial t = -(\partial y/\partial x)\sqrt{1+(\partial y/\partial x)^2} \quad (6)$$

where x is the abscissa (distance) and y the height of the slope. As it stands (and as it was used in 1961), it is assumed that the slope height y is monotonously increasing (or at least: not decreasing) with increasing x; if it is decreasing, $(\partial y/\partial x)$ becomes negative and the slope would become higher by erosion, which is manifestly impossible under natural conditions. If Eq. (6) is to be applied to slopes decreasing with increasing x, it must be applied by replacing the partial differential by its absolute value:

$$\partial y/\partial t = -|\partial y/\partial x|\sqrt{1+(\partial y/\partial x)^2} \quad (7)$$

This is a highly nonlinear partial differential equation. When solving it on a computer by the method of finite differences, questions of mathematical stability are of utmost importance. Firstly, the conditions resulting from the characteristics have to be observed (for background see e.g. Collatz, 1951); however, this is only a necessary, not a sufficient condition. Thus, it turns out, secondly, that, to ensure stability, the differentials have always to be taken in the direction of decreasing y. Thirdly, the iterations must only be continued to a point where the boundary conditions (y at $x=0$, 100) do not make themselves felt. The solutions obtained in 1961 only involved slopes increasing (or at least not decreasing) in the direction of $+x$, so that Eq. (6) could be used. We have now modified the program to be applicable also to slopes decreasing in the direction of $+x$, observing the conditions mentioned above. All solutions obtained were mathematically stable, but an knick or hole in a smooth slope led now to a positive-feedback mechanism. As an example, we show here profiles of the regular non-hazardous development of a smooth slope bank (Fig. 1), and the development of the same bank with a ditch (through which all eroded material is thought to be removed) at the foot (Fig. 2). Figs. 1 and 2 differ only by the initial conditions (top curve); the units are normalized, i.e. $0 \leqslant x \leqslant 100$; $\Delta x = \Delta y$; $\Delta t = \Delta x$ (at least for the representation; for the iterations, smaller time steps had to be taken). It is clear that the development of a slope without a ditch is regular (or stationary), if there is a ditch, the latter becomes rapidly deeper, until the critical height (Fellenius, 1940) of a stable slope bank (at least for a small angle of internal friction of the slope material) is inexorably (since the slope *angle* stays more or less the same) exceeded and a slide occurs. Thus, a minute change in the initial conditions (i.e. introducing a small ditch) causes a fundamentally different long-term behavior. This is characteristic for complex systems.

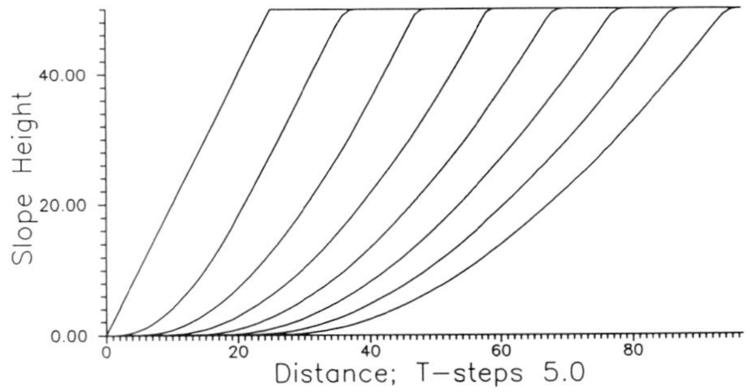

Fig. 1. Regular uniform recession of a smooth slope bank. Regarding units: see text.

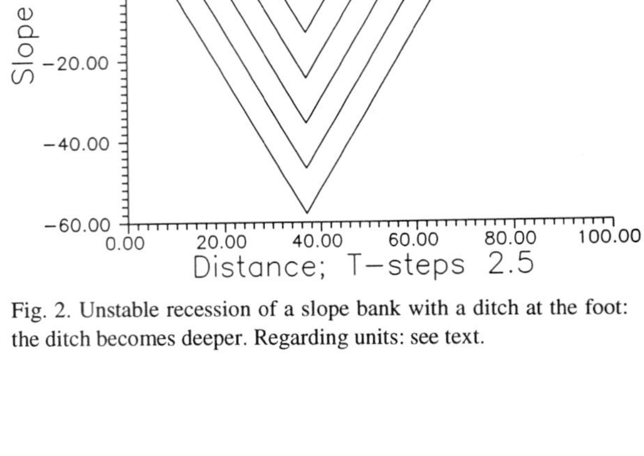

Fig. 2. Unstable recession of a slope bank with a ditch at the foot: the ditch becomes deeper. Regarding units: see text.

The slope bank considered above has been assumed as somehow a priori given. Evidently, it is at the edge of criticality. An initial slope bank can also be the result of *self-organized* criticality (regarding this concept cf. Nicolis and Prigogine, 1977; Haken and Wunderlin, 1988 and Kauffman, 1993); the best known example in geomorphology is that of a sand or scree slope that has been formed by the steady addition of material from above (Bak and Chen 1991; see also Waldrop, 1992, p. 305). In this case the mass-movement is caused by gravity alone; the (temporary) stability is due to the interlocking of the sand or scree particles. Further additions of particles cause intermittent slides at irregular intervals; the number of slides in a given time interval (assuming the particle-addition rate as constant) as a function of size follows a power-law distribution (the average frequency of a slide of a given size is inversely proportional to some power of its size); as noted a feature of complex systems. It may be assumed to hold for the landslide size versus frequency distribution if the latter are due to a constant mass input from endogenic (increase of height of relief due to tectonic uplift as a result of the tectonic plate motions) and consequent denudation by exogenic processes.

4.3. Planes and sink holes

As a second example we refer to the development of a plain with a hole in it (without a hole, a horizontal plain does not denude at all; i.e. its profile remains a horizontal straight line forever). Fig. 3 shows the profile of such a plane with a small knick in it, using the usual slope-development equation (Eq. 7): this solution is for a linear case but is qualitatively the same for circular coordinates. The original knick can be assumed to have been caused by random fluctuations in the denudation processes; a good example would be the formation of a small solution-hole in a karst plateau which develops into a huge sink-hole. Such cases are well known from the limestone platforms of Yucatan and Florida: in Yucatan, deep "Cenotes" (water-filled holes) represent characteristic landscape features, and in Florida sinkholes with houses disappearing in them (cf. the Winter Park event: Foose, 1981) are notorious. Thus, again, small changes in the initial conditions cause large effects in the long run which are experienced as hazards.

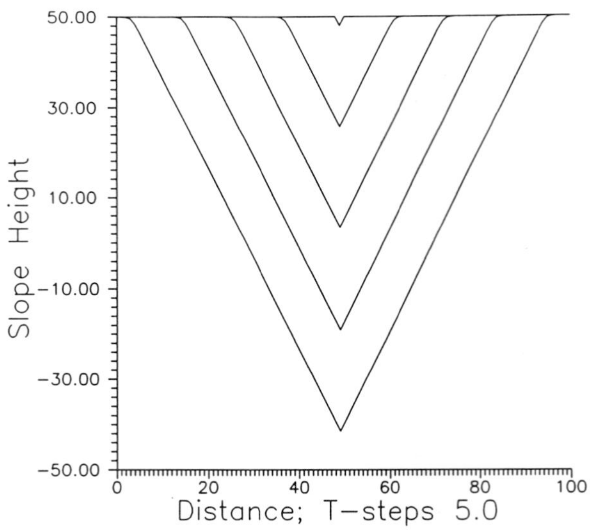

Fig. 3. Deepening of a hole in a plane.

5. Conclusions

The above analysis shows that geomorphic systems are principally open and highly nonlinear with the possibility of positive feedback. They cannot be treated by ordinary system theory; one has to use concepts from the science of complexity and non-equilibrium thermodynamics (i.e. its system-analog). "Stationary" landscapes are in reality self-organized patterns (cf. stone stripes) at the edge of chaos; small changes in the initial conditions can cause very different long-term behavior-patterns: thus the "process–response" concept is suspect *in principle*. "Hazards" are the result of sudden changes in long-term behavior caused by minute changes in the initial conditions. These can be the result of input from tectonic processes into the system and the resulting establishment of self-organized criticality. In this case, the hazard events follow generally a power-law distribution for size versus frequency, provided the input occurs at a constant rate. Hazards are therefore simply part and parcel of the normal geomorphic evolution: the steady-rate build-up of stress (input into the geomorphic system) by the plate-tectonic motions causes intermittent catastrophes with a power law (Eq. 3) characterizing the size-frequency distribution.

References

Bak, P. and Chen, K., 1991. Self-organized criticality. Sci. Am., 1991(1): 46–53.

Bak, P., Tang, C. and Wiesenfeld, K., 1988. Self-organized criticality. Phys. Rev., A38: 364–374.

Çambel, A.B., 1993. Applied Chaos Theory, A Paradigm for Complexity. Academic Press, Boston, 246 pp.

Carson, M.A. and Kirkby, M.J., 1972. Hillslope Form and Process. University Press, Cambridge, 475 pp.

Chorley, R.J., 1962. Geomorphology and General Systems Theory. U.S. Geol. Surv. Prof. Pap., 500B: 10 pp.

Collatz, L., 1951. Numerische Behandlung von Differentialgleichungen. Springer, Berlin, 458 pp.

Fellenius, W., 1940. Erdstatische Berechnungen mit Reibung und Kohäsion (Adhäsion) und unter Annahme kreiszylindrischer Gleitflächen; 2. erg. Aufl. W. Ernst, Berlin, 48 pp.

Foose, R.M., 1981. Sinking may be fast or slow. Geotimes 26(8): 21–22.

Gerber, E.K., 1969. Bildung von Gratgipfeln und Felswänden in den Alpen. Z. Geomorphol. Suppl., 8: 94–118.

Gerber, E.K. and Scheidegger, A.E., 1974. On the dynamics of scree slopes. Rock Mechanics, 6: 25–38.

Gerber, E.K. and Scheidegger, A.E., 1984. Eine chronische Rutschung in tonigem Material. Vierteljahrsschr. Naturforsch. Ges. Zürich, 129(3): 294–315.

Gutenberg, B. and Richter, C.F., 1949. Seismicity of the Earth and Associated Phenomena. Princeton Univ. Press, 211 pp.

Hack, J.T., 1960. Interpretation of erosional topography in humid temperate regions. Am. J. Sci., 258A: 80–97.

Haken, H. and Wunderlin, A., 1991. Die Selbststrukturierung der Materie. Vieweg, Braunschweig, 466 pp.

Kauffman, S.A., 1993. The Origins of Order: Self-Organization and Selection in Evolution. Oxford Univ. Press, 709 pp.

Langbein, W.B. and Leopold, L.B., 1966. River meanders — Theory of minimum variance. U.S. Geol. Surv. Prof. Pap., 422H: 1–15.

Milne, G., 1935. Some suggested units of classification and mapping, particularly for East African soils. Soil Res., 4: 183–198.

Milne, G., 1947. A soil reconnaissance journey through parts of Tanganyika Territory. J. Ecol., 27: 192–265.

Morisawa, M., 1975. Tectonics and geomorphic models. In: W.N. Melhorn and R.C. Flemal (Editors), Theories of Landform Development. Allen and Unwin, London, pp. 199–216.

Nicolis, G. and Prigogine, I., 1977. Self-Organization in Non-Equilibrium Systems. 8th printing. Wiley, New York, 491 pp.

Prigogine, I. and Stengers, I., 1984. Order out of Chaos. Bantam Books, Toronto, 349 pp.

Scheidegger, A.E., 1961. Mathematical models of slope development. Bull. Geol. Soc. Am., 72: 37–50.

Scheidegger, A.E., 1979. The principle of antagonism in the Earth's evolution. Tectonophysics, 55: T7–T10.

Scheidegger, A.E., 1983. Instability principle in geomorphic equilibrium. Z. Geomorphol., 27(1): 1–19.

Scheidegger, A.E., 1986. The catena principle in geomorphology. Z. Geomorphol., 30(3): 257–273.

Scheidegger, A.E., 1987. The fundamental principles of landscape evolution. Catena Suppl., 10: 199–210.

Scheidegger, A.E. and Ai, N.S., 1986. Tectonic processes and geomorphological design. Tectonophysics, 126: 285–300.

Slingerland, R., 1981. Qualitative stability analysis of geologic systems with an example from river hydraulic geometry. Geology, 9: 491–493.

Strahler, A.N., 1957. Quantitative analysis of watershed geomorphology. Trans. Am. Geophys. Union, 38: 913–920.

Strahler, A.N., 1980. Systems theory in general geography. Phys. Geogr., 1:1–27

Taylor, G.I., 1950. The instability of liquid surfaces when accelerated in a direction perpendicular to their planes. Proc. R. Soc. Lond. Ser. A, 201: 192–196.

Terjung, W.H., 1982. Process–Response Systems in Physical Geography. F. Duemmler, Bonn, 65 pp.

Thom, R., 1972. Stabilité stucturelle et morphogénèse; essai d'une théorie générale des modèles. Benjamin, Reading PA, 362 pp.

Turcotte, D.L., 1992. Fractals and Chaos in Geophysics. University Press, Cambridge, 221 pp.

Twidale, C.R., 1971. Structural Landforms. MIT Press, Cambridge, MA, 247 pp.

Von Bertalanffy, L., 1950. An outline of general systems theory. Br. J. Philos. Sci., 1: 134–165.

Waldrop, M.M., 1992. Complexity. Simon and Schuster, New York, 380 pp.

Werner, B.T. and Hallet, B., 1993. Numerical simulation of self-organized stone stripes. Nature, 361: 142–145.

ELSEVIER

Geomorphology 10 (1994) 27–36

GEOMORPHOLOGY

The geomorphological dimensions of natural disasters

Charles L. Rosenfeld*

Department of Geosciences, Oregon State University, Corvallis, OR 97331-5506, USA

Received March 31, 1994; accepted April 6, 1994

Abstract

The International Decade for Natural Disaster Reduction (IDNDR) was designed to focus the attention of scientists, engineers and economists on the increasing losses and deaths occurring globally as a result of natural hazards. Geomorphologists have considerable expertise in process studies and the mapping of precursors and antecedent conditions of surficial phenomena, and important understanding of "magnitude and frequency" concepts applied to the earth sciences. This paper examines several projects where geomorphologists provide significant contributions to interdisciplinary research, and suggests a framework for a comprehensive approach to natural hazards studies.

1. Introduction

The dramatic increase in losses and casualties due to natural disasters during the past three decades has prompted a major international scientific initiative into the probable causes and possible mitigation strategies. The damage has increased three-fold from the 1960's through the 1980's, leaving more than three million dead and causing the displacement of more than 800 million persons during that period. The U.S. National Academy of Sciences (NAS), prompted by the ominous trend of disaster losses, presented an initiative to the United Nations in December of 1987. The UN subsequently established a secretariat for the IDNDR in April of 1989, located in Geneva, Switzerland along with the UN Disaster Relief Organization (UNDRO). Many of the crucial questions are related to geomorphological processes, therefore the International Council of Scientific Unions (ICSU) has given tasks to several of its member unions with specific scientific themes during the International Decade of Natural Disaster Reduction (IDNDR) that include geomorphology.

As the population of the Earth has doubled from the three billion of 1960, the annual losses due to disaster has grown almost ten-fold (Bruce, 1993). At least part of the problem can be traced to the increase in population density of many regions, which often results in the human occupance of geomorphically hazard-prone lands that were previously avoided. But many states and cultures have not emphasized "flood-proofing" buildings in floodplains or coastal areas, earthquake design standards in seismic hazard areas, or building codes to withstand windstorm or fire dangers. Often it has not been for lack of knowledge or vigilance, but for lack of resources or unwillingness to divert limited national wealth to such causes, especially in developing economies. Even in developed countries, governments have been reluctant to impose costly building standards and zoning regulations in the most vulnerable areas. We might well ask: what if Hurricane Andrew had focused a few kilometers to the north and struck downtown Miami?

*Chairman IGU Commission on Natural Hazards Studies.

SSDI 0169-555X(94)00029-Q

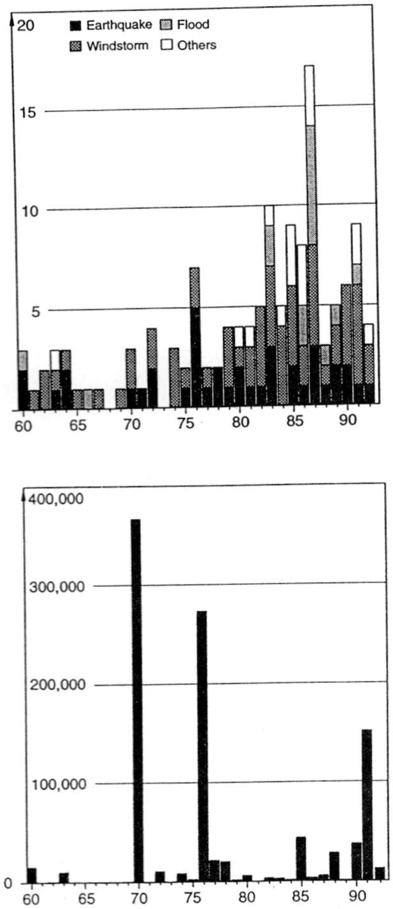

Fig. 1. Annual total of fatal hazard events with deaths and missing persons directly attributed to natural disasters, 1960–1992 (after Berz, 1992).

Another perplexing question centers upon the apparent increase in frequency of natural disasters. Accurate statistical analyses of such infrequent occurrences requires an observation period well in excess of a century (Berz, 1993), while the consistent reporting of most disaster types has a much shorter history. Monitoring techniques, measurement scales (e.g. Richter, Beaufort), and communications have only recently allowed the global reporting of events in comparable terms. A study of volcanic eruptions by Simpkin et al. (1981) concludes that the reported increase in volcanic activity over the past 120 years is almost certainly due to improved reporting and communications technology; they even cite a reduction in "apparent" activity during the two world wars. Fig. 1 (after Berz, 1993) shows the trends of earthquakes, floods, windstorms and other casualty producing natural events since 1960. Despite the growing influence of global databases, scientific consortia and the news media, additional factors may be influencing the growing number of reported disasters.

While considerable scientific debate lingers around the issue of global climatic change, geomorphologists are well aware of other indirect effects of human activity. Certainly the deforestation of large areas has caused landslides (Swanson and Dyrness, 1975) and increased both the frequency and peak flows of flood events in many areas (Kaith, 1960; Harrold, 1971; Kahn and Nazem, 1988), while over-grazing has accelerated drought effects and erosion (Berry, 1984). Groundwater withdrawal and irrigation diversions have affected the natural vegetation and micro-climates of some regions (Tweedie, 1966), and has even induced earthquakes in some cases (Evans, 1966).

As global climate models become increasingly realistic, their mathematical results consistently point to a more hazardous world in the future (Lohman, 1993; Office of Technology Assessment, 1993). That increasing concentrations of greenhouse gases in the atmosphere, primarily resulting from the burning of fossil fuels, is changing the radiation balance and perhaps the climate is consistent with recent disaster experience. There is general agreement among atmospheric scientists that a "warmer" world would be a "wetter" world, with no increase in the number of days with rain, but with more intense rainfall (Hively, 1988). Combined with the hydrologic effects of landuse changes, the frequency and effect of floods would surely increase, especially in the monsoon climates of South Asia where flooding already reaches catastrophic proportions. Drought effects in sub-Saharan Africa, South America and Australasia could occur more frequently and be more severe as a result of intensified El Niño–Southern Oscillation events (UNESCO, 1963). Resultant sea level rise could pose additional storm surge or tsunami risk to low-lying coastal regions such as Egypt or Bangladesh, and many Pacific islands, with the loss of most fresh groundwater resources in the latter case (Reynolds, 1990). Regardless of the causes, the impacts of anticipated changes in extreme weather hazards as a result of global climatic change, and their implications for human activity, demands the immediate attention of geomorphologists.

Recent experience has shown that disaster planning and early warning systems have resulted in variable successes. In Japan, major landslides (primarily debris flows) triggered by heavy typhoon rains were responsible for the loss of more than 130,000 homes and nearly 500 lives in 1938. But in 1976, following the government's comprehensive landslide control program, the worst storm-produced landslide year in over two decades cost less than 2000 homes and fewer than 125 lives — and this reduction trend is still improving (Watari, 1988). In Bangladesh a cyclonic storm pushed a 3 to 8 meter storm surge through Sandwhip and other coastal islands killing more than 10,000 people in 1985. However, a satellite-based early warning system aided evacuation of island and coastal dwellers to points further inland, which reduced the toll by 30 to 50 times from that of a similar storm in 1970 (Rasid and Pramanik, 1990). Despite these successes, and more than 50 years of comprehensive flood control development, the 1993 floods along the Mississippi valley in the American mid-west were the costliest flood events in U.S. history (Bottorff, 1994). In the Philippines, flooding triggered by Typhoon Uring in November 1991 struck Ormoc City in Leyte Provence with a 3 meter flood crest. Although the storm itself was monitored, the lack of hazard zoning, and poor infrastructure design in the urban area left nearly 7000 people dead or missing (Pearson and Oliver, 1992).

In some instances, disaster mitigation strategies and international relief efforts may actually be partially responsible for the rising losses. In many developed countries, state-backed hazard insurance programs are designed to encourage the use of hazard zoning and the implementation of damage resistant building codes and a reduction in the demand for structural control measures; however, this may have actually encouraged the development of hazard-prone areas through the combined effect of lower land costs and cheap indemnity. Thus the ''insured'' actually transfers the risk to the ''insurer'', and may dismiss the risk or reduce the individual concern for loss prevention measures. Fig. 2 shows the rapid increase in both economic losses and in insured losses since 1960 (Berz, 1993). One critical problem is that insurance rates are calculated based upon the loss experience of the past; however, given the escalating nature of both frequency and magnitude of disaster losses this may not be appropriate. Likewise, many lesser developed countries have developed a

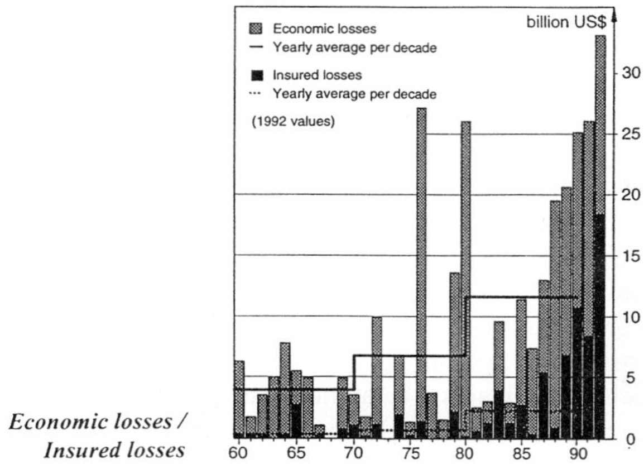

Fig. 2. Annual total of economic losses, showing the insured losses, from 1960–1992, with averages shown for each decade (after Berz, 1992).

dependance on international disaster relief, whereby they become less willing to divert their limited national resources to immediate relief or recovery efforts.

2. Role of the geomorphologist

Geomorphologists are increasingly engaged in the mapping of geophysical, hydrological and surficial materials characteristics which expose areas to rupture, failure, fire, inundation, drought, erosion or submergence. Coupled with landuse and infrastructure analyses they examine the location, value, exposure, and vulnerability of the structural environment to hazard damage, failure, loss of function, and release of hazardous materials. When population density and demographics are added, potential casualties and emergency services needs can be forecast. This, of course, involves the integration of social scientists whose focus upon social, technical, administrative, political, legal and economic forces which structure a society's strategies and policies for risk management (i.e., prevention, mitigation, preparedness, prediction and warning, and recovery), public awareness, training, regulation and social insurance. Such a comprehensive approach would have been nearly inconceivable in the past, but with the advent of computerized geographic information systems (GIS) such mapping, modeling, and decision support systems are becoming more common place. The United Nations Institute for Training and

Research (UNITAR) has sponsored several training workshops entitled "Explorations in Geographic Information Systems Technology", operated by the Graduate School of Geography at Clark University (USA). A similar program is operated by the International Institute for Aerospace Survey and Earth Sciences (ITC) in the Netherlands.

While satellite remote sensing has become a routine tool for land surface classification and mapping, the more recent fusion of these methods with the Global Positioning System (GPS), and integration with GIS technology marks a catalytic change in our approach to geographic data collection. Modern field-portable (laptop) computers can now be equipped with GPS receivers to accurately provide field locations anywhere on earth from a "constellation" of 26 Navstar satellites. As the computer displays the location on a digital map or remotely sensed image, the researcher can map or update a GIS data base in the field. The use of such technology will undoubtedly expand rapidly in the near future as more applications develop and as the costs of the required equipment drop. A "pen-based" laptop computer, appropriate software and an attached GPS receiver currently costs about US$8000. Six computer manufacturers have announced plans to release similar computers with "built-in" GPS receivers within the next 18 months (Penmetrics, 1993). The author has operated one of these GPS-equipped computers while mapping flood refuge sites in Bangladesh. This computer did not require a keyboard as it was designed to "recognize" mapping symbols and handwriting from an electronic "pen". Field data from numerous field research teams can quickly be merged electronically into a comprehensive GIS. Other aspects of digital data technology are emerging as significant future contributors to natural hazards studies; the Digital Chart of the World (Defense Mapping Agency, 1992) makes digital cartographic databases available on low-cost compact discs, and many other geophysical and hydrologic data resources are also available in this format, enabling researchers to access long-term records of hazard activity.

3. A comprehensive approach to hazards studies

A significant "bridging function" by the ICSU is resulting in the collaboration of diverse scientific unions under IDNDR. Unfortunately, other UN global scientific programs, such as the International Geosphere–Biosphere Program (IGBP) and the Human Dimension of Global Change Program (HDGP) were independently structured and funded, resulting in separate program functions with limited interaction. IDNDR considers both the physical aspects of hazards as well as their social consequences, and involves geomorphologists from several member unions including the International Geographical Union (IGU), the International Union of Geological Sciences (IUGS), the International Union of Geodesy and Geophysics (IUGG), the International Association of Engineering Geology (IAEG) and the International Commission for the Lithosphere, Habitat (UN).

This comprehensive approach to natural hazards studies may be graphically portrayed as a pyramid (Fig. 3) incorporating a temporal scale of the duration of hazard events, a spatial scale proportional to the areas affected by each hazard event, an economic scale relative to the level of development of the society affected, proportional casualties and economic losses, and levels of response required to meet the impacts. The duration, recurrence interval and areal extent of natural hazards affects both our observation base and our understanding of such events. For example, while the mechanics of mass wasting are well understood, their "triggers" are often less clear and the antecedent conditions preceding their occurrence may be too widespread and numerous to monitor effectively. However, adequate safety measures are possible, as seen in Japan's example, provided that economic and social circumstances warrant them. The ability of a society to deal with the effects of disasters to some extent varies with the size of the area affected and the degree to which the infrastructure can retain its function. The 1972 earthquake which struck Managua, Nicaragua caused 10,000 deaths, 20,000 injuries and caused such damage to the infrastructure that three quarters of the city's 400,000 inhabitants were left homeless (National Academy of Sciences, 1987). Given Nicaragua's state of economic development, the losses exceeded the nation's entire gross national product.

To explore the role of economic development upon the effects and response to natural disasters, we divide states into lesser developed countries (LDC's), transitioning economies (TE's), and developed countries (DC's). While some economists have observed a

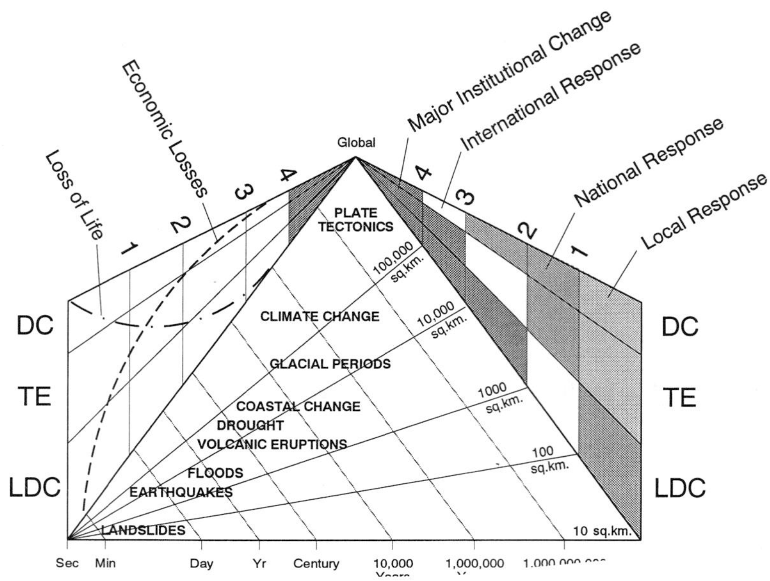

Fig. 3. A comprehensive approach to natural disaster analysis may be graphically portrayed as a pyramid: the front face depicts the relative duration and areal extents of different hazards, the left side shows that casualties escalate with hazard severity in lesser developed countries, while developed countries suffer greater economic losses. The right face illustrates the level of response usually required to meet disaster conditions as a function of the economic development of the affected area (computer graphic by A. Jon Kimerling).

growing gap between the rich and poor nations of the world, there has been an increase in the number of industrializing transitional economies. The losses sustained by states at varying levels of economic development is shown on the left face of our pyramid: as the severity of disasters increase, there is an exponential rise in the number of casualties among the poorer nations, while the developed countries, with better warning systems, planning and construction regulations, and public services suffer fewer fatalities. At the same time, property losses can be extreme under similar circumstances in those developed countries. The 1994 San Fernando valley earthquake which centered in urban Los Angeles caused billions of dollars of property damage, with the loss of less than 70 lives (Schweig, 1994). The casualty and economic loss curves converge in transitioning economies, where new investment in industrialization often takes higher priority than the costs of hazard mitigation. Awareness of the risks of natural hazards to economic development is essential among the leaders of these nations, as the general population has a heritage of enduring such events, with recovery left to personal initiative — this does not work in the industrial world. There is, however, a threshold of extreme events beyond which cataclysmic losses occur, regardless of a nation's wealth. Geomorphologists, familiar with magnitude and frequency concepts, have little difficulty envisioning this model.

The opposite face of our pyramid depicts the levels of emergency response that are generally required for recovery at the various severity levels of disaster. Once again, our relative severity scale ascends from (1) significant events, moderately exceeding "normal" conditions, (2) exceptional events, beyond the required design standards for routine construction, but below those set for "key" facilities, (3) catastrophic events which exceed any cost-effective design strengths, and (4) extreme events yielding almost total destruction of structures and support systems. Beyond the obvious relation between response containment and level of economic development there is an institutional dimension that is frequently overlooked by major national and international hazards research programs: geologic and geophysical hazards can be identified at large scales, affecting nations or regions, but vulnerability and risk assessments need to be carried out at the detail of human activities, generally at scales of 1:1000 to 1:50,000. Geomorphologists have traditionally concentrated their mapping at these scales (Tricart, 1965;

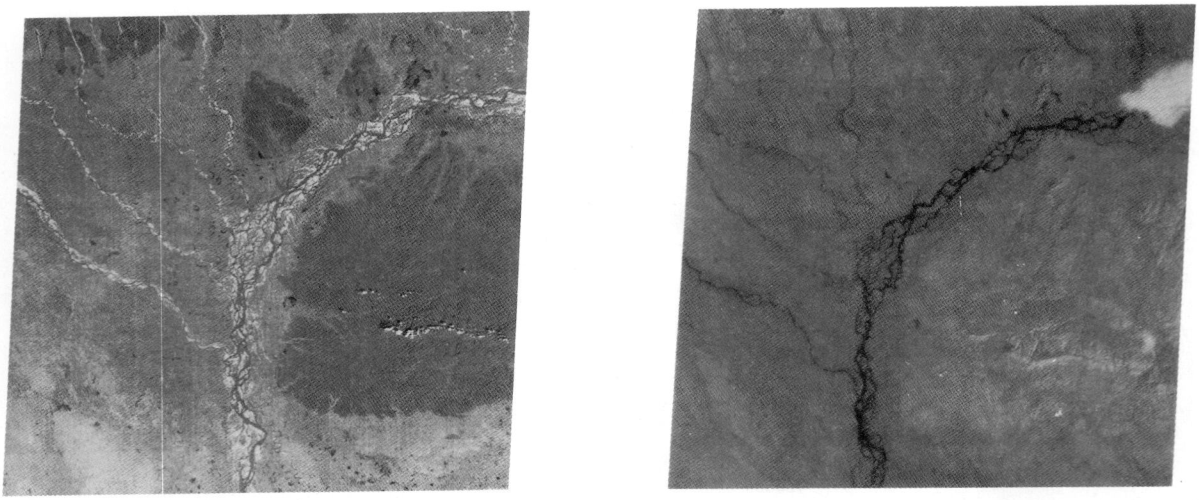

Fig. 4. Landsat MSS imagery of the Brahmaputra–Jamuna river system before and after the 1977 floods (courtesy of NASA–Ames Research Center).

Demek, 1972; Verstappen and Van Zuidam, 1989) where many other fields of the earth sciences have taken a smaller scale "synoptic" approach. Priority for hazards assessment needs to center on those impacts and response options at the local level, as they are the most immediate, and potentially the most crucial at severity levels 2 and 3. Level 1 events can usually be accommodated within "normal" risk management capabilities, except in the least aware societies, while level 4 is beyond the maximum recovery capability of even the wealthiest of states. Major social and institutional changes often result when losses far exceed the best available mitigation or recovery mechanisms.

The IGU Commission on Natural Hazards Studies in its flood hazard project in Bangladesh has endeavoured to incorporate this model. The channel changes brought about by the catastrophic flooding of major rivers of Bangladesh were analyzed (Fig. 4) using Landsat Multi-spectral Scanner imagery (MSS) and were subsequently studied in the field by geomorphologists (Huh et al., 1985; Kubo, 1993) to determine the implications of lateral bank cutting and sedimentation upon land use and control flood structures. Using regional climatic forcing models and hydrologic data, the physical scale of flood severity can be established. This is then combined with geomorphological mapping and satellite imagery to delineate the hazard regions. Flood forecasting and flood-alert schemes are constructed using weather satellites and monitoring stations. Field surveys delineate the human occupance and the flood responses of the population, and assess the depth of flood inundation. Alternative flood strategies may be evaluated using the combination of field data, imagery and geomorphological classifications of site characteristics (Rosenfeld, 1994). Vulnerability and potential damage mapping techniques are taught at training workshops to local field mapping technicians, with the idea that local involvement in risk management fosters an increased awareness of the flood hazard among authorities and the population, which inevitably leads to local "adjustments" in landuse patterns and population densities. Evacuation strategies, flood refuge sites, and relief distribution points are identified using GIS technology, and key facilities and major urban areas are evaluated as candidates for cost-effective flood control structures. The resulting analyses are tested against the field data; for example, Islam (1990) determined from interviews that cropland is often protected by dikes, and people place their residences near the rivers or atop the dikes. When flooding occurs, the dikes may raise the water levels near the rivers, encouraging residents to intentionally cut the dikes to protect their homes. Finally, since flood mitigation is cost-limited at certain severity levels, planning consideration must be given to events which could possibly exceed the mitigation efforts (Verstappen, 1993).

4. Remote sensing

The role of remote sensing has increased both in the frequency of its use and in its influence upon the mon-

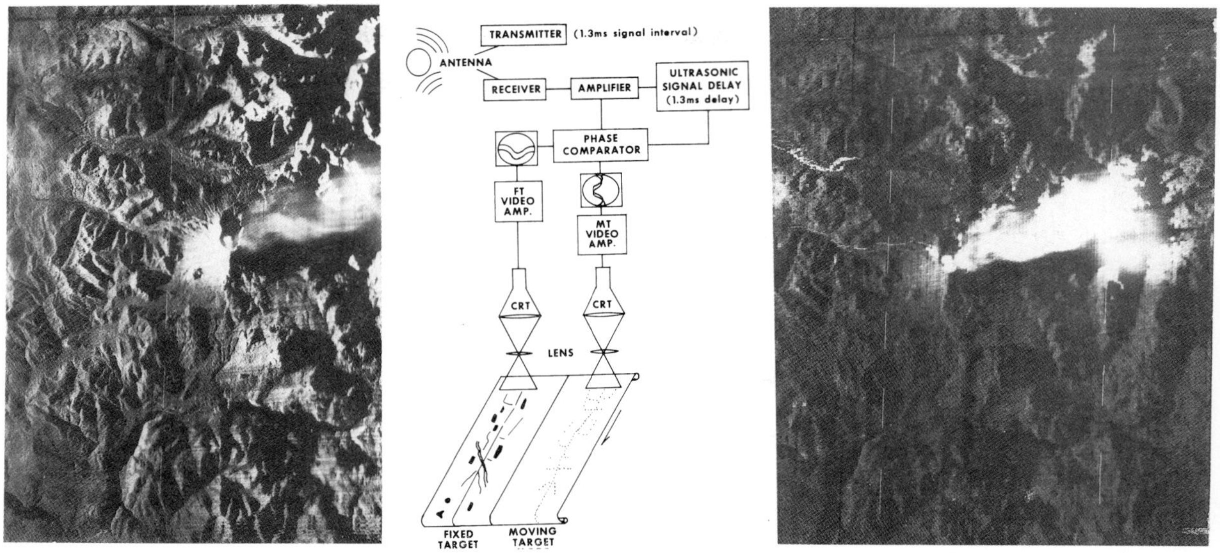

Fig. 5. Airborne RADAR imagery used to study the geomorphic processes associated with the May 18, 1980 eruption of Mount St. Helens, Washington (courtesy Oregon Army National Guard).

itoring of large geomorphological events (Rosenfeld, 1984, 1987). A growing number of studies have successfully utilized remote sensing to monitor geomorphological process activity, and subsequently have concentrated on large scale investigations of hazard areas to provide hazard zoning maps, or process studies to assist in structural mitigation.

Fig. 5 is an airborne RADAR image acquired during the May 18, 1980 eruption of Mt. St. Helens. This image electronically separates moving objects from "fixed" objects, thus allowing analysis of active geomorphic events in the scene (Rosenfeld, 1980). The "fixed" objects include the new crater and debris avalanche emanating from it, and the "moving" objects include the vents on the crater floor, the Plinean plume, and mudflows rushing down the river valleys to the west. When this process information was combined with a multi-spectral Landsat image of the entire disaster area, digital image classification techniques were "trained" to delineate geomorphic zones with distinctive origins, surficial materials and erosional sensitivities. These zones were used to establish ground and aerial photo monitoring transects to gauge the sediment production resulting from post-eruption processes (Rosenfeld and Pearson, 1982). This geomorphic approach produced sediment yield estimates that were used to design sediment retention structures, flood control levees and dredging programs downstream of the volcanically affected area (Fig. 6). In contrast to the more traditional "hydraulic" methods, using stream gauges and sediment sampling techniques, the geomorphic approach permitted rapid assessment of sediment volumes related to process activity, and maximized the use of remote sensing in a very dangerous field environment.

MSS has also been used to track ice advances. The Bering and Stellar glaciers, in Alaska, were imaged at three year intervals (Fig. 7), when the two images are viewed stereoscopically the regions of the glaciers experiencing most rapid ice advance appear to have elevated surfaces. The author is taking part in a research project to study the current surge of the Bering glacier, where advances of 6 to 50 meters a day have been observed.

This study employs an aircraft equipped with a GPS and high-resolution video camera to map the terminus at 30 day intervals. Field observations of shear mechanisms and ground deformation at the leading edge are yielding new insight to the operation of wet-temperate ice sheets and may prove to be a good analogue for Pleistocene glacial erosion and deposition in Puget Sound and southern New England. With the speed of this glacial advance and the difficulty of access to this remote site, many types of observations are being

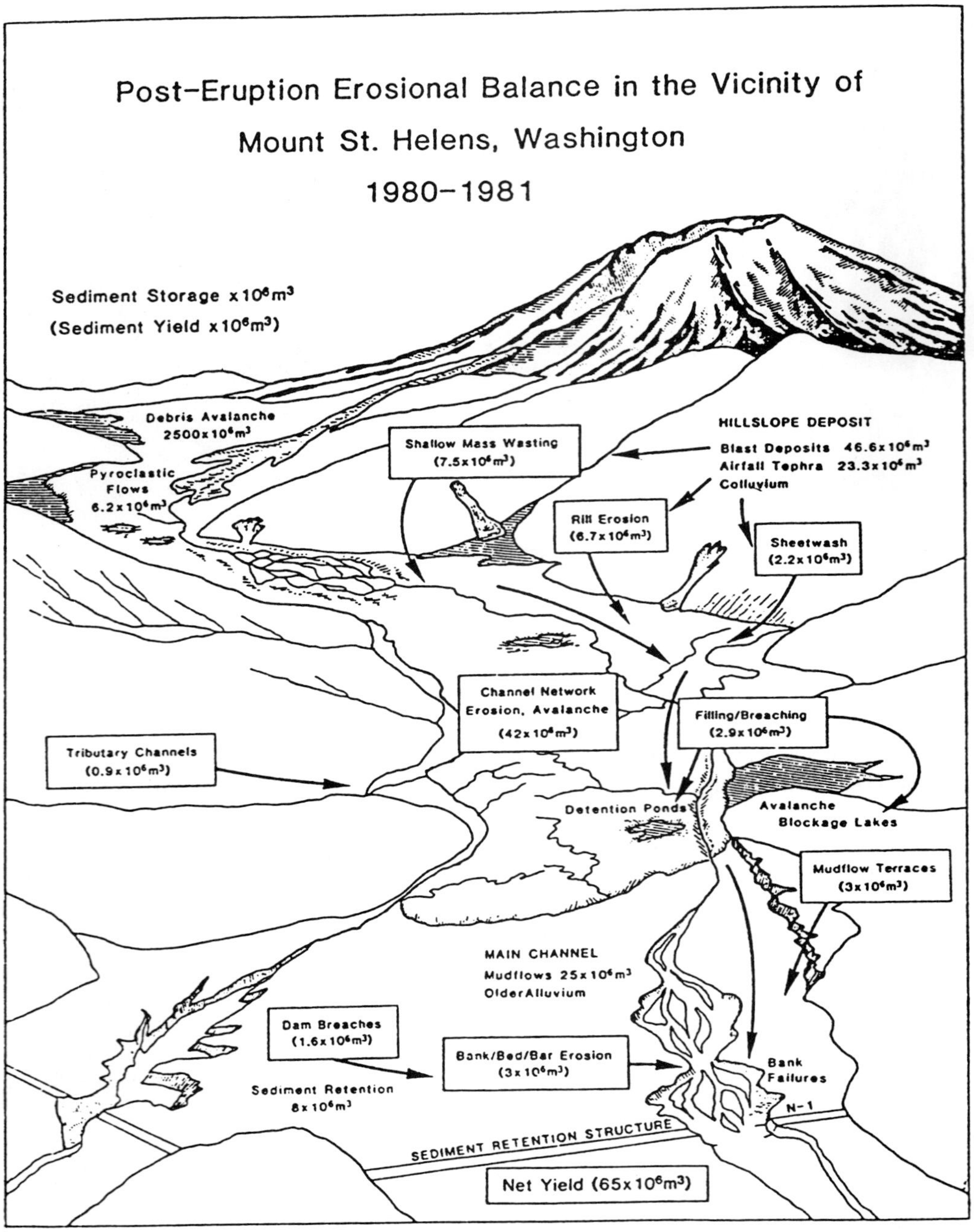

Fig. 6. Erosion budget estimated from field measurements of geomorphic process, photogrammetry of surface features, and mapping of surface extent of affected areas by the eruption of Mount St. Helens (Rosenfeld, 1987).

recorded as part of a multi-layered GIS. Data from crevasse patterns and terminus locations to ice-penetrating radar sections and carbon-14 dates are stored for future analysis. Rapid-onset, short duration phenomena can be subjected to intense observation using a combination of remote sensing and traditional field methods, with GIS used as an archive and retrieval system for long-term analysis — this technique may be

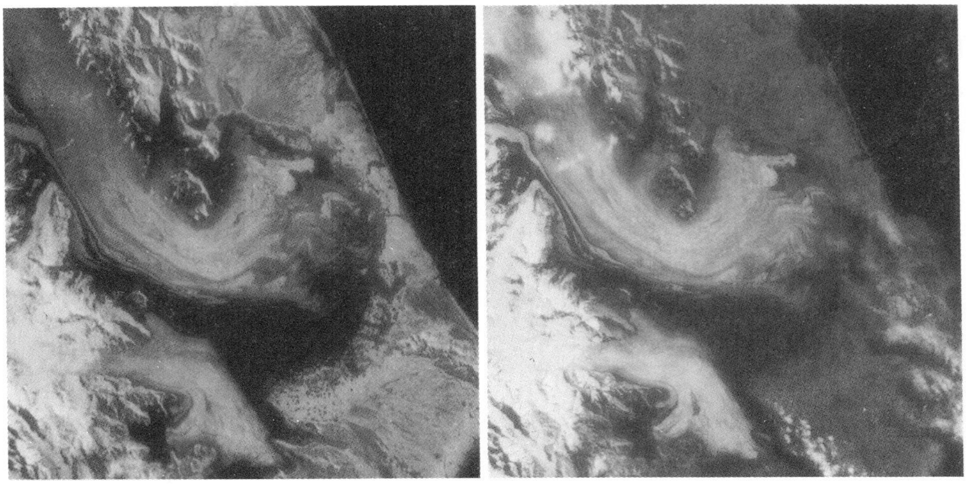

Fig. 7. Landsat MSS imagery of the Bering and Stellar glaciers along the Gulf of Alaska, imaged at three year intervals. When viewed stereoscopically, the areas of highest ice velocity appear to be elevated (courtesy of NASA–Ames Research Center).

appropriate for many other ''hazard-producing'' processes.

The 1993 floods along the Mississippi valley were monitored closely using satellite imagery, not only by the disaster relief agencies, but also by the insurance industry, urban and land use planners, flood control engineers and, hopefully, geomorphologists — undoubtedly many lessons will be learned for the future (Meyers and White, 1993). A major benefit derived from any detailed analysis of episodic geomorphic events is the enhanced process–response insights that results. The increased awareness of processes linked to rapid-onset, or short duration events (such as earthquakes, floods. volcanic eruptions, or severe storm effects) enable us to better anticipate the far reaching effects of long-term changes to the global geomorphic system (i.e; climatic change, sea-level adjustments, anthropogenetic effects).

5. Conclusions

The international cooperation being applied to the development of the Earth Observing System promises a coordinated approach from multiple space platforms giving us unprecedented monitoring capabilities from space. Existing technologies such as GPS and field-portable GIS computers offer vast improvements in interdisciplinary research capabilities. Automated monitoring networks and advanced computer modeling techniques are giving us new tools to test alternative hazard mitigation strategies. Geomorphologists must be willing to embrace new technologies which will permit them to exercise their specialized talents globally, interface more readily with professionals outside the earth science community (for example engineers and social scientists), and associate more closely with monitoring networks and scientific unions to assure that major events are anticipated by identifying their physical precursors, and adequate disaster planning efforts.

Natural hazards research is obviously an interdisciplinary field involving a range of natural scientists with social scientists assessing the human dimension of the problems. Given the limited observed record of hazard events in most regions, a geomorphological approach, where the areal extent (and possibly the frequency) of events can be determined from the landscape, is essential. The geomorphological approach may also encourage the ''nature knows best'' approach to designing hazard mitigation strategies in balance with the dynamics of processes within the region. In the final analysis, occupance of hazard-prone areas is both physically and economically self-regulating. It is the function of science, as a servant of society, to identify those limitations and point the way toward minimizing the disastrous consequences of ''learning our lessons'' in nature's way.

References

Berry, L., 1984. Desertification in the Sudano–Sahelian region, 1977–1984. Desertification Control Bulletin 10, UNEP.

Berz, G., 1993. The insurance industry and IDNDR: common interests and tasks. IDNDR Newsletter, #15, Osservatorio Vesuviano, pp. 8–11.

Bruce, J.P., 1993. Natural disasters and global change. IDNDR Newsletter #15, Osservatorio Vesuviano, p. 3.

Bottorff, H., 1994. Rapid relief: GPS helps assess Mississippi River flood damage. GPS World, 5(5): 22–28.

Coleman, J.M., 1969. Brahmaputra River: channel processes and sedimentation. Sediment. Geol., 3: 129–239.

Cooke, R.U. and Reeves, R.W., 1976. Arroyos and Environmental Change in the American South-West. Clarendon Press, Oxford, 276 pp.

Defense Mapping Agency, 1992. Digital Chart of the World. Washington. 6 compact disks.

Demek, J., 1972. Manual of Detailed Geomorphological Mapping. IGU Commission on Geomorphological Survey and Mapping, Prague, 344 pp.

Evans, D.M., 1966. Man-made earthquakes in Denver. Geotimes, 10: 11–18.

Harrold, L.L., 1971. Effects of vegetation on storm hydrographs. In: Biological Effects in the Hydrological Cycle. 3rd Intl. Seminar for Hydrology Professors, NSF, pp. 332–346.

Hively, W., 1988. Global Change. Am. Sci., 76(2): 127–130.

Huh, O.K., Ali, A. and Quadir, D.A., 1985, Use of NOAA satellite AVHRR data to monitor river flood hydrology in Bangladesh. Report to UNFAO, Louisiana State University, 36 pp.

Islam, M., 1990. Environmental Perceptions and Agriculture: Environmental Aspects of Agricultural Development in Bangladesh. University Press, Dhaka, 321 pp.

Kahn, A.R. and Nazem, N.I., 1988. Abundance and scaricty of water in Bangladesh: Issues revisited. BIISS J., 9(4).

Kaith, D.C., 1960. Forest practices in control of avalanches, floods, and soil erosion in the Himalayan Front. Proc. Fifth World Forestry Congr., Seattle (3): 1640–1643.

Kubo, S., 1993. Geomorphological features of northwestern Bangladesh and some problems on flood mitigation. GeoJournal, 31(4): 313–318.

Lohmann, E., 1993. Strategy design and policy making for extreme weather events as a result of climatic changes. IDNDR Newsletter #15, Osservatario Vesuviano, pp. 6–10.

Meyers, M.F. and White, G.F., 1993. Floods... and floods of studies. Nat. Hazards Observer, 18(2): 1–2.

National Academy of Sciences, 1987. Confronting Natural Disasters. National Academy Press, 60 pp.

Office of Technology Assessment, 1993. Preparing for an Uncertain Climate. GPO, 2 Vols., 768 pp.

Pearson, M.L. and Oliver, J.G., 1992. Reconnaissance report: flooding resulting from typhoon Uring in Ormoc City, the Philippines. Waterways Experiment Station, Misc. Paper GL-92, Corps of Engineers, Washington, 49 pp.

Penmetrics, Inc., 1993. Fact sheet on pen-based computers with integral GPS capabilities. Unpub. Memorandum, Corvallis, 2 pp.

Rasid, H. and Pramanik, M., 1990. Visual interpretation of satellite imagery for monitoring floods in Bangladesh. Environ. Manage., 14(6): 815–821.

Reynolds, W., 1990. Personal communication concerning Corps of Engineers study of groundwater availability in Pacific atolls due to sea-level rise.

Rosenfeld, C.L., 1980. Observations on the Mount St. Helens eruption. Am. Sci., 68(5): 494–509.

Rosenfeld, C., 1984. Remote sensing for geomorphologists. In: J. Costa and P. Fleisher (Editors), Developments and Applications of Geomorphology. Springer, Berlin, pp. 1–37.

Rosenfeld, C.L., 1987. Observation and measurement of erossion following the 1980 eruptions of Mount St. Helens, Washington, USA. In: A. Goddard and A. Rapp (Editors), Processus et Mesure de l'Erosion. 25th Int. Geographical Congr., Paris. CNRS, pp. 565–571.

Rosenfeld, C.L., 1992. Watershed management: fighting the effects of drought in west Africa. Geo Info Systems, 2(3): 29–39.

Rosenfeld, C.L., 1994. Flood hazard reduction: GIS maps survival strategies in Bangla Desh. Geo Info Syst., 4(5): 30–37.

Rosenfeld, C. and Pearson, M., 1982. Aerial monitoring of the erosional characteristics of the North Fork Toutle River: a geomorphic approach to sediment yield. Final Report, Portland District, Corps of Engineers, 99 pp.

Simpkin, T., Seibert, T., McClelland, L., Bridge, D., Newhall, C., Latter, J., 1981. Volcanoes of the World. Hutchinson-Ross, Stroudsburg, 233 pp.

Schweig, E.S., 1994. A wake-up call from Northridge. Nat. Hazards Observer, 18(4): 2–3.

Swanson, F.J. and Dyrness, C.T., 1975. Impact of clearcutting and road construction on soil erosion by landslides in the western Cascade Range, Oregon. Geology, 3: 494–496.

Tricart, J., 1965. Principes et Méthodes de la Géomorphologie. Legende des Cartes Géomorphologique. Masson, Paris, pp. 185–242.

Tweedie, A., 1966. Water and the World. Nelson, Sidney, 317 pp.

UNESCO, 1963. Changes of Climate. Arid Zone Research, Vol. 20. UNESCO, Paris, 488 pp.

Verstappen, H.Th., 1993. The International Decade for Natural Disaster Reduction and the IGU flood hazard research programme. GeoJournal, 34(4): 309–312.

Verstappen, H. and Van Zuidam, R., 1989. The ITC system of geomorphologic survey. Int. Inst. for Aerospace Survey and Earth Sciences (ITC), Enschede, 41 pp.

Watari, M. (Editor), 1988. Landslides in Japan. Nat. Conf. on Landslide Control. Japan Society of Landslides, 54 pp.

ELSEVIER

Geomorphology 10 (1994) 37–63

GEOMORPHOLOGY

Geologic effects of hurricanes

Nicholas K. Coch

Geology Department, Queens College, CUNY, Flushing, NY 11367, USA

Received February 10, 1994; revised March 27, 1994; accepted April 2, 1994

Abstract

Hurricanes are intense low pressure systems of tropical origin. Hurricane damage results from storm surge, wind, and inland flooding from heavy rainfall. Field observations and remote sensing of recent major hurricanes such as Hugo (1989), Andrew (1992) and Iniki (1992) are providing new insights into the mechanisms producing damage in these major storms. Velocities associated with hurricanes include the counterclockwise vortex winds flowing around the eye and the much slower regional winds that steer the hurricane and move it forward. Vectorial addition of these two winds on the right side of the storm gives a higher effective wind speed than on the left side. Coast-parallel hurricane tracks keep the weaker left side of the storm against the coast, whereas coast-normal tracks produce a wide swath of destruction as the more powerful right side of the storm cuts a swath of destruction hundreds of kilometers inland. Storm surge is a function of the wind speed, central pressure, shelf slope, shoreline configuration, and anthropogenic alterations to the shoreline. Maximum surge heights are not under the eye of the hurricane, where the pressure is lowest, but on the right side of the eye at the radius of maximum winds, where the winds are strongest. Flood surge occurs as the hurricane approaches land and drives coastal waters, and superimposed waves, across the shore. Ebb surge occurs when impounded surface water flows seaward as the storm moves inland. Flood and ebb surge damage have been greatly increased in recent hurricanes as a result of anthropogenic changes along the shoreline.

Hurricane wind damage occurs on three scales — megascale, mesoscale and microscale. Local wind damage is a function of wind speed, exposure and structural resistance to velocity pressure, wind drag and flying debris. Localized extreme damage is caused by gusts that can locally exceed sustained winds by a factor of two in areas where there is strong convective activity.

Geologic changes occurring in hurricanes include beach erosion, dune erosion, inlet formation from flood and ebb surge, landscape changes through tree destruction by wind and nearshore channeling and sedimentation resulting from ebb surge.

Multi-decadal wet and dry cycles in West Africa seem to be associated with increases (wet periods) and decreases (dry periods) in the frequency of Atlantic Coast landfalling hurricanes. Coastal zone population and development has increased markedly in a time of low hurricane frequency in the 24 year dry cycle from 1970 to the present. However, no previous climatic cycle in this century has exceeded 26 years. We may be entering a multi-decadal cycle of greater hurricane activity, placing these highly urbanized shorelines in considerable danger.

1. Introduction

Hurricanes Hugo (1989) in South Carolina, Andrew (1992) in South Florida, and Iniki (1992) in Hawaii have provided valuable insights into the geologic processes and damage patterns associated with major storms. New technological advances such as satellite imaging, computer simulation and prediction of storm surge, and Doppler Radar analysis of approaching storms have yielded new data useful in analyzing damage intensities and types. In addition, analysis of these storms have been increasingly interdisciplinary with meteorologists as well as geologists, biologists, engi-

SSDI 0169-555X(94)00016-K

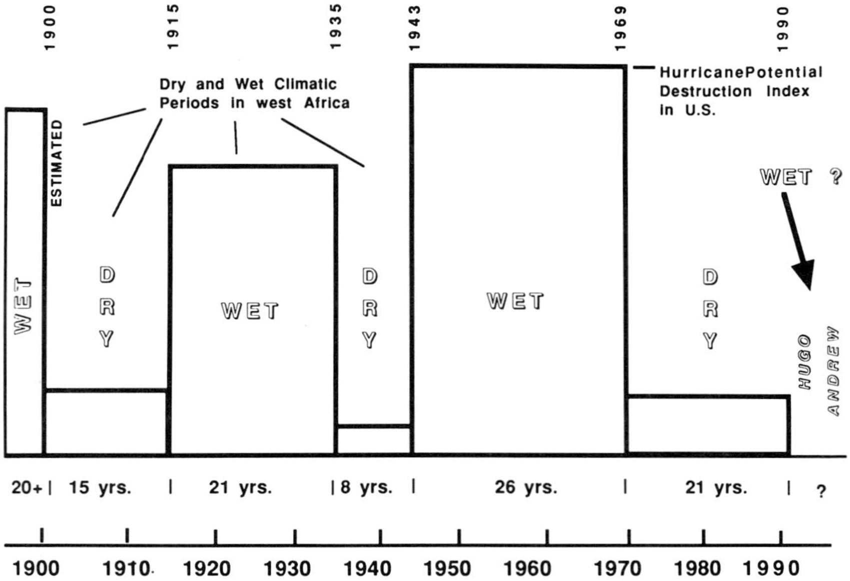

Fig. 1. Hurricane damage potential related to multi-decadal wet and dry cycles in West Africa. The hurricane damage potential (vertical scale of diagram) is a measure of a hurricane's potential for wind and storm surge destruction. It is defined as the sum of the square of a hurricanes maximum wind speed (in 10^4 knots^2) for each 6-hour period of a storm's existence. (After Gray and Landsea, 1991.)

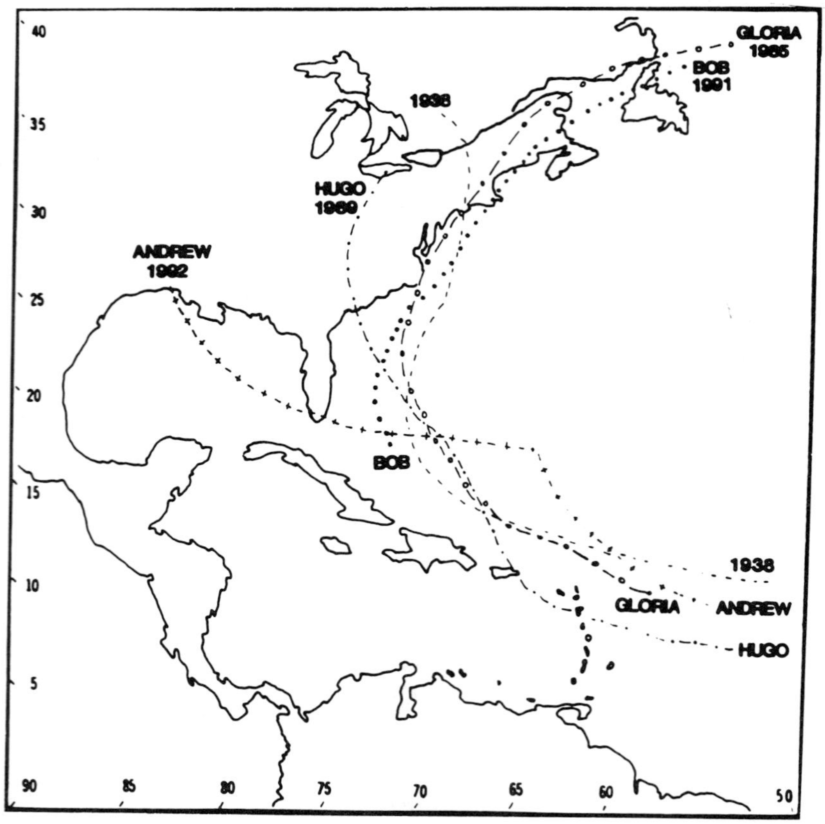

Fig. 2. Map of tracks of hurricanes described in the text.

neers, and emergency managers providing a *holistic* picture of hurricane effects for the first time.

Gray (1990) has shown that the frequency of Atlantic Coast landfalling hurricanes is related to multi-decadal ''wet'' and ''dry'' cycles in West Africa. Gray and Landsea (1991) have delineated these wet and dry cycles and the hurricane damage done during each of them for the last century (Fig. 1). During wet cycles, such as 1945–1969 (Fig. 1), Atlantic Coast landfalling hurricanes were more frequent. In contrast, during dry cycles (1970–to ?present) they have been less frequent (Fig. 1). Unfortunately, much of our coastal development has occurred in the 24 years since 1970 when few major hurricanes hit the Atlantic Coast of the United States. This period of relative quiescence has done little to raise hurricane consciousness among coastal inhabitants. No previous wet or dry cycle in this century has lasted more than 26 years (Fig. 1), and many experts believe we soon will be entering a new ''wet cycle'' in which greater numbers of hurricanes will wreak catastrophic damage on our highly populated and developed shorelines.

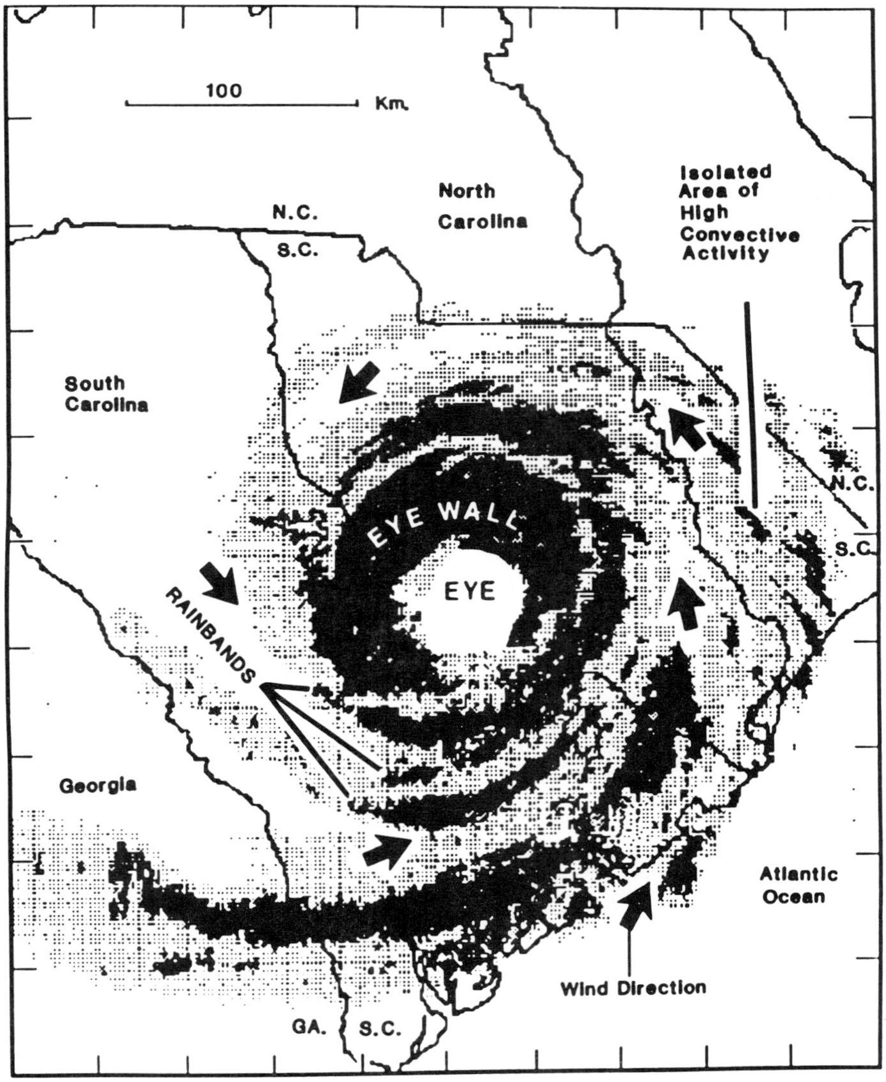

Fig. 3. Doppler Radar image of Hurricane Hugo (1989) showing storm components mentioned in the text. Modified from radar image by Hurricane Research Division, NOAA.

Table 1
The Saffir–Simpson scale for classification of hurricanes based on their wind speeds and storm surge. Some examples of typical damage for each category are included here. Actual damage patterns depend not only on wind speed and surge levels but other facrors described in the text. NOAA version of scale (1990)

Category	Wind velocity		Surge height[a,b]		Damage
	(km/h)	(mi/h)	(m)	(ft)	
1	119–153	74–95	1.2–1.5	4–5	No real damage to building structures. Damage primarily to unanchored mobile homes, shrubbery and trees. Also, some coastal road flooding and minor pier damage
2	154–177	96–110	1.8–2.4	6–8	Some roofing material, door and window damage to buildings. Coastal and lowlying escape routes flood 2–4 hours before arrival of center. Small craft in unprotected anchorages break moorings
3	178–209	111–130	2.7–3.6	9–12	Some structural damage to small residences and utility buildings with a minor amount of curtainwall failures. Mobile homes are destroyed. Flooding near the coast destroys smaller structures with larger structures damaged by floating debris. Terrain continuously lower than 5 feet above sea level may be flooded inland as far as 6 miles
4	210–249	131–155	3.9–5.5	13–18	More extensive curtainwall failures with erosion of beach areas. Major damage to lower floors of structures near the shore. Terrain continuously below 10 feet above sea level may be flooded requiring massive evacuation of residential areas inland as far as 6 miles
5	> 249	> 155	> 5.5	> 18	Complete roof failure on many residences and industrial buildings. Some complete building failures with small utility buildings blown over or away. Major damage to lower floors of all structures located less than 15 feet above sea level and within 500 yards of the shoreline. Massive evacuation of low areas within 5–10 miles of the shoreline may be required

[a]Actual storm surge height will vary depending on local coastal configuration and other factors.
[b]Surge heights given above normal water levels.

This paper describes the styles and intensities of hurricane damage and provides information on the factors that increase damage from wind, storm surge and inland flooding. It is based on both published studies of Hurricane Andrew (1992), Iniki (1992), Bob (1991) and Hugo (1989) plus ground and aerial studies of the damage effects of those storms by the author. The tracks of the hurricanes described in this paper are shown in

Fig. 2. Information is synthesized from a number of different cognate fields because a *holistic* view of hurricane damage is essential if we are to minimize loss of life and structural damage in future hurricanes.

2. Hurricanes

United States hurricanes are formed in the warm tropical waters of the North Atlantic, the Caribbean and the eastern Pacific and migrate westward and northward until they make landfall. Details on hurricane formation, intensification and migration are given in the work by Risnychok (1990) and Coch (1994).

2.1. Structure

A mature hurricane consists of a calm central area of low pressure referred to as the eye, surrounded by spiral bands of thunderstorm activity (Fig. 3). Note that the rainbands are not continuous, isolated areas of high convective activity occur in the outer parts of the storm, and the winds flow in a counterclockwise direction around the eye (Fig. 3).

The most violent winds and strongest convective activity occur in the eyewall, the section just outside of the eye (referred to as the eyewall). Maximum winds occur in the right front portion of the eyewall in a zone varying from 32–56 km (20–35 miles) wide in a typical hurricane such as Hugo (Fig. 3).

2.2. Power

Hurricanes are classified based on their sustained wind speeds into 5 categories on the Saffir–Simpson Scale (Simpson, 1974). The wind speeds, central pressures, storm surge and typical damage associated with each category is given in Table 1.

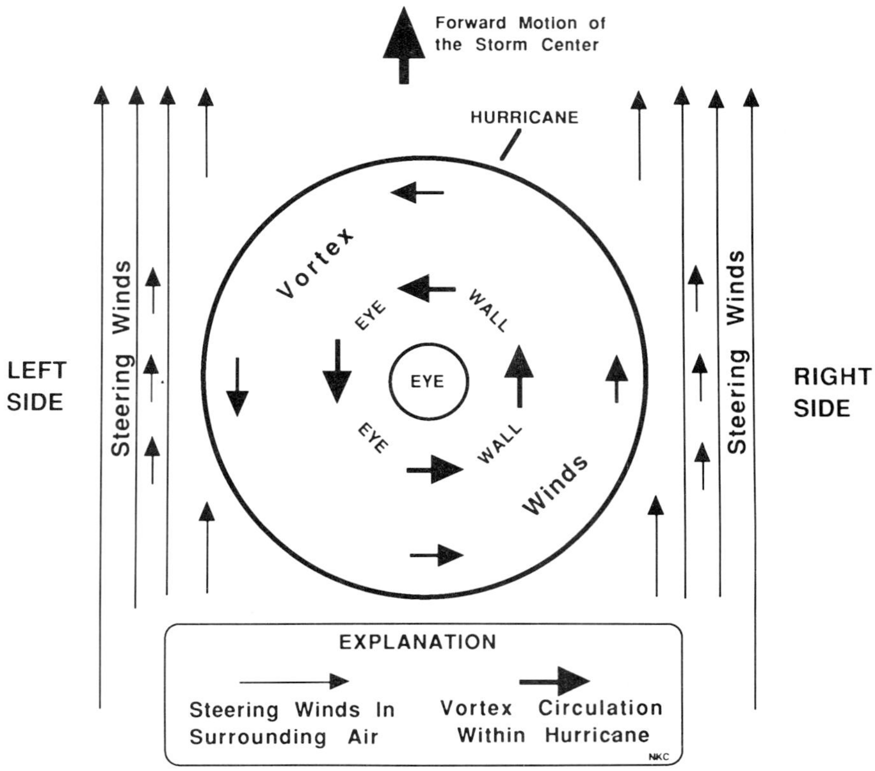

Fig. 4. Diagrammatic view (horizontal section) showing the velocity components described in the text. The high velocity vortex winds flow counterclockwise around the eye. At the same time the storm center is being moved forward by the regional steering winds. Note that on the right side of the storm (looking forward) the two velocities are in the same direction and so are additive. Consequently, the winds on the right side of a hurricane are always stronger.

While the Saffir–Simpson category is defined by sustained wind velocity ranges, other parameters such as the forward velocity of the storm are very important in determining the destructive power of a hurricane. Given two hurricanes of similar power (Saffir–Simpson category), the faster moving one will cause significantly more damage. The regional significance of this is discussed later in this paper.

2.3. Velocity

Two different velocities are associated with hurricanes and their interplay determines the damage intensities and gradients along the diameter of the storm as it makes landfall. The vortex winds are the hurricane force winds (Table 1) that flow counterclockwise around the eye (Fig. 4). The storm system itself, hundreds of kilometers across, is embedded in a regional wind system that guides the storm system forward. This storm center forward velocity is far less than the hurricane force vortex winds flowing around the eye.

The winds on either side of a hurricane are not equal. Note that on the right side of the storm, looking in the direction of movement (Fig. 4) the steering winds and the vortex winds are in the same direction and are thus additive to some degree. On the left side, the winds are in opposite directions and are subtractive. *In short, the winds on the right side of a hurricane are always stronger.*

Forward velocities of hurricane systems also vary with latitude. Southern landfalling hurricanes are under the influence of easterly (trade) wind systems and generally move less than 32 km/h (20 mile/h). Once a hurricane moves northward along the Carolinas, it begins to be affected more by westerly wind systems whose upper level flows are faster. Thus, hurricanes speed up as they move toward the northeastern United States. A more typical speed for a northern hurricane is 48–64 km/h (30–40 mile/h). According to Pierce (1939), the extremely destructive Long Island–New England hurricane (1938) moved forward at a velocity of 97 km/h (60 mile/h).

The northward increase in forward speed of a hurricane has a very important role in increasing damage at, and after landfall. In short, it increases the effective wind on the right side of the storm and this contributes to greater wind damage and the development of a higher storm surge. The increase in speed also affects the power of the storm when it approaches landfall. If a hurricane is moving faster than 56 km/h (35 mile/h) as it passes over the cooler waters south of Long Island, it will not begin to begin to weaken and break up and will release its full power on the land (R. Sheets, pers. commun., 1993). The great forward speed of the 1938

Fig. 5. Hurricane Bob (1991) damage on Cape Cod, Mass. Elevated surge undermined the cottages in the foreground. Note the wind-induced cladding loss on the side of the building (dark area) on the left of the photo. Photograph courtesy of Ana Butler.

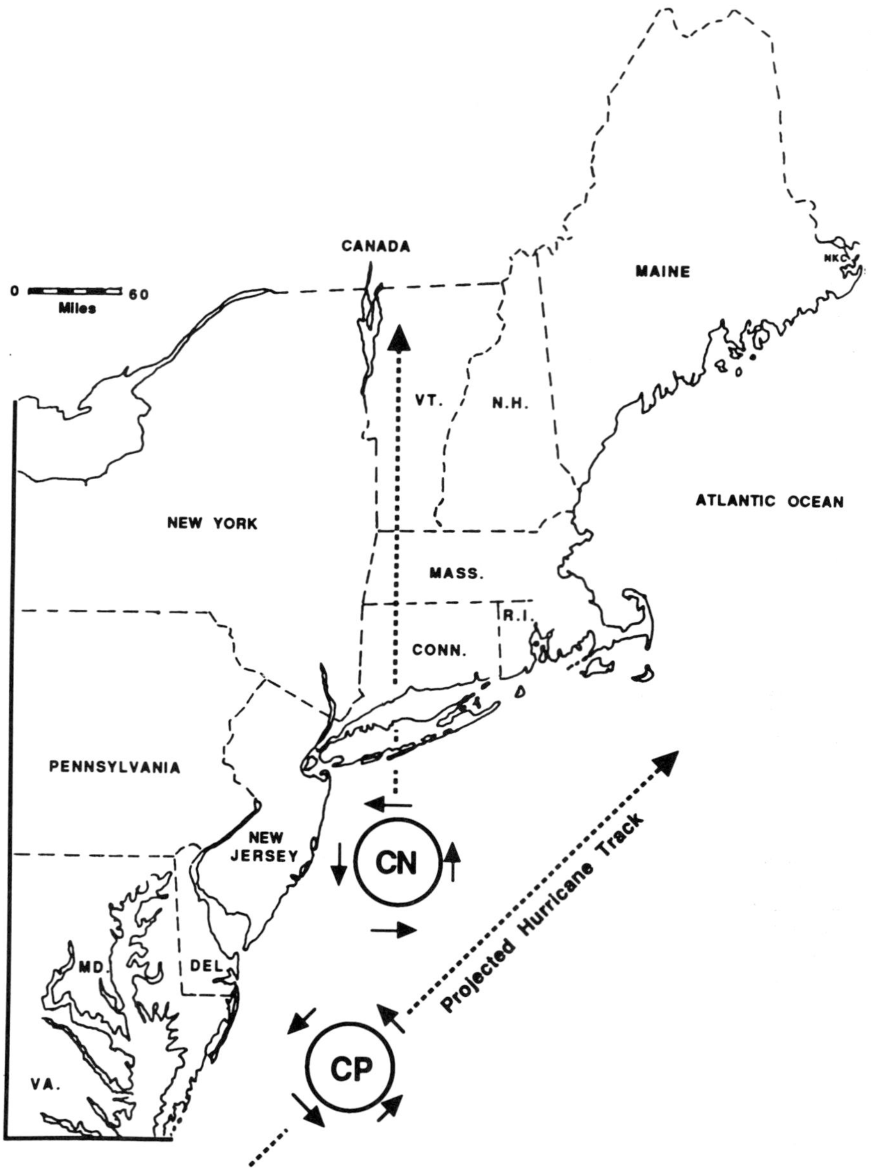

Fig. 6. Tracks of Hurricanes relative to the coastline. In a coast-parallel track (*CP*), the weaker left side of the storm is against the coast. In a coast-normal track (*CN*) the more powerful right side cuts a wide swath of devastation deep inland as it crosses the coast. The problem is especially serious in Long Island and southern new England where most hurricanes have a coast-normal track.

New England Hurricane (Fig. 2) resulted in very little reduction in storm power until it was well into central New England. Hurricane force winds persisted to the Canadian border. In contrast, Hurricane Bob (1991), a potentially dangerous Category 3 hurricane, was moving only 48–56 km/h (30–35 mile/h) as it approached Long Island and southern New England (Fig. 2), and this weakened the storm (Fig. 2). Yet, it still caused considerable destruction in the Cape Cod area of Massachusetts (Fig. 5).

2.4. Tracks

As massive and powerful as a hurricane is (Fig. 3), it is easily blocked by local weather systems and steered by regional upper level wind systems (Coch, 1994). For example, Hurricane Andrew moved steadily northwestward (Fig. 2) until it encountered a high pressure system blocking its northward path and this diverted it sharply westward (Fig. 2) to cut a swath of destruction across South Dade County, Florida. When Andrew moved into the Gulf of Mexico from South Florida, it had a potential northwest track towards the highly developed Texas Coast. However, a low pressure system developed in the western part of the Gulf of Mexico and that prevented the hurricane from drifting westward. The winds associated with that low pressure center steered the hurricane into the relatively unpopulated area of the western Mississippi Delta (Coch, 1994).

One of the most important aspects of hurricane paths is the relationship of the track of the storm to the coastline. In a coast-parallel track (Fig. 6) the storm keeps its weaker left side against the coast and damage drops off rapidly inland. In contrast, in a coast-normal track (Fig. 6) the hurricane crosses the coast and the powerful right side cuts a wide 80–160 km (50–100 miles) swath of heavy destruction deep inland. In the 1938 Long Island–New England Hurricane (Fig. 2), significant damage occurred almost to the Canadian border (Federal Writers Project, 1938). In Hurricane Hugo (Fig. 2) damage extended across South Carolina and deep into North Carolina.

A major hurricane with a coast-normal track striking the highly urbanized northeastern United States could have catastrophic consequences. The path scenario for such a storm is shown in Fig. 7. This meteorological set up occurred in the 1938 Hurricane and in several

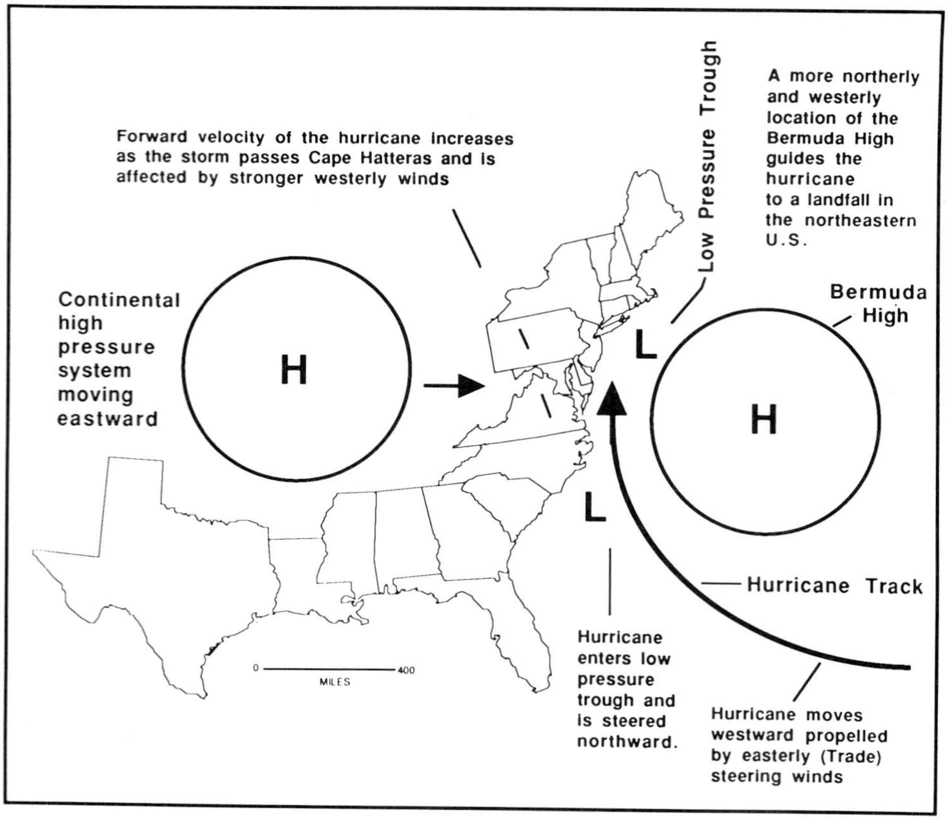

Fig. 7. Meteorological setup favoring landfall of a major hurricane in the northeastern United States. A low pressure zone develops between a continental high and the semi-permanent Bermuda High. If a major hurricane rounds the Bermuda High and enters the low pressure trough, it is directed northward. the storm forward velocity increases as it passes the Carolinas and is affected by stronger upper level westerly winds.

subsequent hurricanes such as Hurricane Bob (1991). The scenario involves a well developed Bermuda High and a high pressure system moving eastward across the United States. In between the two air masses is a low pressure trough (Fig. 7). This set-up can occur as much as ten times during a hurricane season (from June–October). If the passage of a Category 3 hurricane (Table 1) coincides with such a set up, the hurricane can move around the western edge of the Bermuda High and enter the low pressure trough. It will then speed up as it moves northward and be steered into a heavily developed coast of the northeastern United States with catastrophic consequences (Coch and Wolff, 1990).

It takes only a slight change in hurricane track landfall position to make a great difference in the resulting damage. It is only in the last few hours before landfall that the official forecasts are both accurate and consistent to allow local decision makers to use them as a sure guide to preparatory actions (Carter, 1983). The damage caused by Hurricane Andrew as it made landfall at Homestead, Florida (Fig. 2) was staggering. However, if it had made landfall just 32 km (20 miles) to the north, over the highly developed Miami–Ft. Lauderdale area, the results would have been truly catastrophic.

3. Hurricane damage

The damage caused by a hurricane is the sum total of storm surge, wind, and inland flooding caused by high precipitation prior to and during the storm. The wide variety of damage associated with a coast-normal hurricane track crossing a low lying shoreline is shown in generalized fashion in Fig. 8.

The relative percentage of the three types of storm damage at any given locality is a function of three factors: (1) distance from the coast; (2) hurricane side and distance from the eye center and; (3) the strength of isolated conductive activity centers in the outer rainbands (Fig. 3).

Field observations of damage patterns during several recent major hurricanes, as well as remote sensing analysis by governmental agencies during those storms, have provided new insights into the mechanisms which create this damage.

3.1. Storm surge

The elevated sea surface resulting from a storm is referred to as storm surge; this is a topic fraught with misconceptions. People commonly visualize the process as a series of massive tidal waves hitting the coast and think the major factor in elevating the sea surface is the low pressure associated with the eye of the hurricane. In reality, it is the *wind shear* that raises the sea surface and the storm surge is really a dome of water on which high waves are superposed. The doming occurs as the hurricane winds drive water across the continental shelf towards land.

Two types of storm surge are distinguished here. A common misconception is that surge comes only from the ocean. In reality it first moves landward from the ocean (flood surge) and then seaward from both the rivers and bays behind the coast (ebb surge). The flood surge occurs later as the hurricane moves onshore and drives nearshore waters inland. The ebb surge occurs as the hurricane moves across the coast, permitting the entrapped surge waters and rivers engorged from rainfall to move seaward. Ebb surge is significantly less intense than flood surge, but is capable of proportionately greater destruction because it affects structures already weakened by waves, flood surge, and windborne debris.

Measuring and predicting surge levels

Techniques of measuring storm surge levels in the field have been described by Coch and Wolff (1991) from investigations in Hurricane Hugo. Three criteria are used: (1) abrasion marks made on structures and trees by floating debris; (2) heights of debris trapped or abandoned as flood waters receded; (3) stain lines from muddy waters and finely divided organic materials on building walls. It is important to exclude marks made by waves (higher than the surge level). In this regard, debris and stain lines in protected areas and enclosed rooms are the best indicators of maximum surge levels (Nelson, 1991).

The SLOSH (Sea, Lake, and Overland Surge from Hurricanes) model of the National Weather Service now provides an accurate prediction of expected surge levels along major coastal segments (Jarvinen and Lawrence, 1985). Mathematical analysis of a number of factors generates contours showing the expected height of storm surge for a given a set of conditions on

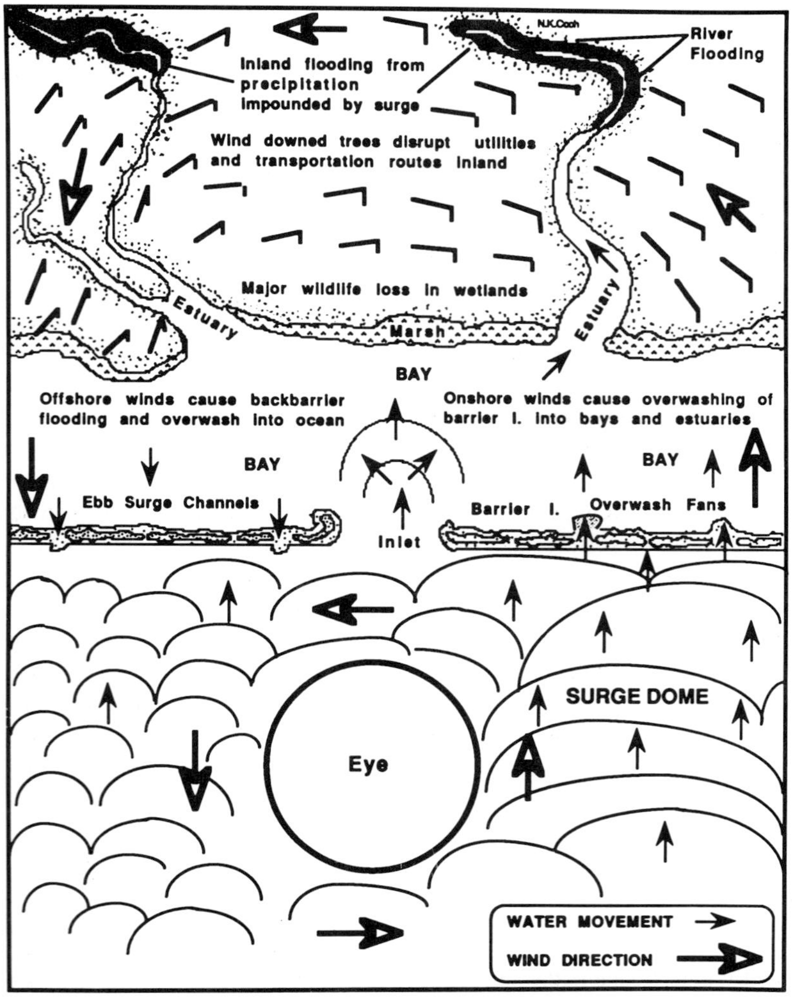

Fig. 8. Generalized map showing the variety of damage patterns associated with a coast-normal hurricane track across a barrier island-low lying coastal plain shoreline segment.

a specific shoreline segment. The pre-hurricane SLOSH analysis done for the South Carolina coast proved amazingly accurate in predicting actual surge levels in Hurricane Hugo (Coch and Wolff, 1991). One of the SLOSH maps drawn by the National Hurricane Center for the apex of the New York Bight is shown in Fig. 9. The map contours give the predicted storm surge for a Category 3 hurricane moving northwest across northern New Jersey at a forward speed of 65 km/h (40 mile/h). The reasons for the high surge values and their significance are discussed in a subsequent section of this paper.

Flood surge

The major factors determining the height of the flood surge are: (1) wind speed, (2) central pressure, (3) slope and width of the continental shelf, (4) tidal stage, (5) shoreline configuration, and (6) anthropogenic changes along the coast. All these factors aid in raising the water surface into a dome upon which wind shear will develop high waves.

The wind speed, rather than the low pressure in the eye, is the most important factor in generating flood surge. The sea surface is drawn up in the low pressure within the eye but this accounts for only a small portion

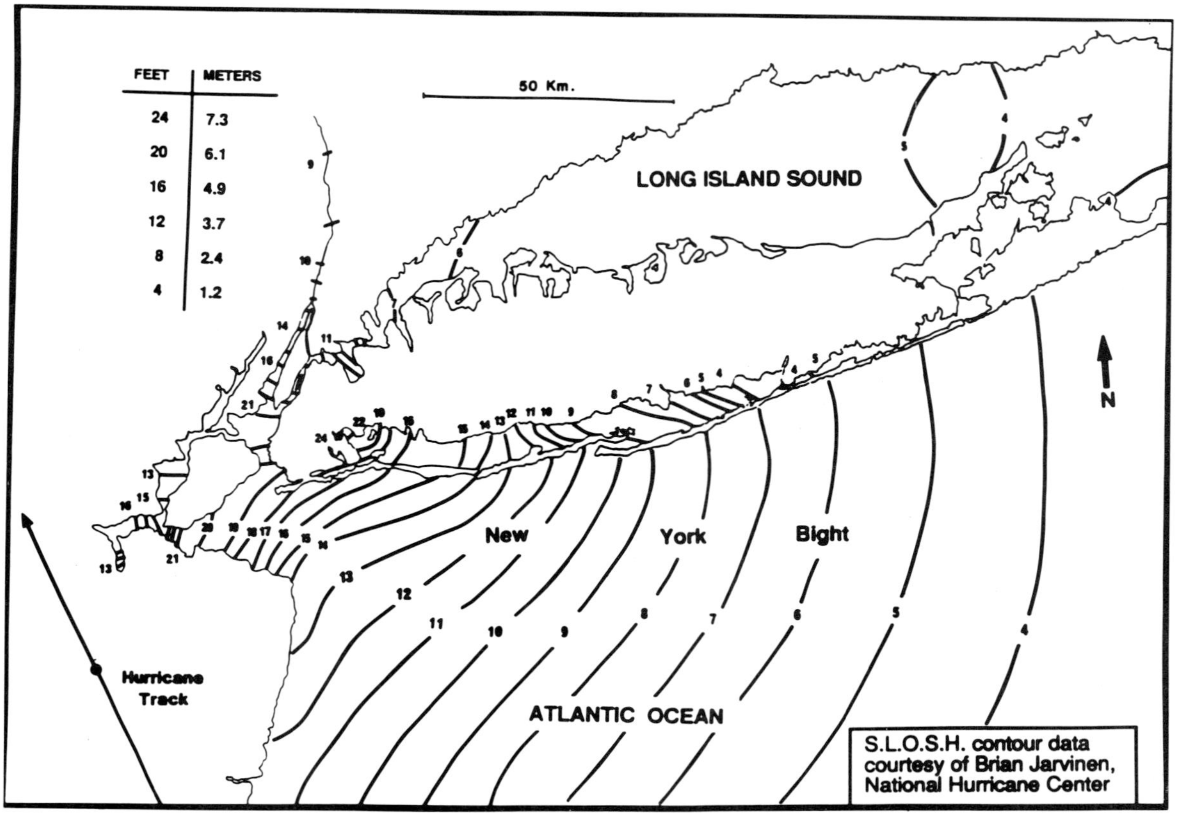

Fig. 9. SLOSH map for the apex of the New York Bight showing the predicted storm surge levels (in feet) resulting from passage of a category 3 hurricane moving NW across northern New Jersey at a forward speed of 65 km/h (40 mile/h). SLOSH contours courtesy of Brian Jarvinen, National Hurricane Center.

of the surge level because the rise is only about 0.3 m (1 ft) for an atmospheric pressure drop of 33.9 mbar (1 inch of mercury) according to Pore and Barrientos (1976). In all recent hurricanes the maximum surge has not been under the eye (where pressure is lowest) but to the right of the eye at the radius of maximum winds, where winds are greatest (Penland et al., 1989; Coch and Wolff, 1991; Coch, 1994).

The slope and width of the continental shelf exert an effect on both surge and wave height. In general, the gentler the slope the greater the surge, but the lower the waves. Conversely, the steeper the offshore slope, the lower the surge, but the higher the waves.

The steep offshore slope south of the island of Kauai produced a relatively low surge during Hurricane Iniki in 1992. The storm was a Category 4 hurricane (Table 1) but surge levels were only about 1.5–2 m (4.5–6 ft) as Iniki made landfall on the south coast of Kauai (Central Pacific Hurricane Center, 1992). However, massive 6.1–10.6 m (20–35 ft) waves pounded the many resort complexes along the coast causing extensive damage (Fig. 10). Debris lines marked surf inundation to an altitude of 6.7 m (22 ft) inland. These high values were very close to the U.S. Army Corps of Engineers "worst case scenario" for a storm in Hawaii (Central Pacific Hurricane Center, 1992).

The configuration of the coast is also important and surge height increases in concave seaward (bay) coastal sections. In Hurricane Hugo, gentle shelf slope, location at the radius of maximum winds, and a concave seaward coastal segment, combined to raise surge levels of 6 m (20 ft) at Bulls Bay, about 22 km (14 miles) northeast of Charleston (Coch and Wolff, 1991; Fig. 10).

Tidal stage at landfall can also significantly increase the height of storm surge. Surge levels can increase 1–

Fig. 10. Hurricane Iniki (1992) damage to homes along the Poipu coast of southern Kauai, Hawaii. Photograph taken one year after the storm. Massive waves destroyed the second story of this home. The surge and waves broke the stone wall, deeply eroded the lawn and deposited basalt boulders inland. Photograph by Nicholas K. Coch.

3 m (3–9 ft) at high tide depending on tidal range along a coastal segment. The path of Hurricane Gloria (1985) across highly urbanized western Long Island (Fig. 2) provides an example. Millions of people in the region were warned before landfall that this Category 3 storm could cause major damage. However, it slowed down to 48 km/h (30 mile/h) as it moved northward and it made landfall on the south shore of Long Island at low tide, resulting in a surge of only 1.5 m (5 ft). Thus, after a big media build-up the surge damage was minimal (although winds wreaked extensive tree damage and other havoc inland). Consequently, inhabitants of the northeast United States resumed their complacency, believing major hurricanes do not affect northern coastal areas. In reality, they were just lucky. If Gloria had hit at high tide, with all the other factors the same, the surge would have been 3 m (10 ft) above sea level at the Battery in New York City. Each borough of New York City would have experienced tidal flooding on an average distance of 61 m (200 ft) inland from the shore. The expected damage and dislocation of regional transport in the New York Metropolitan area is presented by Gigi and Wert (1986, pp. 3–6).

The worst case scenario for storm surge is approximated by the 1938 New England Hurricane (Fig. 2). This high Category 3 storm (B. Jarvinen, pers. commun., 1993) was moving forward at greater than 96 km/h (60 mile/h) and hit the south shore of Long Island at just about high tide. Surge levels as recorded by high water marks (Pore and Barrientos, 1976) reached from 2.1 m (7 ft) to 4.8 m (15.6 ft) along the south shore of Long Island.

The geographic orientation and shape of a coastal segment can also exert a major effect on amplifying storm surge. Coastal segments along the Atlantic Coast of the United States that amplify storm surge are right angle segments and funnel-shaped estuaries that open to the east. Both of these types of shoreline segments exist close together in the apex of the New York Bight and in Long Island Sound to the north (Fig. 11). When a hurricane enters the New York Bight, the counterclockwise (easterly) flow of winds around the front of the storm center drive the waters into the right angle made by the New York–New Jersey juncture, increasing surge levels westward. This surge amplification was the cause of a great deal of damage in the shore communities of western New York and northern New Jersey in the ''nor'easter'' storms of December 1992 (Coch, 1994).

The worst case scenario for surge damage in the New York Bight region (B. Jarvinen, pers. commun., 1993) is for a hurricane passing over northern New Jersey. The SLOSH prediction for that scenario would have water levels rising to 6.4 m (21 ft) at the entrance to

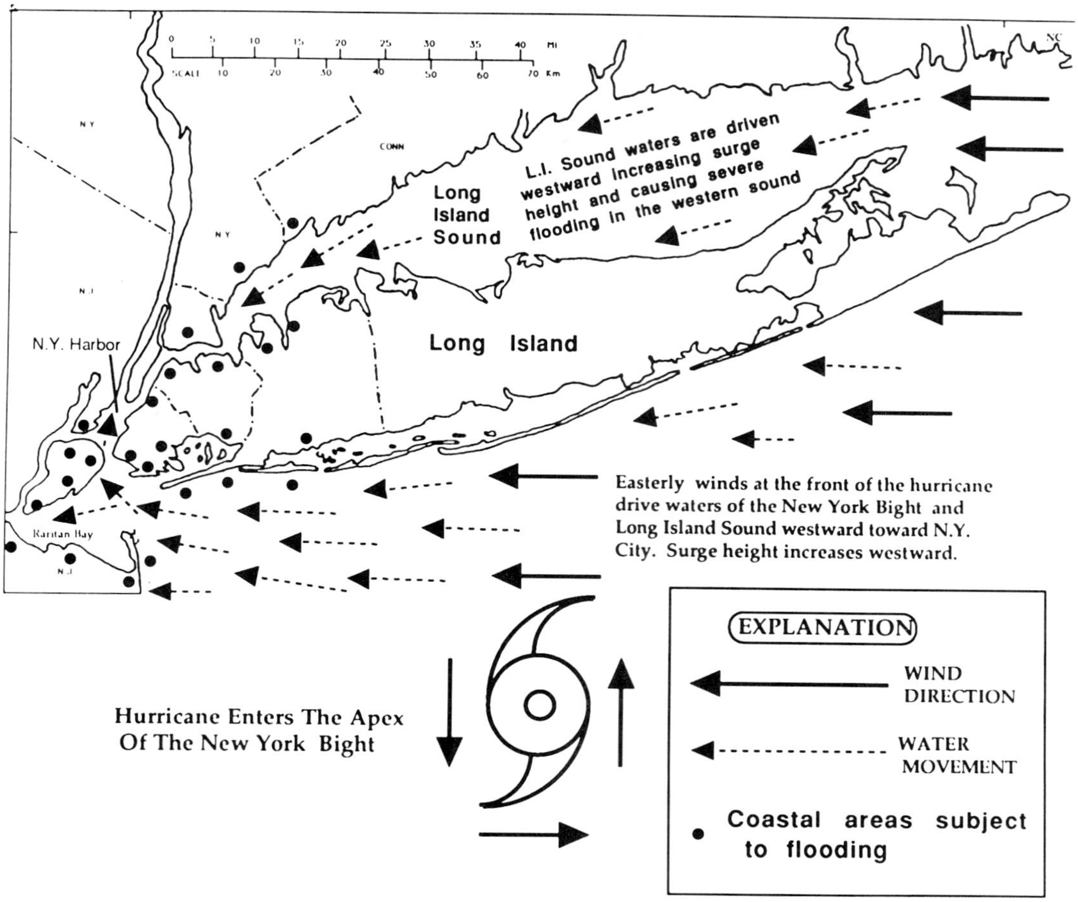

Fig. 11. Storm surge amplification in the apex of the New York Bight and in Long Island Sound. The easterly winds at the front of a hurricane entering this region drive both the waters of the New York Bight and Long Island Sound westward causing severe flooding during landfall and passage across Long Island.

New York Harbor at the Narrows and 7.3 m (24 ft) in Jamaica Bay at the western end of Long Island (Fig. 9).

The same hurricane winds that drives New York Bight Apex waters westward also drive the waters of Long Island Sound westward. Long Island Sound is an east–west oriented funnel-shaped estuary opening to the east. This is the direction from which hurricane winds come as the storm moves into New England (Fig. 11). Long Island Sound varies in width from 27 km near its eastern end to less than 3 km at its western end in New York City. The easterly winds at the front of the hurricane drive the waters of the Sound westward, through a decreasing cross section, towards New York City as the hurricane approaches the south shore of Long Island. However, the progressive wave takes about three hours to move westward across the Sound so that maximum surge levels are attained in the western end of the Sound when the Hurricane is in central New England (B. Jarvinen, pers. commun., 1993).

Some indication of the degree to which surge is amplified in a funnel shaped estuary like Long Island Sound is available from measurements made in the 1938 Hurricane (Pore and Barrientos, 1976). Surge in the eastern part of the sound was about 2.3 m (7.5 ft) increasing steadily to a maximum of 3.9 m (12.7 ft) at the entrance to the East River in New York City (Fig. 12).

This hurricane had made landfall on the south shore of Long Island at 3:30 p.m. and moved inland at a

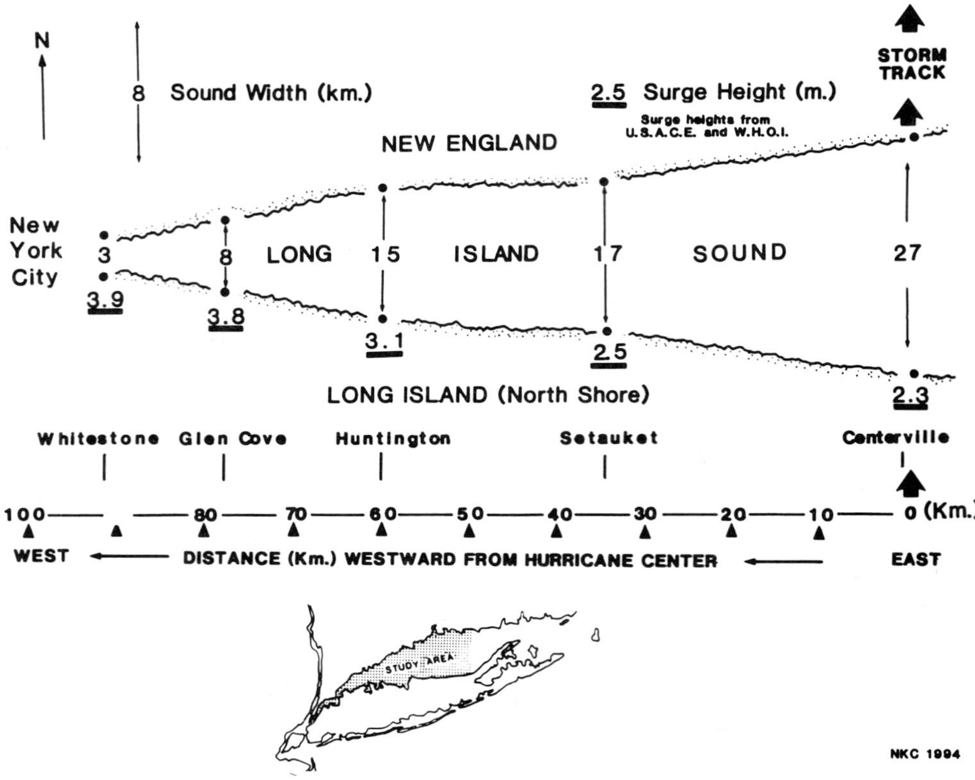

Fig. 12. Map of surge levels in Long island Sound in the 1938 New England Hurricane. Surge heights increased steadily westward as the estuary narrowed. The maximum surge level at the western end of the sound (New York City) was reached three hours after the hurricane made landfall on the south shore of Long Island. Based on data in Pore and Barrientos (1976). The shoreline of Long Island Sound has been generalized, but widths and distances are accurate.

velocity of 80–96 km/h (50–60 mile/h). However, the maximum storm surge in western Long Island and the East River in New York City was reached *three hours later*, about 7:30 p.m. Rising surge waters flooded lowlying areas of the Borough of Queens in New York City and inundated hospitals and power plants on islands in the East River. The inundation of the power plant near Hell Gate in the East River plunged parts of the Boroughs of the Bronx, and Manhattan into darkness, as well as bringing part of New York City's subway system to a halt. At this time (7:30 p.m.) the hurricane center was in northern Vermont (Bricker, 1988).

Ebb surge

Ebb surge is the seaward return of water impounded on land as the hurricane moves inland. The height of the ebb surge is a function of the flood surge, the amount of precipitation preceding and during the hurricane and the local obstructions that block flow back into the ocean. Ebb surge is lower than flood surge, but can still cause significant damage because it affects structures that were partially destroyed by flood surge, wave action, and wind borne debris (Fig. 13).

A number of factors increase the possibility of ebb surge damage at a given location. These factors include; (1) failure or absence of shore protection structures (rip-rap aprons, bulkheads); (2) streets that run perpendicular to the shoreline; (3) areas of open land with little vegetation and; (4) presence of beach access paths, dune walkovers, and locations on narrow parts of barrier islands seaward of bays (Lennon, 1991).

The damage done by ebb surge is more elusive than the visible "dams" of debris washed inland by flood surge (Fig. 13). Side-scan sonar studies of the South Carolina nearshore zone by Gayes (1991b) after Hur-

Fig. 13. Ebb surge cut channel at Pawleys Island, South Carolina formed in Hurricane Hugo (1989). Two homes were lost and bulkheading was destroyed by the ebb surge. Note the flood-surge deposited debris at the edges of the trees at the top of the photo. Photograph by Nicholas K. Coch.

ricane Hugo have provided new information on the erosional and depositional dynamics of ebb surge. The study detected 30–50 m (98–164 ft) wide channels perpendicular to the shoreline that cut through shore oblique bar systems parallel to the shore. Side-scan sonar sweeps and diver observations documented a wide variety of debris scattered over the nearshore area (Gayes, 1991b; Fig. 13) up to 100 m (328 ft) offshore in water depths of 2–4 m (7–13 ft). Specific types of debris recognized during the study included portions of fishing piers, seawalls, and mobile homes (P. Gayes, pers. commun., 1991a). Development along the shore of the study area correlated well with locations of many scour channels. Massive multi-storied structures not raised on pilings had clearly obstructed, diverted and channeled the storm surge and there was frequently a noticeable absence of a well-developed nearshore bar and the presence of steeper beachface slope in those areas.

3.2. Wind

Wind and storm surge cause the major damage along the coast. Inland, wind is the dominant cause of damage. In a hurricane with a coast-parallel track (Fig. 6), wind damage drops off rapidly inland. However, in a hurricane with a coast normal track (Fig. 6), winds can cut a wide swath of devastation far inland.

Scales of wind damage

A common conception about hurricane wind damage is that the storm cuts a continuous swath across the land. Remote sensing, and post-hurricane ground and aerial damage analysis suggest that the damage is far more uneven. It is clearly more intense near the eyewall area (Fig. 3), but in outlying areas the intensity of damage may simply be a matter of chance.

Three scales of wind damage are used in this paper to describe wind damage effects (Fig. 14). The sum total of all damage over the diameter of the storm makes up the *megascale* pattern (Fig. 14). The second scale of damage is the *mesoscale* pattern and involves differential convective activity within the rainbands spiraling around the eye and the areas between the rainbands (Fig. 3). The higher Doppler Radar reflectivities (Db values in Fig. 14) associated with the rainbands reflect the greater degree of radar impulse return from the raindrops in the bands. Simply put, under an area of higher radar reflectivity such as in the rainbands in Fig. 14, there will be more severe damage than to either side. An actual example will be described later in this paper.

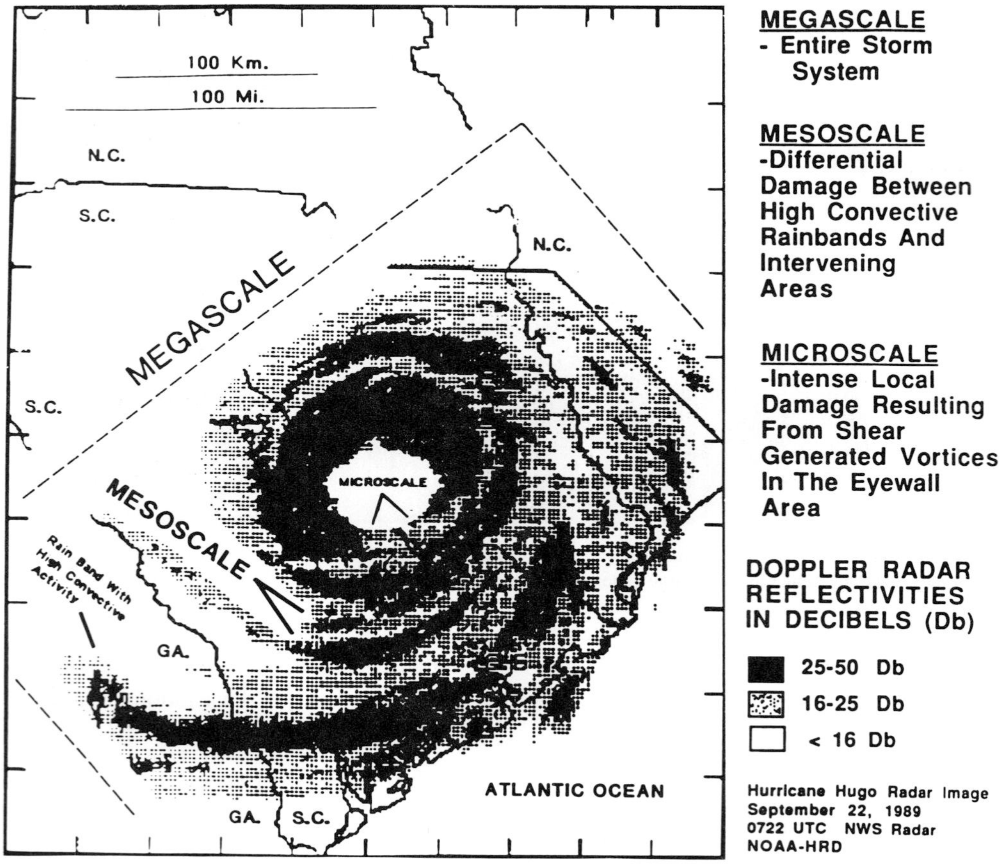

Fig. 14. Three scales of wind damage in a hurricane. The megascale pattern is the sum total of all damage across the diameter of the storm. The mesoscale pattern describes differential damage between highly convective areas on the rainbands and the areas in between. Microscale patterns of damage are narrow, local paths of extreme destruction caused by spin-up vortices described by F. Fujita in Hurricane Andrew.

The smallest scale of damage is the *microscale* pattern that is most common in the eyewall area (Fig. 14). Microscale damage was described by Prof. F. Fujita, of the University of Chicago, from aerial photograph analysis and field surveys after Hurricane Andrew. Aerial photographs of Andrew's impact zone in South Dade County, Florida showed scattered, narrow, short and curving paths of extreme damage within an overall severely devastated area. Fujita (1993) believes that these damage patterns result from *spin up vortices* generated in the eyewall. According to Fujita (1993), slowly rotating vortices 60–152 m (200–500 ft) in diameter form as a result of slight variations in wind speed along the eyewall. If these vortices migrate inward, into the calmer area in the eye, they dissipate. However, if the vortex migrates outward, under an area of high convective activity, the results are quite different. The convection stretches the vortex upward, reducing its diameter and increasing its speed in the process. Vortex velocity now may reach 129 km/h (80 mile/h). However, the small spin up vortex is imbedded in the faster hurricane winds moving at 193 km/h (120 mile/h), leading to *effective* winds of 322 km/h (200 mile/h). This type of microscale activity caused extreme damage in the Naranja Lakes and Pine Woods Villa developments in South Dade County, Florida in Hurricane Andrew (Fujita, 1993).

Sustained versus peak velocities

Hurricane wind velocities are generally reported as *sustained winds*, persisting at that level for several minutes. However, *gust velocities* persisting for short

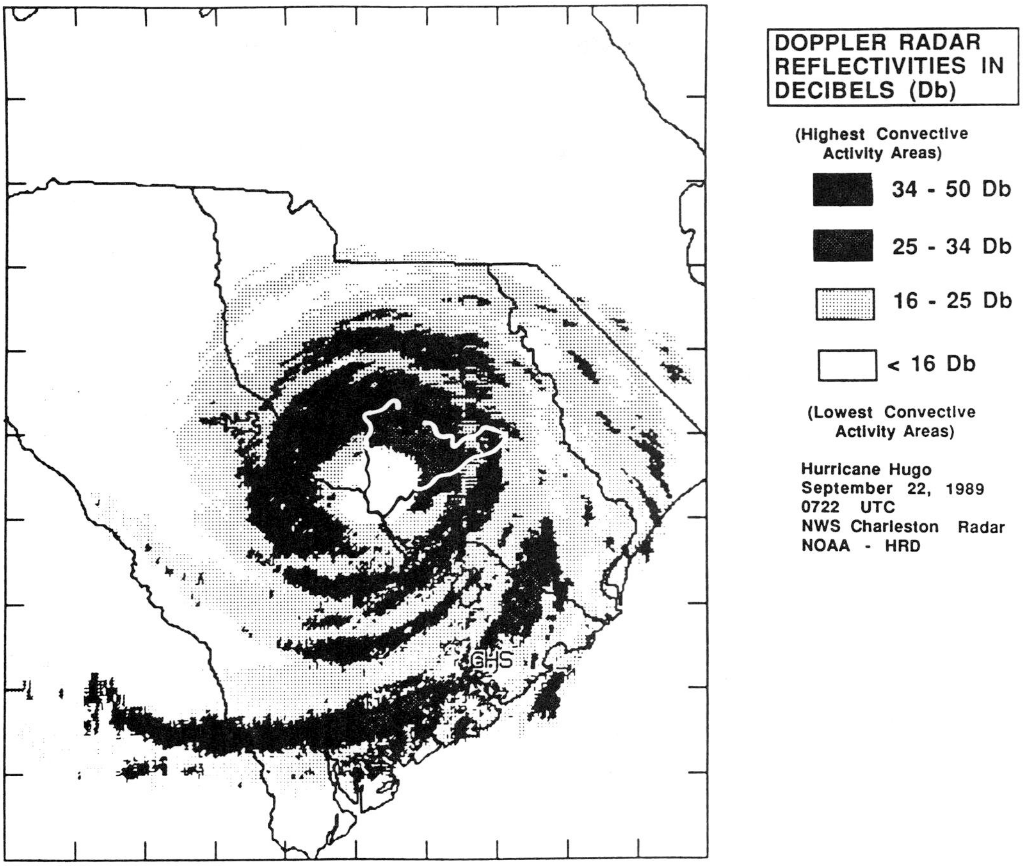

Fig. 15. Doppler Radar image of Hurricane Hugo when the eye and eyewall area was located over inland Sumter County (outlined area). Areas of extreme convective activity (34–50 Db areas on the diagram) resulted in peak gust winds far greater than the mean winds of the weakening hurricane.

duration are important because gusts can exceed sustained winds by up to 100% in highly convective situations. Gust speeds are estimated by applying a gust factor to the mean wind of a given averaging period. Krayer and Marshall (1991) derived a gust factor of 1.5 based on a study of major hurricanes. As a hurricane moves inland, its sustained winds drop as a result of frictional effects and a filling-in (weakening) of the storm. Studies done after Hurricane Hugo (Powell et al., 1991) have shown that isolated convective centers inland (Fig. 3) can generate wind gusts as strong as the hurricane's earlier sustained winds at landfall. This results in unexpectedly severe damage 60 km (100 miles) or more inland from the coast.

The effect of peak gust winds inland was well demonstrated during Hurricane Hugo in Sumter County, South Carolina (Fig. 15). As Hurricane Hugo passed over Sumter County, its winds had decreased from Category 4 at the coast to just above Category 1 (Table 1). However, the right eyewall of the storm containing areas of high convective activity, passed over a good part of the county and was to create mesoscale-pattern damage (Fig. 14). The 10-minute gust factor approached a value of 2.0 because of the presence of extremely convective rain band features (Fig. 15) during eyewall passage (Powell et al., 1991). Thus, while the mean wind speed was about 121 km/h (75 mile/h), or a Category 1, a good part of the county was being battered by *Category* 4 level winds of up to 225 km/h (140 mile/h) and destruction was severe in a county far inland from the ocean (Fig. 15). The inhabitants of the county had expected to shelter refugees from coastal destruction but instead required massive help themselves. Damage in Sumter County exceeded 790 million dollars and 75% *of all injuries* during Hurricane Hugo occurred in Sumter County (V. Jones, pers. commun., 1993).

Wind damage factors

Wind damages structures and vegetation by its force, pressure differential effects and by creating and transporting debris that destroys other structures. The force exerted by the wind varies with the square of the velocity. Thus a 3× increase in wind speed from gale force winds of 55 km/h (34 mile/h) to Category 2 hurricane winds of 164 km/h (102 mile/h) increases the wind force 9 *times* (3^2). The inverse relationship between air flow and fluid pressure creates areas of differential pressure around a structure. Where the wind flows faster, the pressure is lower, and vice versa. The differential pressure fields around a structure may cause it to fail (Fig. 16). The side of a structure facing the wind is subject to compressive stresses that tend to push the wall and windows in. The faster moving air flowing over the roof and around the sides of the structure generate areas of low pressure that act as suction forces,

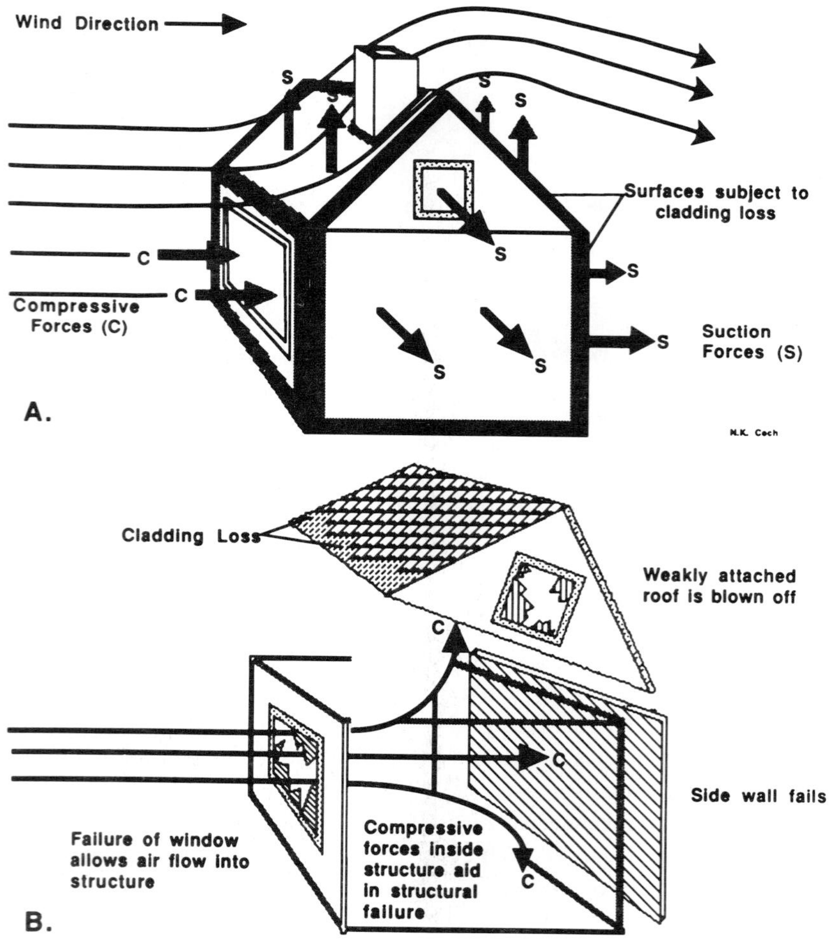

Fig. 16. (A) Wind forces acting on a structure in a hurricane. The side facing the wind is subject to compressive forces that tend to push it in. The faster moving air flowing around the sides and roof create a low pressure zone that results in suction forces tending to pull the structural walls outward. As long as the structure remains intact it may resist these forces. (B) the integrity of the structure has been destroyed by debris that broke a window. the free flowing air inside the structure now pushes outward on the walls and roof and the structure is destroyed. Diagram by the author suggested by descriptions in Sparks (1991).

Fig. 17. Hurricane Andrew surge and wind destruction at the national headquarters of the Burger King Corporation on Biscayne BayW, Florida. This location, near the radius of maximum winds of Hurricane Andrew, experienced the highest surge and winds. The lower story was gutted by surge while the upper ones were destroyed by winds. Aerial photograph by Nicholas K. Coch.

tending to pull the walls and roof outward (Fig. 16A). The structure can therefore fail if: (1) structural walls are weak; (2) the roof, walls, and foundation are not held together by combinations of reinforcing plates and straps; or (3) if wind can gain access to the inside of a structure through structural failure or airborne debris damage to windows, home doors, or garage doors. If wind enters the structure, it exerts a compressive force on the interior walls and roof that acts with the suction forces outside to tear the structure apart (Fig. 16B). In Hurricane Andrew, rolling garage doors were shifted off their tracks by sustained winds and gusts, allowing

Fig. 18. Homes gutted and trees damaged by the high winds of Hurricane Iniki (1992) at Princeville, Kauai. Photograph taken one year after the storm, these homes were on a ridge fully exposed to winds that were channeled through the mountains in the distance. The channeling increased the wind speed, resulting in great damage on the lee side of the island. Photograph by Nicholas K. Coch.

wind to gain access to many homes (FEMA, 1993a).

Exposure is also an important factor in wind damage. Homes in South Dade County, Florida that had open spaces (parks, lakes, vacant sites, etc.) upwind of them sustained greater damage in Hurricane Andrew. (R. Sheets, pers. commun., 1993). Exposure along a coast is also a factor. Homes nearest the coast are affected by unimpeded flow that decreases rapidly inland as a result of frictional drag over topographic features, structures, and vegetation. An extreme case of wind velocity decrease inland was reported by SethuRaman (1979) before Hurricane Belle (1976) made landfall on the south shore of Long Island. Anemometer readings 10 km inland were 50% of those at the coast, and readings 20 km inland were 25% of the coastal values (SethuRaman, 1979)

The effects of wind increase away from the ground surface where friction with the topography, structures and vegetation slow down air flow. The problem is especially serious along urbanized coasts where high

Fig. 19. Changes in the landscape accompanying Hurricane Andrew (1992) at the Coral Gables Canal, Dade County, Florida. (a) Before Andrew (1987) the cross bedded oolite outcrop is well exposed and lush vegetation covers the area in the distance. (b) After Andrew (1992) the trees are broken and defoliated and the house in the distance is fully exposed. Photographs by Nicholas K. Coch.

rise buildings are subject to increasing wind velocities on their upper floors while the lowest floors are being destroyed by storm surge (Fig. 17).

Hurricane Iniki (1992) made landfall on the island of Kauai, Hawaii and provided a good view of the effects of a hurricane on a mountainous island. Wind speeds increased as the flow was channeled through mountainous passes to cause unexpected devastation on the *lee side* of the island (FEMA, 1993b). Destruction of homes and vegetation on ridge crests was especially severe (Fig. 18). Debris mobilized by hurricane winds caused significant damage. Pieces of structures, such as the corrugated roofs widely used on Kauai, served as projectiles that damaged other structures. Loss of glazing (windows), allowed storm winds to enter the structure (Fig. 16B) and assist in its disintegration (FEMA, 1993b).

4. Geologic changes in hurricanes

Dramatic changes in the landscape occur after a major hurricane (Fig. 19). Structural damage may be restored in weeks or months, but vegetation may take far longer to recover. The vegetation loss may significantly increase the danger of forest fires as well as increasing soil erosion and gullying. However, the most dramatic geologic changes occur near the coast and involve beach and dune erosion, barrier island breaching and nearshore channeling by ebb surge.

4.1. Beach erosion

Elevated storm surge and high waves cause severe beach and dune erosion. Stauble et al. (1991) noted that Hurricane Hugo (1989) reduced most of the dunes over a 160 km (110 miles) stretch on the right side of the storm to a flat planar surface while others were severely eroded. Where winds and surge were highest in Hurricane Hugo, barrier sands were eroded away to expose the underlying peat (Fig. 20). The sand was transported landward as overwash fans built into water bodies behind the shoreline.

Nelson (1991) utilized before and after topographic surveys to describe beach and dune alteration (Fig. 21) from Hurricane Hugo (1989) in South Carolina. In most cases, there was significant recession of the beach face. Some examples of beach face recession are: North Myrtle Beach, 19 m (62 ft); Myrtle Beach, 8.5 m (28 ft); Surfside Beach, 8.5 m (28 ft); and Garden City Beach, 12.8 m (42 ft). The profile at Surfside Beach (Fig. 21) showed a flattening and recession of the beach face, and the development of an overwash apron and nearshore bar as a result of the storm.

Fig. 20. Flood surge erosion and deposition on the barrier island south of Bulls Bay, South Carolina. This area was at the radius of maximum winds in Hurricane Hugo (1989). Sand was stripped from the beach and deposited as overwash fans (light areas) into the bay behind the barrier island. An underlying peat deposit (dark layer) is now exposed along the ocean shoreline (right side of photo). Aerial Photograph by Nicholas K. Coch.

4.2. Dune erosion

Dune erosion in hurricanes depends on dune height, width, and vegetation. The best documentation of the extent of dune erosion in hurricanes comes from studies done by several groups after Hurricane Hugo (1989) in South Carolina. This Category 4 event leveled dunes, reducing them to planar surfaces or severely eroding

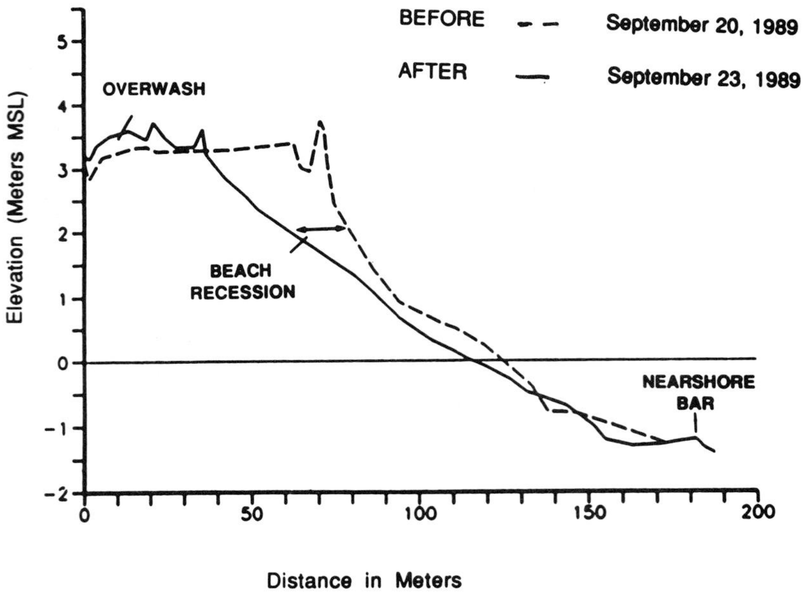

Fig. 21. Before and after topographic profile at Surfside Beach, South Carolina showing erosion and deposition from Hurricane Hugo (1989). The beach face was flattened and receded, and an overwash apron and a nearshore bar were formed. Modified from Nelson (1991).

them over a 160 km (110 miles) stretch from north of the eye at Myrtle Beach to south of the eye at Folly Beach (Stauble et al., 1991). Thieler and Young (1991) found that dunes survived when high enough to prevent their being overwashed and wide enough to prevent being completely eroded. A good example of this were the dunes at Litchfield, South Carolina that were scarped and cut back but not destroyed. Prior to the storm, the dunes reached up to 7 m (23 ft) and dune width was 15 m (49 ft) according to Stauble et al. (1991). The other situation where dunes survived was when they were low but well vegetated so that they could be rapidly submerged without significant erosion such as occurred at Sullivans Island and the Isle of Palms near the eye of Hugo (Thieler and Young, 1991). The only problem with this type of ''dune survival'' is that it provides little protection for structures behind the dunes.

4.3. Vegetation loss and soil erosion

Loss of vegetation provides a striking change in the landscape after a hurricane (Fig. 19) and has a great impact on birds and animals (Coch, in prep.). In this paper only the geological aspects of vegetation loss are considered.

Hurricanes Hugo, Andrew, and Iniki showed the wide range of damage that results from the uprooting, bending and breaking of trees Falling trees cause structural damage and bring down utility lines. They also block roads needed for access by emergency vehicles. The roots of overturned trees break underground water and utility lines. This type of tree damage caused major disruption when Hurricane Hugo (Fig. 15) crossed over Sumter County, South Carolina (V. Jones, pers. commun., 1993).

The patterns of tree fall are not random but are relatively uniform *at any one place*. There is usually a significant difference in tree fall orientation on either side of a hurricane (Fig. 8). This has been documented by detailed plots based on analysis of aerial photographs (Penland et al., 1989, fig. 6). In general, the pattern of tree fall orientation (Fig. 8) in landfall areas is controlled by the wind directions at the front of the hurricane (Fig. 4). However, the tree fall orientation in inland areas may be much more complex and variable.

The number of trees downed and the type of failure (bending, breaking, loss of upper canopy only, etc.) is a function of both tree type and location with respect to the eyewall. Hook et al. (1991) compared tree loss

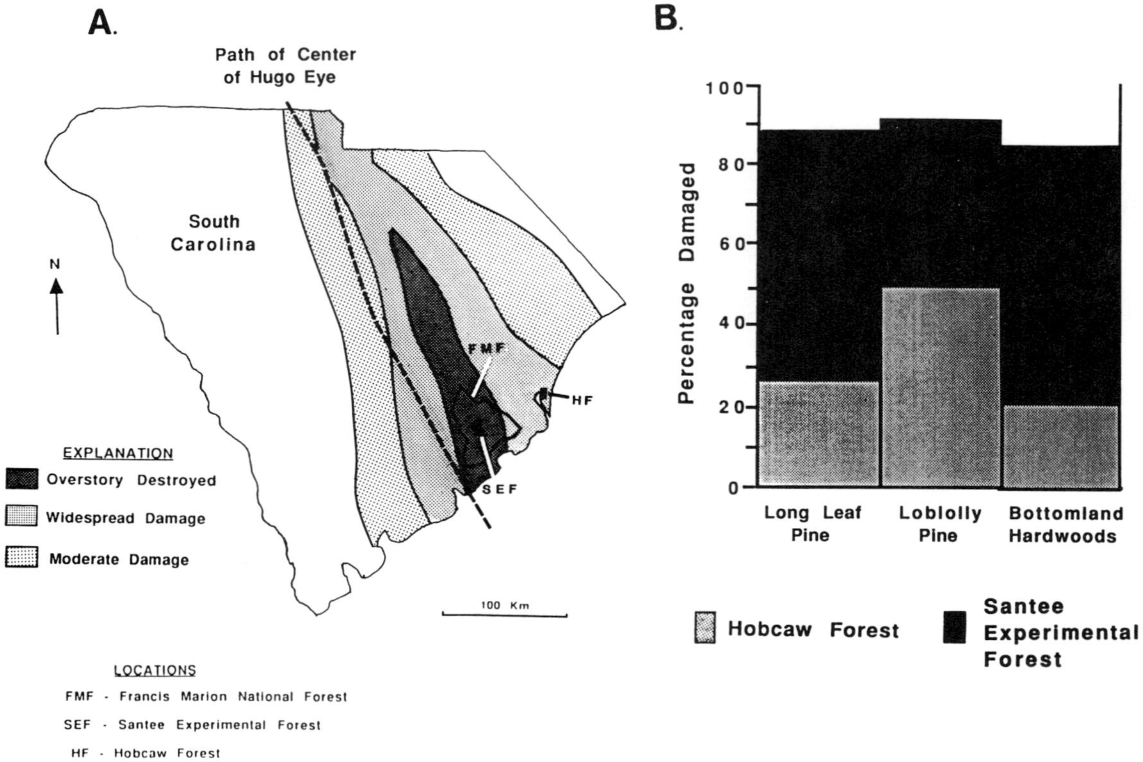

Fig. 22. Comparison of tree damage in two South Carolina Forests during Hurricane Hugo (1989). Damage to all species studied was severe in the Santee Experimental Forest close to the eye of the storm. Damage was far less, and species dependent, in the Hobcaw Forest on the outer part of the storm. (Modified from Hook et al., 1991).

Fig. 23. Photograph of extensive tree breakage and deforestation on Kauai, one year after Hurricane Iniki (1992) struck the area. Massive tree destruction occurred as a result of wind acceleration, near the ridge crest in Waimea State Park, Kauai. Photograph by Nicholas K. Coch.

Fig. 24. Gullying and soil erosion resulting from Hurricane Iniki (1992) on Kauai Hawaii. Photograph taken one year after hurricane. The high rainfall and steep slopes in Waimea State Park on Kauai accelerated soil loss and gullying when vegetation was destroyed at higher altitudes. Photograph by Nicholas K. Coch.

in two South Carolina forests during Hurricane Hugo. The Santee Experimental Forest was in the Francis Marion National Forest within the right eyewall of Hurricane Hugo (Fig. 22) 41–58 km north of the eye center. Winds there during the storm were 43–66 km/s (27–41 mi/h). The Hobcaw Forest was 83–108 km north of the eye center and was subjected to considerably less wind force than the Santee Forest. In the Santee Forest there was little difference between the wind resistance of different species. The percent damaged were as follows (Fig. 22); long leaf pine (89%); loblolly pine (91%); and bottom land hardwoods (86%). In the Hobcaw Forest the damage was considerably less and was species dependent. Damage percentages were as follows: long leaf pine (27%); loblolly pine (52%); and bottom land hardwoods (20%).

Loss of vegetation can have significant geological consequences, especially in mountainous areas with high rainfall. Hurricane Iniki (1992) a Category 4 storm, caused widespread tree damage on the island of Kauai, Hawaii (Fig. 23). Uprooted trees exposed the soil to erosion. The high rainfall plus the steep slopes substantially increased the rate of soil erosion and gullying (Fig. 24).

Re-vegetation after a hurricane in tropical and subtropical areas is much faster than in temperate climates. Data from the 1938 New England Hurricane shows that a major hurricane hitting the mountainous areas of the northeastern United States will cause significant deforestation (Federal Writers Project, 1938). Re-growth will take years and high rates of soil erosion as well as the threat of forest fires will persist during the recovery period.

5. Summary and conclusions

Hurricanes cause a wide variety of damage including flood and ebb surge, wind destruction and inland flooding. Hurricanes with coast-parallel tracks cause significant damage along a long stretch of coast, but the effects drop off rapidly inland. Hurricanes with coast-normal tracks affect a shorter section of shoreline but far more severely. In addition, they cut a wide swath of destruction hundreds of kilometers inland. The destructive power of hurricane winds is not only a function of the vortex velocity but also the forward velocity of the storm center. On the right side of the hurricane, the two wind speeds are vectorially added, greatly increasing the effective wind speed in that sector. The increase in hurricane center speed as the storm passes north of Cape Hatteras and becomes affected by stronger westerly winds, gives northeast United States landfalling hurricanes far more destructive power than storms of similar Saffir–Simpson categories that impact southern coasts.

Storm surge is the rise in the ocean surface resulting from hurricane winds driving shelf waters into shallower areas nearshore as well as the rise in the surface resulting from the low pressure in the eye of the storm. Storm surge is more accurately pictured as a broad dome of water on which high waves usually develop. Maximum surge levels occur on the right side of the storm at the radius of maximum winds and levels rise higher along coasts that have gently sloping continental shelves and embayed sections. Potential surge is highest along segments that make right angles opening eastward, such as the New York Bight and along funnel-shaped estuaries opening east, such as Long Island Sound. Surge amplification has caused especially serious damage along these two highly urbanized coastal segments both in the great ''nor'easter'' storms of the 1990's (Coch, 1994) and in every hurricane that has made a landfall along the coast of Long Island, Rhode Island or southern Massachusetts (Pore and Barrientos, 1976). Ebb surge occurs as impounded flood surge waters and previous rainfall move seaward as the hurricane passes inland. Although ebb surge is much lower in height than flood surge, ebb surge can cause considerable erosion because it affects structures that have already been weakened by winds and flood surge as well as debris carried by both agents. Ebb surge overwashes barrier islands from the bayward side, disperses debris across the nearshore zone to create navigational hazards, and cuts inlets across the barrier islands (Fig. 13)

Hurricane winds are the major destructive force inland. Wind is especially destructive along exposed coasts and the effects drop off inland as frictional effects and hurricane filling decrease wind speed. While the overall power of a hurricane decreases inland, local convective centers on the rainbands can cause severe local damage. Wind damage occurs on storm-wide levels (megascale), associated with local high convective centers on the rainbands (mesoscale) and in very local, but extremely intense, ''spin-up vortices'' (microscale features) associated with the right eyewall of the storm.

Hurricanes result in considerable geologic change along the coast and inland. Beaches are stripped of sand that is deposited as overwash fans into bays. Dunes that are not high enough, or well vegetated are eroded and breached in many cases. Low, but well vegetated dunes may be overstepped with little erosion, although this provides little protection for structures landward of the dunes. The great tree destruction inland disrupts transportation and communication lines as well as buried utilities, and creates a severe fire hazard. Hurricane-induced vegetation loss in humid mountainous terrain can significantly increase soil erosion until cover is restored.

Atlantic-landfalling hurricanes seem to be more frequent during multi-decadal wet cycles in west Africa, and far less frequent in dry cycles. No wet or dry cycle has lasted longer than 26 years, and we are now in the 24th year of a dry cycle. Unfortunately, this period of quiescence has coincided with great urbanization along the Atlantic Coast and the development of public apathy toward the hurricane threat. If past climatic patterns continue, we will soon be entering a cycle of far more frequent Atlantic Coast landfalling hurricanes. These future storms will affect a heavily developed and populated coast where inhabitants have made little preparation to minimize destruction and loss of life. The result could be devastation on a scale last experienced in the northeast during the New England Hurricane of 1938. Unfortunately, it has been 56 years since that regional catastrophe, and its many lessons have been largely ignored or forgotten.

Acknowledgments

A number of colleagues and government officials were of great help in the author's four post-hurricane field studies. My special thanks to Mel Nishihara, hurricane program manager for the State of Hawaii, who set up meetings with emergency management officials on Oahu and Kauai and briefed the author on Hurricane Iniki damage. Glenn Trapp, National Weather Service, Honolulu, provided satellite images of Iniki and a preliminary science report on the storm that was of great help in field studies. Tom Batey, Administrator, County of Kauai, suggested damage study sites that facilitated field work on Kauai and he graciously discussed the results and interpretations of the observations afterwards with the author.

Sam Houston, Hurricane Research Division NOAA, discussed his preliminary findings on Hurricane

Andrew and suggested field study sites to the author three days after that hurricane devastated south Dade County, Florida. Billy Wagner and Andy Eads, Monroe County Emergency Management, gave valuable insights into damage patterns and recovery problems in the Homestead-Florida City areas.

Don Thompson, Long Island Weather Observers, provided information on Hurricane Bob meteorological effects and damage types in easternmost Long Island and accompanied the author in the field. Ana Butler conducted Hurricane Bob damage studies on Cape Cod, Mass. and graciously provided access to the data and photographs for this paper.

Hurricane Hugo field studies (Coch and Wolff, 1991) were aided by the information and logistical support provided by Paul Gayes and Doug Nelson of the Center for Marine and Wetlands Studies, Coastal Carolina College. In late 1993, Gayes and Nelson also provided an update and further information on Hugo ebb surge and dune destruction, respectively. Mark Powell, Hurricane Research Division NOAA, supplied the Hugo radar images used in this paper. Victor Jones, Director of Public Safety in Sumter County, South Carolina, described the inland effects of Hurricane Hugo and the recovery problems that ensued after the passage of the storm.

The staff of the National Hurricane Center, particularly Robert Sheets, Brian Jarvinen, Miles Lawrence, Jack Bevan and John Pavone, deserve special thanks for providing information on a number of different aspects of hurricanes over the past few years. William Gray, Colorado State University, has discussed his research on hurricane frequency with me on a number of occasions. Mike Wylie and Tony Gigi, National Weather Service, New York City, gave valuable insights into the damage done by ''nor'easter'' storms and by hurricanes impacting the northeastern United States. Fred Wolff, Hoffstra University provided valuable additions and corrections to the manuscript.

This research was sponsored in part by PSC-BHE grants #663177, #664048 and #771762. Drafting and word processing assistance was given by Henry Mesa, Sara Woolley and Katherine Andrade. Nancy Bareis and Matt Dinzie, Photographic Unit, CIMS at Queens College, produced the photo plates for this paper. My special thanks to Katherine Andrade for her overall administrative help in the production of this paper.

References

Bricker, R.K., 1988. The Long Island Express — Tracking the Hurricane of 1938. Hodgins, Batavia, NY, 125 pp.

Carter, T.M., 1983. Probability of hurricane/tropical storm conditions: A user's guide for local decision makers, U.S. Dept. Commerce, NOAA, NWS Severe Weather Branch, 25 pp.

Central Pacific Hurricane Center, 1992. Preliminary Science Report, Hurricane Iniki, September 5–13, 1992, Honolulu, Hawaii.

Coch, N.K., 1994. The hurricane hazard in the northeastern U.S. In: C.W. Finkl (Editor), Coastal Hazards. Perception, Susceptibility and Mitigation. J. Coast. Res., Spec. Vol., 12, in press.

Coch, N.K. and Wolff, M.P., 1990. Probable effects of a storm like Hurricane Hugo on Long Island N.Y. Northeast. Environ. Sci. 9 (1/2): 1–14.

Coch, N.K. and Wolff, M.P., 1991. Effects of Hurricane Hugo storm surge in coastal South Carolina. In: C.W. Finkl and O.H. Pilkey (Editors), Impacts of Hurricane Hugo: September 10–22, 1989. J. Coast. Res. Spec. Issue, 8: 201–228.

Federal Writers Project, Works Progress Administration in the New England States, 1938. New England Hurricane — A Factual Pictorial Record. Hale Cushman and Flint Co., Boston, 220 pp.

FEMA, 1993a. Building Performance in Hurricane Andrew in Florida. FIA-22: 93.

FEMA, 1993b. Building Performance in Hurricane Iniki in Hawaii. FIA-23: 100.

Fujita, T., 1993. Origin of spin-up vortices. Am. Meteorol. Soc. address, Summarized in Earth, 2(4): 11.

Gayes, P., 1991a. Personal communication, Center for Marine and Wetland Studies, Coastal Carolina College, Conway, SC 29526, USA.

Gayes, P., 1991b. Post-hurricane Hugo nearshore side scan sonar study: Myrtle Beach to Folly Beach, South Carolina. In: C.W. Finkl and O.H. Pilkey (Editors), Impacts of Hurricane Hugo: September 10–22, 1989. J. Coast. Res. Spec. Issue, 8: 95–112.

Gigi, A.F. and Wert, D.A., 1986. Hurricane Gloria's potential storm surge. NOAA Tech. Mem. NWS ER-70, July 1986.

Gray, W.M., 1990. Strong association between West African rainfall and U.S. landfall of intense hurricanes. Science, 249: 1251–1256.

Gray, W.F. and Landsea, C.W., 1991. Predicting U.S. hurricane spawned destruction from west African rainfall. Unpubl. report, Dept. Atmospheric Sciences, Colorado State University, Fort Collins, CO, 40 pp.

Hook, D.L., Buford, M.A. and Williams, T.M., 1991. Impact of Hurricane Hugo on the South Carolina Coastal Plain Forest. In: C.W. Finkl and O.H. Pilkey (Editors), Impacts of Hurricane Hugo: September 10–22, 1989. J. Coast. Res. Spec. Issue, 8: 291–300.

Jarvinen, B.R., 1993. Personal communication, National Hurricane Center, 1320 South Dixie Highway, Coral Gables, FL 33146, USA.

Jarvinen, B.R. and Lawrence, M.B., 1985. An evaluation of the SLOSH storm- surge model. Am. Meteorol. Soc. Bull., 66(11): 1408–1411.

Jones, V.C., 1993. Personal communication, Director, Office of Public Safety, Sumter County, SC 29150, USA.

Krayer, W.R. and Marshall, R.D., 1991. Gust factors applied to hurricane winds. Preprints: Eight Int. Conf. on Wind Engineering. Assoc. for Wind Engineering, London, Ont., Canada.

Lennon, G., 1991. The nature and causes of hurricane-induced ebb scour channels on a developed shoreline. In: C.W. Finkl and O.H. Pilkey (Editors), Impacts of Hurricane Hugo: September 10–22, 1989. J. Coast. Res. Spec. Issue, 8: 237–248.

Nelson, D., 1991. Factors effecting beach morphology changes caused by Hurricane Hugo, Northern South Carolina. In: C.W. Finkl and O.H. Pilkey (Editors), Impacts of Hurricane Hugo: September 10–22, 1989. J. Coast. Res. Spec. Issue, 8: 163–179.

Penland, S., Suter, J.R., Sallenger, A.H., Williams, S.J., McBride, R.A., Westphal, K.E., Reimer, P.D. and Jaffe, B.E., 1989. Morphodynamic signature of the 1985 hurricane impacts on the northern Gulf of Mexico. Proc. Sixth. Symp. on Coastal and Ocean Management. Am. Soc. Civ. Eng., pp. 4220–4234.

Pierce, C.H., 1939. The meteorological history of the New England Hurricane of September 21, 1938. Mon. Weather Rev., 67(August): 237–285.

Pore, N.A. and Barrientos, C.S., 1976. Storm Surge, Mesa N.Y. Bight Atlas Mon. 6, N.Y. Sea Grant Institute, Albany, NY, 44 pp.

Powell, M.D., Dodge, P.P. and Black, M.L., 1991. The landfall of Hurricane Hugo in the Carolinas — Surface wind distribution. Weather Forecast., pp. 379–399.

Risnychok, N.T., 1990. Hurricane — A familiarization booklet. National Hurricane Center, NOAA, 30 pp.

SethuRaman, 1979. Atmospheric turbulence and storm surge due to Hurricane Belle (1976). Mon. Weather Rev., 107: 314–321.

Sheets, R.C., 1993. Personal communication, National Hurricane Center, 1320 South Dixie Highway, Coral Gables, FL 33146, USA.

Simpson, R.H. 1974. The hurricane disaster potential scale. Weatherwise, 27: 169–186.

Sparks, P.B., 1991. Wind conditions in Hurricane Hugo and their effect on buildings in South Carolina. In: C.W. Finkl and O.H. Pilkey (Editors), Impacts of Hurricane Hugo: September 10–22, 1989. J. Coast. Res. Spec. Issue, 8: 13–24.

Stauble, D.K., Seaburgh, W.C. and Hales, L.Z., 1991. Effects of Hurricane Hugo on the South Carolina coast. In: C.W. Finkl and O.H. Pilkey (Editors), Impacts of Hurricane Hugo: September 10–22, 1989. J. Coast. Res. Spec. Issue, 8: 129–162.

Thieler, E.R. and Young, R.S., 1991. Quantitative evaluation of coastal geomorphological changes in South Carolina after Hurricane Hugo. In: C.W. Finkl and O.H. Pilkey (Editors), Impacts of Hurricane Hugo: September 10–22, 1989. J. Coast. Res. Spec. Issue, 8: 187–200.

ELSEVIER

Geomorphology 10 (1994) 65–81

GEOMORPHOLOGY

Natural hazards of shoreline bluff erosion: a case study of Horizon View, Lake Huron

Patrick L. Lawrence

Department of Geography, University of Waterloo, Waterloo, ON N2L 3G1, Canada

Received February 7, 1994; revised March 26, 1994; accepted April 2, 1994

Abstract

Assessment of hazard of erosion of shoreline bluffs requires an understanding of the natural processes and human use in these high risk areas. As an important component associated with erosion and natural evolution of bluff shorelines, alongshore sediment transport is an essential process in defining the hazard associated with bluff erosion. The objective of this study is the integration of an understanding of alongshore sediment transport into a proposed management framework for cohesive bluff shorelines. A study of a ten kilometre area along the southeastern Lake Huron shoreline, at the cottage community of Horizon View, north of Goderich, Ontario examines the natural and human components which have led to the development of bluff erosion hazards. Although dominant sediment transport is north to south alongshore, the pattern of wave refraction during south and southwest waves, suggests frequent northward transport of material from the 30 meter high bluffs at the site. Examination of beach sediments and bluff conditions support this finding and indicate long term sediment supply and bluff instability. Planning of future development at this location must appreciate the potential for continued bluff erosion and the study results suggest that when combined with other factors such as wave energy, bluff morphology, groundwater conditions, vegetation cover, land uses and impacts of fluctuating lake levels, studies of alongshore sediment transport can be a useful tool to assist in management of cohesive coastlines and understanding the natural hazard of bluff erosion in developed coastal areas. A management framework for cohesive bluffs attempts to integrate an understanding of natural processes with human uses and land use policies in order to identify, assess and select a preferred management option reflecting this understanding.

1. Introduction

With increasing pressure for development, many coastal regions are experiencing rapid alteration of the natural landscape, resulting in widespread impacts on natural systems. A basic understanding of geomorphic processes such as sediment transport is therefore necessary in order to facilitate future land use management within coastal regions. Knowledge of key geomorphological processes is integral to the understanding of the evolution of any landscape. In a coastal environment bluff recession represents the unique and complex dynamic responses of landforms. Wave action, subaerial weathering, fluvial processes, and human impacts can all interact to cause instantaneous, short and long term removal of bluff material and subsequent recession of the bluff face.

The purpose of this paper is to examine the physical and human components that comprise the erosion hazard of cohesive bluff shorelines. A case study of the Horizon View site on southeastern Lake Huron is presented where computer modelling of wave energy has indicated the potential for reversals in alongshore sediment transport which would have important implications for bluff erosion at this site where residential development is on the increase. A series of field tech-

SSDI 0169-555X(94)00015-J

niques were used to assess the sediment dynamics and relative risks from shoreline erosion and the results are discussed in the context of current land uses and planning regulations. The objective is to develop a management model for eroding bluff shorelines based on understanding wave energy and sediment transport and human uses.

1.1. Great Lakes shoreline

The shoreline of the lower Great Lakes (Lakes Huron, Erie and Ontario) is comprised of large sections of cohesive shoreline bluffs, totalling 40% of the total shoreline length (Davidson-Arnott, 1988). These bluffs consist of thick deposits of glacial till and post-glacial lacustrine clay, often with interbedded silt and sand. Wave erosion and associated mass wasting processes along cohesive shorelines on the Great Lakes have resulted in long term mean annual bluff recession rates up to 2.0 metres per year at some sites (Boyd, 1981). Storms events, particularly those associated with periods of high lake levels, most recently in 1951–52, 1972–73 and 1985–86, have resulted in extensive damage to property through nearshore and bluff erosion in many shoreline communities (Kreutzwiser, 1987).

Over the last thirty years, a series of provincial and federal studies have examined the issue of the impact of fluctuating Great Lakes water levels and the resultant shoreline flooding and erosion hazards (Ontario Shoreline Management Review Committee, 1986; International Joint Commission, 1989). Recommendations to the province focused on the need to alleviate the long term impact of fluctuating water levels, shore erosion, and subsequent shoreline damages. The International Joint Commission (1989) noted that "wave action and composition of shore materials are the most significant determinants of long term changes in the [Great Lakes] shorelines" (p. 27). Provincial studies following high water levels on the Great Lakes in 1985 led to the development of a shoreline management program in which a major component is the prevention of flooding and erosion hazards by assessment of the risk to human use and development and the use of land use controls (Ontario Ministry of Natural Resources, 1987).

These studies point out the need for a more comprehensive understanding of the processes of shore erosion, as controlled by wave energy, sediment transport and bluff conditions, for individual site specific cases. In order to provide solutions to long term hazards associated with bluff erosion and for development of management policies, examination and integration of coastal process studies is required. With the Ontario provincial shoreline management program in existence, local conservation authorities with their limited staff expertise and funding require shore process information, in particular assessment of shore hazards, in order to develop shore management plans.

1.2. Erosion of cohesive shoreline bluffs

Long term evolution of cohesive shorelines includes severe erosion of nearshore and bluff material and continuing readjustment of the shore form. The mode of bluff failure is controlled by direct toe erosion due to wave action, causing instability and successive shallow slips upslope (Boyd, 1981). Steep offshore profiles are also typical of eroding areas, resulting in increased direct wave impact on the shorefront. The irregular distribution of erosion rates alongshore can be considered a function of bluff stratigraphy, offshore bathymetry, local groundwater and surface drainage conditions, vegetation cover and land use management practices.

The key factors controlling erosion at the toe of the bluffs have been identified by Sunamura (1983) as: (1) the assailing force of the waves at the bluff foot; and (2) the resisting force of the bluff forming material. Water levels, offshore bathymetry and beach profile, and beach sediment also indirectly influence the assailing force of the waves. Wave activity is controlled by fetch length (distance prevailing winds travel over water body), orientation to shore, wind climate (direction of prevailing waves, height and period), water levels and ice cover (length of season). Subaerial processes (slope processes) are considered a third factor in the evolution of cohesive shorelines. When waves erode a bluff toe, the slope becomes unstable due to increases in the slope angle and in the slope stresses. This instability then can induce mass movements of various types including slumps, debris slides, mudslides, mudflows and debris falls (Sunamura, 1983). The occurrence of mass movement is a function of several contributing factors including lithology, geological structure, geotechnical properties of the materials, precipitation and the groundwater level.

All these processes, however, appear to have little effect on the actual rate of toe erosion and the equilibrium that develops. Rather this is due to higher potential alongshore sediment transport than actual sediment supply (Davidson-Arnott, 1988). If the production of large amounts of debris due to slope instability, subaerial processes, or mass movements exceeds the removal rate by wave action, then the slope angle decreases. These two conditions, the production and removal of debris, reflect the processes of shoreline evolution which lead to slope evolution over the long term. The form of the slope and shoreline will continue to alter until conditions change or the slope reaches a stable level.

The processes of the transport of sediment alongshore in the nearshore coastal zone are complex and reflect the interaction between the dynamics of alongshore currents, erosion by wave action and available sediment. The predominant wave and wind climates, nearshore and bluff characteristics, and the offshore conditions are the primary controls on the rate of transport and amount of sediment moved at any given location. Sediment transport is associated with the location of eroding shore reaches and depositional environments, formation of offshore bars, and the distribution of nearshore sediments. The determination of alongshore sediment transport is primarily predicted by the use of empirical models of wave climate, the calculation of available sediment, and estimation of sediment transport by various field measurement techniques (for examples refer to Komar and Inman, 1970; Galvin and Vitale, 1976; Dean et al., 1982).

1.3. Bluff erosion hazards

An understanding of sediment transport, identification of sediment sources and sinks, and estimating sediment budgets is essential to understanding the spatial distribution of erosion hazards in the coastal zone. Wave energy dissipated at the base of a cohesive bluff leads to instability, bluff erosion, and in areas that have been developed for human use, identification of a hazard to land uses. Long term deposition at a site will promote the establishment of beaches and dunes which can provide a barrier to bluff erosion by acting as a source of sediment to be removed by wave energy. However, sediment dynamics alone are not hazards, they can be considered essential natural processes within coastal ecosystems which allow for evolution of the landscape.

Erosion and sediment accumulation are also greatly influenced directly and indirectly by a variety of human activities. Bluff erosion supplying sediment to the nearshore and alongshore sediment transport can be interrupted by dredging, shoreline protection structures, docks, cottage developments, and harbour structures. Evaluating and understanding natural hazards requires the assessment of human actions and perceptions in land uses and development in coastal areas characterized by erosion. In many cases it is important also to note the institutional arrangements in society which often operate both to keep people in the same place and to protect existing interests by reinforcing the status quo. Development pressures, opportunity, incentives, investment, assistance and planning often contribute to increasing hazards and related costs by encouraging and supporting increased occupance of known hazard zones.

In understanding natural hazards, one requires an appreciation for the complexity of human and natural interactions in the environment. Assessing the natural processes leading to the development of conditions conducive to flooding and erosion and understanding the sociological, economical, behavioral, and attitudinal characteristics of human activity are essential. The key components of a human ecological approach to hazards have been identified as (modified from Burton et al., 1993):

(1) assessment of the extent of human occupance by hazards zones,

(2) identification of the full range of possible human adjustments to the hazard,

(3) study of how man perceives and estimates the occurrence of hazard,

(4) description of the process of adoption of damage-reducing adjustments in their social context,

(5) estimation of the optimal set of adjustments in terms of anticipated social consequences.

2. Study area

Horizon View is located along the southeastern shoreline of Lake Huron within Ashfield Township, Huron County, Ontario (Fig. 1). The area is located in the physiographic region of southwestern Ontario clas-

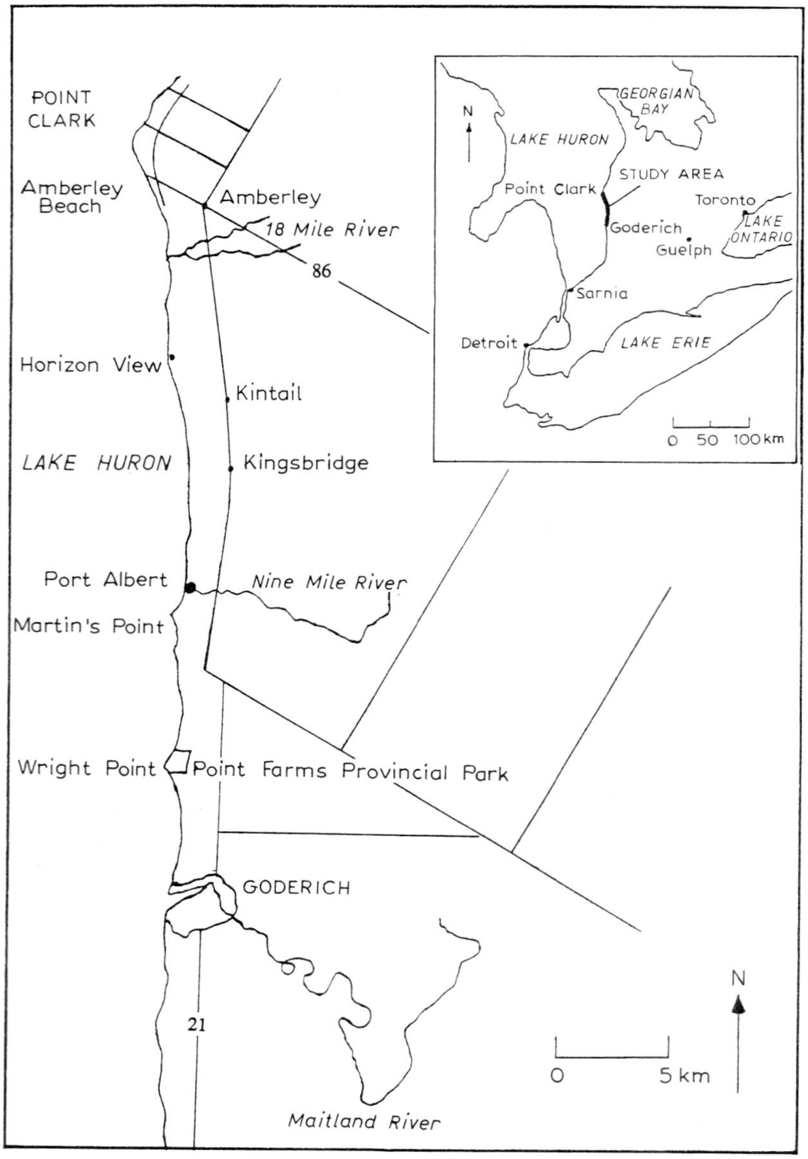

Fig. 1. Southeastern Lake Huron shoreline.

sified as the Huron slope, which consists of a clay plain of glacial Lake Warren overlying a clay till gently sloping westward forming shoreline bluffs. The bluffs are 20 to 40 metres in height and are comprised of cohesive sediments including glacial till, clay and silts (Reinders Ltd., 1989). Land uses consist of agriculture, with several small urban communities and extensive cottage development along the shoreline.

2.1. Shoreline description

The St. Joseph till is the main surficial deposit along the southeastern Lake Huron shoreline. The till is highly resistive to erosion, except in the wet, exposed, weathered state in which surface runoff, mass wasting and direct abrasion by wave action leads to rapid removal. This type of clayey till material has a unconfined compressive strength of 1.0 to 4.0 kg/cm^2 when dry and 0.25 to 1.0 kg/cm^2 with a high water content (Boyd, 1981). Bedrock consists of Devonian medium grained dolomite and limestone of the Hamilton or Detroit River Groups and is located 28 to 33 metres below the surface, overlain by the clay/clay till deposits (Reinders Ltd., 1989). A review of offshore bathymetry from hydrographic charts indicates that the two metre contour is generally 150 to 200 metres offshore while in the vicinity of Point Clark, shallower offshore slopes occur as a function of the offshore stony till shelves. Examination of surficial sediments of the Lake Huron basin has revealed a dominance of sand, till and bedrock offshore within the study area. Erosion rates of the offshore platform and nearshore till deposits along this shoreline are at present unknown.

The climate is continental with January and July average temperatures −4 and 22°C. The Lake Huron water mass has a moderating effect on temperatures, also resulting in increased local snowfall, and an average annual precipitation of 900 mm (Reinders Ltd., 1989). Ice begins to form along the coast by mid-December and persists until early April. The ice cover on Lake Huron is usually 40 to 70 percent, inhibiting wave generation and sediment transport along the shore. Mean monthly wind speeds vary from 12 to 21 knots, with dominance of south winds at 19% occurrence and east and southeast at 13 and 14%, respectively. The northwest fetch is longer than the west allowing larger waves to be generated by storms and causing high transport rates during storm events.

Examination of wave hindcasted frequency distribution tables completed by Philpott Ltd. (1988) shows that 42% of the total hourly frequency are northwest waves, with south and southwest waves totalling 48%, indicating the importance of these waves in potential sediment transport. Wave heights from all directions have been recorded less then 3 metres for 95% of the monthly mean records. Long period wave classes (5 to 7 seconds, 1 to 2 metre waves) are noted in the northwest waves, as a function of the fetch. From the south and southwest, 3 to 5 seconds, less than 1.5 metre waves are most common.

Lake Huron experiences a variety of cycles of fluctuating lake levels; long term, seasonal and short term. The long term mean level (1860 to 1989) for Lake Huron is 176.4 metres with extreme variations of 1.5 to 2 metres from the mean occurring at irregular intervals due to climatic variations within the Great Lakes basin (International Joint Commission, 1989). Seasonal variation, as a function of precipitation, drainage and runoff is in the order 0.4 metres, with water levels typically reaching annual peaks in early June or July (International Joint Commission, 1989). Short term fluctuations as a result of storm surge, seiches and wave runup, lasting for a few hours can account for as much as 1 metre local change in water levels.

The net direction of littoral transport along southeastern Lake Huron is from north to south as a result of large northwesterly storm waves. Total sediment transported alongshore from Point Clark south to Goderich has been estimated at 26,770 m^3 per year (Reinders Ltd., 1989). Sediment is frequently intercepted and deposited as beaches by a variety of mechanisms including natural headlands formed by glacial till or bedrock shelves and possibly at several artificial groyne fields. Harbour structures at Goderich act as a total barrier to continued southward transport and serves as the termination of the littoral cell.

The computer program WAVENRG (May, 1974) has been utilized to simulate wave refraction and to model wave energy and potential sediment transport within the Lake Huron shoreline study area (Lawrence, 1991). The examination of the plots for the total mean annual wave energy as well as the individual wave refraction patterns, particularly the south and southwest waves, has suggested the possibility for frequent reversals of the north–south alongshore transport of sediment within the study area (Fig. 2). Of interest is the northern ten kilometre section from Horizon View to Point Clark. As has been suggested by several researchers (Boyd, 1981; Reinders Ltd., 1989), sediment transport in this reach may be subject to periodic northward transport as a function of dominant south waves, often evident in the wave climate regime.

2.2. Horizon View

The study site (Fig. 3) consists of a ten kilometre shoreline bordered on the north by the Point Clark headland and extends south to a cottage community. South, in the lee of the Point Clark boulder till promontory, the shoreline is at first characterized by narrow beaches with sand and gravel and considerable boulders offshore. The beaches rapidly increase in width

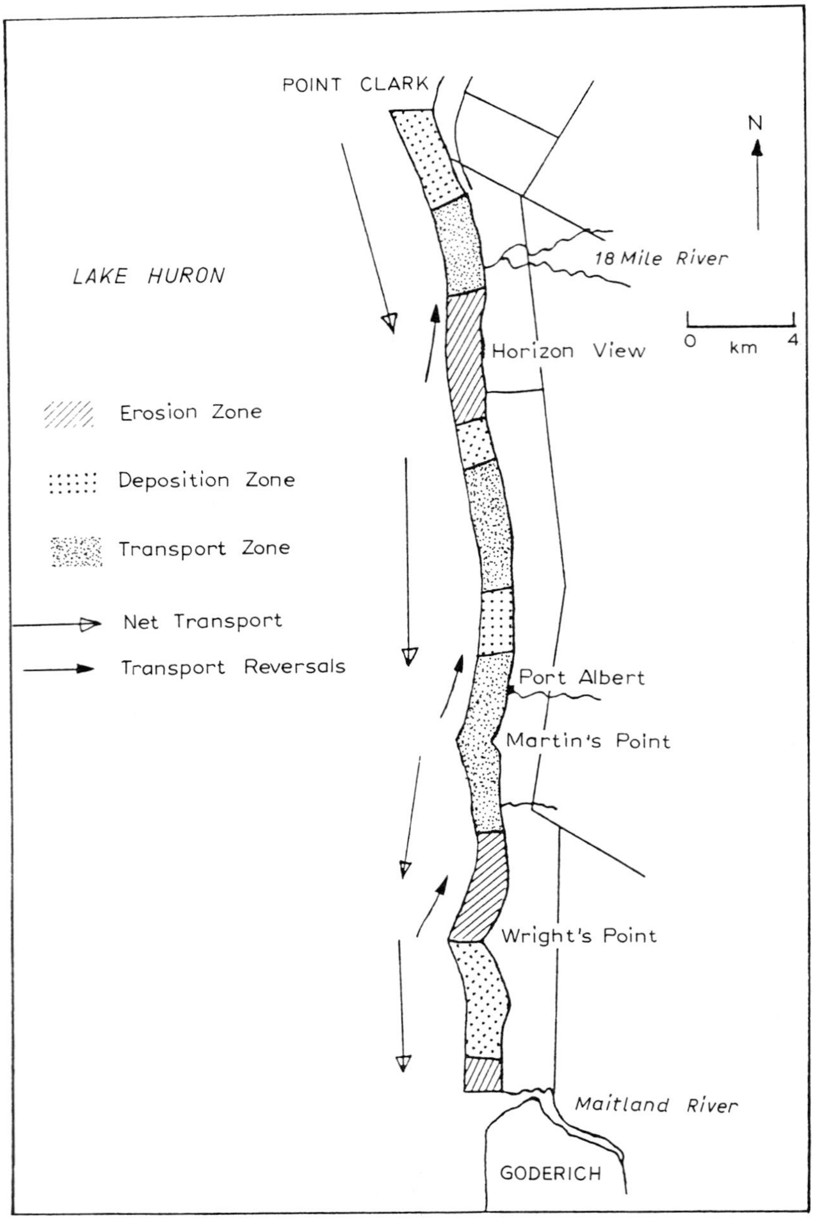

Fig. 2. Alongshore sediment transport patterns.

and sediment supplied either from the south or encircling around the headland from the north, has led to the establishment of wide sandy beaches with discontinuous foredune development. The backshore is dominated by 20 to 30 metre high clay till post-glacial Lake Algonquin bluffs, located several hundred metres inland. Extensive cottage development occurs at Amberley Beach with eighty seasonal (cottage) and permanent residences.

Continuing south, the beach narrows in width to less than 20 metres and the Lake Algonquin bluffs intersect with the contemporary shoreline. The bluffs are 15 to 20 metres in height, with complete vegetation cover consisting of pines, cedars and ground shrubs. Erosion

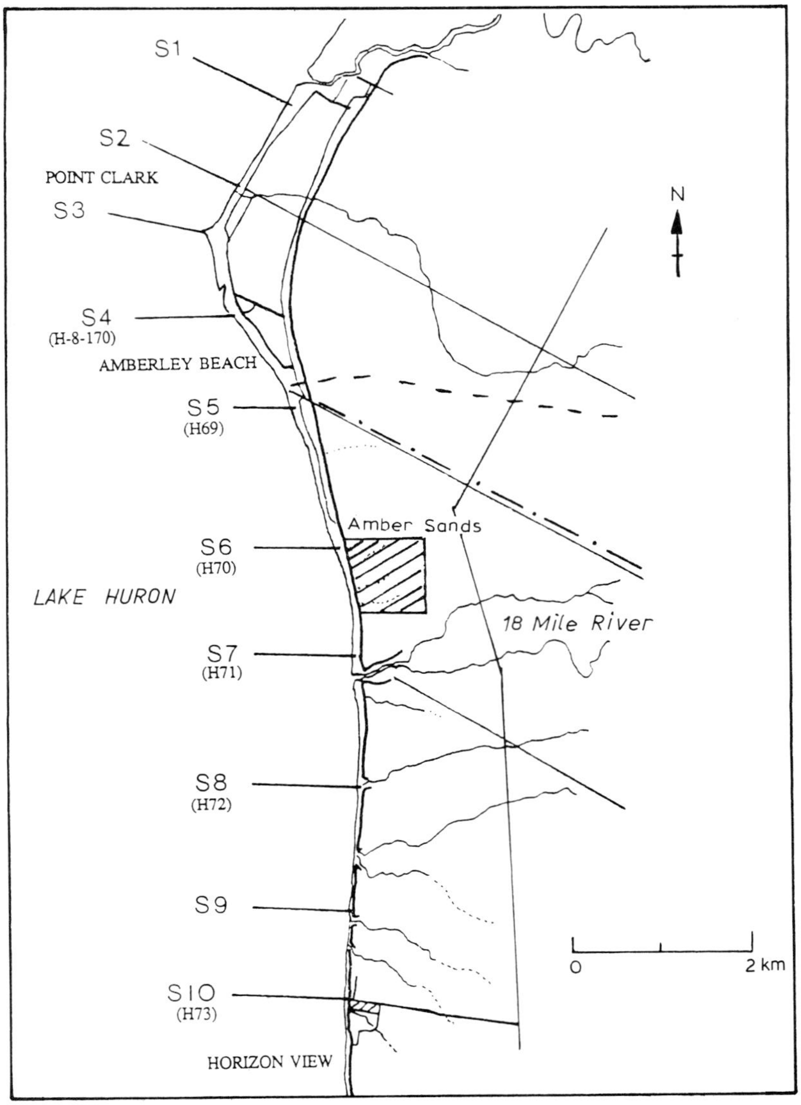

Fig. 3. Horizon View study site.

rates of up to 0.11 metres per year have been estimated from available data (Reinders Ltd., 1989). Development is restricted to a series of small cottages along the base of the bluffs. Shoreline protection is limited to a series of six concrete groynes and breakwalls protecting only four cottages.

The 18 Mile River runs into Lake Huron, providing an influx of sediment. South of the river, the shoreline begins a gradual transition towards high eroding bluffs, with substantial slumping and mass wasting, toe erosion, limited vegetation cover, steep offshore profiles and narrow sandy beaches. The area has been assessed as low risk with limited flooding and beach erosion during high lake level periods (Lawrence, 1991). Several small cottages are located on top of the bluff, just south of the 18 Mile River, with the remainder of the shoreline dominantly woodland or agricultural land.

At the southern extent of the shoreline section, the cottage development of Horizon View occurs with twenty cottages located on top of steep 30 metre bluffs, undergoing recession in the order of 0.5 metres per year from direct wave erosion and severe slumping. Exten-

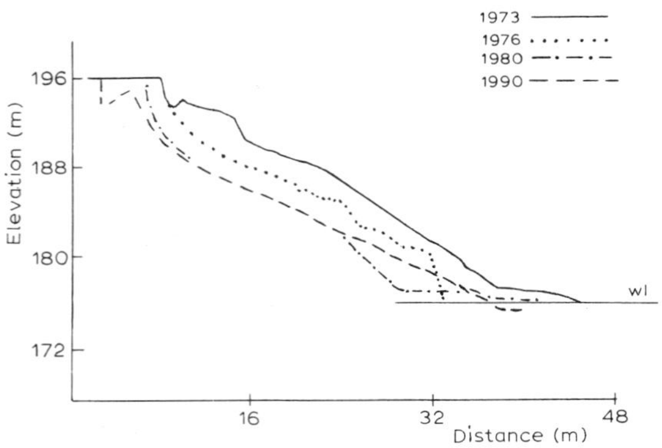

Fig. 4. Horizon view erosion monitoring station.

sive mass wasting, including slumping, debris flows and gullies, leading to removal of vegetation cover occurs along this section of shoreline. With many seasonal residences located at the toe of the bluff, severe damage to buildings occurred during high water levels in 1972–73 and 1985–86 but limited use of shore protection structures has been noted at this site. Wave erosion of the beach and toe of the bluff, especially during high water levels, is considered the most important hazard to development, with bluff slumping and gully erosion also severe (Lawrence, 1991).

An erosion monitoring station, located at Horizon View, Ashfield Township has been surveyed infrequently since 1973 (Fig. 4). The clay till bluff profile exhibits typical gradual slope decline as material moves from the top, in major and minor slump events, to the base, where it is removed by wave action. The top of the bluff at this site has experienced recession of approximately five metres in the last 17 years, for an average rate of 0.3 metres per year. However, as is common on many of these bluffs, recession is an episodic event, with the majority occurring since 1986, with a series of slumps effectively removing 2.5 metres from the top of the bluff.

3. Geomorphic studies

It was decided that a multi-faceted technique based on geomorphic and sedimentological indicators would be utilized to examine variability of beach form and long term alongshore sediment transport along a section of shoreline from Point Clark to Horizon View, within the southeastern Lake Huron study area. This would allow for examination of the natural processes which occur along the shoreline and relate to the identification of the severity of the hazard from bluff erosion. The objective is to determine the alongshore variations in beach and bluff form that can indicate whether the site is prone to continued erosion or deposition. The variations between sites can be used to indicate patterns of long term sediment transport and suggest the degree of stability and degree of hazard to human use and development.

Taggart and Schwartz (1988) suggest the systematic use of geomorphic and sedimentological indicators to determine the direction of net shore-drift as a quick and accurate method. The key indicators examined include beach width, sediment size graduation, beach slope, bluff morphology, headland beaches, sediment trapped by shore structures, stream month diversions, inlet migrations, spit growth and the development of offshore features (e.g. nearshore bars). Trends in beach and bluff conditions alongshore can be used a key indicators in determining littoral drift patterns. In the zone of erosion, the shoreline is characterized by narrow beaches and near-vertical to vertical eroding bluffs which are devoid of vegetation cover. Bluff undercutting by wave action and vertical slumping due to readjustment of slope form are common.

3.1. Methods

Six techniques were employed during the field research; beach profiles, offshore echo sounding, beach sediment analysis, sediment wedge thickness, stream diversions and geomorphological characteristics. One kilometre profile stations were measured alongshore within the study site. Onshore beach profiles were completed using abney level, stadia rod and standard surveying procedures. Offshore profiles to a distance approximately 800 metres offshore were completed at each station, using a Raytheon echo sounder mounted on the stern of a nine metre boat. Horizontal position control was obtained by use of a 200 metre beach parallel offset and subsequent angle correction communicated to the boat via radio or hand signals. Raw data were corrected for water level and draft.

At each of the ten profile stations (S1–S10), samples of beach sediment were also collected within the swash zone and at midbeach for analysis of size and sorting parameters. The average thickness of beach material should increase in the direction of transport, reflecting deposition of material as wave energy decreases alongshore (Taggart and Schwartz, 1988). Measurements of the thickness of beach sediment overlying the till material were attempted alongshore at each station, and at 200 metre intervals in between, measurements of beach width, slope and depth to till were determined. The sediment thickness was determined by measuring the depth of sediment overlying till.

The mouths of the 18 Mile River and several smaller streams were examined in the field and on aerial photographs to assess the direction of stream diversion within the study area. Beach form can also be examined to assist in the assessment of littoral drift. Jacobsen and Schwartz (1981) state that gradual widening of beaches, leading to development of backshore dune complexes often occurs in the direction of predominant transport. Beach slope is influenced by particle size (which affects permeability), slope of any underlying platform material, and the thickness of the beach sediment (Taggart and Schwartz, 1988). In general, beach slope decreases in the direction of transport as a function of sediment accumulation and decreasing wave action. As a result, beach profiles should become more gentle in slope, wider and develop foredunes in the direction of littoral drift.

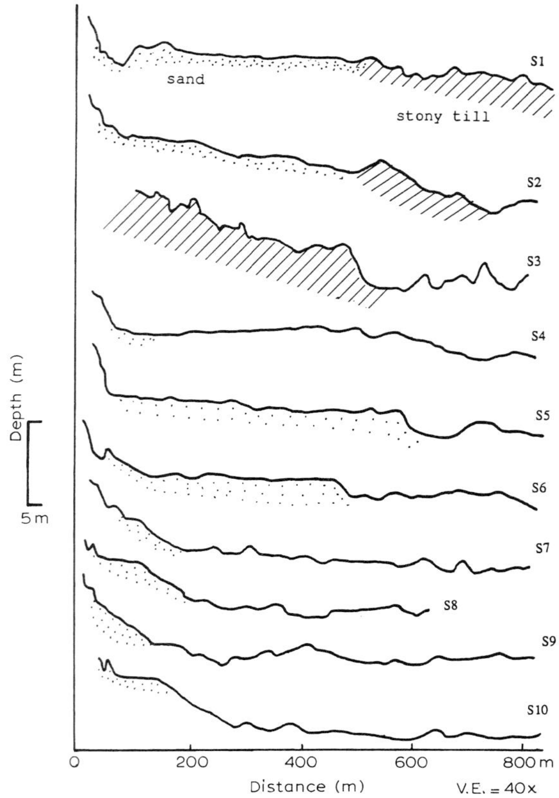

Fig. 5. Nearshore profiles S1–S10.

3.2. Results

Nearshore profiles along the eight kilometre shoreline section indicate the importance of the offshore stony till platform at Point Clark in controlling southward transport around the headland (Fig. 5). At site S3 (Point Clark), the offshore profile exhibits a concave slope to a distance of 600 metres. At this point, the profile drops sharply, indicating the end of the platform. It can be seen that sediment transport north to south around this barrier would be impeded. A large amount of sediment would be diverted offshore and deposited in deeper water. When combined with the insignificant amount of sediment coming towards Point Clark from the north, 449 m^3 per year (Reinders Ltd., 1989), it can be stated that very little sediment passes around the point. South of Point Clark, in the lee of the headland, gentle nearshore profiles continue out beyond of 800 metres offshore. Boulders mixed with pockets of coarse sediment are gradually replaced by sand and silt sediments. At site S6, sediment influx from the river has produced a wide, shallow, sandy profile out to 500 metres.

Southward, a decreasing influence of the platform is noticed, with typical cohesive concave offshore profiles developing. Within the last few kilometres, the nearshore profiles drop off to depths of 7 to 10 metres very rapidly, within a hundred metres. Occasional alongshore migrating bars consisting of fine grained sands and silt lie south to north from Horizon View towards the 18 Mile River. These conditions would seem to support northward transport from Horizon View, with offshore and bluff erosion supplying sediment, towards Amberley Beach, with its gentle sandy profile and beaches.

Beach conditions at each of the ten sites tend to exhibit development reflecting northward transport of material (Fig. 6). Narrow steep beaches occur from S10 at Horizon View, north to S8, south of the 18 Mile River. Wide gentle sloping beaches are found northward from the 18 Mile River (S7) towards Amberley Beach (S5). In the lee of Point Clark (S4), narrow,

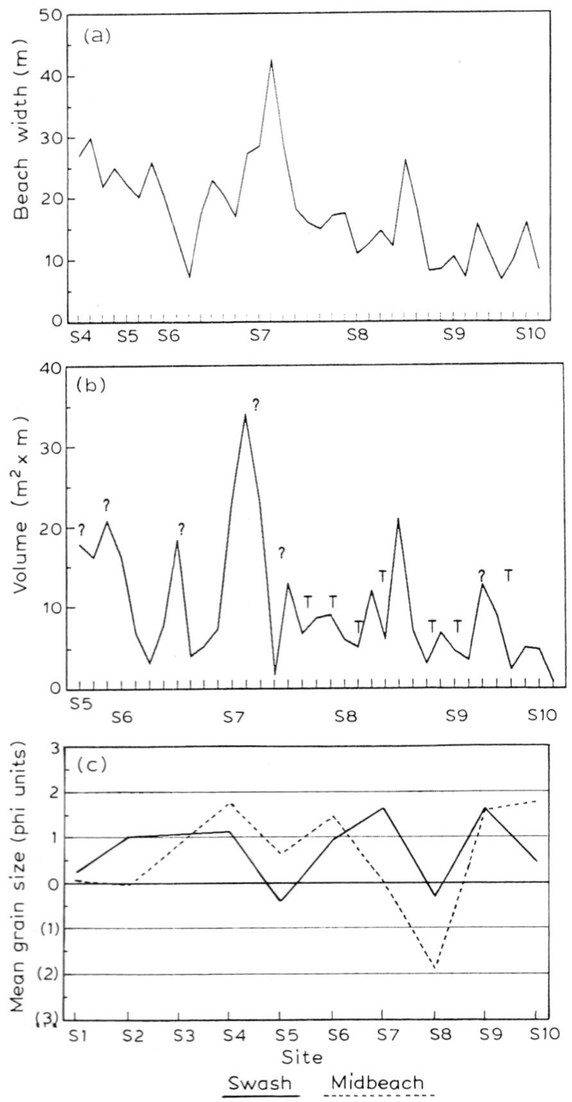

Fig. 6. Beach conditions S4–S10. (a) Beach width; (b) sediment wedge thickness, Amberley Beach (S5) to Horizon View (S10); (c) mean grain size.

level beaches have developed. Beaches increase in width towards the north, from less than 10 metres at Horizon View to greater than 20 metres at Amberley. A very large beach at the 18 Mile River (43 metres) is the result of the influx of river sediment and accumulation along the south bank of the river channel. At several locations, a layer of coarse sand and gravel was noted overlying the till, and as such was occasionally used as an indicator. South of the 18 Mile River (S7), till was confirmed at a majority of locations. The calculated volume of material at stations S8 to S10 was less then 20 m^3. North from the 18 Mile River (S7) towards Amberley Beach (S5), beach width became the determining factor within the calculation of the sediment wedge. Overlying beach sediment was too thick to measure accurately, therefore inconclusive results were obtained.

Along the shoreline area changes in shoreline conditions, which reflect the sediment conditions, can be characterized to assist in this interpretation (Fig. 7). In the south at Horizon View (S10), eroding bluffs with steep near-vertical slopes and limited vegetation cover and narrow beaches comprise the shoreline. Severe subaerial mass wasting, often in the form of rotational or vertical slumps hundreds of metres in length, and extensive wave erosion at the toe of the bluff, are common characteristics for this shoreline, and extending northward a few kilometres. The shoreline begins a rapid transition towards wider beaches (S9), increasing vegetation cover, stable slopes with periodic evidence of wave erosion, perhaps during recent storm events or high water levels in 1985–86. Northward at site S8, vegetation becomes established, subaerial mass wasting decreases, with slumping at the bluff toe a function of wave undercutting.

At the 18 Mile River (S7) a wide stable beach has developed, due to the influx of sediment provided by the river. Slopes are stable and well vegetated. Just north of this location at site S6, lies a small section of narrowing beaches. Erosion of the toe of the bluff occurs at rates of 0.11 metres, perhaps indicating changes in offshore profile conditions and sediment supply. Continuing northward, the beaches gradually widen, reaching in excess of 30 metres at Amberley (S5), with well established, vegetated foredunes. Several hundred metres north at S4, however, the sandy beaches are replaced by 15 to 25 metre coarse sand beaches with boulders offshore. These conditions continue northward towards the Point Clark headland (S3).

On the basis of the geomorphic evidence provided, supported by the WAVENRG model results (Lawrence, 1991), a proposed littoral drift pattern for the shoreline from Point Clark to Horizon View is suggested (Fig. 8). Sediment eroded from the bluffs, beach, and offshore at Horizon View is transported frequently in both north and south directions. Towards the north this mate-

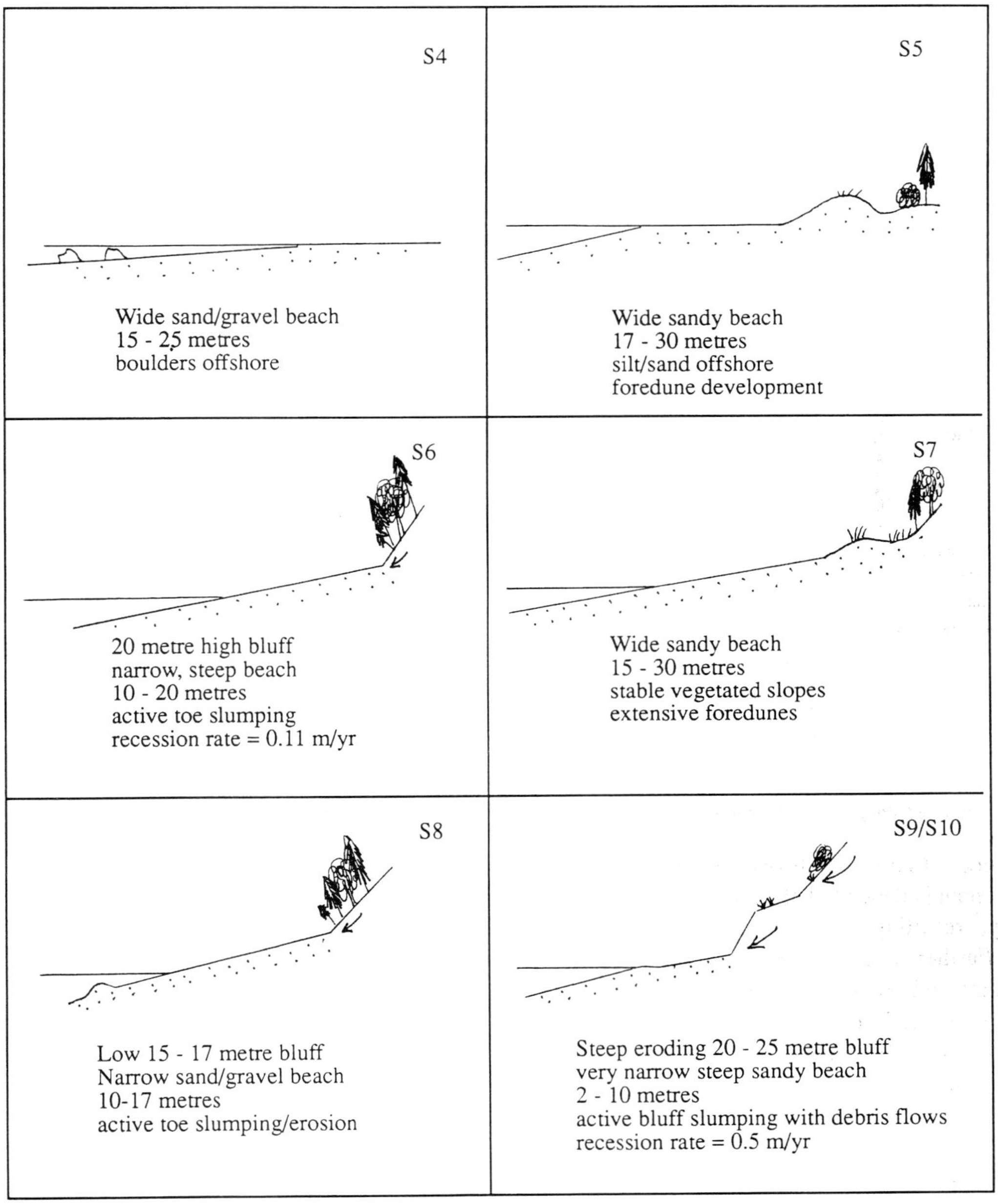

Fig. 7. Shoreline characteristics S4–S10.

rial, continuing to be supplied by bluff erosion, maintains a northward alongshore sediment transport pattern, as a function of frequent south and southwest waves. South of the 18 Mile River, stable slopes are protected by a wide beach supplied from this northward transport and sediment influx from the river.

A decrease in sediment immediately north of the 18 Mile River reflects offshore conditions and the potential for erosion due to northwest waves bypassing or refracting around the Point Clark headland. Gradually towards the north, sediment is accumulated along wide sandy beaches, as potential transport as a function of

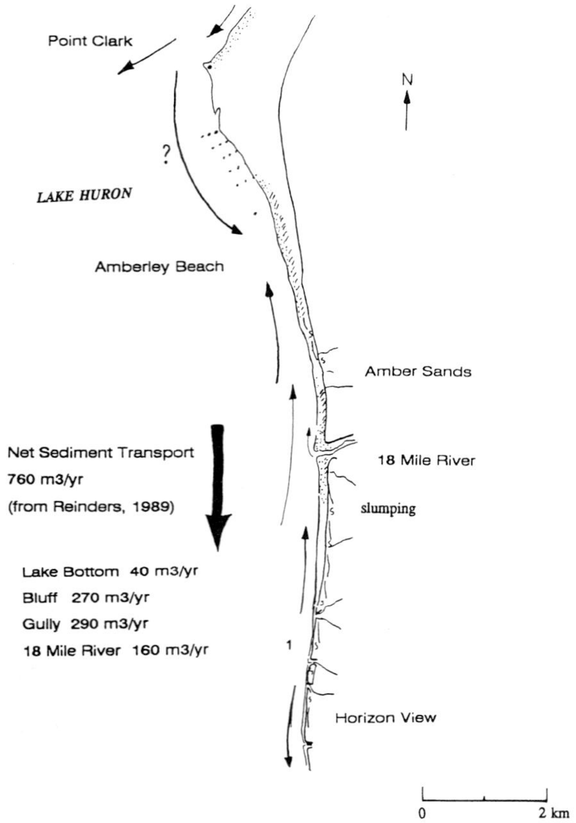

Fig. 8. Proposed alongshore sediment transport patterns.

wave energy (south/southwest waves) decreases. Most sediment is deposited along extensive beaches at Amberley, resulting in the progradation of wide beaches. Further north, the beach sediment becomes coarse (sand and gravel), reflecting decreasing transport. In lee of the Point Clark headland, very coarse beach sediments, with offshore boulders and pockets of fine silt characterize the shoreline.

4. Discussion

The identification of a northward alongshore sediment transport pattern towards Point Clark, and the assessment of Horizon View as a major source of material to the alongshore sediment transport regime, has major implications for land use management practices along this shoreline. Maintenance of this sediment transport system is essential for the continued natural evolution of this section of Lake Huron shoreline. New development and any proposed shoreline protection schemes must appreciate the importance of sediment supply from Horizon View, transported north and south, supplying and maintaining natural beaches.

4.1. Land use and management

The current land use zoning along the Lake Huron shoreline in Ashfield Township, Huron County includes a Natural Environment (NE1) designation which restricts development. The NE1 zone includes the area from the shoreline to the top of the bank. At Horizon View, the zoning designation east of the top of the bank is presently zoned for Recreational Residential (RC1). This means that development can currently legally locate right up to the edge of the bank, even in highly eroding areas like Horizon View with approval of the local township and county planning departments and elected councils.

The Maitland Valley Conservation Authority (MVCA) has had an Erosion Risk Limit setback established since 1986 for Ashfield Township. Recent improvements in recession rate data has allowed refinement of the definition and position of this setback. Boyd (1993) used 1934 Ontario Land Survey data to provide over 50 years of recession rate data when the position of top of bank is compared to recently prepared digital survey mapping completed at 1 : 2000 under the Federal Flood Damage Reduction Program. In addition, geotechnical studies done for the MVCA and by the Ontario Ministry of Natural Resources (OMNR) in preparing technical guidelines for a draft Provincial Shoreline Policy Statement, confirms that a stable slope factor of 3 : 1 is appropriate for eroding cohesive bluff shorelines in further defining a development setback.

However, even with all of this information, neither the municipality, nor the Conservation Authority, can restrict development within the Erosion Risk Limit until the setback has been established in the Township's Comprehensive Zoning Bylaw with revisions to the existing recreational residential zoning. Even though Horizon View may be an inappropriate place to develop, the MVCA can currently only advise individuals of the risk and hope that they develop behind the setback. Experience has shown, though, that people will still locate a building in a high risk area, with the full knowledge that the property has a limited life expectancy. They appear to be willing to accept the risk as long as they can enjoy the property for a few years at least. What tends to happen, however, is that after a time, the owner will sell the property to get back his investment, and the next owner may be faced with the prospect of losing the cottage due to continued bluff erosion. The "buyer beware" maxim holds very true here. Some people will have the sense to contact the Conservation Authority to determine what constraints lie with the property before they make an offer to purchase. However, this does not happen often enough.

At Amber Sands, north of the 18 Mile River, a planned subdivision consisting of 106 lots located on 128 acres, with 900 metres of shoreline, is currently under development. This particular plan of subdivision has been on file at MVCA for several years, predating recent shoreline management initiatives. Proposed site plans include a sufficient setback from top of bluff, anticipating bluff recession. Concerns for this development include beach access, removal of bluff face vegetation, accelerated gully development due to increasing surface drainage runoff and alteration of groundwater conditions. Recent indications are, however, that the development is experiencing difficulty in selling sufficient lots in order for construction to begin. Whether this situation results from difficult economic conditions or from an improved public awareness of the risk associated with this type of investment is unknown.

A similar development is currently under construction at Horizon View where several lots are advertised for sale, although they are located directly along a bluff experiencing recession in the order of 0.5 metres per year and with a history of successive slumping and mass wasting. Recent field surveys have also indicated a lack of activity and investment at this location. At present there are two structures located along the lakefront lots of this subdivision. One is a very large house originally built behind the erosion risk limit. Since built, the owner has begun to encroach into the risk area by building additions. The planned subdivision at Horizon View was developed in the 1970's before the MVCA had adequate bluff erosion data on which to prepare a setback. In fact, most of Ashfield Township was outside of MVCA jurisdiction until the mid-1980's. The subdivision is just now being developed with several residences completed or under construction. Again, the Conservation Authority, can only advise that the development should be setback behind the Erosion Risk Limit.

The only way to prevent development from locating in high risk areas is to establish the setback in the local land use planning bylaw. The MVCA has attempted to encourage Ashfield Township to incorporate the Erosion Risk Limit land use zoning setback into their bylaw. Public meetings were held on two separate occasions since 1992. The public reaction was not overly receptive because the setback would apply to the whole of Ashfield's shoreline, and those individuals at sites at low risk due to limited bluff erosion perceived that the setback would be too onerous and invasive of property rights. Ironically, residents of Horizon View were most understanding of the need for the setback.

Ashfield Township Council has taken a very weak position on this issue by deferring the matter until a draft Shoreline Policy Statement currently in development has been released by the provincial government. Their feeling was that the Township might be in a stronger position for a legal challenge if there were a government endorsed policy under the provincial Planning Act which gives the legal mandate for municipalities to regulate land use and development within their jurisdiction. However, in the meantime, development in high risk areas continues, and it is the position of the MVCA staff that with the quality of technical information available, any land use decision would be supported in case of a legal challenge.

4.2. Management framework

Given the extent of coasts, especially in the mid-latitude environments, which consist primarily of erodible unconsolidated bluffs, general application of the study results and management implications is possible to assist in long term planning of eroding bluff shorelines. Using the southeastern Lake Huron shoreline as a case study, it is possible to consider the key issues regarding management and development along bluff shores. The formation of a comprehensive management framework, integrating coastal morphology and processes, land use management practices and the range of human responses to the hazard associated with eroding bluffs, is the essential goal.

In order to fully assess the management of eroding shoreline bluffs a number of concepts must be integrated to formulate a procedure or conceptual frame-

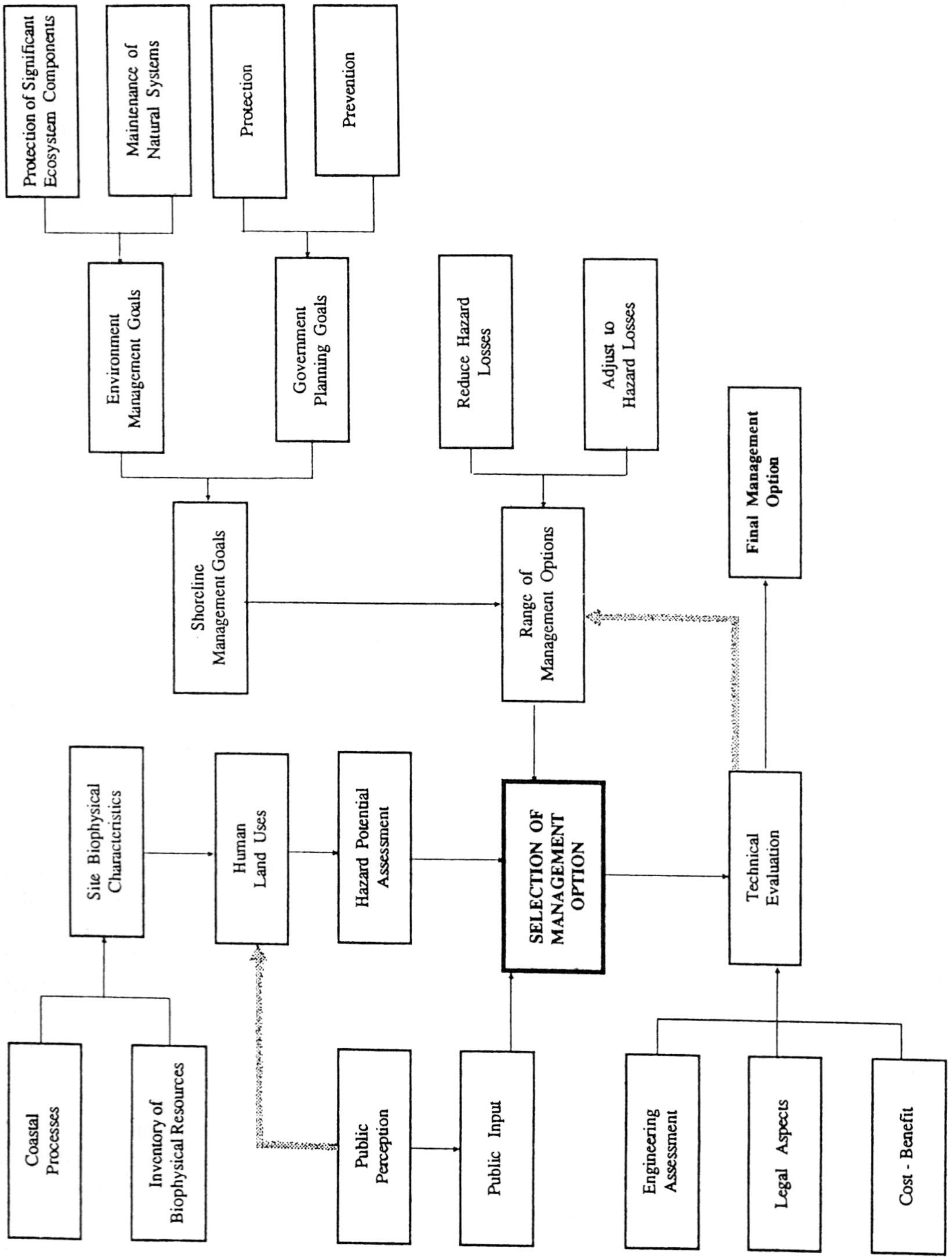

Fig. 9. Management framework for eroding cohesive bluff shorelines.

work for selecting management approaches. A review of literature concerning the appraisal of hazard conditions, government policy, public perception and management options has provided for the development of a management framework for eroding shoreline bluffs. A number of authors (Bokuniewicz and Tanski, 1980; Amir, 1983; Fischer, 1985) have discussed multi-system approaches to management of eroding bluffs.

The framework (Fig. 9) attempts to combine site biophysical inventory and human land uses into a hazard potential assessment for which a management option can be selected respecting overall shoreline management goals. Each component requires individual consideration as to how it is integral to the management framework. The individual management option is then evaluated based on public policy and site specific considerations before determination of a final management approach.

Assessment of the hazard potential requires resource classification, evaluation, impact forecasting and monitoring of coastal processes (Amir, 1983). The stability of eroding bluff shorelines should also be examined in assessment of hazard zones, planning of stabilization works and local land use planning. Techniques employed include aerial photography, historical and present mapping and observations, field studies (profile surveys, wave and climate measurements) and lab analysis (bluff material and beach sediment analysis), with both short term (seasonal) and long term studies (Bokuniewicz and Tanski, 1980). The procedures for assessing the hazard must contain four main components: (1) delineation of areas potentially affected; (2) estimating vulnerability, recognizing that human land uses involves beneficial resource use and risk of potential loss; (3) identification of ranges of responses to a hazard; and (4) assessment of present and future adjustment impacts. The designation of hazard areas must be accompanied by reasonable use of available scientific information and provision for public examination and review of the designations.

The overall management framework is driven by a set of shoreline management goals. McKenzie (1991) discusses the need for shoreline planning policies based on an ecosystem approach. In the context of the Ontario Shoreline Management Program, he argues that a shoreline policy should strive to maintain the ecological integrity of coastal systems, promote development that is sustainable and reduce the potential for future property damage. These principles are underlined by a need to understand and appreciate the complexity of physical and biological processes operating within the shore zone environment, that littoral drift (sediment transport) is a essential coastal element to be preserved and that management policies should consider an ecosystem planning approach.

Within the management framework these ideals are combined with government planning goals to create the shoreline management goals. Ontario provincial shoreline management goals (McKeen, 1990) are utilized as typical government goals. The provincial shoreline management Program incorporates three main components: prevention, protection, and emergency response. In order to consider long term management strategies, prevention and protection are considered as fundamental goals within the management framework. The selection of a management option must then reflect a consideration and assessment of the individual shoreline management goals, in association with the hazard potential assessment. There may exist an apparent contradiction between the management goals. It is then the responsibility of the management agency to assess the relative importance of each, if a preferred management option is in conflict.

The management options are then evaluated on the basis of their technical, legal and cost–benefit merit. Technical and legal aspects of any site specific management approach may also need to be reviewed in order to determine if there exists any obstruction to implementation including shore ownership questions, impacts on adjacent properties, right to public access and planning for long term impacts. On the failure of any of the evaluation criteria, the management option would be rejected and a new management option would then need to be considered.

The final management option represents a site specific approach to managing the hazard of bluff erosion on the basis of individual criteria and overall shoreline management goals. The management decision can be provided to prospective purchasers of waterfront property and integrated into more comprehensive regional management strategies, reflecting spatial variations in shoreline characteristics and land use concerns. In this manner, individual concerns of the hazard associated with eroding bluffs can be regulated within a comprehensive management framework, utilizing local and regional management goals and hazard assessment procedures.

5. Conclusions

The integration of coastal process studies into management applications can be considered a recent phenomenon. The utilization of such technical information can be made difficult by an inability to directly apply results to assessment of resources and hazards, thus hampering the evaluation and application processes. Consideration must be given to the appropriate types of information required by land use managers or planners. Researchers need to develop a keen awareness for application of potential research results, while planners must appreciate the need for physical morphology and process information and foster an understanding of the utilization of such data into the planning process.

It has been the primary purpose of this research to assess one of the main components of the evolution of cohesive bluffs, sediment transport, in order to assist future management initiatives along the southeastern Lake Huron shoreline in Ontario. These factors play a major role in the erosion of cohesive shorelines and alongshore sediment regime conditions, including development of sediment sources through material removal, creation of alongshore sediment deposits, and beach stabilization. The alongshore variations of wave energy and sediment transport also have important human impacts through the development of varying shoreline stability conditions. Eroding bluffs create a direct hazard to the maintenance and safety to existing and future structures along these shores, while stable beach environments provide welcome environments for human habitation.

For many eroding bluff shorelines, management plans need to place a greater importance on coastal processes studies to assist in the evaluation of alternative management practices. Management agencies must realize the importance of long term monitoring of erosion processes and incorporating extensive data sets, including wave hindcasting, wind climate, erosion rate data and historical shoreline observations into the planning process. The need for complete, accurate and comprehensive data on shore resources and processes is essential to the development of a successful and thorough plan.

There exists a need to assimilate and communicate this information to both the management agencies involved in shore planning and to the public concerned with development in hazard areas. Utilization of sediment transport studies can provide basic information to assist in the site specific assessment of the erosion hazard, provide a basis for restrictions on development, and aid potential land developers and property owners in assessing the practicality of extensive bluff development within these regions. Continued examination of these and related shoreline characteristics such as wave energy, groundwater, bluff stability, beach development and nearshore erosion would provide an ability to accurately assess site specific and regional hazard conditions with the goal of assisting development practices and land use management strategies.

Acknowledgments

This research was conducted as partial fulfilment of requirements for the degree of Master of Science in the Department of Geography at the University of Guelph, Ontario, Canada from 1989 to 1991. Funding was provided by support from the Department of Geography and Faculty of Graduate Studies at the University of Guelph and in part from a National Sciences and Engineering Research Council of Canada (NSERC) grant to Dr. Robin Davidson-Arnott who acted as academic supervisor. Geoff Peach, Maitland Valley Conservation Authority and Dr. Gary Boyd, Fisheries and Oceans provided comments and suggestions. Gillian Novaseic, Jeff Ollerhead, Anthony Gabriel, Heather Conliffe-Reid, Trevor Chandler and Jan Slaats assisted with field studies and research.

References

Amir, S., 1983. Tools for a coastal resource management framework with multiple objectives. J. Environ. Manage., 17: 121–131.

Bokuniewicz, H. and Tanski, J., 1980. Managing localized erosion of coastal bluffs. In: Coastal Zone '80, pp. 1883–1898.

Boyd, G.L., 1981. Great Lakes Erosion Monitoring Program — Final Report 1973/1980. Unpublished manuscript, Bayfield Lab for Marine Science and Surveys, Department of Fisheries and Oceans, Burlington, Ont., 200 pp.

Boyd, G.L., 1993. Nearshore control of coastal landform change: Late Wisconsinan to Present, Lake Huron, Canada. Ph.D. Thesis,

Department of Geography, University of Waterloo, upubl.

Burton, I., Kates, R.W. and White, G.F., 1993. The Environment as Hazard. Guilford Press, New York, 290 pp.

Davidson-Arnott, R.G.D., 1988. An overview of coastal processes in the Great Lakes. In: Impact of Great Lakes Water Levels on Shore Processes. International Joint Commission, Ottawa, Canada, 22 pp.

Dean, R.G., Berek, E.P., Gable, C.G. and Seymour, R.J., 1982. Longshore transport determined by an efficient trap. In: 18th Coastal Engineering Conf. Proc. Am. Soc. of Civil Engineers, pp. 954–968.

Fischer, D.W., 1985. Shoreline erosion: a management framework. J. Shoreline Manage., 1: 37–50.

Galvin, C. and Vitale, P., 1976. Longshore transport prediction — SPM 1973 equation. In: 15th Coastal Engineering Conf. Proc. Am. Soc. of Civil Engineers, pp. 1133–1148.

International Joint Commission, 1989. Living with the Lakes Challenges and Opportunities Phase I Project Management Team, Burlington, Ontario, 108 pp.

Jacobsen, E.E. and Schwartz, M.L., 1981. The use of geomorphic indicators to determine the direction of net shore drift. Shore and Beach, October 1981: 38–43.

Komar, P.D. and Inman, D.L., 1970. Longshore sand transport on beaches. J. Geophys. Res., 75: 5914–5927.

Kreutzwiser, R.D., 1987. Managing the Great Lakes shoreline hazard. J. Soil Water Conserv., 42: 150–154.

Lawrence, P.L., 1991. Wave energy and sediment transport patterns on southeastern Lake Huron and Implications for shoreline management. M.Sc. thesis. Department of Geography, University of Guelph, 151 pp., unpubl.

May, J.P., 1974. WAVENRG: A computer program to determine the distribution of energy dissipation in shoaling water waves with examples from coastal Florida. In: W.F. Tanner (Editor), Sediment Transport in the Nearshore Zone. Florida State University, pp. 22–61.

McKeen, P., 1990. Shoreline management: a provincial perspective. In: Proc. Canadian Coastal Conf. Short Coarse Notes — Coastal Management: Experiences, Lessons and Needs. National Research Council of Canada, pp. 65–82.

McKenzie, D.I., 1991. Shoreline erosion on Ontario's Great Lakes in Ontario. In: B. Mitchell (Editor) Geographical Perspectives on Economy and Environment. Department of Geography, Univ. of Waterloo Publ. Series No. 34, pp. 253–268.

Ontario Ministry of Natural Resources, 1987. Guidelines for Developing Great Lakes Shoreline Management Plans. Report to the Ontario Conservation Authorities, Toronto, Ontario, 32 pp.

Ontario Shoreline Management Review Committee, 1986. Report of the Shoreline Review Committee to the Ministers of Natural Resources and Municipal Affairs. Toronto, Ontario, 75 pp.

Philpott Ltd., 1988. Wave climate data for Ontario's Great Lakes: Lake Huron/Georgian Bay. Final Report to the Ontario Ministry of Natural Resources, Toronto, Ontario, 122 pp.

Reinders Ltd, 1989. Lake Huron processes study — Final Report and Appendices. Final Report to the Maitland Valley Conservation Authority, Wroxeter, Ontario, 102 pp.

Sunamura, T., 1983. Processes of sea cliff and platform erosion. In: P. Komar (Editor), CRC Handbook of Coastal Processes and Erosion, pp. 233–265.

Taggart, B.E. and Schwartz, M.L., 1988. Net shore-drift determination: a systematic approach. J. Shoreline Manage., 3: 285–309.

ELSEVIER

Geomorphology 10 (1994) 83–94

GEOMORPHOLOGY

Geomorphological effects of tsunami run-up and backwash

A.G. Dawson

Centre for Quaternary Research, Division of Geography, School of Natural and Environmental Sciences, Coventry University, Coventry, CV1 5FB, UK

Received January 31, 1994; revised March 17, 1994; accepted March 30, 1994

Abstract

Geomorphological processes associated with tsunami run-up and backwash are highly complex. To date, most studies have considered the sedimentology of palaeo-tsunamis while almost no detailed studies have been undertaken in areas known to have been affected by tsunamis in recent times. Research on contemporary tsunamis show that coastal landscapes may be greatly altered not only by direct tsunami run-up orthogonal to the shoreline, but also by episodes of vigorous backwash and by water flow sub-parallel to the coastline. The combined effect of these processes produces hitherto unrecognised coastal landforms that are dominated by the effects of high-magnitude erosion and deposition. Deposition of large boulders is also associated with severe tsunamis, while in adjacent areas run-up processes often result in the deposition of continuous and discontinuous sediment sheets. In many instances the upper limit of sediment deposition lies well below the upper limit of wave run-up which is frequently marked by a well-defined zone of stripped vegetation and soil. The processes described have considerable implications for models of coastal evolution. It may even be the case that the evolution of particular stretches of coastline are dominated by such high-magnitude low-frequency events.

1. Introduction

The word tsunami is derived from the Japanese word meaning harbour wave. They are often described as tidal waves but this view is incorrect since they have nothing to do with tides. Tsunamis are frequently generated by fault displacement on the seabed caused by earthquakes. However, many tsunamis are generated by submarine landslides and are therefore unrelated to seismic activity. In ocean floor areas where sediment or rock has been displaced, water rushes in from the surrounding sea to restore the water level. When this happens, there is a rapid lowering of sea level along nearby coasts, occasionally below the level of the lowest tides. Thereafter, large kinematic waves travel outwards from the area located above the zone of seabed displacement. Tsunamis are long period ocean waves with very long wavelengths and very high velocities, the latter being mostly a function of the depth of the water over which the wave trains pass. Areas of very deep water generate the fastest waves that may often exceed 480km h^{-1}. The height of the waves in the open ocean is usually quite small (often no more than 1–2 m). At the coast, however, the flood levels (run-up) are partly a function of the dimensions of the propagated waves, but are also greatly influenced by the topography and bathymetry of the coastal zone. It may often be the case that relatively small tsunami waves may generate the greatest run-up as a result of the effects of wave resonance within confined estuaries. It is well known, however, that most tsunami waves deposit sediment. For example, after the Chilean earthquake and tsunami of May 1960, "...white sand was deposited on valley slopes wherever the (tsunami)

0169-555X/94/$07.00
SSDI 0169-555X(94)00017-L

wave passed, 6–30 cm thick, up to 4.25 km inland.'' (Wright and Mella, 1963).

Despite geomorphological observations such as that described above, the geomorphological impact of tsunamis on coasts is an area of research that has been neglected in coastal geomorphology. Indeed, there is no detailed paper on the subject — a fact made all the more astonishing since such high-magnitude low-frequency events are likely to have played an important role in the evolution of many coasts. In the following sections, the principal geomorphological processes associated with tsunami erosion and deposition are described. However, discussion of the processes of tsunami sediment transport lie beyond the scope of this paper.

2. Tsunami erosion

There are many historical accounts of coastal erosion caused by tsunamis. In most instances, the accounts document the destruction of man-made structures with little consideration of geomorphological changes. In general, the principal erosive processes include the erosion of bedrock blocks in the foreshore zone, beach sediment removal and the lowering of coastal surfaces, rapid coastline retreat as well as vegetation removal and soil stripping (cf. Bourrouilh-Le Jan and Talandier, 1985; Ota et al., 1985; Shi et al., 1993). In numerous cases, processes of coastal erosion are accelerated by earthquake ground cracking and by earthquake-induced landslide activity. Tsunami waves that cross the coastal zone are normally rapidly-moving long-period breaking waves that frequently travel considerable distances inland. According to the local hydraulic regime, tsunamis do not always run-up orthogonal to the coastline. Instead water movement can frequently take place as Mach stem flow oblique, or parallel, to the coast and cause highly complex patterns of run-up and backwash.

Photo 1. Coastal area of eastern Flores inundated by the tsunami. The upper run-up limit here was ca. 26 m (top centre). Note areas of boulders, discontinuous sediment sheets, stripped soil and vegetation. Note also that the orieintation of felled trees varies considerably, thus demonstrating complex patterns of tsunami current flow. (Photo courtesy H. Yeh.)

3. Tsunami deposition of boulders

Observations of geomorphological processes associated with modern tsunamis indicate that tsunami runup may often cause widespread deposition of large boulders. For example, the Flores tsunami of December 12th, 1992, was locally associated with the destruction of sections of coral reef and the deposition of numerous coral boulders farther inland (Yeh and Shi, pers. commun.) (Photos 1 and 2). Several former tsunamis are also considered to have been associated with widespread boulder deposition. For example, Bourrouilh-Le Jan and Talandier (1985) have described several areas of coral reef at Rangiroa, Tuamoto archipelago, SE Pacific that are characterised by large numbers of giant blocks (up to 750 m^3). The pattern of boulder distribution indicates that there is often a progressive decrease in block size landward. In addition, examples are cited where individiual blocks have been transported across the atoll rim and into the lagoonal areas (Bourrouilh-Le Jan and Talandier, 1985). They also consider palaeo-tsunamis as likely depositional agents for the boulders but also speculate that storm surges associated with former hurricanes may also have been responsible for block deposition.

Similar boulder fields have been observed in the southern Ryukyu Islands, Japan (Ota et al., 1985; Kawana and Pirazzoli, 1990; Nakata and Kawana, 1993). In this area, numerous giant blocks, individually up to circa 100 m^3 in size and occurring up to 30 m above present sea level are unequivocally interpreted as having been deposited by tsunamis during the Holocene (Fig. 1). Radiocarbon dating of marine mollusca cemented onto the surfaces of individual boulders has provided information on the ages of individual tsunamis some of which are also documented in historical

Photo 2. Section of coastline of northeastern Flores after December 1992 tsunami. Note accumulations of coral boulders (in foreground) deposited as a result of coral reef erosion. Upper limit of tsunami runup corresponds to upper limit of soil and vegetation stripping. Discontinuous sediment accumulations deposited by the tsunami are also evident. Note helicopter for scale. (Photo courtesy H. Yeh.)

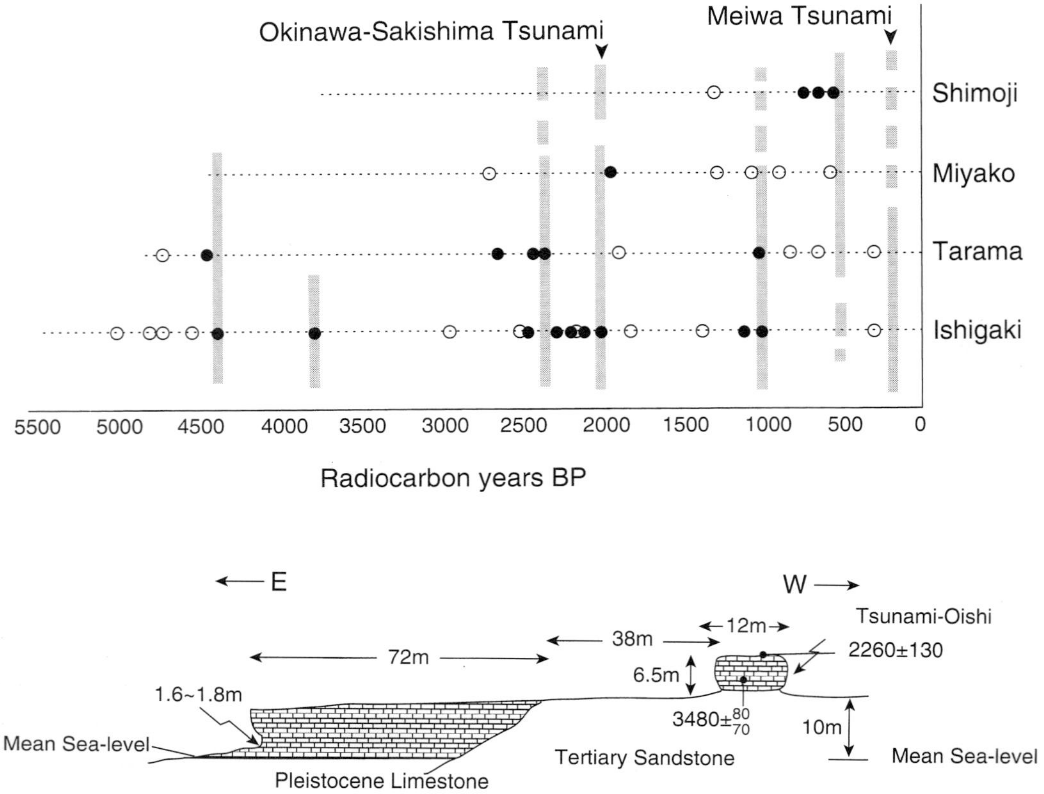

Fig. 1. (Top) Radiocarbon ages of tsunami boulders in southern Ryukyu Islands, Japan. Dots denote samples taken from upper part of the boulders. The thick vertical bars indicate likely tsunami ages. (Bottom) The tsunami boulder "Tsunami-oishi" located on the lower terrace in the southern part of Ishigaki island, Ryukyus (after Nakata and Kawana, 1993).

accounts. Boulder deposits of last interglacial age and attributed to tsunami inundation have also been identified in areas of the eastern Mediterranean (Wood and Rose, pers. commun.).

On the islands of Lanai and Molokai, Hawaii, boulders considered to have been deposited by a tsunami during the last interglacial occur within a remarkably thick (up to 8 m) deposit (Moore and Moore, 1984, 1988). Here, limestone blocks occur within a gravel bed in which most of the clasts are composed of basalt. Skeletons of coral and other reef organisms also occur as individual clasts (Fig. 2). The deposits, known as the Hulopoe Gravel, occur up to 375 m above sea level and were first interpreted as sediments associated with Quaternary emerged shorelines (Stearns, 1940) and later as sediments deposited by a tsunami (Moore and Moore, 1984). The Hulopoe Gravel is unique among deposits inferred to have been deposited by tsunamis since it is remarkably thick. Moore and Moore (1988) describe sediment exposures 200 m inland from the Lanai coastline as consisting of three distinctive sedimentary units, successively 2, 4 and 2 m thick that are considered to have been deposited by succesive waves in the tsunami wave train. Each bed consists of two sub-units: a lower sub-unit of basalt and limestone boulders (circa 0.5 m in diameter), cobbles and sand, and an upper bimodal sub-unit of large basalt boulders in a sandy matrix (Fig. 2). Moore and Moore (1984) have argued that the lower sub-units are indicative of sedimentation associated with tsunami run-up while the upper sub-units are considered to have been deposited during episodes of tsunami backwash.

Although the upper limit of tsunami run-up (indicated by soil stripping) is here believed to be 375 m and the upper limit of sediment deposition is 326 m, most boulders with diameters of circa 0.5 m occur no

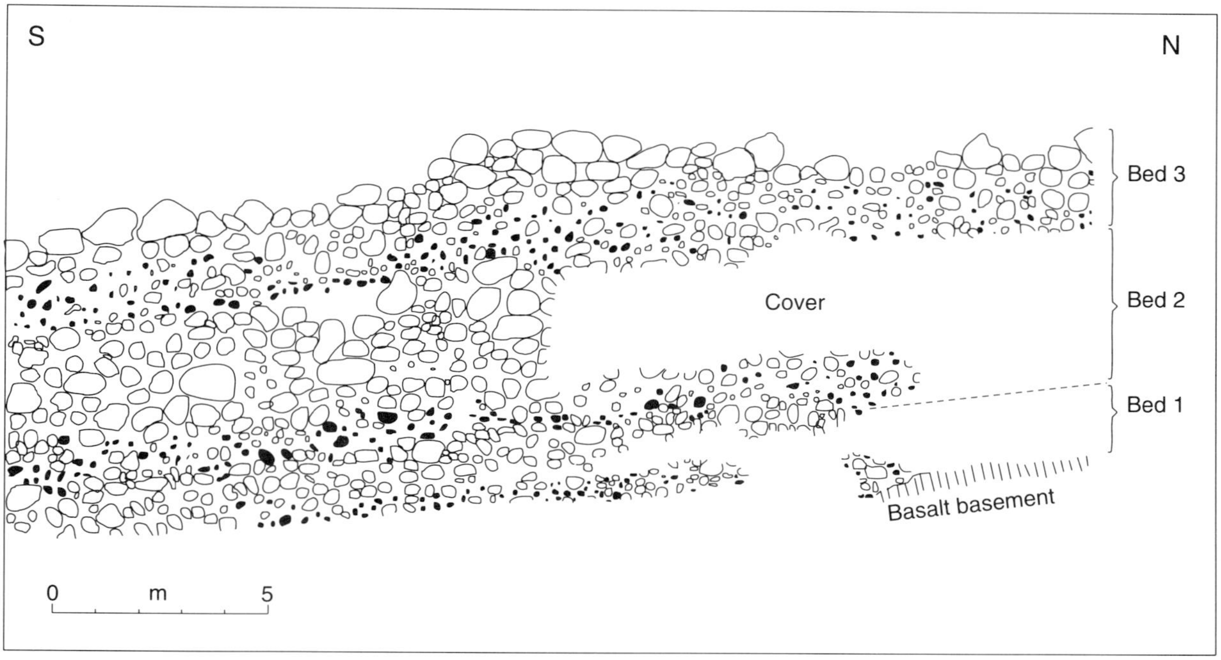

Fig. 2. Section of Hulopoe Gravel at head of Kapihaa Bay, Lanai, Hawaii. The altitude of the base of the section in the centre of the profile is 22 m. Basalt clasts are shown as white and limestone clasts as black. Note the graded nature of the beds as indicated by the distribution of limestone and basalts clasts within the exposure (after Moore and Moore, 1988).

higher than circa 50 m above sea level (Moore and Moore, 1984). These astonishing run-up elevations are believed to have been caused by a tsunami generated by a large submarine slide/debris avalanche (the Alika phase 2 event) that took place circa 105,000 years ago on the western flank of the Manua Loa volcano (Moore et al., 1989, 1992). This tsunami is also believed to have caused 20 m high tsunami waves to strike the coastline of New South Wales, Australia where numerous giant blocks are reported to have been removed from pre-existing emerged coastal rock platforms (Young and Bryant, 1992).

The field evidence of boulder deposition from Lanai and from Tahiti appears to indicate that tsunami deposition is associated with a progressive decrease in boulder size from seaward to landward. Farther inland, boulder deposits are replaced by sheets of finer-grained sediment that progressively rise in altitude and taper inland. In many instances, however, tsunami flooding is not associated with the deposition of boulders but instead is characterised solely by the deposition of continuous and discontinuous sediment sheets (Shi et al., 1993). The above discussion also highlights the need to distinguish debris and sediment deposited by tsunamis from that associated with hurricane-induced storm surges. For example, Teissier (1969) has described from Rangiroa, Tahiti, the transport of coralline reef blocks weighing several tonnes as a result of the cyclone of 15 January 1903. Similarly, Pirazzoli et al. (1988) disagrees with the conclusions of Bourrouilh-Le Jan and Talandier (1985) regarding tsunami boulder transport and deposition in French Polynesia and argues on several grounds that most reef boulders have been deposited by cyclone-induced storm surges.

4. Other large-scale geomorphological changes

There is some evidence to indicate that large-scale tsunamis are often associated with profound reorganisation of coastal barrier systems. For example, in the Algarve region of Portugal, Andrade (1992) has shown that the Ria Formosa barrier island system was greatly modified by the Lisbon tsunami of 1755 AD. Here, tsunami inundation appears to have been accompanied by widespread submergence, barrier breaching and the deposition of large washover fans as well as the complete reorganisation of backbarrier drainage systems. The washover delta fans are very similar to features described from northeastern Japan and attributed to the effects of tsunamis (Minoura and Nakaya, 1991). Similar processes of barrier breaching have also been described from the Tuamoto archipelago, French Polynesia where they are associated with the formation of breached sections of atoll rim (Bourrouilh-Le Jan and Talandier, 1985).

On the island of Lanai, Hawaii, parts of the Hulopoe gravel form dunelike ridges that are aligned approximately transverse to the hillslopes and appear to have been produced during episodes of exceptionally rapid water movement (Moore and Moore, 1988). Similar ridges attributed to tsunamis, occur in the South Ryukyu islands, Japan where they are up to 40 m wide, several metres high and are aligned parallel to the coastline (Ota et al., 1985).

5. Tsunami sedimentation, flood run-up and backwash

Recent investigations of coastal sediments deposited by ancient tsunamis have shown that tsunami sediment deposition is associated with the development of sheets of sediment that rise in altitude inland as tapering sediment wedges (Dawson et al., 1988). Tsunami deposits generally consist of accumulations of sands and coarse silts that occur as sets of fining upwards sequences. In addition the base of many tsunami sediment accumulations rest unconformably on the underlying strata and, in many cases, intraclasts from the substrate are incorporated within the overlying tsunami sediments. The nature of tsunami sedimentation is greatly influenced by offshore sediment supply and rock lithology with the thickest and most extensive sediment sheets being deposited in areas where coastal dunes are present. Geomorphological investigations of the recent tsunami in Flores, Indonesia of 12 December 1992 showed that the tsunami was associated with the deposition of extensive sheets of sediment, up to 1 m thick, that are continued landward by discontinuous sediment accumulations. The highest of these sediment accumulations always occurs below the upper limit of tsunami run-up (Photos 1 and 2) (Shi et al., 1993). In Flores, the areas of coastline affected by the highest flood run-up are characterised by a zone of erosion in which soil and vegetation stripping has been the dominant process caused by repeated episodes of run-up and backwash. The Flores study has therefore demonstrated that the upper limit of sediment deposition does not necessarily correspond with the upper limit of run-up. Thus, the upper limit of sediment sheets deposited by palaeo-tsunamis represents a minimum estimate of flood run-up.

The most detailed sedimentological information available for a palaeo-tsunami is for the tsunami caused by the Second Storegga submarine landslide that took place in the Norwegian Sea and North Atlantic at circa 7000 ^{14}C yr BP (Dawson et al., 1988). Most of the field evidence for the tsunami occurs along the coastline of Scotland although supplementary information is available for Norway and Iceland (Fig. 3).

5.1. Scotland

Evidence for a regionally extensive layer of marine sediment within isostatically-uplifted Holocene coastal sequences in eastern Scotland has been known for over 20 years (Sissons and Smith, 1965) and, for a long time, was attributed to a major storm surge (Smith et al., 1985). More recently, it was suggested that the marine sediments were deposited by a tsunami associated with the Second Storegga submarine landslide (Dawson et al., 1988) (Fig. 3; Photo 3). Later research has shown that the sediments deposited by the tsunami are, in fact, very complex both in terms of their composition (ranging from a coarse gravel to a clay/silt) and sedimentary structures. In most cases the base of the sediment layer rests unconformably on peat depos-

Fig. 3. Bathymetric map of Norwegian Sea and North Sea regions showing area of the Storegga submarine landslides (after Bugge, 1983 and Bugge et al., 1987). Dots show principal locations where coastal deposits exist that have been attributed to deposition by the tsunami generated by the Second Storegga Slide (for more information see Long et al., 1989; Svendsen and Mangerud, 1990; Hansom and Briggs, 1991 and Bondevik and Svendsen, 1993). The graphs for sites 1, 2 and 3 are numerical model predictions of tsunami run-up according to Harbitz (1992). Note the pronounced initial drawdown and the extreme and astonishingly rapid water-level fluctuations at the Norwegian offshore site.

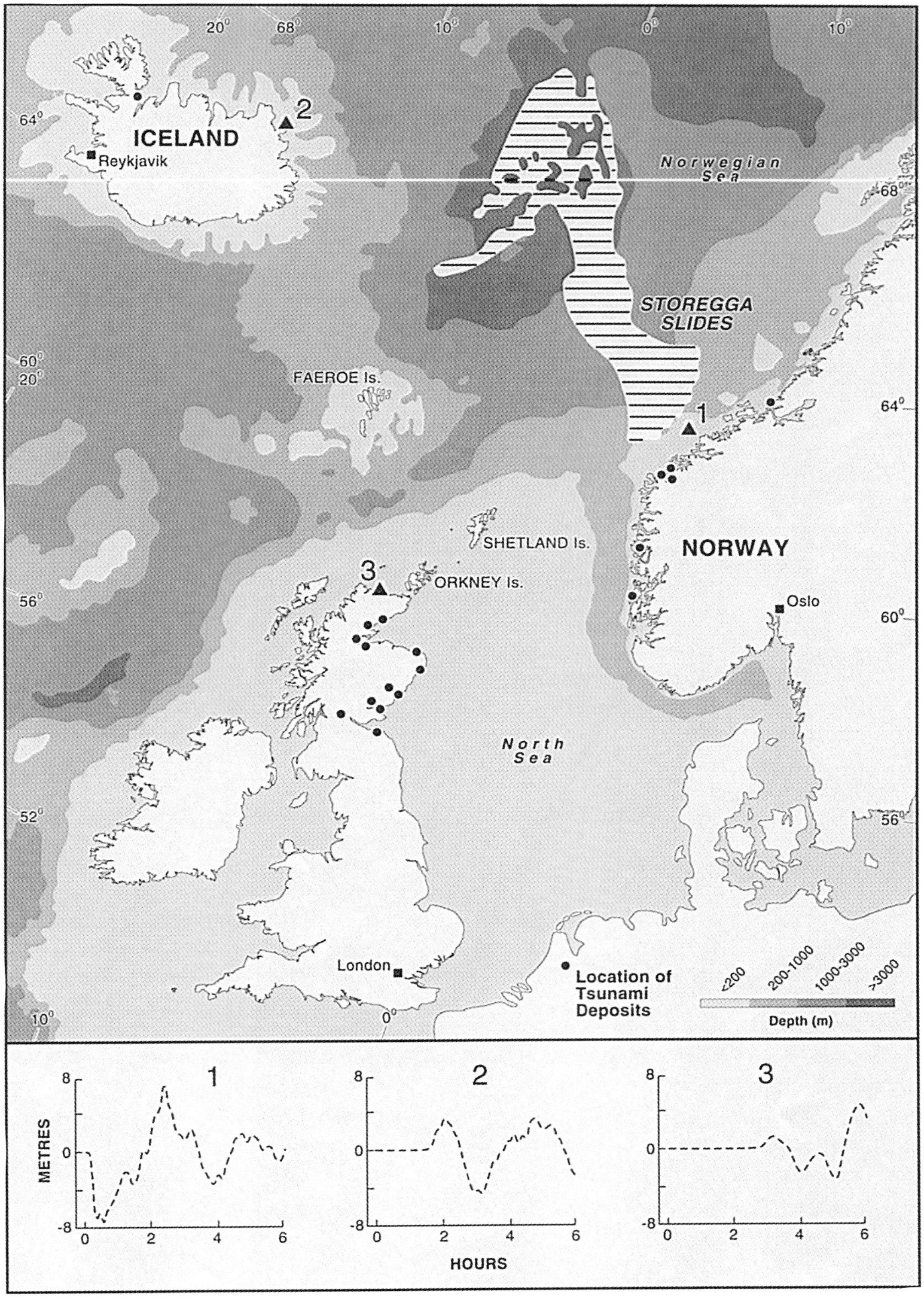
ICELAND
Reykjavik
Norwegian Sea
STOREGGA SLIDES
FAEROE Is.
SHETLAND Is.
ORKNEY Is.
NORWAY
Oslo
North Sea
London
Location of Tsunami Deposits
<200
200-1000
1000-3000
>3000
Depth (m)
METRES
HOURS

Photo 3. Sediment exposure showing tsunami layer (upper white layer) associated with second Storegga landslide at Montrose, eastern Scotland. Note unconformity between base of sand and top of underlying peat. (Photo courtesy D. Smith.)

its. Frequently, the sediment accumulations contain intraclasts of peat that have been eroded and transported in suspension during tsunami inundation. In addition, the deposit is often characterised by the presence of large numbers of marine and brackish water diatom species of which the most common is the marine diatom *Paralia sulcata* (Ehrenberg Cleve). Very large numbers of the diatoms (60–80%) are broken and may indicate sedimentation under extremely high energy conditions.

In Scotland, the tsunami sediment layer has been radiometrically dated at over 15 separate sites (Smith et al., 1991) with the ^{14}C ages showing a distinct clustering of ages at circa 7000 yr BP (Fig. 3). Stratigraphic

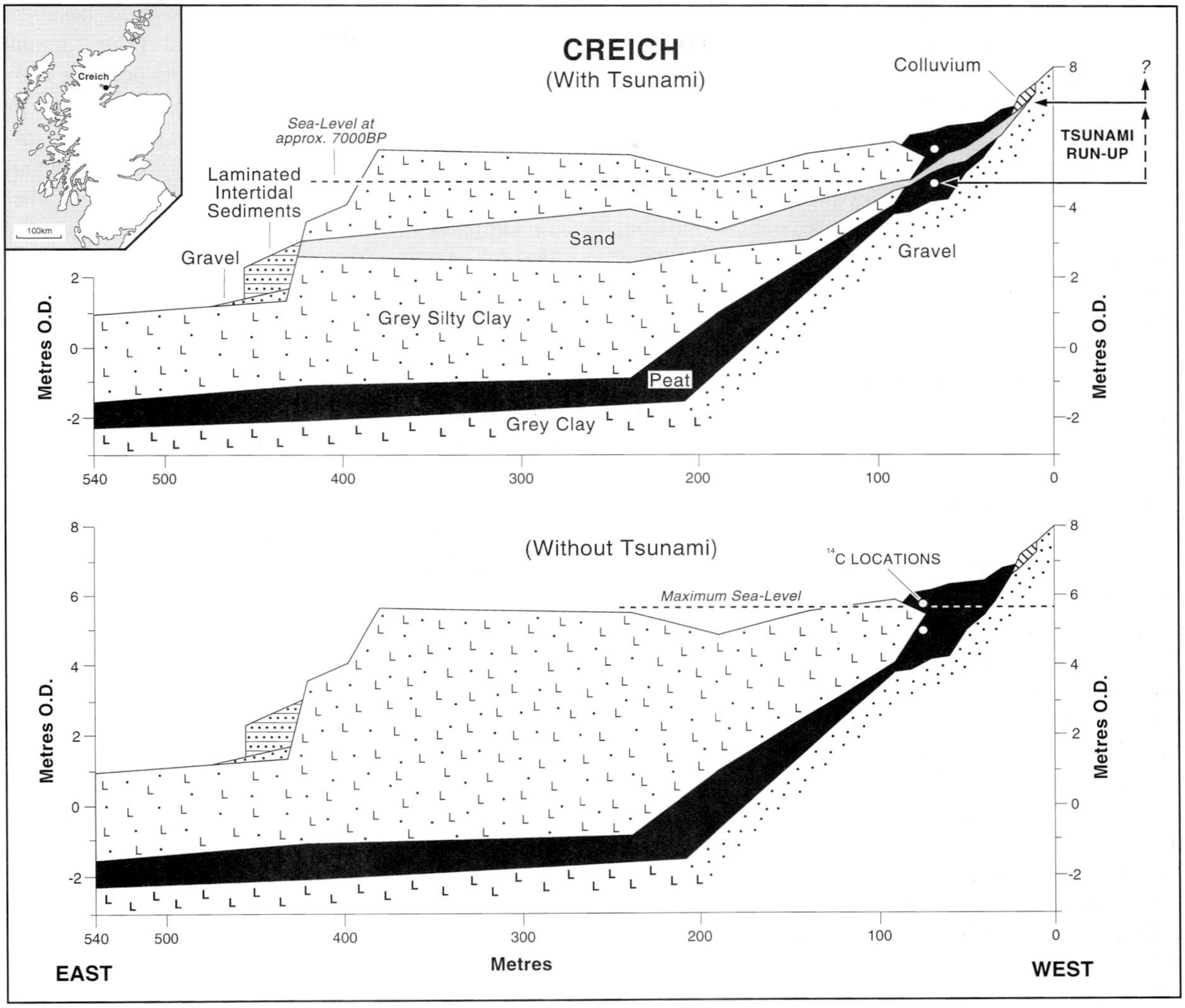

Fig. 4. Schematic diagram of a site at Creich, north of Inverness, Scotland where Storegga tsunami sediments are preserved. The bottom diagram is a theoretical reconstruction of the nature of the Holocene coastal stratigraphy if there had not been a tsunami at ca. 7000 yr BP. The top diagram shows the way in which a sediment layer was deposited by the tsunami. Also shown is an estimate of the tsunami run-up at this site based on stratigraphic evidence (see also Smith et al., 1991).

investigations have shown that it is possible to calculate the minimum run-up of the tsunami for each of the sites investigated (Fig. 4). These show typical values of between +1 and +4 m (cf. Smith and Dawson, 1990). Large local variations in coastal flood run-up are to be expected, however, due to the effects of wave resonance and amplification within individual inlets. In general, however, the estimates of minimum tsunami run-up based on the maximum altitudes of marine sediment layers exhibit a good agreement with theoretical calculations of flood run-up based on the numerical models of the Storegga tsunami made by Harbitz (1991, 1992) and Henry and Murty (1992).

Detailed investigations of the coastal sediments deposited by the circa 7000 yr BP tsunami have been undertaken by Shi et al. (1991). These studies have shown that individual tsunami sediment deposits appear to be moderately well-sorted and characterised by a series of fining-upwards sediment sequences that are frequently also accompanied by a progressive fining-upwards of the whole deposit. The sets of fining-upwards sequences within individual deposits have been interpreted as possibly due to deposition by individual tsunami waves (Dawson et al., 1991). However, the hydrodynamics of tsunami coastal sedimentation processes are very poorly known. For example, considerable sediment deposition may take place during the backwash (ie. from landward to seaward) following each coastal flood inundation. Furthermore, the role of sediment erosion and reworking during periods of backwash has not been studied. Moreover, it is likely that local variations in the configuration and bathymetry of individual inlets as well as variations in nearshore sediment supply may result in very complex patterns of coastal run-up and sediment deposition.

5.2. Norway and Iceland

More recently, numerous sites showing possible former tsunami inundation during the early/middle Holocene have been reported from western Norway (Svendsen and Mangerud, 1990; Bondevik and Svendsen, 1993) (Fig. 3). The former authors describe a lake isolation basin at Almestadmyra at 13.7 m a.s.l. where a sand layer (attributed to a tsunami) containing brackish diatoms is interbedded in the lower part of a Holocene lacustrine gyttja. The threshold of the basin is located circa 4m above the highest level reached by the Holocene sea in this area (the Tapes transgression) and hence this altitude can be considered as a minimum estimate of tsunami run-up. Svendsen and Mangerud (1990) have also reported a site in western Norway where possible tsunami deposits exist. They describe (pp. 120–122) an isolation basin at Skolemyra the threshold of which occurs circa 5 m above the highest Tapes marine level. Here, a disturbed layer, which includes terrestrial turf, rests unconformably upon volcanic ash of Late Glacial age. The disturbed layer occurs in the deepest part of the lake and is apparently the result of slumping. Svendsen and Mangerud argue that because the surrounding area is flat and not conducive to slumping, an external cause (for example a tsunami) may be a plausible triggering mechanism. Preliminary research by Bondevik and Svendsen (1993) suggests that the Norwegian sites were indundated by a tsunami at circa 7200 yr BP, an age slightly older than that indicated for tsunami deposits of the same age in Scotland.

A separate site of tsunami deposition has been reported from northwestern Iceland (Hansom and Briggs, 1991) (Fig. 3). These authors describe a site at Bitrufjorour, Hvitahlio, in Vestfirdir where marine silts at circa 8.5 m a.s.l. are overlain by two separate accumulations of terrestrial peat upon which is a raised beach ridge composed of cobbles and sand. Radiometric dating of the associated peat deposits suggest that the high energy marine event which deposited the beach ridge took place between 8830 ± 60 yr BP (GrN-15844) and 6910 ± 100 yr BP (GrN-15843) (Hansom and Briggs, 1991, pp. 84–87). The full significance of the Vestfirdir site must await the discovery of other circa 7000 yr BP sites in Iceland that exhibit unequivocal evidence of a severe episode of coastal flooding at this time. No information is presently available on tsunami run-up for the Vestfirdir site. The site at Vestfirdir is particularly unusual since tsunami activity in this area appears to have been associated with the construction of a single beach ridge.

Such problems highlight the almost complete absence of sedimentological research on tsunami deposits. Indeed, apart from the studies that have been undertaken on the sediments deposited by the Storegga tsunami, there are only a limited number of studies known to the author in which detailed analyses of tsunami sediments have been undertaken (cf. Darienzo and Peterson, 1990; Atwater and Yamaguchi, 1991; Foster et al., 1991; Minoura and Nakaya, 1991; Atwater and Moore, 1992).

5.3. Oregon and Washington

Intercalated sequences of tsunami-deposited sediments and intercalated subsurface saltmarsh deposits have been described from northern Oregon (Netarts Bay) by Darienzo and Peterson (1990). Here, six intervals of sudden marsh burial have been documented for the last several thousand years. These authors observed that five of the buried marsh surfaces exhibit sharp, non-erosional upper contacts with either anomalous sand layers (tsunami deposits) or tidal flat mud deposits. Their results appear to indicate the former occurrence of tsunamis generated by earthquakes in the Cascadia subduction zone that, in turn, were associated with episodes of abrupt coastal coseismic subsidence in the order of 1–1.5 m followed by slow interseismic uplift. Field evidence for similar late Holocene palaeo-tsunamis has been described for areas of coastal Washington State (Reinhardt and Bourgeois, 1989; Atwater and Yamaguchi, 1991; Atwater and Moore, 1992). In particular, two major tsunamis appear to have taken place at ca. 300 and 1000–1100 yr BP and were similarly related to tectonic activity in the Cascadia subduction zone. In the case of the 300 yr BP event, numerous stems of salt marsh vegetation are buried in growth position as a result of an extraordinary landward-directed surge (probably a tsunami) while elsewhere, tree ring data indicates the ocurrence of sudden coseismic submergence (Atwater and Yamaguchi, 1991). The 1000–1100 yr BP tsunami is believed to have been generated by an earthquake along the Seattle fault in the Puget Sound and was not only associated with widespread sediment deposition and coastal subsidence but also with subaerial landsliding.

5.4. Southwest England

Foster et al. (1991) have undertaken detailed analyses of marine sediments in the Scilly Isles, SW England, considered to have been deposited by the Lisbon tsunami of 1755 AD. The sediments occur within an infilled lagoon and consist of up to 1 m of marine sands that taper landward and consist of a series of fining-upwards units. The sand accumulation rests with unconformity upon an underlying terrigenous peat and contains numerous small clasts of reworked peat within it. Marine sediments associated with the Lisbon tsunami of 1755 AD are also believed to have been deposited in areas of the Algarve region, Portugal (Hindson and Andrade, pers. commun.).

5.5. Japan

In Japan, Minoura and Nakaya (1991) have described several high-energy marine sand layers intercalated with organic muds from lacustrine sediments in the Tsgaru and Sendai Plains region. In total, thirteen laterally persistent sand layers have been identified. Many of the sand layers rest unconformably upon underlying muds while in some cases, flame structures locally occur at the base of the layers and are interpreted as the result of differential loading caused by the rapid deposition of sand. The distribution of sand layers indicate that in some cases the tapering marine layers often extend inland as far as 3km beyond the position of the former coastline. Each sediment unit is believed to represent an episode of tsunami deposition indicated by radiometric dating (^{210}Pb) to represent evidence for 13 major tsunamis in this area since 1710 AD.

6. Summary

The study of the geomorphological processes associated with tsunami run-up and backwash is very much in its infancy. Until recently, most research has considered patterns of sediment deposition associated with former high-energy flood events and has concluded that the observed geomorphological and stratigraphic evidence is consistent with the view that palaeo-tsunamis have been the causal mechanisms. The Flores study of Shi et al. (1993) was the first geomorphological investigation of a modern tsunami. In addition to Flores, additional investigations of recent tsunamis that took place in 1992–1993 have been undertaken in Nicaragua (Bourgeois, unpubl.) and Hokkaido (Takahashi and Shuto, unpubl.). It is hoped that these and other studies will provide important information that will be of value in the identification of palaeo-tsunamis in coastal sediment sequences. This, in turn, will help contribute to our knowledge of the significance of such high-magnitude low-frequency events in the evolution of individual coastal landscapes.

Acknowledgements

The author is grateful to Shaozhong Shi, Prof. David Smith, Sue Dawson, Dale Howes, Robert Hindson, Dr. Ian Foster, Phil Wood and Prof. Jim Rose and an anonymous referee for comments on the draft of this paper. Cartographic assistance was kindly provided by Ruth Gaskell and Shirley Addleton.

References

Andrade, C., 1992. Tsunami generated forms in the Algarve barrier islands, In: A.G. Dawson (Editor), European Geophysical Society 1992 Tsunami Meeting. Sci. Tsunami Hazards, 10: 21–34.

Atwater, B.F. and Moore, A.L., 1992. A tsunami about 1000 years ago in Puget Sound, Washington. Science, 258: 1614–1617.

Atwater, B.F. and Yamaguchi, D.K., 1991. Sudden, probably coseismic submergence of Holocene trees and grass in coastal Washington State. Geology, 19: 706–709.

Bondevik, S. and Svendsen, J.I., 1993. Palaeotsunamis in the Norwegian Sea and the North Sea. In: S. Tinti (Editor), Genesis and Impact of Tsunamis on the European Coasts — GITEC. Mid-Term Scientific Report, Contract EV5F-CT92-0175, 7. 2-7. 27.

Bourrouilh-Le Jan, F.G. and Talandier, J., 1985. Sédimentation et fracturation de hautes énergie en milieu récifal: tsunamis, ouragans et cyclones et leurs effets sur la sédimentologie et la géomorphologie d'un atoll: motu et hoa, à Rangiroa, Tuamotu, Pacifique SE. Mar. Geol., 67: 263–333.

Bugge, T., 1983. Submarine slides on the Norwegian continental margin, with special emphasis on the Storegga area. Continental Shelf and Petroleum Technology Research Institute A/S publ. 110, 152 pp.

Bugge, T., Befring, S., Belderson, R.H., Eidvin, T., Jansen, E., Kenyon, N., Holtedahl, H. and Sejrup, H.-P., 1987. A giant three-stage submarine slide off Norway. Geo-Marine Lett., 7: 191–198.

Dawson, A.G., Long, D. and Smith, D.E., 1988. The Storegga Slides: evidence from eastern Scotland for a possible tsunami. Mar. Geol., 82: 271–276.

Dawson, A.G., Foster, I.D.L., Shi, S., Smith, D.E. and Long, D., 1991. the identification of tsunami deposits in coastal sediment sequences. Sci. Tsunami Hazards, 9: 73–82.

Darienzo, M.E. and Peterson, C.D., 1990. Episodic tectonic subsidence of Late Holocene salt marshes, northern Oregon central Cascadia margin. Tectonics, 9: 1–22.

Foster, I.D.L., Albon, A.J., Bardell, K.M., Fletcher, J.L., Jardine, T.C., Mothers, R.J., Pritchard, M.A. and Turner, S.E., 1991. High energy coastal sedimentary deposits: an evaluation of depositional processes in southwest England. Earth Surface Process. Landforms, 16: 341–356.

Hansom, J.D. and Briggs, D.J., 1991. Sea-level change in Vestfirdir, north west Iceland. In: J.K. Maizels and C. Caseldine (Editors), Environmental Change in Iceland: Past, Present and Future. Kluwer, Dordrecht, pp. 79–91.

Harbitz, C.B., 1991. Model simulations of tsunami generated by the Storegga slide. Inst. of Mathematics, Univ. of Oslo, Series No. 5, 30 pp.

Harbitz, C.B., 1992. Model simulations of tsunamis generated by the Storegga Slides. Mar. Geol., 105: 1–21.

Henry, F. and Murty, T.S., 1992. Model studies of the effects of the Storegga Slide tsunami. Sci. Tsunami Hazards, 10: 51–62.

Kawana, T. and Pirazzoli, P., 1990. Re-examination of the Holocene emerged shorelines in Irabu and Shimoji Islands, the South Ryukyus, Japan. Quat. Res., 28: 419–426.

Long, D., Smith, D.E. and Dawson, A.G., 1989. A Holocene tsunami deposit in eastern Scotland. J. Quat. Sci., 4: 61–66.

Minoura, K. and Nakaya, S., 1991. Traces of tsunami preserved in inter-tidal lacustrine and marsh deposits: some examples from northeast Japan. J. Geol., 99: 265–287.

Moore, G.W. and Moore, J.G., 1988. Large–scale bedforms in boulder gravel produced by giant waves in Hawaii. In: H.E. Clifton (Editor), Sedimentologic Consequences of Convulsive Geologic Events. Geol. Soc. Am. Spec. Pap., 229: 101–110.

Moore, J.G. and Moore, G.W., 1984. Deposit from a giant wave on the island of Lanai, Hawaii. Science, 226: 1312–1315.

Moore, J.G., Clague, D.A., Holcom, R.T., Lipman, P.W., Normark, W.R. and Torresan, M.E., 1989. Prodigious submarine landslides on the Hawaiian Ridge. J. Geophys. Res., 94: 17465–17484.

Moore, J.G., Normark, W.R. and Gutmacher, C.E., 1992. Major landslides on the submarine flanks of the Manua Loa Volcano, Hawaii. Landslide News, 13–16.

Nakata, T. and Kawana, T., 1993. Historical and prehistorical large tsunamis in the southern Ryukyus, Japan. In: Tsunami '93, Proc. IUGG/IOC Int. Tsunami Symp., Wakayama, Japan, August 23–27, 1993, pp. 297–307.

Ota, Y., Pirazzoli, P.A., Kawana, T. and Moriwaki, H., 1985. Late Holocene coastal morphology and sea-level records on three small islands, the South Ryukyus, Japan. Geogr. Rev. Jpn., 58B: 185–194.

Pirazzoli, P.A., Montaggioni, L.F., Salvat, B. and Faure, G., 1988. Late Holocene sea level indicators from twelve atolls in the central and eastern Tuamotos (Pacific Ocean). Coral Reefs, 7: 57–68.

Reinhardt, M.A. and Bourgeois, J., 1989. Tsunami favored over storm or seiche for sand deposit overlying buried Holocene peat, Willapa Bay, Washington. EOS, October 24: 1331.

Shi, S., Dawson, A.G. and Smith, D.E., 1991. Sedimentology of a Holocene tsunami deposit. Abstr. XIII Int. Congr. of the International Union for Quaternary Research (INQUA), Beijing, China, August 2–9, 1991, p. 329.

Shi, S., Dawson, A.G. and Smith, D.E., 1993. Geomorphological impact of the Flores tsunami of 12th December, 1992, In: Tsunami '93, Proc. IUGG/IOC Int. Tsunami Symp., Wakayama, Japan, August 23–27, 1993, pp. 689–696.

Sissons, J.B. and Smith, D.E., 1965. Peat bogs in a post-glacial sea and a buried raised beach in the western part of the Forth valley. Scott. J. Geol., 1: 247–255.

Smith, D.E. and Dawson, A.G., 1990. Tsunami waves in the North Sea. New Sci., 4th August 1990: 46–49.

Smith, D.E., Cullingford, R.A. and Haggart, B.A., 1985. A major coastal flood during the Holocene in eastern Scotland. Eiszeitalter Ggw., 35: 109–118.

Smith, D.E., Turbayne, S.C., Dawson, A.G. and Hickey, K.R., 1991. The temporal and spatial variability of major floods around European coasts, Final Report for European Commission, Contract EV4C 0047 UK (H), 128 pp., unpubl.

Stearns, H.T., 1940. Geology and ground-water resources of the islands of Lanai and Kahoolawe, Hawaii. Hawaii Div. Hydrography Bull. 6, 177 pp.

Svendsen, J.I. and Mangerud, J., 1990. Sea-level changes and pollen stratigraphy on the outer coast of Sunmore, western Norway. Norsk Geol. Tidsskr., 70: 111–134.

Teissier, R., 1969. Les Cyclones en Polynesie Francaise, Bull. Soc. Etudes Oceaniennes, 14, 20–21.

Wright, C. and Mella, A., 1963. Modifications to the soil patterns of south central Chile resulting from seismic and associated phenomena during the period May to August 1960. Bull. Seismol. Soc. Am., 53, 1367–1402.

Young, R.W. and Bryant, E.A., 1992. Catastrophic wave erosion on the southeastern coast of Australia: impact of the Lanai tsunamis ca. 105ka? Geology, 20: 199–202.

ELSEVIER

Geomorphology 10 (1994) 95–106

GEOMORPHOLOGY

Distribution of hazard types in a drainage basin and its relation to geomorphological setting

Hiroo Ohmori[a], Hiroshi Shimazu[b]

[a]*Department of Geography, Graduate School of Science, The University of Tokyo, 7-3-1 Hongo, Bunkyo-ku, Tokyo 113, Japan*
[b]*Department of Geography, Faculty of Letters, Kanazawa University, Kakumacho, Kanazawa City, Ishikawa Prefecture, 920-11, Japan*

Received January 1, 1994; revised March 20, 1994; accepted March 30, 1994

Abstract

Hazards along a river course in a drainage basin are characterized by three types of events: debris flow, turbidity flow and flood. Each has its own channel segment with different sediment transport processes. The sediment transport process is controlled by transportation force: tractive force and slope-direction component of sediment weight, both of which are significantly affected by channel slope. Along large rivers in Japan the boundary between the upstream turbidity flow segment and the downstream flood segment is located at the position with a channel slope of about 1/1000. Along steep, small rivers the boundary between the upstream debris flow segment and the downstream turbidity flow segment is located at the position with a channel slope of about 80/1000. The channel slope depends on the shapes of longitudinal profiles of rivers. The longitudinal profiles of Japanese rivers, main rivers and tributaries, can be described by an exponential, power or linear function. The transportation force of the rivers fitted with exponential functions markedly decreases downstream due to the large curvature of the profiles, causing noticeable sediment deposition in the middle courses. The transportation force of the rivers fitted with power or linear functions maintains its strength through the whole river courses due to the small curvature, causing sediment transportation down to the lower courses. The function types are strongly affected by relief in the drainage basins. The rivers flowing in small relief areas are fitted with exponential functions and those flowing in large relief areas are fitted with power or linear functions. Thus, the distribution of hazard types along a river course in a drainage basin is controlled by the distribution of relief in the drainage basin.

1. Introduction

Hazards occurring in a drainage basin are classified into three types of events characterized by different transportational and/or depositional processes of sediment. The first type is a debris flow which is defined here as a high-density flow containing numerous gravels of boulder size, the second type is a turbidity flow which is defined as a torrent containing much tractional load of cobble and pebble size, and the third type is a flood which is defined as an overflow and ponding of muddy water without gravel. The debris flow causes markedly serious destruction of land by transportation and deposition of boulders, and debris control dams and ground sills are required to prevent the occurrence. The turbidity flow also causes serious destruction by transportation and deposition of gravels, and dykes and spurs are required as defense from the attack. The flood causes widespread damage primarily by ponding of muddy water, and levees and reservoirs are required to prevent flooding. As each requires different control works for protection, distinguishing where and which types of hazard occur along a river course is important for planning land use and flood control projects.

SSDI 0169-555X(94)00019-N

As defined above, hazard types are characterized by the grain size of transported sediment which is controlled by transport processes: debris flow, traction and suspension. The grain size of debris flow sediment depends on the channel slope (Shimazu, 1991) because the transportation force is determined mainly by the slope-direction component of sediment weight itself. The grain size of the sediment transported by traction and suspension depends on the tractive force of stream flow (Hjulstrom, 1935; Plumley, 1948; Iwagaki, 1956; Hack, 1957; Brush, 1961; Egiazaroff, 1965; Bagnold, 1966; Inokuchi and Mezaki, 1974; Knighton, 1975, 1980, 1982; Schumm, 1981; Brierley and Hickin, 1985; Ohmori, 1991; Inoue, 1992). Tractive force is also determined by the channel slope for Japanese rivers (Ohmori, 1991; Inoue, 1992). The channel slope can be evaluated based on the longitudinal profiles of rivers. The characteristics of the longitudinal profiles are strongly affected by the distribution of relief in the drainage basins (Ohmori, 1993). This paper examines first the segmentation of a river course based on the hazard types as characterized by sediment transport processes. Second, the sediment transport processes controlled by the shape of longitudinal profile of the river are discussed. Third, the relationship between the shape of the longitudinal profile and the relief in the drainage basin is examined. And finally, how the distribution of relief in a drainage basin controls the distribution of hazard types is discussed. Forty-five large Japanese rivers with drainage areas of 200 to 14,300 km^2 are examined for the hazards in a large river and 7 steep Japanese small rivers with drainage areas of 50 to 950 km^2 are examined for the hazards in a steep, small river.

2. Segments with different sediment transport processes in a large river

In relatively large rivers in Japan, the river courses are divided into two segments: the upstream fans (or gravel-bed channels) and the downstream natural levee and delta systems (Yatsu, 1954, 1955; Oya, 1977; Saito, 1988; Ohmori, 1988). The boundary between the segments is clearly recognized at the position where the channel slope is about 1/1000 (Fig. 1; Ohmori, 1991). The position indicates the front of the depositional area of gravels and is called the "FDG" (Ohmori, 1991). Using the equation expressing the tractive force, τ (dyn/cm^2):

$$\tau = \rho g R I \quad (1)$$

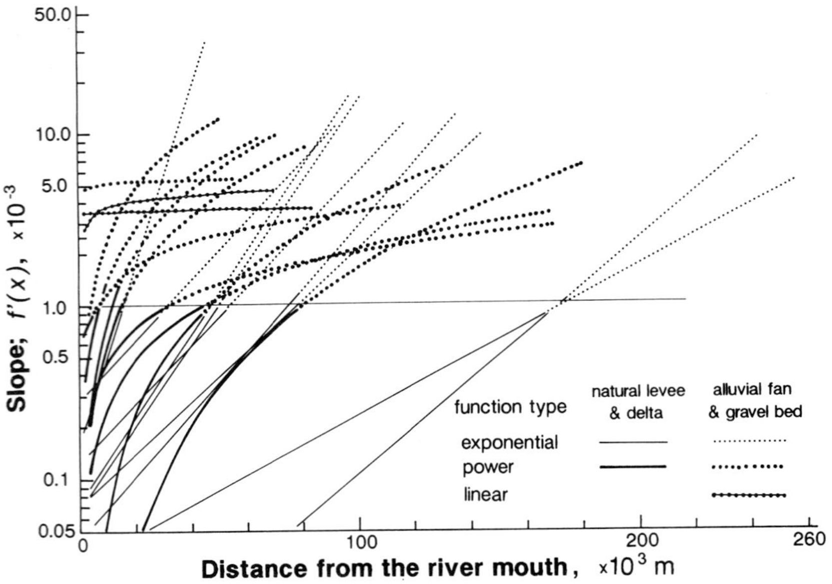

Fig. 1. Segmentation of large rivers in Japan. The boundaries between the gravel bed channels and the natural levee and delta systems are located at the positions with a channel slope of about 1/1000 which is evaluated by $f'(x)$ of the best fit functions (after Ohmori, 1991).

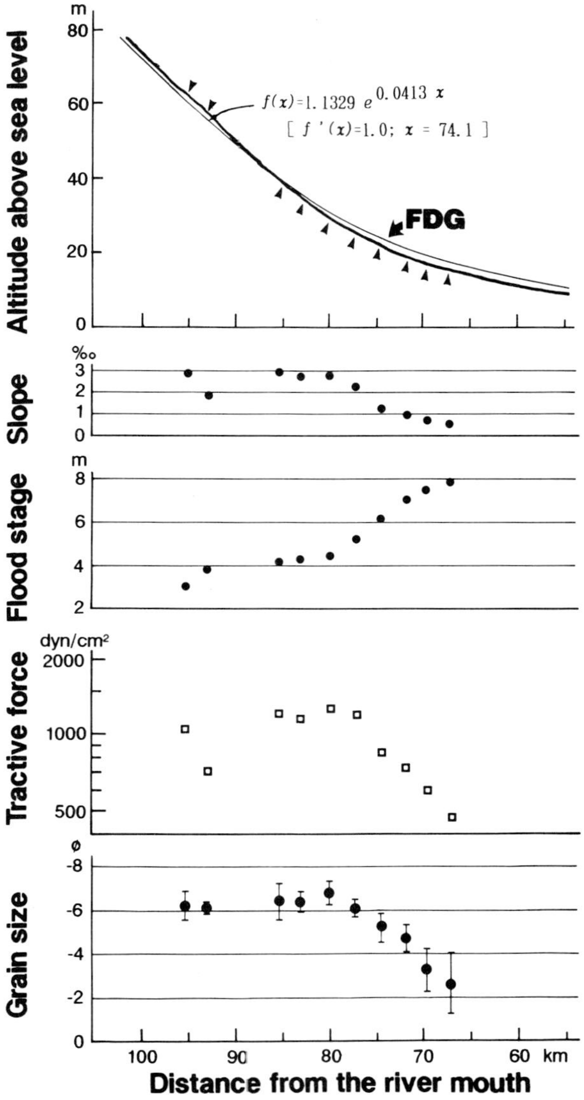

Fig. 2. Example of a part of a longitudinal profile of the River Arakawa, in the Kanto Plain, central Japan, which shows part of a regression line of the best fit function, the observed channel slope, the observed flood stage of a ten-year recurrence interval, the calculated tractive force and the observed grain size. The grain size indicates the mean diameter together with the standard deviation of the tractional sediment. The sediment is divided into groups such as tractional sediment and suspended sediment, each of which shows a log-normal distribution of grain size (based on Inoue, 1992).

and the equation expressing the critical tractive force, τ_c (dyn/cm^2):

$$\tau_c = 0.05(\sigma - \rho) g d \qquad (2)$$

the maximum diameter of sediment, d (cm), is given by:

$$d = RI/0.05/(\sigma/\rho - 1) \qquad (3)$$

where ρ is the fluid density (g/cm^3), g is the acceleration of gravity (cm/s^2), R is the hydraulic radius (cm), which can be substituted by water depth or flood stage, I is channel slope and σ is the sediment density (g/cm^3). Because the maximum flood stages frequently observed in Japanese rivers are in the range of 100–500 cm, and $\sigma/\rho = 2.65$, the sediment with a maximum diameter of 1 to 6 cm is transported downstream to the position with a channel slope of about 1/1000. At this position, the sediment changes from gravel to sand, forming FDG (Ohmori, 1991).

Based on the relationships between the tractive force and the grain size distribution of sediment observed in 5 large rivers in the Kanto Plain, central Japan, Inoue (1992) showed that the alongstream change in grain size is determined by the tractive force which changes downstream. Gravels larger than cobble size (about 6 cm in diameter or -6 in ϕ scale), which comprise 80% or more of bed material in weight, are transported by traction and most of them are deposited in the reach with a channel slope equal to or larger than 1/1000 (Fig. 2; Inoue, 1992). It means that the downstream end of cobble-size gravel deposition which is equivalent to FDG is located at the position with a channel slope of about 1/1000. Because the tractive force markedly decreases downstream due to the abrupt decrease in channel slope, the downstream course gentler than 1/1000 is comprised mainly of sandy material. Thus, a large river course is classified into two segments: the upstream turbidity flow segment and the downstream flood segment.

3. Segments with different sediment transport processes in a steep small river

Viewed in the drainage basin, steep small rivers are equivalent to the upper courses and tributaries of the large rivers examined above. They are frequently attacked by debris flows occurring in their uppermost steep channels and tributaries. Based on the grain size distribution along the channels of main courses and tributaries of steep small rivers in Japan, Shimazu (1990, 1991) pointed out that the mean diameter of

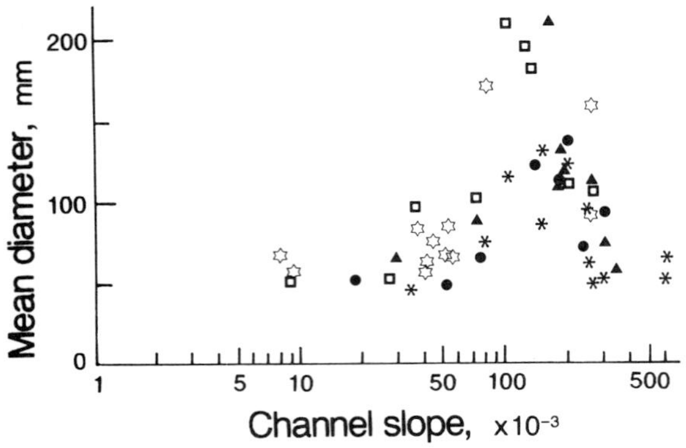

Fig. 3. Relationship between channel slope and mean diameter of sediment in steep small rivers and tributaries. Different symbols indicate different drainage basins (after Shimazu, 1991).

sediment increases with the increase in channel slope for the lower courses gentler than about 80/1000, whereas the mean diameter decreases with the increase in channel slope for the steeper upper courses (Fig. 3; Shimazu, 1991). Along the river courses steeper than 200/1000, bedrock is often exposed continuously and a marked sedimentation is not recognized. Sediment deposition becomes conspicuous downstream, forming clearly shaped debris flow lobes in the courses with a channel slope of 200/1000 to 80/1000. The debris flow lobes contain a number of boulders with a maximum diameter larger than 100 cm. There are no huge boulders in the downstream courses gentler than 80/1000 where the sediment diameter decreases with the decrease in channel slope. In such gentle gradients, gravels are transported by turbidity flows and the upstream huge boulders are also moved into the gentle courses by turbidity flows after they are diminished in size due to weathering (Schumm and Stevens, 1973; Benda, 1990).

The values of the critical slopes agree well with those for debris flows observed in Japan: 11–15° (195/1000–270/1000) for the upstream end of the debris flow deposition and 3–4° (50/1000–70/1000) for the downstream end of the deposition (Takahashi and Yoshida, 1979; Kobashi et al., 1980). Thus, a steep small river course is divided into two segments: the debris flow segment which is steeper than 80/1000 and the turbidity flow segment which is gentler than 80/1000.

Using the channel slope at the lower end of a tributary, represented by the mean slope of the channel between 200 and 400 m upstream from the junction with its main river, tributaries are classified into two categories: the debris flow tributary which has a lower-end channel slope larger than 80/1000 and the turbidity flow tributary which has a lower-end channel slope less than 80/1000. In the debris flow tributary, debris flows run down through the whole course into the main river, whereas in the turbidity flow tributary, debris flows, even if they occur, cannot reach into the main river because they are deposited in the middle course, and turbidity flows are dominant in the lower course.

4. Longitudinal profiles of rivers controlling sediment transport processes

Based on regression analysis, the longitudinal profile of a large river in Japan can be described by one of three mathematical functions: exponential, power and linear functions (Ohmori, 1988, 1991; Inoue, 1992; Ohmori and Saito, 1993). It was noted that the tractive force, which is calculated by Eq. (1), significantly decreases downstream in the rivers fitted with exponential functions due to the large curvature (Fig. 4a), even if flood stage increases downstream (Ohmori, 1991; Inoue, 1992; Fig. 2). Downstream decrease in tractive force is small along the river courses fitted with power functions due to the relatively small curvature, and the tractive force in the rivers fitted with linear functions maintains the same strength through the whole river courses because the change in channel gradient is negligible (Fig. 4a).

As the channel slope is estimated by the first derivative of the best fit functions, the location of FDG is also estimated based on the best fit functions. The locations of FDG estimated by the first derivative of the best fit functions, $f'(x) = 1/1000$, are not different from the locations of FDG observed in Japanese rivers, as shown in Fig. 1, where the observed FDG is indicated by the boundary between the alluvial fan and gravel bed and the natural levee and delta (Ohmori, 1991). Thus, it can be said that the degree of development of the upstream turbidity flow segment and/or the downstream flood segment depends on the function type.

Utilizing linear, power, exponential and logarithmic functions, regression analysis between altitude and distance was performed for the longitudinal profiles of

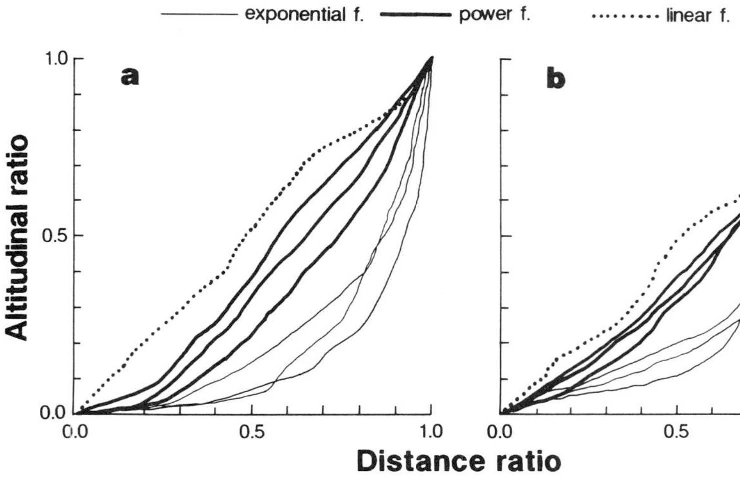

Fig. 4. Examples of the longitudinal profiles of rivers described by different functions. (a) For large rivers, (b) for small, steep rivers. Distance and altitude are expressed by ratios, where both the horizontal distance and relative altitude between the lowest and highest points of a measured river course are 1.0.

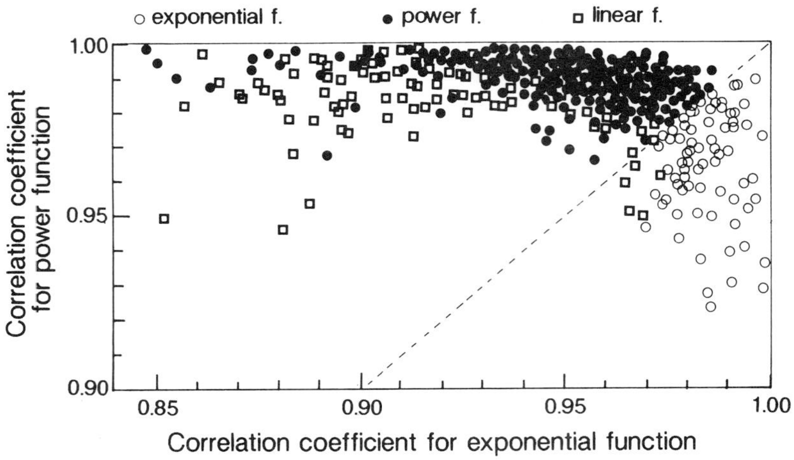

Fig. 5. Distribution of correlation coefficients of exponential and power regression functions; for steep, small rivers and tributaries. The function types are determined based on the least mean error among the utilized exponential, power, linear and logarithmic functions.

main courses and tributaries of steep small rivers in Japan. The longitudinal profiles also show a good fit to both exponential and power functions. The correlation coefficients of individual rivers are larger than 0.85 for both functions and larger than 0.95 for either of the functions (Fig. 5). Based on the mean error of estimate, some of them, which have also large correlation coefficients larger than 0.95 for linear function as well for exponential and power functions, are best described by a linear function. Thus, the longitudinal profiles of steep small rivers in Japan are expressed by an exponential, power or linear function, as also noted by Shepherd (1985) for bedrock rivers in Texas, USA. Based on the frequency of the function type for each class of the lower-end channel slope (Fig. 6), tributaries with an exponential function type profile are limited to those with a lower-end channel slope less than about 200/1000 (about $10^{2.3}$x10^{-3}). Tributaries with a power function type profile increase in number with the increase in slope and the increment is remarkable for a channel slope larger than about 60/1000 (about $10^{1.8}$x10^{-3}). Tributaries with a linear function type

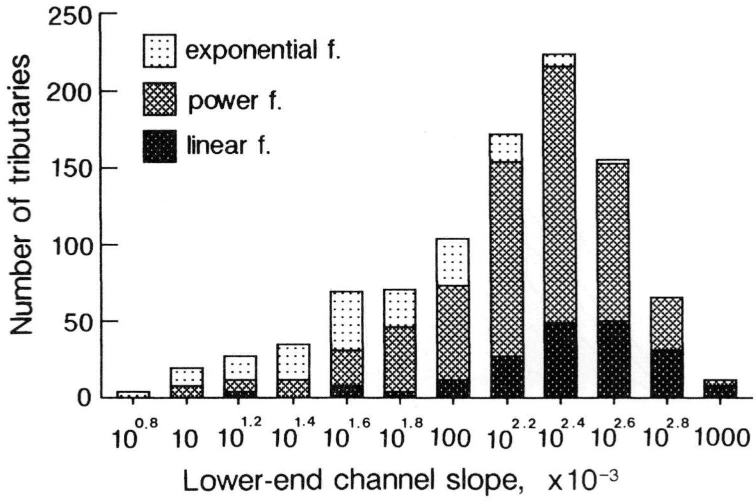

Fig. 6. Frequency distribution of function types for each class of the lower-end channel slope of tributaries. The lower-end channel slope indicates the mean slope of the 200 m long channel just above the junction with its main river.

profile are limited to those with a gardient larger than about 100/1000.

As noted previously, using the lower-end channel slope of 80/1000, tributaries are classified into either a turbidity or debris flow tributary. Most of the tributaries fitted with exponential functions are classified as turbidity flow tributaries. The transportation force of debris flows, which is principally caused by the slope-direction component of weight of the debris, markedly decreases downstream due to the large curvature (Fig. 4b), causing the deposition of boulders in the middle courses, with the lower courses dominated by turbidity flows. The tributaries fitted with power or linear functions are classified as debris flow tributaries where the downstream decrease in transportation force is small due to the small curvature (Fig. 4b) and debris flows run through the whole courses. Because the above characteristics are representative of steep, small rivers, hazard types of a steep, small river also depend on the function type.

5. Relief controlling the function types for large rivers

The relief condition in a drainage basin is expressed by many indicators. For large rivers, the relief ratio defined by Schumm (1956) and the hypsometric integral defined by Strahler (1952) are used for examining the relationship between function types and relief. Both indicators represent the relief condition of a drainage basin by a single value. As suggested by the relationship between the hypsometric integral and the relief ratio (Fig. 7), small hypsometric integrals indicate small-relief drainage basins and large hypsometric inte-

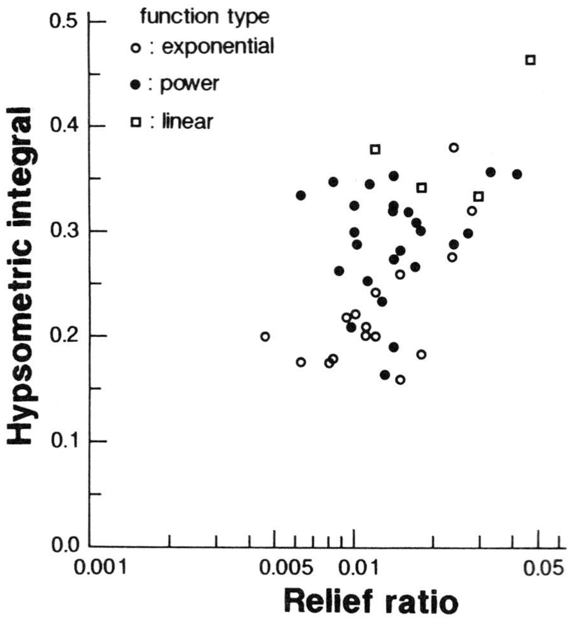

Fig. 7. Distribution of function types of large rivers in relation to the relief in drainage basins (after Ohmori, 1993).

grals the large-relief drainage basins for Japanese rivers (Ohmori, 1993). From Fig. 7, the distribution of a function type is limited in a range of hypsometric integral. Most rivers fitted with exponential functions have integrals less than 0.25, those fitted with power functions have integrals ranging from 0.20 to 0.35, and those fitted with linear functions have integrals larger than 0.30. Thus, the function types describing the longitudinal profiles of rivers are strongly affected by the relief in the drainage basins and it can be said that long flood segments are developed in the drainage basins with integrals less than 0.25, and long turbidity flow segments are developed in the drainage basins with integrals larger than 0.25.

6. Relief controlling the function types for steep small rivers

Because the distribution of tributaries with different hazard types in a drainage basin affects the distribution of hazard types along their main river courses, a more detailed examination was performed for the steep, small rivers. Fig. 8 shows the relationship between the relief ratio and the lower-end channel slope of tributaries. Channel slope increases with the increase in relief ratio. Based on Fig. 8, where each function type is expressed by a different symbol, exponential functions generally show small relief ratios, and power and linear functions usually show larger relief ratios. This suggests that the turbidity flow tributaries occur in small relief areas and the debris flow tributaries in large relief areas. When a drainage basin of a main river is covered with a grid system and divided into unit squares in which several lattice points are included, we can calculate the mean altitude and the standard deviation for each unit square based on the altitudinal data of the lattice points. Mean altitude and the standard deviation of a unit square are given by:

$$\bar{x}=\left(\sum_{i=1}^{n}x_i\right)/n \tag{4}$$

and:

$$D=\left\{\left[\sum_{i=1}^{n}(x_i-\bar{x})^2\right]/(n-1)\right\}^{1/2} \tag{5}$$

where $\bar{x}$ is the mean altitude, D is the standard deviation, x_i is the altitude of the lattice point i, and n is the number of lattice points in a unit square. The standard deviation, which is called the "dispersion of altitude" (Ohmori, 1978), is, statistically, an absolute measure of the dispersion of surface altitude, and indicates physically the dispersion of potential energy of surface material. It has been noted that the dispersion of altitude is proportional to both relief energy and surface gradient (Ohmori, 1978, 1981; Ohmori and Sohma, 1983; Ohmori and Hirano, 1984), and is in a functional relation with erosion rate (Ohmori, 1978, 1981, 1983). Using a grid system with intervals of 250 m on 1:50,000 scale top-

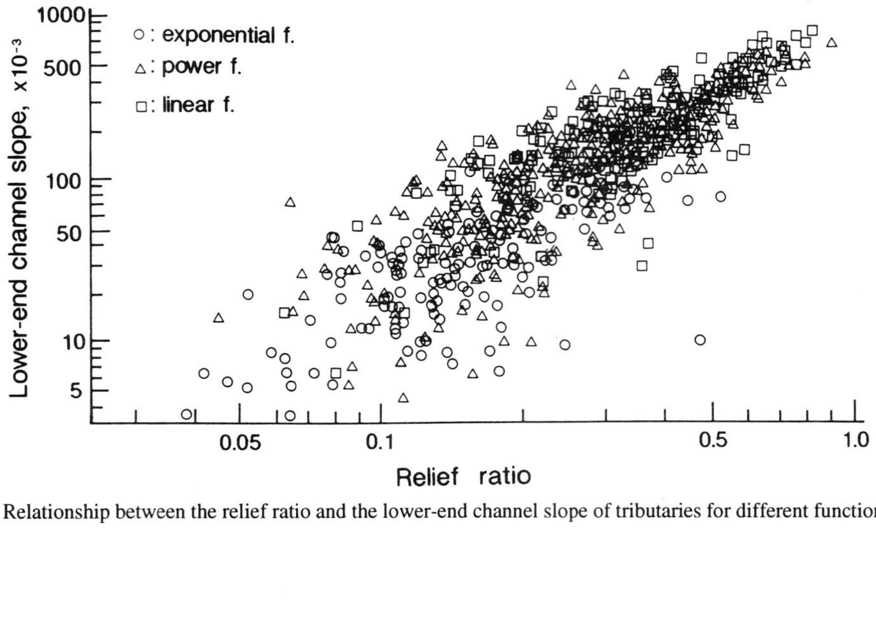

Fig. 8. Relationship between the relief ratio and the lower-end channel slope of tributaries for different function types.

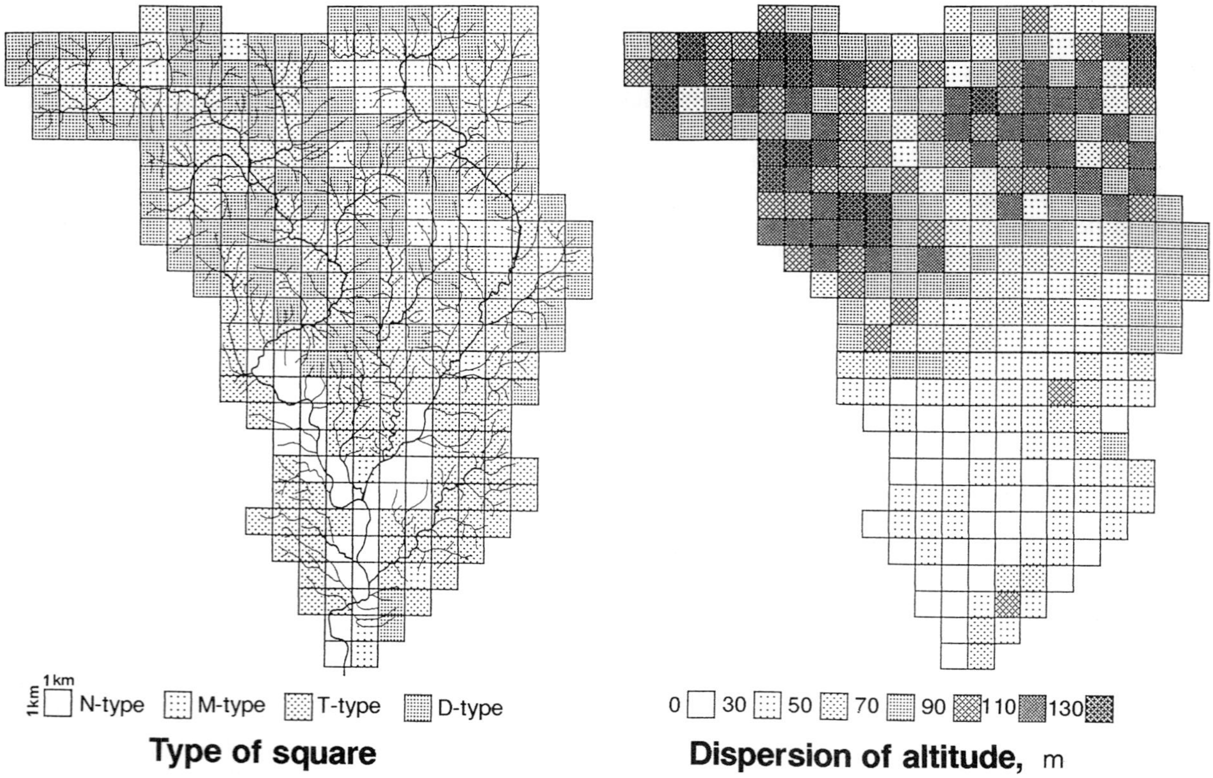

Fig. 9. Example of the distribution of square types and the dispersion of altitude for a drainage basin of a steep, small river: the Fujikoto River with a drainage area of 285 km^2, in northeast Japan.

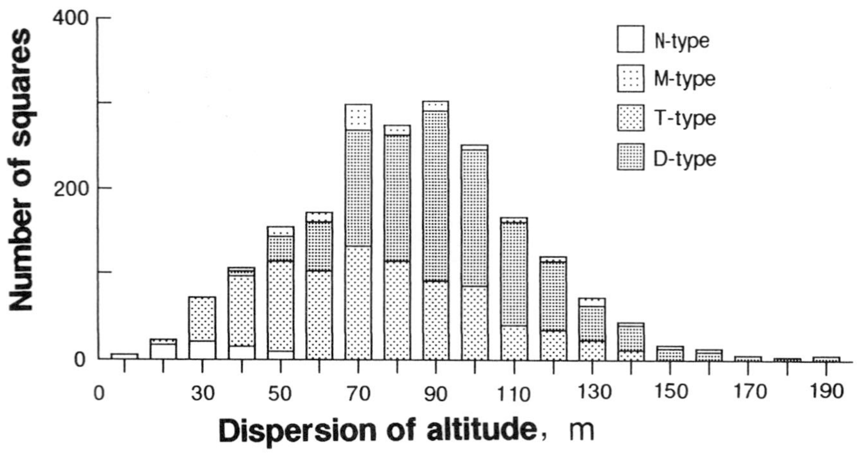

Fig. 10. Frequency distribution of square types for each class of the dispersion of altitude for 7 steep, small rivers.

ographic maps whose contour interval is 20 m, each drainage basin of 7 steep small rivers in Japan was divided into squares of 1 km^2 with 25 lattice points and the dispersion was calculated by Eq. (5) for individual squares.

Then, using the channel slope measured on 1:25,000 scale topographic maps with a contour interval of 10 m, channels of 7 main rivers and their tributaries with a drainage basin larger than 0.1 km^2 were divided into turbidity flow segments and/or debris flow segments. Because the critical channel slope is 80/1000, the boundary between the two segments is located on the channel where the distance between two neighbouring contour lines is 5 mm on the maps. Based on the segment types dominantly distributed in each square, all squares in the drainage basins examined were classified into the following four square types:

N-type square is a square where turbidity flow segments are dominant together with wide flood plains and/or river terraces over half of the square. *M-type square* is a square with poor development of river channels themselves where landslide blocks and smooth soil-creep slopes are well developed. Excluding N-type squares, the residual squares dominated by turbidity flow segments are classified as *T-type square. D-type squares* are occupied mainly with debris flow segments. An example of the distribution of square types in a drainage basin is shown in Fig. 9a. The distribution of the dispersion of altitude for the same drainage basin is shown in Fig. 9b. These figures suggest that N-type squares correspond to a small dispersion and D-type squares to a large dispersion.

The frequency distribution of square types for each class of the dispersion is shown in Fig. 10 for all squares totalling 2079 squares of the 7 steep small rivers examined. N-type squares are limited to an altitudinal dispersion less than about 50 m, where most tributaries have no debris flow segment. M-type squares have a low frequency and a wide range from 40 to 160 m. D-type squares are dominant for dispersions larger than about 70 m.

T-type squares which are dominated by turbidity flow tributaries also have a wide range. All of the turbidity flow tributaries have turbidity flow segments at least in their lower courses, but they are classified into two categories: a tributary with a debris flow segment in its upper course and a tributary without a debris flow segment. For T-type squares, the squares occupied with turbidity flow tributaries with upstream debris flow segments have a large dispersion, larger than 100 m, whereas the squares occupied with turbidity flow tributaries without a debris flow segment have a small dispersion, less than about 60 m.

Based on the above relationship between the dispersion and the square types, the squares with a dispersion larger than about 100 m are occupied with rivers which are often attacked by debris flows. Although the squares with a dispersion of 60–100 m cannot be clearly divided into either of the two categories, the squares with a dispersion less than about 60 m are occupied with rivers where turbidity flows are dominant. Especially, most squares with a dispersion less than 40 m are occupied with rivers without a debris flow segment. Thus, the distribution of hazard types along a river course is controlled by the distribution of relief in the drainage basin.

7. Conclusions

Each of the hazard types; debris flow, turbidity flow or flood, occupies its own segment with a different channel slope in a river course. Along a large river in Japan, gravels of cobble size are transported down to a position with a channel slope of about 1/1000. The position indicates the boundary between the upstream turbidity flow segment and the downstream flood segment. Along a steep small river, debris flows move down to the position with a channel slope of about 80/1000. This position is a boundary between the upstream debris flow segment and the downstream turbidity flow segment. Based on the relationships between function types and segment types, between segment types and square types and between square types and relief, the distribution of hazard types along a river course is summarized in Fig. 11.

A river course flowing through an area with an altitudinal dispersion larger than about 100 m is attacked by debris flows occurring in the tributaries whose longitudinal profiles are fitted with power or linear functions (Fig. 11a). Due to the small curvature of the longitudinal profile, the transportation force maintains the same strength through the whole course and boulders are transported into the main river (Fig. 11b). The river course flowing through an area with a dispersion less than about 60 m is attacked, not by debris flows, but by turbidity flows occurring in the tributaries whose

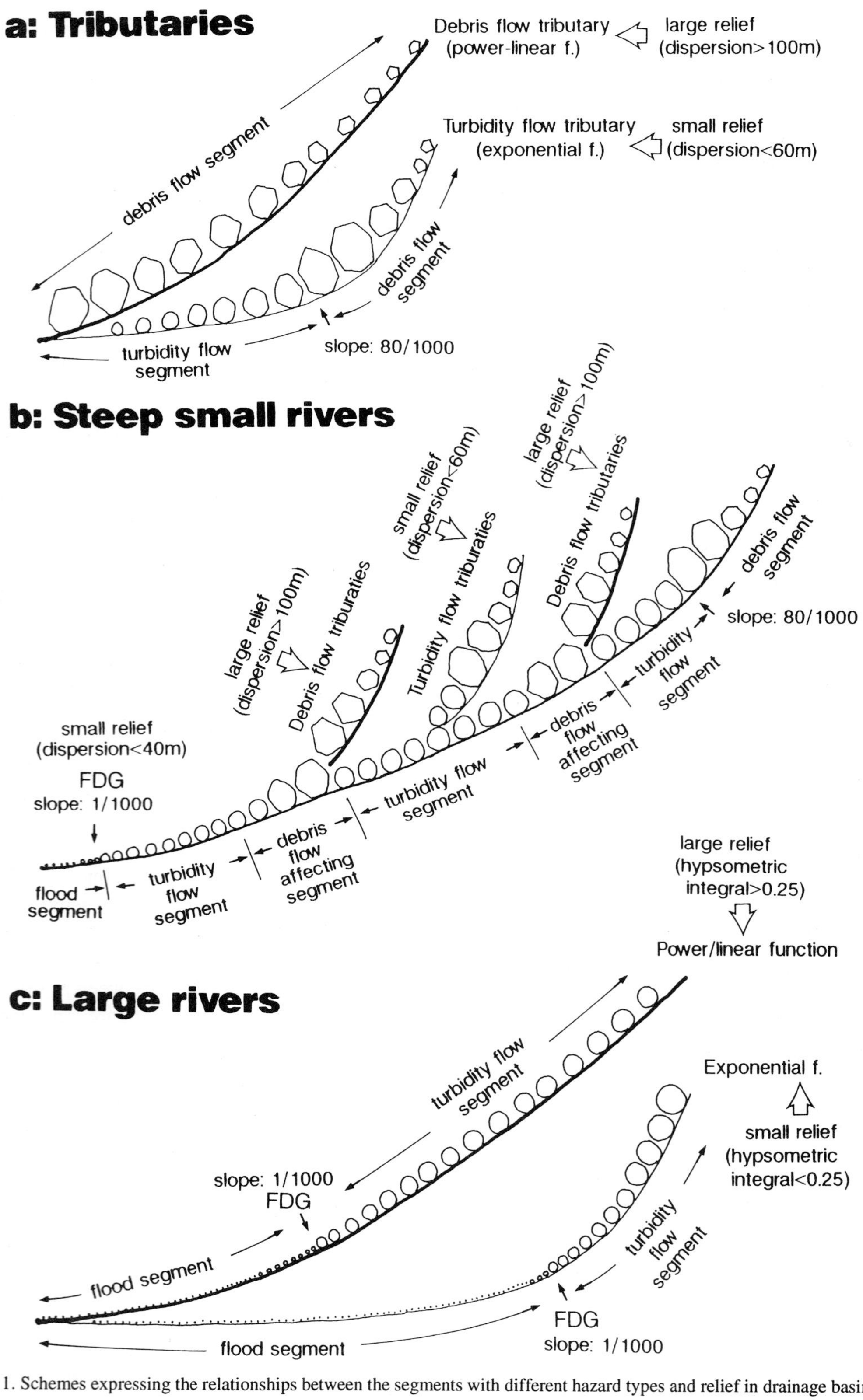

Fig. 11. Schemes expressing the relationships between the segments with different hazard types and relief in drainage basins.

profiles are fitted with exponential functions (Fig. 11a). Due to the large profile curvature, transportation force markedly decreases downstream, and, even if debris flows occur in the upstream segments, they cannot run down to the downstream gentle slope segments where turbidity flows are dominant (Fig. 11b). When a river flows through an area with a dispersion of altitude less than about 40 m, it has a segment with a channel gradient less than 1/1000 and hazards are caused only by floods (Fig. 11b). In a large river course, the hazards mentioned above occur through the whole river course, but the development of a turbidity flow segment and/or a flood segment also depends on the relief in the drainage basin (Fig. 11c).

Acknowledgements

The authors would like to thank Prof. M. Morisawa, State University of New York at Binghamton, for her kind invitation to the 25th Annual Binghamton Geomorphological Symposium on Natural Hazards and Geomorphology, and for her critical comments on the manuscript.

References

Bagnold, R.A., 1966. An approach to the sediment transport problem from general physics. U.S. Geol. Surv. Prof. Pap., 422-I: 1–37.

Benda, L., 1990. The influence of debris flows on channels and valley floors in the Oregon Coast Range, U.S.A. Earth Surface Process. Landforms, 15: 457–466.

Brierley, G.J. and Hickin, E.J., 1985. The downstream gradation of particle sizes in the Squamish River, British Columbia. Earth Surface Process. Landforms, 10: 597–606.

Brush, L.M., Jr., 1961. Drainage basins, channels and flow characteristics of selected streams in central Pennsylvania. U.S. Geol. Surv. Prof. Pap., 282-f: 145–181.

Egiazaroff, I.V., 1965. Calculation of nonuniform sediment concentrations. Proc. Am. Soc. Civ. Eng., 91(HY4): 225–247.

Hack, J.T., 1957. Studies of longitudinal profiles in Virginia and Maryland. U.S. Geol. Surv. Prof. Pap., 294-B: 45–97.

Hjulstrom, F., 1935. Studies of morphological activity of rivers as illustrated by the River Fyris. Bull. Geol. Inst. Uppsala, 25: 221–527.

Inokuchi, M. and Mezaki, S. 1974. Analysis of the grain size distribution of bed material in alluvial rivers. Geogr. Rev. Jpn., 47: 545–556 (in Japanese, with English abstr.).

Inoue, K., 1992. Downstream change in grain size of river bed sediments and its geomorphological implications in the Kanto Plain, central Japan. Geogr. Rev. Jpn., 65B: 75–89.

Iwagaki, Y., 1956. Fundamental study on critical tractive force. I. Hydro-dynamical study on critical tractive force. Trans. Jpn. Soc. Civ. Eng., 41: 1–21 (in Japanese, with English abstr.).

Knighton, A.D., 1975. Channel gradient in relation to discharge and bed material characteristics. Catena, 2: 263–274.

Knighton, A.D., 1980. Longitudinal changes in the size and sorting of stream-bed material in four English rivers. Geol. Soc. Am. Bull., 91: 55–62.

Knighton, A.D., 1982. Longitudinal changes in the size and shape of stream bed material: evidence of variable transport conditions. Catena, 9: 25–34.

Kobashi, S., Nakayama, M. and Imamura, R., 1980. Movement of surface material. In: A. Takei (Editor), Landslides, Slope Collapses and Debris Flows. Kajima Press, Tokyo, pp. 28–64 (in Japanese)

Ohmori, H., 1978. Relief structure of the Japanese mountains and their stages in geomorphic development. Bull. Dept. Geogr. Univ. Tokyo, 10: 31–85.

Ohmori, H., 1981. Simulation of change of landform from a geomorphometric method. Trans. Jpn. Geomorphol. Union, 2: 95–100.

Ohmori, H., 1983. Characteristics of the erosion rate in the Japanese mountains from the viewpoint of climatic geomorphology. Z. Geomorphol. Suppl., 46: 1–14.

Ohmori, H., 1988. The relationship between the mathematical function types describing longitudinal profiles of rivers and the types of alluvial plains in Japan. In: Research for the Formations of Alluvial Plains and Alluvial Deposits, and Their Relations to Natural Environment during the Late Quaternary in Japan. Dept. Geogr., Nagoya Univ., Nagoya, pp. 6–15 (in Japanese).

Ohmori, H., 1991. Change in the mathematical function type describing the longitudinal profile of a river through an evolutionary process. J. Geol., 99: 97–110.

Ohmori, H., 1993. Changes in the hypsometric curve through mountain building resulting from concurrent tectonics and denudation. Geomorphology, 8: 263–277.

Ohmori, H. and Hirano, M., 1984. Mathematical explanation of some characteristics of altitude distribution of landforms in an equilibrium state. Trans. Jpn. Geomorphol. Union, 5: 293–310.

Ohmori, H. and Saito, K., 1993. Morphological development of longitudinal profiles of rivers in Japan and Taiwan. Bull. Dept. Geogr. Univ. Tokyo, 25: 29–41.

Ohmori, H. and Sohma, H., 1983. Landform classification in mountain region and geomorphic characteristic values. J. Jpn. Cart. Assoc., 21(3): 1–12 (in Japanese, with English abstr.).

Oya, M., 1977. Comparative study of the fluvial plain based on the geomorphological land classification. Geogr. Rev. Jpn., 50: 1–31.

Plumley, W.J., 1948. Black hills terrace gravels: a study in sediment transport. J. Geol., 56: 526–577.

Saito, K., 1988. Alluvial Fans in Japan. Kokon-shoin, Tokyo, 280 pp. (in Japanese).

Schumm, S.A., 1956. Evolution of drainage systems and slopes in badlands at Perth Amboy, New Jersey. Geol. Soc. Am. Bull., 67: 597–646.

Schumm, S.A., 1981. Evolution and response of the fluvial system, sedimentologic implications. SEPM Spec. Publ., 31: 19–29.

Schumm, S.A. and Stevens, M.A., 1973. Abrasion in place: a mechanism for rounding and size reduction of coarse sediments in rivers. Geology, 1: 37–40.

Shepherd, R.G., 1985. Regression analysis of river profiles. J. Geol., 93: 377–384.

Shimazu, H., 1990. Segmentation of mountain rivers based on longitudinal change in gravel size of riverbeds in Tohoku District, northeastern Japan. Geogr. Rev. Jpn., 63A: 487–507 (in Japanese, with English abstr.).

Shimazu, H., 1991. Downstream change in gravel size and bed form along tributaries and gravel supply for the main stream in the Japanese mountains. Geogr. Rev. Jpn., 64A: 569–580 (in Japanese, with English abstr.).

Strahler, A.N., 1952. Hypsometric (area–altitude) analysis of erosional topography. Geol. Soc. Am. Bull., 63: 1117–1141.

Takahashi, T. and Yoshida, H., 1979. Study on the deposition of debris flows. 1. Deposition due to abrupt change of bed slope. Ann. Disaster Prevention Res. Inst. Kyoto Univ., 22(B-2): 315–328 (in Japanese, with English abstr.).

Yatsu, E., 1954. On the longitudinal profile of the graded river. Res. Inst. Nat. Resour. Jpn. Misc. Rep., 33: 15–24, 34: 14–21, 35: 1–6 (in Japanese, with English abstr.).

Yatsu, E., 1955. On the longitudinal profile of the graded river. Am. Geophys. Union Trans., 36: 655–663.

ELSEVIER

Geomorphology 10 (1994) 107–128

GEOMORPHOLOGY

Recent climatic change and catastrophic geomorphic processes in mountain environments

Stephen G. Evans[a], John J. Clague[b]

[a]*Geological Survey of Canada, 601 Booth Street, Ottawa, Ont. K1A 0E8, Canada*
[b]*Geological Survey of Canada, 100 West Pender Street, Vancouver, B.C. V6B 1R8, Canada*

Received March 25, 1994; revised April 22, 1994; accepted April 22, 1994

Abstract

Climatic warming during the last 100–150 years has resulted in a significant glacier ice loss from mountainous areas of the world. Certain natural processes which pose hazards to people and development in these areas have accelerated as a result of this recent deglaciation. These include glacier avalanches, landslides and slope instability caused by glacier debuttressing, and outburst floods from moraine- and glacier-dammed lakes. In addition, changes in sediment and water supply induced by climatic warming and glacier retreat have altered channel and floodplain patterns of rivers draining high mountain ranges.

The perturbation of natural processes operating in mountain environments, caused by recent climatic warming, ranges from tens of decades for moraine-dam failures to hundreds of years or more for landslides. The recognition that climatic change as modest as that of the last century can perturb natural alpine processes has important implications for hazard assessment and future development in mountains. Even so, these effects are probably at least an order of magnitude smaller than those associated with late Pleistocene deglaciation ca. 15,000 to 10,000 years ago.

1. Introduction

Climatic warming during the last 100–150 years has resulted in widespread destabilisation of many mountain geomorphic systems and accelerated certain catastrophic processes, largely as a result of dramatic glacier ice loss (Evans and Clague, 1993). These processes include glacier avalanches, landslides and slope instability caused by glacier debuttressing, and catastrophic outburst floods from moraine- and glacier-dammed lakes. Large floods, debris flows, and landslides have altered the supply of sediment and water to streams draining glacier forelands, in some cases inducing major changes in channel and floodplain patterns.

The catastrophic events under discussion are initiated within the glacier foreland, defined by the limits of the Little Ice Age[1] advance, and their effects are often restricted to this zone. Frequently, however, they impact on sites beyond the Little Ice Age limit and their effects may be felt far downstream, as in the case of highly mobile rock avalanches and outburst floods. Such catastrophic processes pose significant hazards to human habitation and economic infrastructural elements in mountain areas, and have been the cause of some of the worst natural disasters of this century. In the last 150 years, the total loss of life from glacier-related catastrophic events, many of which are linked to climatic warming, has been in excess of 30,000; damage to the economic infrastructure of the affected

[1]The Little Ice Age is the most recent period of cooler climate and expanded alpine glaciers during the Holocene (Matthes, 1939); it ended in the nineteenth century.

SSDI 0169-555X(94)00034-O

regions probably has been more than one billion dollars (Evans and Clague, 1993).

A major objective of this paper is to review the most important catastrophic geomorphic processes linked to climate-induced glacier ice loss, with examples mainly from the mountains of western Canada. Further, we explore how recent glacier ice loss has affected the frequency of catastrophic events and stability thresholds, and we speculate on the amplitude and wavelength of perturbations to mountain geomorphic systems caused by climatic warming.

2. Climatic change and glacier ice loss

Hansen and Lebedeff (1987) have identified the spatial and temporal components of the change in mean atmospheric temperature between 1880 and 1985. In general, there was pronounced warming from 1880 until about 1940, followed by cooling from 1940 until 1965, and renewed warming from 1965 to 1985. The net average increase in global temperature over the entire period is about 1°C, but there is significant regional variation. For example, parts of the Peruvian Andes experienced temperature increases of 3–5°C between 1880 and 1940, a period, as noted below, of great glacier ice loss and numerous related catastrophic events. In contrast, in the Canadian Cordillera of western Canada, the net temperature increase from 1880 to 1985 was about 1°C; warming of 1–2°C between 1880 and 1940 and 1–3°C between 1965 and 1985 was substantially offset by marked cooling between 1940 and 1965.

While the relationship between climate and glacier mass balance is complex, Oerlemans and Fortuin (1992), among others, have demonstrated that alpine glaciers are very sensitive to increases in air temperature, a response that is not offset by increases in precipitation. In the present context, the most important effect of recent warming is the substantial loss of glacier ice that has occurred in all mountainous regions of the world, including the Canadian Cordillera (e.g. Mathews, 1951). Most glaciers have thinned and their margins have retreated since the end of the Little Ice Age (Intergovernmental Panel on Climatic Change, 1990). This is manifested in glacierised mountain landscapes by well developed end and lateral moraines that lie outside present glacier margins, and by well defined

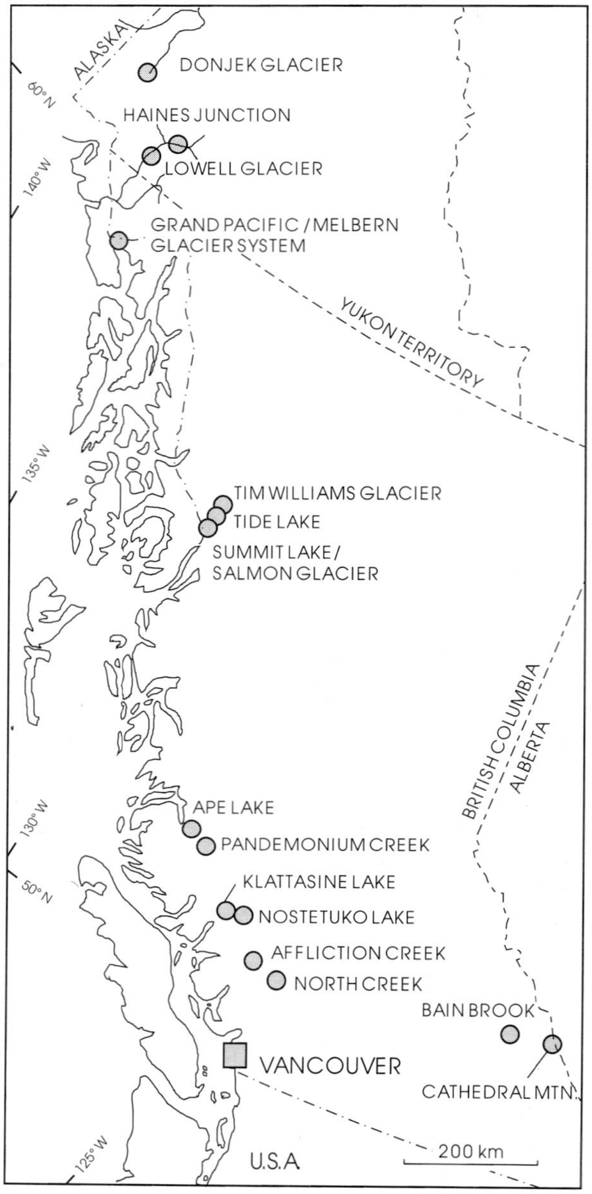

Fig. 1. Map showing localities in the Canadian Cordillera mentioned in the text.

trimlines that mark the upper limit of Little Ice Age glaciers.

In some parts of western Canada, the magnitude of the ice loss has been extraordinary. In the southern St. Elias Mountains, for example, the Grand Pacific/Melbern Glacier system (Fig. 1) has lost over 300 km^3 of ice, or more than 50% of its mass, in the last 200 years in the catastrophic deglaciation of the region (Clague and Evans, 1993). The scale and pattern of historic deglaciation in this area are comparable to those of late Pleistocene time.

3. Glacier avalanches

A glacier avalanche (e.g. Hanke, 1966; Röthlisberger, 1978) is a sudden, rapid, downslope movement of ice following its detachment from the terminus of a glacier. Conditions favourable for ice avalanching are created when the terminus of a glacier retreats up a steep slope, the Type I starting zone of Alean (1985). Glacier avalanches are common in mountainous areas throughout the world and are potentially very hazardous. For example, at least 124 people were killed by glacier avalanches between 1901 and 1983 in the Swiss Alps alone (Alean, 1985), including 88 people in the Allalin Glacier avalanche in August 1965 (Fig. 2; Röthlisberger, 1978). Glacier avalanches usually occur during summer and result from a destruction of tensile strength in the ice mass through progressive fragmentation associated with crevasse development, melting of parts of the glacier that may be frozen to the substrate, and reduction of frictional resistance at the ice–rock interface due to increased water pressures (Röthlisberger, 1978).

Glacier avalanches are not well documented in the Canadian Cordillera, largely because of the short-lived nature of the deposits and the low population density of the region. However, they appear to be common, as their effects are captured on government aerial photographs. Photographs taken in 1975 of Bain Brook in Glacier National Park, British Columbia (Fig. 1), for example, show the deposits of a recent glacier avalanche (Fig. 3). The toe of the glacier below Virtue Mountain became destabilised as it retreated up a steep rock slope. A large mass of ice ($>10^5$ m^3) broke away from the toe of the glacier along a front of 500 m and travelled 1.75 km across the Little Ice Age glacier foreland. The ice fell 670 m (from 2130 to 1460 m) over this distance, corresponding to a fahrboschung of 21°. Little evidence of the event was visible when the site was visited during 1984.

4. Landslides and debris flows

Slopes adjacent to glaciers that have significantly thinned and retreated since the Little Ice Age are particularly prone to landslides (Fig. 4). Glacial erosion and oversteepening of the slopes, in combination with subsequent debuttressing due to glacier retreat, have caused instability, evidenced by progressive mountain slope deformation, rock avalanches, and other landslides.

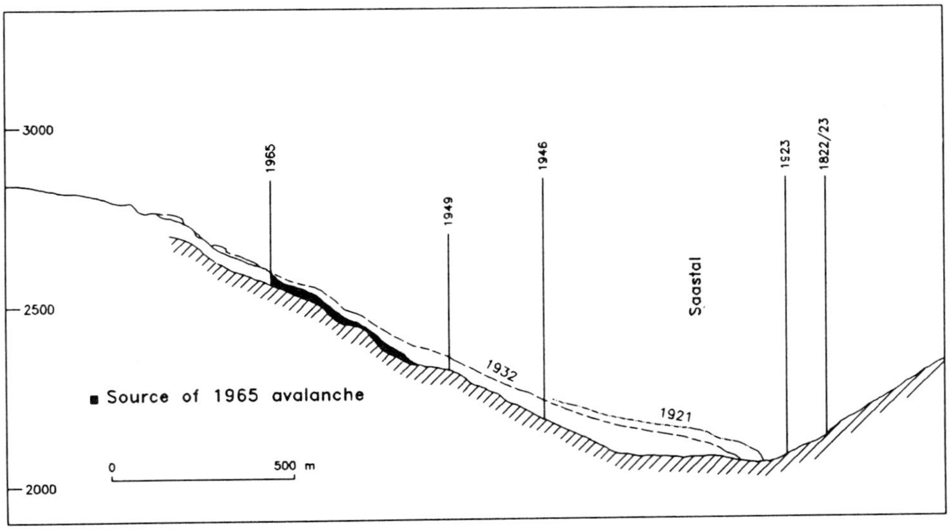

Fig. 2. Profile showing historical retreat of Allalin Glacier, Switzerland, and source of the 1965 glacier avalanche. The avalanche killed 88 construction workers at the Mattmark Dam construction site in the Saas valley. (Modified from Röthlisberger, 1978, fig. 11.)

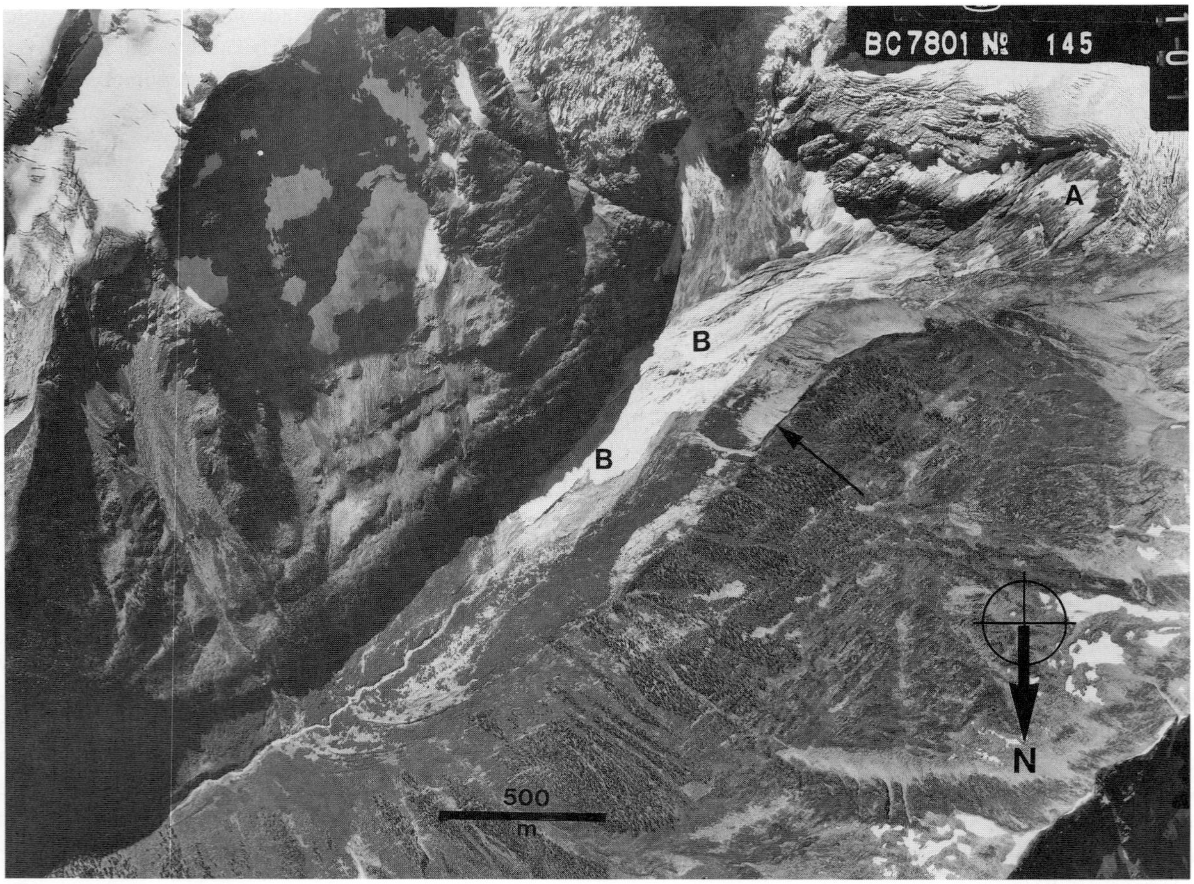

Fig. 3. Aerial photograph (BC 7801;145) of a glacier avalanche at Bain Brook, Glacier National Park, British Columbia, taken in September 1975. *A*: source area; *B*: avalanche debris. Note the Little Ice Age trimlines (arrowed).

4.1. Rock avalanches

Particularly good examples of landslides caused by glacier oversteepening and recent debuttressing are the 1992 rock avalanches ($5–10\times10^6$ m^3) on the slopes of Mount Fletcher above Maud Glacier in the Southern Alps of New Zealand (McSaveney, 1992, 1993). Maud Glacier had thinned approximately 250 m since the Little Ice Age maximum. This thinning, which accelerated after the middle of the present century, exposed a steeper toe slope through which failure occurred (Fig. 5).

Two highly destructive landslides from the north peak of Nevados Huascaran in the Cordillera Blanca of Peru in 1962 and 1970 (Morales, 1966; Ericksen et al., 1970; Plafker and Ericksen, 1978) were probably caused, in part, by rapid glacier thinning and retreat from Little Ice Age limits. Between 1886 and 1942, a period of marked local warming in the region (Hansen and Lebedeff, 1987), the firn limit on Huascaran rose in elevation from 4320 m to 5100 m, including a 500 m rise between 1932 and 1942 (Broggi, 1943). In addition, during the same period, the canopy of ice on the summit ridge of Huascaran thinned dramatically (Clapperton, 1983).

In 1962, approximately 13×10^6 m^3 of rock and glacier ice detached from the north peak of Huascaran and travelled 16 km down the Rio Shacsa at an average velocity of 47 m/s. The landslide overwhelmed several towns and villages and killed about 4000 people, mostly in the town of Ranrahirca. A debris flow continued downstream in the Rio Santa valley to at least Huallanca, 52 km from Ranrahirca, destroying bridges and sections of road (Morales, 1966).

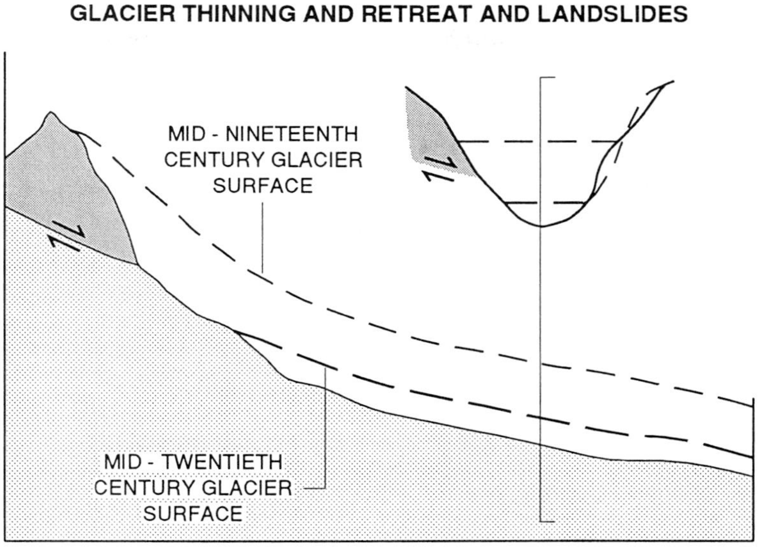

Fig. 4. Schematic diagram showing the relation between glacier thinning/retreat and landslides. (Evans and Clague, 1993, fig. 4.)

In 1970, 50–100 × 10^6 m^3 of rock and ice, undermined by the 1962 event, fell away from a similar position on Huascaran's north peak during a large (M7.7) earthquake centred 130 km to the west. The landslide exhibited spectacular mobility: in the upper part of the path, boulders were hurled into ballistic trajectory and impacted 4 km from their launch positions. The debris travelled a vertical distance of about 4200 m over a horizontal distance of 16 km at a mean velocity of 75 m/s. The landslide caused an estimated 18,000 deaths, mainly in the town of Yungay. It blocked the Rio Santa, and continued downstream as a debris flow at an average velocity of 10 m/s to at least Huallanca. There, residents described the flow as a dark-coloured turbulent wave containing abundant blocks of ice (Ericksen et al. 1970). The debris flow caused extensive damage to transportation and communication facilities downstream of Yungay and deposited up to 10 m of debris in the channel of the Rio Santa (Browning, 1973).

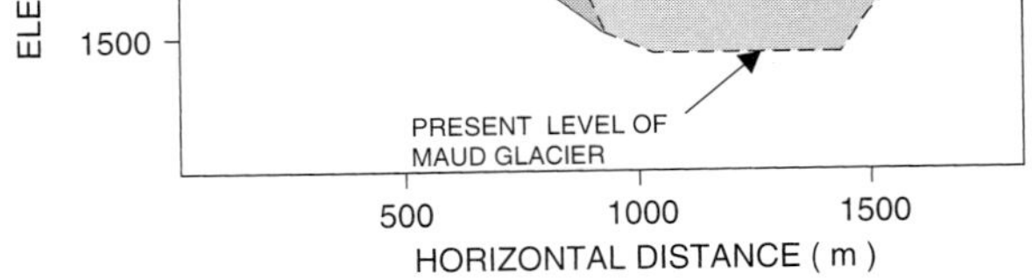

Fig. 5. Profile of the Mount Fletcher rock avalanches, New Zealand. (Modified from McSaveney, 1992, 1993.)

Landslides caused by glacier downwasting and retreat are common on steep slopes adjacent to glaciers in western North America (Evans and Clague, 1988). For example, at least three, large, twentieth-century rock avalanches at Mt. Rainier, Washington, occurred on valley and cirque walls that were partially supported by glacier ice during the Little Ice Age (O'Connor and Costa, 1992), including the Little Tahoma Peak event in 1963. Of the 30 known, large ($>1 \times 10^6$ m^3), historic rock avalanches in the Canadian Cordillera, 16 have occurred on glacially debuttressed slopes (Figs. 1, 6, 7). Field investigations have shown that detachment surfaces of many of these landslides intersect the slopes below Little Ice Age trimlines and were thus exposed during recent glacier retreat (Evans and Clague, 1988, 1990, 1993; Evans et al., 1989). Rock avalanches in this environment may run-out on the glacier without travelling beyond its margins (e.g. Tim Williams Glacier, Fig. 6; Evans and Clague, 1990), be contained within the glacier foreland (e.g. North Creek,

Fig. 6. Oblique aerial view of the debris from the 1956 Tim Williams Glacier rock avalanche, Coast Mountains, British Columbia, taken in 1989. In the period 1956 to 1989 the debris had been transported 1 kilometre down glacier (Evans and Clague, 1990). The source of the rock avalanche is arrowed and the conspicuous Little Ice Age trimline is evident in the centre-left foreground of the photograph.

Fig. 7; Evans and Clague, 1988), or travel a considerable distance beyond the Little Ice Age limit (e.g. Pandemonium Creek, Evans et al., 1989).

Rock avalanches which cross glaciers and glacier forelands are highly mobile and travel farther than rock avalanches of the same volume occurring without interaction with glaciers (Evans and Clague, 1988). This is thought to be due to the low friction of the debris–ice interface, funnelling of flow by Little Ice Age lateral moraines (Evans et al. 1989), and entrainment of significant volumes of snow and ice, which when melted, contribute to the fluidity of the debris (Ericksen et al., 1970).

4.2. Deep-seated slope deformation

Glacier thinning and retreat may also cause non-catastrophic slope deformation, manifested by cracking, subsidence at the top of the slope, and bulging at the toe. Spectacular examples have been reported from the St. Elias Mountains of British Columbia (Evans and Clague, 1993) and Alaska (Radbruch-Hall, 1978), regions that have experienced substantial losses of glacier ice in the twentieth century. At Melbern Glacier (Fig. 1), for example, a 400–600 m lowering of the glacier surface since the Little Ice Age maximum has debuttressed adjacent mountain slopes, causing extensive, non-catastrophic slope deformation (Evans and Clague, 1993; Clague and Evans, 1993).

Although few mountain deformation sites in the Canadian Cordillera have been studied in detail, Bovis

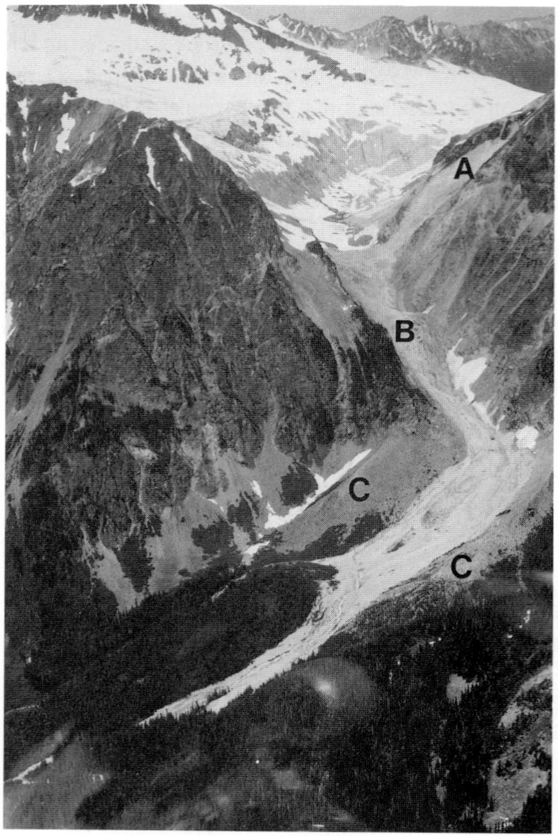

Fig. 7. North Creek rock avalanche, Coast Mountains, British Columbia. A = source area; B = glacier surface; C = Neoglacial lateral moraine. The rock avalanche occurred in 1986.

Fig. 8. Oblique aerial view of tension cracks associated with mountain slope deformation above Affliction Glacier, Coast Mountains, British Columbia.

(1982, 1990) has carried out detailed investigations at Affliction Creek in the southern Coast Mountains of British Columbia (Fig. 1). Tension cracks, uphill- and downhill-facing scarps, grabens, and collapse pits extend for a distance of 1.3 km along Affliction Creek, where local relief is 400 m. The deformation has affected surface Quaternary volcanic rocks as well as underlying quartz monzonite, and, in addition, tension cracks cut a Little Ice Age moraine (Fig. 8). Bovis (1990) has demonstrated that many of the slope movement features developed as a result of debuttressing caused by more than 100 m of downwasting of Affliction Glacier since the Little Ice Age maximum.

The failure geometry at Affliction Creek was determined by Bovis (1990) from field observations and ground movement vectors. We have analysed this failure geometry, using the Morgenstern–Price method, to document the destabilising effect of glacier debuttressing (Fig. 9). Assuming that the existing slope is in limiting equilibrium (i.e., factor of safety = 1.0), the slope, failure surface, and water table configuration shown in Fig. 9 give an effective angle of friction of 32°. Further assuming that cohesion is zero, the analysis shows that the factor of safety against sliding at Affliction Creek, decreases with ice level; thinning of Affliction Glacier since the Little Ice Age has resulted in a decrease of 15% in the factor of safety. Downwasting below elevation 1580 m (Fig. 9), however, does not further affect the stability of the slope, suggesting that the failure threshold is controlled by the location of preexisting slip surfaces in the valley-side slope.

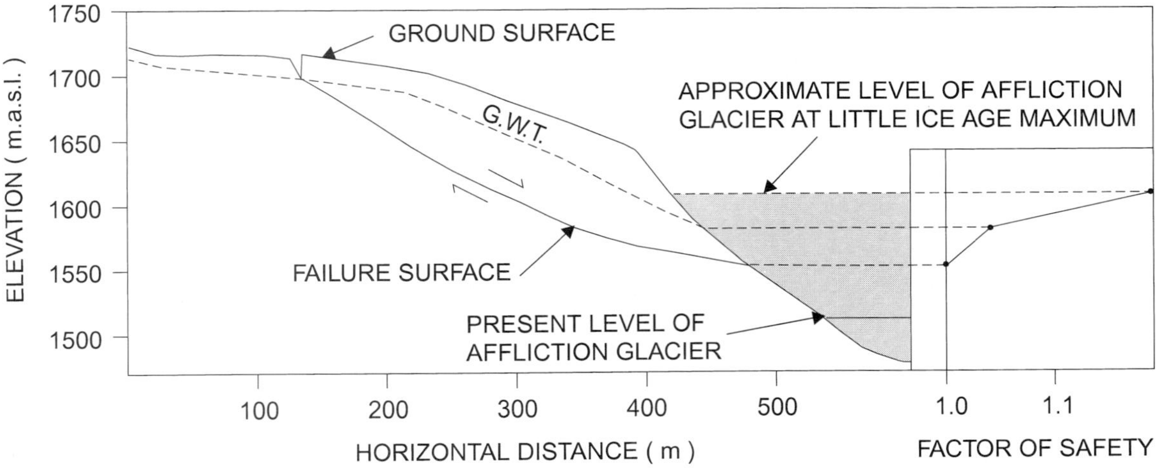

Fig. 9. Cross-section of slope movement at Affliction Creek (modified from Bovis, 1990, fig. 7) and results of stability analysis.

4.3. Debris flows

Debris flow activity has increased in many areas due to recent glacial retreat. In the Swiss Alps, for example, numerous large debris flows were triggered by intense rainfall during the summer of 1987 (Haeberli and Naef, 1988; Zimmerman and Haeberli, 1992). In a large number of cases, the source of the debris is Little Ice Age glacial deposits exposed by recent retreat. Melting of remnant masses of dead ice beneath the drift may also have been a factor in initiating some of the debris flows. Similar explanations have been offered by Jordan (1987) to explain some recent debris flows in the southern Coast Mountains of British Columbia.

5. Outbursts from moraine-dammed lakes

Moraine-dammed lakes are found in high mountains close to existing glaciers (Fig. 10). Most of the moraines impounding these lakes were built during the Little Ice Age. Since then, glaciers have retreated, leaving behind closed basins filled with water (Fig. 11).

Many moraine-dammed lakes have drained suddenly, producing floods and debris flows that have caused considerable damage. Failure commonly occurs when waves generated by a rockfall or glacier avalanche overtop and rapidly incise the moraine dam (Evans, 1987; Clague and Evans, 1994). Melting of ice cores, piping, and earthquakes are other possible failure mechanisms.

Four large moraine dam failures occurred in the Cordillera Blanca of Peru between 1938 and 1950, following the major glacier ice losses of the 1920s and 1930s (Heim, 1948; Lliboutry et al., 1977). The most devastating was the 1941 emptying of moraine-dammed Lake Palcacocha. The moraine was rapidly breached, and the discharged water entered another lake downstream, causing it to drain as well. The event triggered a debris flow (8×10^6 m^3) that destroyed one-third of the city of Huaraz, killing more than 6000 people. Other recent large outbursts from moraine-dammed lakes, which have resulted from the breaching of Little Ice Age moraines, have been reported in other parts of the Andes (e.g. Hauser, 1993), the Himalayas (e.g. Gansser, 1983; Vuichard and Zimmerman, 1987; Ding and Liu, 1992) and the mountains of Central Asia (Niyazov and Degovets, 1975; Yesenov and Degovets, 1979; Popov, 1990). The draining of Laguna del Cerro Largo, Chile, in 1989, reported by Hauser (1993), involved the release of 229×10^6 m^3 of water, and is the largest moraine dam outburst documented in the literature.

Fig. 10. Aerial view of "Iceberg Lake", an informally named moraine-dammed lake in the southern Coast Mountains, British Columbia. The large moraine impounding the lake was constructed during the Little Ice Age. Note other moraines exposed by recent glacier retreat in foreground and distant background beyond Iceberg Lake.

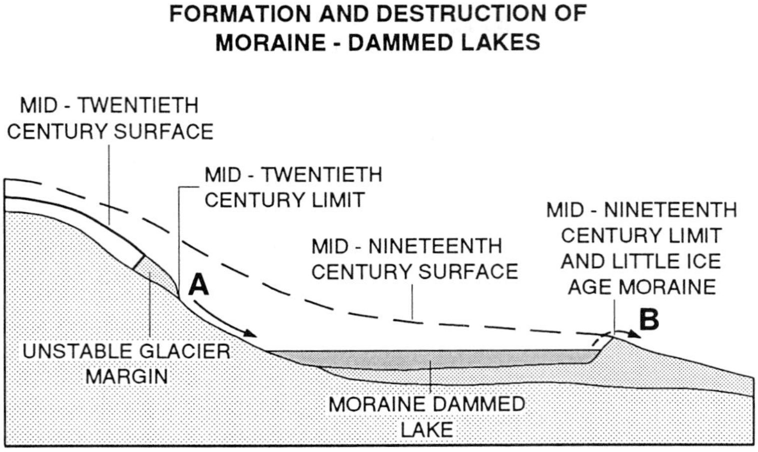

Fig. 11. Schematic diagram showing the relation between glacial thinning and retreat, and the formation and destruction of moraine-dammed lakes. Avalanching of unstable glacier ice (*A*) into the lake generates waves that overtop moraine dam (*B*) initiating catastrophic breaching. (Evans and Clague, 1993, fig. 8.)

5.1. Setting and character

Moraine-dammed lakes typically occur in cirques and steep-walled valleys behind end moraines or at the mouths of tributary valleys blocked by lateral moraines. Most, although not all, are near or above the treeline and may be close to steep rock slopes that are prone to frequent rockfalls and snow and ice avalanches (Fig. 11).

Moraine dams may consist of a single moraine or nested moraines deposited during several Neoglacial advances. Dramatic examples from the Cordillera Blanca, Peru, are illustrated by Heim (1948). Most moraine dams are steep-sided (up to ca. 40°) and have relatively low width-to-height ratios (generally much lower than those of landslide dams; Costa and Schuster, 1988). Heights of a few tens of metres are common, and some moraine dams are more than 100 m high.

The dams comprise poorly sorted, stratified to massive sediment, including: (1) blocky and bouldery deposits with a matrix of sand and gravel, and (2) silty and sandy diamicton. Stratification, where present, commonly dips away from the former ice-contact face towards the distal edge of the moraine. Much of the sediment was deposited at the glacier snout by the dumping of debris carried on and in ice. Some moraines are cored by ice and may have only a thin cover of sediment.

5.2. Historic failures in the Canadian Cordillera

There have been at least eight failures of moraine dams in western Canada since 1945; four in the Coast Mountains, one in the Saint Elias Mountains, and three in the Columbia Mountains (Fig. 1; Evans, 1987; Ryder, 1991; Clague and Evans, 1992, 1994; Clague and Mathews, 1992).

The most spectacular of these is the Nostetuko Lake outburst, which occurred in an uninhabited area of the Coast Mountains in July 1983, 230 km north of Vancouver, British Columbia (Blown and Church, 1985). Nostetuko Lake, located at the head of Nostetuko River, is dammed by a Little Ice Age end moraine (Fig. 12). This moraine was deposited by Cumberland Glacier which has since receded and now terminates on a cliff above Nostetuko Lake. On July 19, 1983, part of the toe of Cumberland Glacier broke away and crashed into the lake. Waves generated by the impact of the glacier avalanche moved down the lake, and overtopped and rapidly incised the moraine at the opposite end. Within four hours, the moraine had been breached to a depth of almost 40 m and about 6×10^6 m^3 of water had been released (Fig. 13). This produced a destructive flood that swept 115 km down Nostetuko and Homathko valleys to the sea (Fig. 14). Over 1×10^6 m^3 of sediment was eroded from the moraine; most of this was deposited as a large fan on top of a former meadow directly downstream from the dam (Fig. 13). Farther

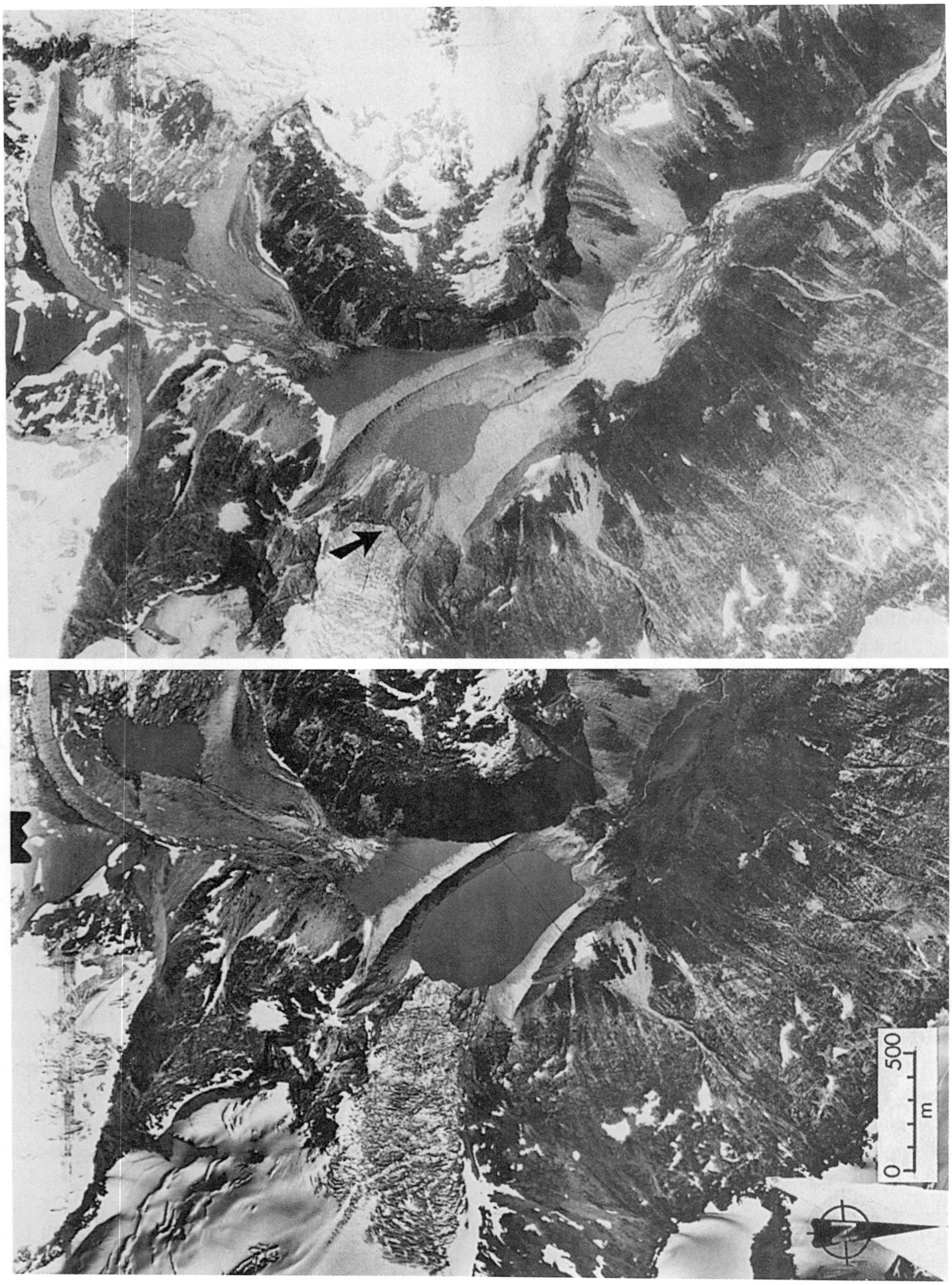

Fig. 12. Aerial photographs of Nostetuko Lake before (left, BC79069-190) and after (right, BC86048-147) the outburst flood of July 19, 1983. Failure occurred when toe of Cumberland Glacier (arrow) collapsed into the lake and generated waves that overtopped the moraine.

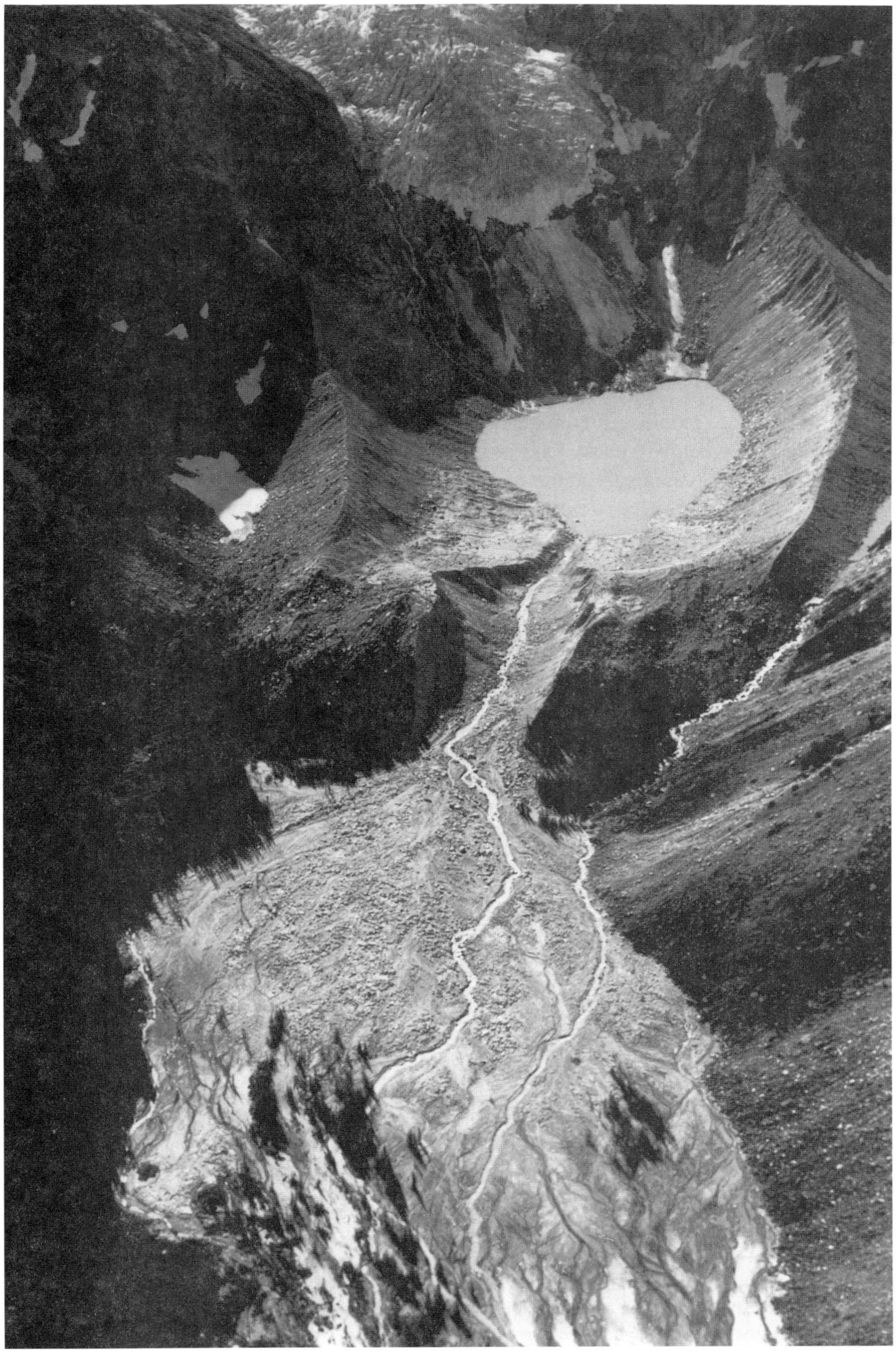

Fig. 13. For caption see overleaf.

downstream, floodwaters extensively eroded unconsolidated deposits in Nostetuko valley, damaged large tracts of forest, and left substantial piles of wood debris and coarse sediment on bars and channel margins. Nostetuko River has not yet recovered its pre-flood channel planform and morphology, suggesting that such rare flood events may affect streams for decades.

Four other historic moraine dam failures produced debris flows rather than water floods. The largest, in Klattasine valley 35 km west of Nostetuko Lake, occurred sometime between June 1971 and September 1973 (Clague et al., 1985). It was triggered by the sudden release of about 1.7×10^6 m^3 of water from moraine-dammed Klattasine Lake. The escaping waters breached the moraine, mobilized large quantities of sediment in the valley below the lake, and generated a debris flow, with an estimated volume of 2–4×10^6m^3, that travelled in one or more surges 8 km to the mouth of Klattasine Creek (Fig. 15). Here, the flow deposited a sheet of coarse bouldery debris up to 20 m thick that temporarily stemmed the flow of Homathko River.

6. Jökulhlaups

Many glacier-dammed lakes are unstable and, on occasion, drain suddenly to produce large floods, termed jökulhlaups (an Icelandic term meaning ''glacier burst'') (Fig. 16). Such floods have caused loss of life and severe property damage in many parts of the world, including Alaska, Austria, France, Iceland, India, Italy, Norway, Pakistan, Argentina, Peru, and Switzerland (see Eisbacher and Clague, 1984, and Costa and Schuster, 1988, for references). Some of the world's largest historic jökulhlaups occurred in the Karakoram Himalayas in the first half of this century (Mason, 1930; Hewitt, 1982). The damming of the Upper Shyok River by Chong Kumdan Glacier formed a lake with an estimated volume of 1.4×10^9 m^3. A sudden outburst from this lake occurred in 1929, and the flood wave travelled down the Shyok River into the Indus River. A rise in the level of the Indus of 8 m was measured at Attock, 740 km downstream from the ice dam.

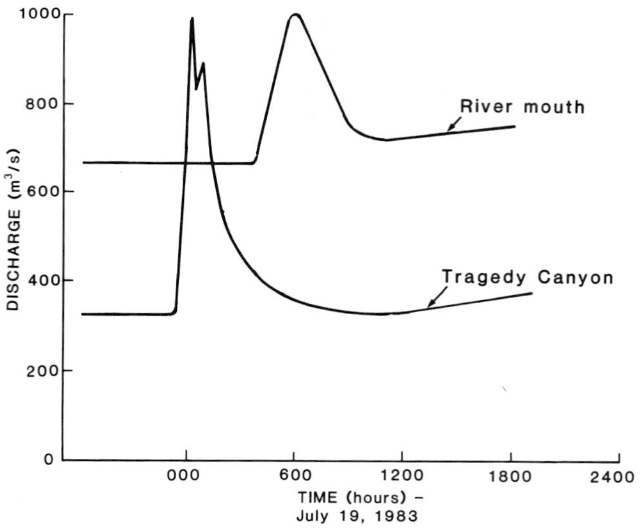

Fig. 14. Hydrographs showing the flood that resulted from the failure of the moraine dam at Nostetuko Lake in July 1983. The lower hydrograph is from a site 67 km downstream from the dam and shows an increase in discharge from 330 m/s to over 900 m/s in one hour; the upper hydrograph, from a site a further 45 km downvalley, shows a somewhat attenuated flood wave. (Clague and Evans, 1994, fig. 17.)

Fig. 13. Oblique view of Nostetuko Lake, a moraine-dammed lake in the Coast Mountains of British Columbia which drained catastrophically in 1983. The bulky, sharp-crested Little Ice Age moraine complex formed during the nineteenth century. Approximately 6×10^6 m^3 of water was released from the lake. A glacier avalanche from the Cumberland Glacier, visible above the remnant lake, probably initiated the breaching event. About 1.5×10^6 m^3 of material was eroded from the moraine dam during the outburst, much of which was deposited in a debris fan immediately downstream.

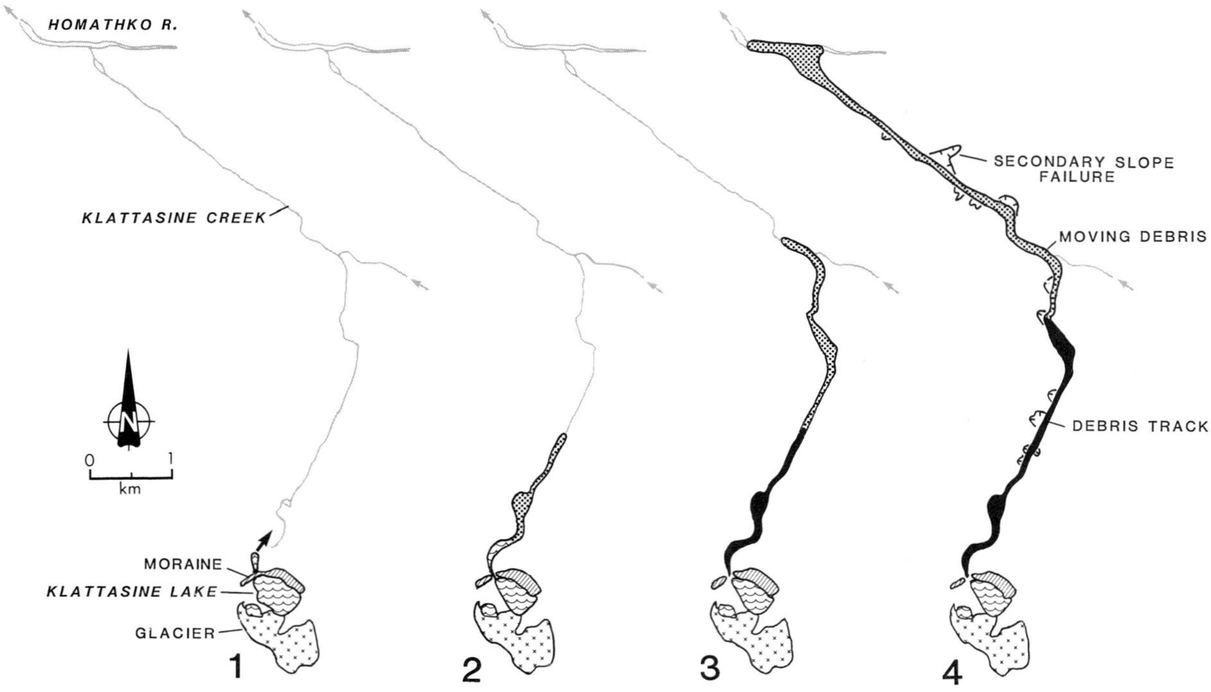

Fig. 15. Stages in the evolution of the Klattasine Creek debris flow. (1) The moraine damming Klattasine Lake begins to fail; (2) escaping waters mobilize large quantities of sediment, initiating a debris flow; (3) the debris flow rapidly moves downvalley and entrains additional sediment; (4) the front of the debris flow reaches Homathko River and temporarily blocks it; secondary landslides occur in Klattasine valley. Stippled area: moving debris; black area: wake of debris flow. (Clague and Evans, 1994, fig. 21.)

6.1. Setting

Glacier-dammed lakes (Fig. 17) are found mainly at the margins of valley glaciers, although some occur within or beneath cirque and valley glaciers and mountain ice caps. Some of the largest lakes are situated in main valleys adjacent to the snouts of tributary glaciers, at the mouths of tributary valleys blocked by trunk glaciers, and in low-lying glacier-marginal areas (Post and Mayo, 1971; Blachut and Ballantyne, 1976).

6.2. Prehistoric and historic failures in the Canadian Cordillera

An extremely large, self-dumping lake formed many times during the Little Ice Age when Lowell Glacier advanced or surged across Alsek River in the Saint Elias Mountains (Fig. 1; Clague and Rampton, 1982). Giant dunes on the floor of Alsek valley (Clague and Rampton, 1982; Schmok and Clarke, 1989) provide evidence of large jökulhlaups from Lake Alsek during recent centuries. Calculations using a paleohydrologic simulation model (Clarke, 1982; Clarke et al., 1984) indicate that the peak discharges of floods from Lake Alsek during the mid-nineteenth century were roughly 3×10^4 m^3/s (Clarke, 1989), which is about one-third of the mean flow of Amazon River at its mouth. A jökulhlaup associated with an earlier phase of the lake discharged approximately 30 km^3 of water and had a peak discharge of about 4.7×10^5 m^3/s (Clarke, 1989); this is perhaps the largest flood of the last 10,000 years on Earth. Although Lake Alsek no longer exists, it would reform if Lowell Glacier were to surge about 1 km. A major blockage of Alsek River might inundate the town of Haines Junction and sections of the Haines and Alaska highways.

Similar, although smaller, jökulhlaups occurred in the 1800s in Donjek River valley, 100 km north of Lowell Glacier, after Donjek Glacier advanced to impound a lake (Clarke and Mathews, 1981). Likewise, Tide Lake (Fig. 1) in the northern Coast Mountains produced several large jökulhlaups in the nineteenth and early twentieth centuries (Clague and Mathews, 1992). Neither of these lakes exists today, but they, like Lake Alsek, would form again if the glaciers that formerly dammed them readvanced. Jökulhlaups from a renewed Lake Donjek would threaten the Alaska Highway (Fig. 18).

There have been many jökulhlaups in western Canada during the historical period (Clague and Evans, 1994). Studies of some of these have laid the foundation for much of our present understanding of the physics of outburst floods. Jökulhlaups from Summit Lake (Fig. 17) in the northern Coast Mountains, menaced the transportation corridor between the Granduc mine site and Hyder, Alaska, and motivated the research of W.H. Mathews (1959, 1964, 1965, 1973) and his student R. Gilbert (1969, 1971, 1972). Mathews (1973) reconstructed the 1965 and 1967 Summit Lake jökulhlaups from observations of falling lake levels, lake areas, and surface outflow, plus estimates of inflow. He showed that outflow discharge increased in proportion to a power (1.5) of the volume of water previously lost from the lake (Fig. 19). From this and from calculations of effective tunnel diameter following the jökulhlaups, Mathews concluded that the tunnels developed by melting due to heat transfer from water, rather than mechanical erosion. The exponential increase in discharge with time indicates a positive feedback loop in outflow from the lake: progressively larger discharges cause more heat production and melting of ice, leading to a larger tunnel, further increases in discharge, more heat, and so on until the water supply is exhausted.

Mathews and Clague (1993) summarized the record of jökulhlaups from Summit Lake and noted the tendency towards smaller and earlier floods in recent years. This suggests a progressive development with time of tunnels within or beneath Salmon Glacier and a corresponding thinning and weakening of the dam (Fisher, 1973). A critical threshold in the thinning of the glacier was reached in 1961 when the seasonal enlargement of a small tunnel exceeded the rate of closure due to flow and the overburden of the ice; this resulted in the first jökulhlaup.

Other examples of recent jökulhlaups that have been extensively studied are those from Ape Lake in the southern Coast Mountains of British Columbia (Jones et al., 1985; Desloges et al., 1989). The first historical outburst flood from this lake occurred in October 1984 when a subglacial tunnel opened in the snout of Fyles Glacier. Following tunnel closure, the lake refilled in 150 days, but it drained again in August 1986. High

Fig. 16. Jökulhlaup from Summit Lake, at Ninemile, Alaska, September 17, 1967. The discharge is almost 3000 m^3/s, which is 100 times larger than the mean discharge of the stream at this site. The steel bridge was washed away a few hours after this photograph was taken. (Photo by J.J. Plummer.)

Fig. 17. Oblique photo of Summit Lake, British Columbia, looking down the Salmon Valley. Salmon Glacier forms the ice dam which impounds the lake. Summit Lake has drained annually since 1970.

discharges during these floods (up to about 1600 m^3/s) damaged forestry roads, bridges, a logging camp, and an airstrip, and caused widespread channel and floodplain erosion. Retreat of Fyles Glacier after 1986 allowed drainage around the north edge of the ice dam and prevented the lake from refilling. No additional jökulhlaups are expected from this basin unless there is a significant readvance of Fyles Glacier. The presence of undamaged trees as old as 300 years on the floodplain below Fyles Glacier before the first flood indicates that the two recent outburst events are unique in the Little Ice Age history of the basin.

The downstream effects of jökulhlaups can be considerable. The channel pattern of Salmon River, for example, is much less stable today than it was prior to the first jökulhlaups from Summit Lake in 1961. Similarly, the Noeick River is still adjusting to the two jökulhlaups from Ape Lake in 1984 and 1986 (Desloges and Church, 1992). Recovery of the Noeick floodplain and channel to a normal (climatic) regime will take on the order of decades to a century, assuming that there are no additional jökulhlaups.

Some jökulhlaups have triggered debris flows (Clague and Evans, 1994). The best documented and most destructive are the jökulhlaups from Cathedral Glacier in the southern Rocky Mountains of British Columbia (Jackson, 1979; Jackson et al., 1989) (Fig. 20). These have resulted from the repeated sudden draining of a small ephemeral lake on the south side of Cathedral Glacier. Several times during this century, water from this lake, augmented perhaps by water stored within the glacier, emptied into a steep ravine and mobilized large volumes of glacial and colluvial sediments. This produced debris flows that travelled up to 3 km and blocked the Canadian Pacific Railway (CPR) mainline and the Trans-Canada Highway in Kicking Horse valley. The volumes of the largest debris flows are about 100,000 m^3. Peak discharges and veloc-

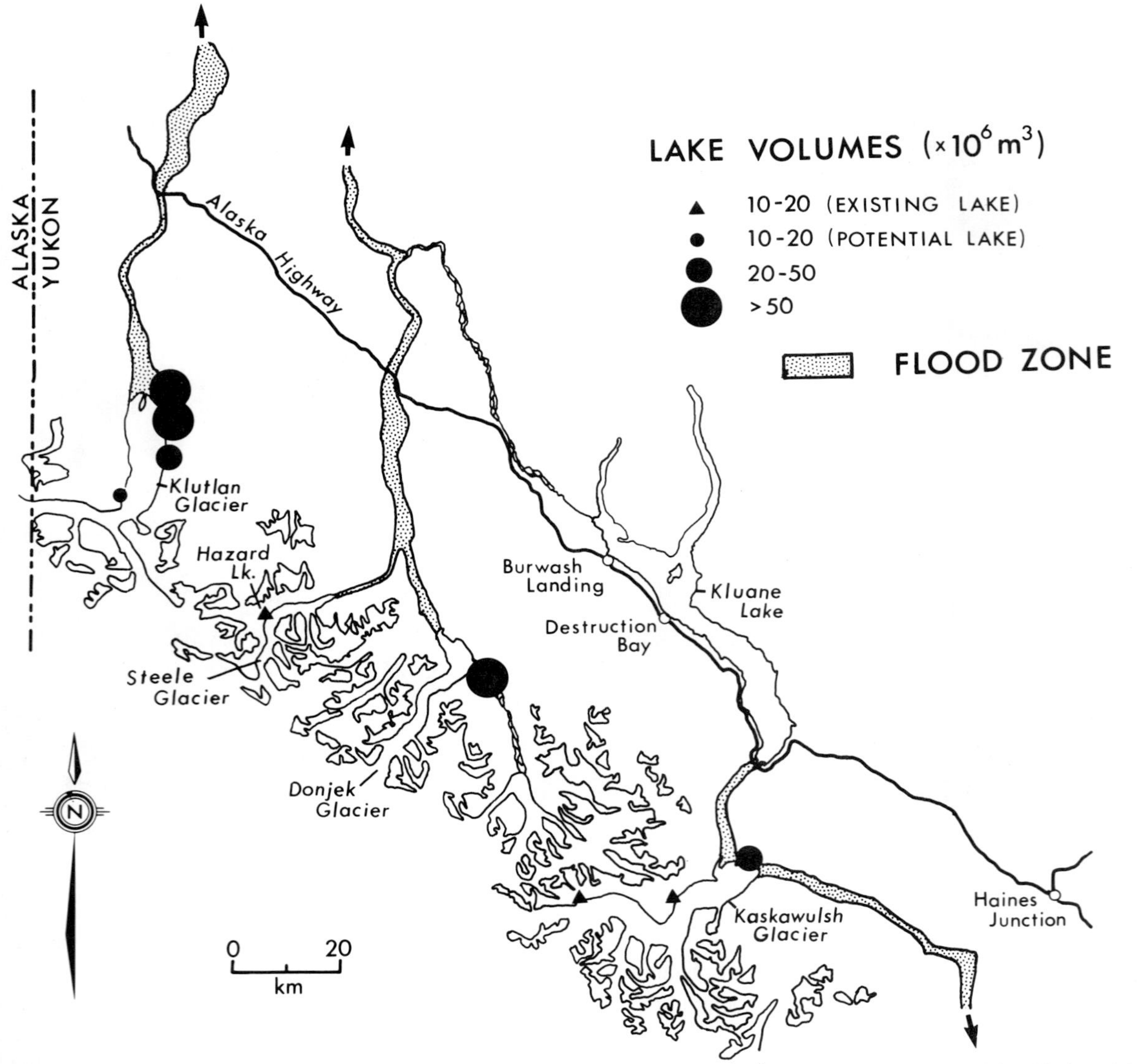

Fig. 18. Jökulhlaup zones and existing and potential glacier-dammed lakes larger than 10×10^6 m^3 in the White, Donjek, and Slims–Kaskawulsh basins, Saint Elias Mountains, Yukon Territory. Potential lakes are basins in which water would be ponded in the event of a glacier advance. Basins near the termini of Klutlan and Kaskawulsh glaciers have not held water for several centuries and are unlikely to do so in the foreseeable future. (Modified from Clague, 1982, fig. 9.)

ities of the flows just above the head of the fan crossed by the Trans-Canada Highway and CPR mainline are 210 m^3/s and 5.5 m/s, respectively. Historical debris flow activity began in 1925 when retreat of Cathedral Glacier left unstable accumulations of sediment supported by melting and collapsing stagnant ice. This probably was accompanied by changes in the morphology and internal hydrology of the glacier, which led to the periodic storage and catastrophic release of water. Debris flows increased in frequency from 1925 to 1985, when CPR began pumping water from Cathedral Glacier. Since then, no jökulhlaups or significant debris flows have occurred in this area.

6.3. The jökulhlaup cycle

Some formerly stable, glacier-dammed lakes in western Canada have gone through a cycle of jöku-

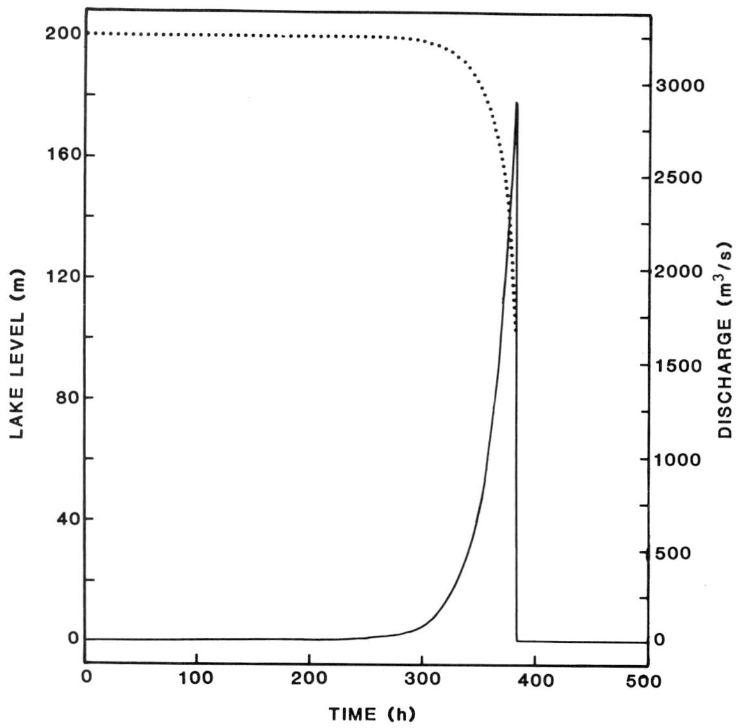

Fig. 19. Variations in lake level (dotted line) and discharge (solid line) for a jökulhlaup from Summit Lake (September 1967), illustrating the characteristics of a typical outburst, namely an exponential increase in discharge followed by an abrupt termination. (Mathews and Clague, 1993, fig. 4.)

lhlaup activity during the twentieth century as glaciers have retreated from maximum positions achieved during the Little Ice Age climax (Fig. 21). As a glacier dam weakens due to downwasting and retreat, a critical threshold is reached when the dam can no longer continuously support the water behind it (e.g. Summit Lake, British Columbia, in December 1961; Mathews, 1965; Mathews and Clague, 1993). Thereafter, the lake drains and refills on either a regular (one or more times a year) or irregular basis. In the words of W.H. Mathews (1965), the lake becomes ''self-dumping''. This continues until either the glacier readvances and forms a stronger dam or, more typically, at least in this century, until it retreats to the point where it no longer impounds water (e.g. Ape and Strohn lakes, British Columbia; Mathews, 1965; Desloges et al., 1989). The frequency of dam failure is also likely to change with time. As a glacier recedes, failures may increase in frequency but decrease in magnitude, ending with the establishment of a permanent outlet (Thorarinsson, 1939; Marcus, 1960).

7. Discussion

The sensitivity of glacierised mountain regions to climatic change is indicated by catastrophic natural events described in this paper. Warming during the last 100–150 years has resulted in large losses of glacier ice in most mountain ranges of the world. This, in turn, has perturbed alpine geomorphic systems and created conditions that favour ice avalanching, slope failure, debris flows, catastrophic failures of moraine and glacier dams; it has also produced significant changes in the regime and planform of some rivers.

Almost all moraine-dammed lakes, for example, owe their existence to glacier retreat since the Little Ice Age. Many of the moraines that impound these lakes are inherently unstable and have failed catastrophically during the twentieth century. Failure commonly has occurred in response to wave overtopping triggered by an external impulse such as a rockfall or icefall. Climatic warming, however, may have played a more direct role by melting ice beneath or within moraines.

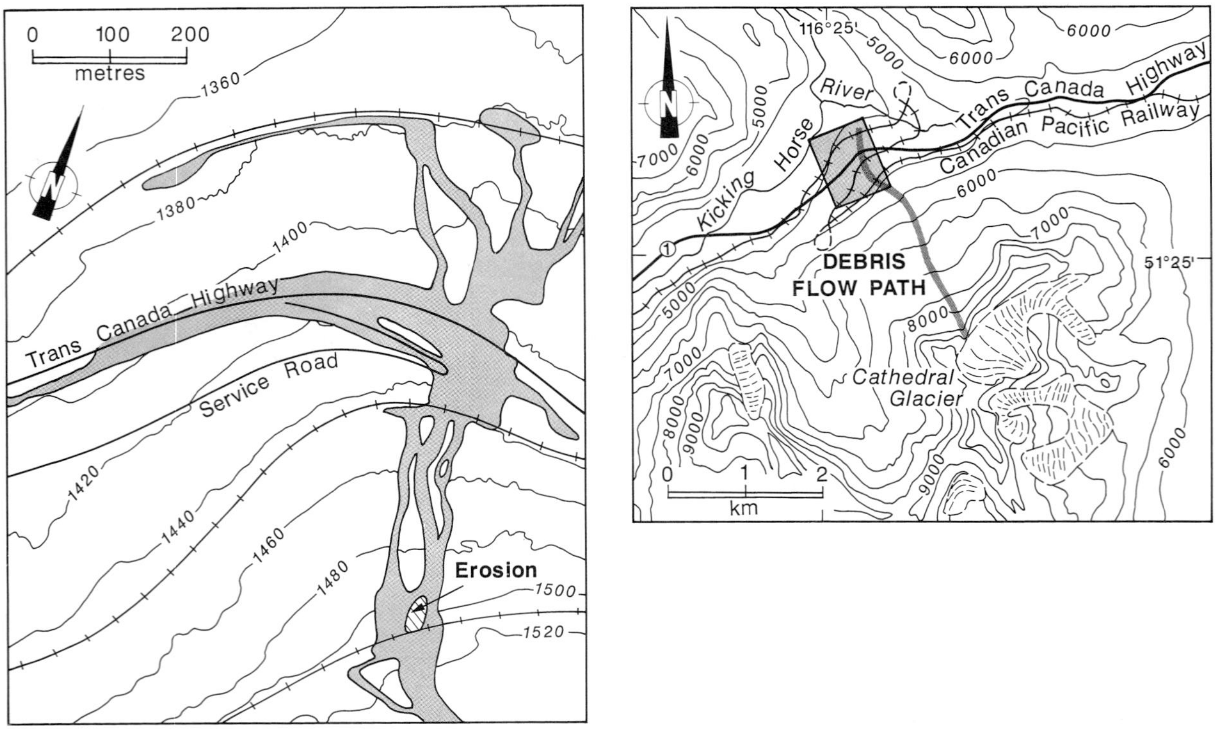

Fig. 20. Extent of the debris flow in Kicking Horse valley resulting from the Cathedral Glacier jökulhlaup of September 6, 1978. Topographic contour interval: left = 20 m; right = 500′ (152 m). (Clague and Evans, 1994, fig. 35.)

In any case, the most unstable moraine-dammed lakes already have drained, and today there are fewer dams capable of failure. In the future, the number of outburst floods from moraine-dammed lakes will diminish exponentially (or in some other nonlinear fashion) until the only ones that remain will be those impounded by moraines that are unlikely to fail under any conditions (e.g. dams with armoured overflow channels, high width-to-height ratios, and no ice cores). The large number of moraine-dam failures between the 1930s and the early 1980s suggests that this phenomenon may have peaked in many mountain ranges; within another 100 years, such failures will probably be much less common than today.

In contrast, the effects of recent glacier thinning and retreat on the stability of high mountain slopes may persist longer, perhaps hundreds of years. There are two main reasons for this. First, slope deterioration, initiated by debuttressing, is likely to occur slowly and progressively over a long period of time. In fact, many large landslides that have occurred during the historical period may have been conditioned by late Pleistocene deglaciation more than 10,000 years earlier, as suggested by Cruden and Hu (1993). Of course, the scale of deglaciation at the end of the Pleistocene, in most areas, was far greater than that of historical time. Second, assuming glaciers continue to retreat in response to the anticipated warmer climate of the next century, additional debuttressing of glacially oversteepened slopes is likely to occur. This will trigger more catastrophic landslides and, in some areas, will initiate deep-seated sagging and cracking of slopes that are now only marginally stable.

There is also a relation between climate and jökulhlaup activity, although it is not as strong as the relation between climate and floods from moraine-dammed lakes. Jökulhlaups can occur as glaciers advance, during a period of climatic cooling, but they are more common during periods of glacier retreat as ice dams progressively thin and weaken. In this context, however, timescale is important; individual glacier-dammed lakes at the threshold of stability are extremely

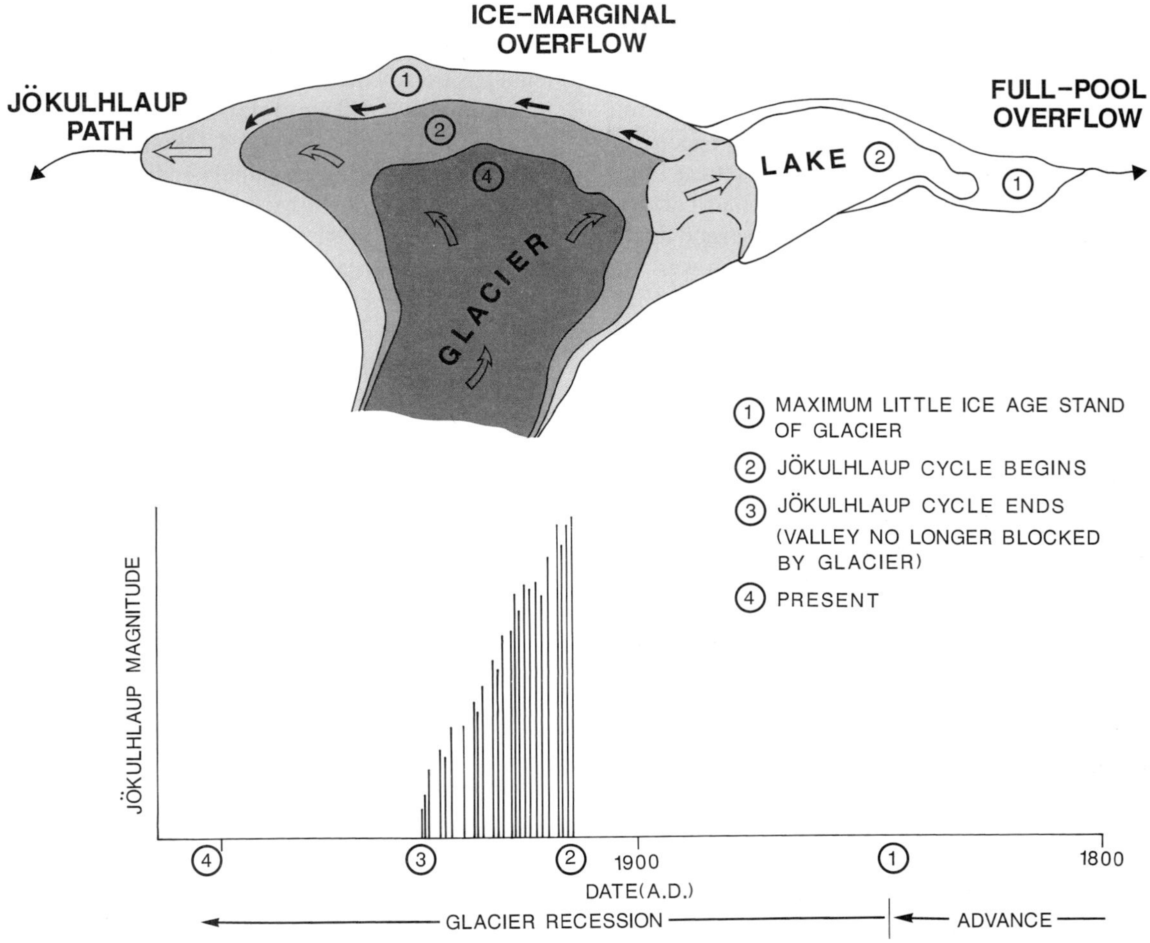

Fig. 21. Pattern of jökulhlaup activity accompanying glacier retreat during the twentieth century. (Top) Plan view of the toe of a hypothetical glacier at three times (1, 2, 4; ice-flow directions are shown by open arrows). At time 1, the lake impounded by the glacier is stable and drains via a stable overflow channel. Between times 1 and 2, the lake overflows along the margin of the glacier (solid arrows). The first jökulhlaup occurs at time 2, and sporadic or cyclic outburst floods continue until time 3 when the glacier has retreated to the point that it can no longer able impound a lake. (Bottom) Plot of jökulhlaup magnitude (relative scale) versus time for the scenario outlined above. (Clague and Evans, 1994, fig. 27.)

sensitive to minor climatic change. Short-lived cooling may cause minor advances of glaciers that are sufficient to form new lakes and trigger new jökulhlaups; alternatively, such cooling may change the frequency of floods from an existing glacier-dammed lake. During a prolonged period of warming, old self-dumping lakes may disappear and new ones form, thus different areas may experience jökulhlaups at different times.

Similar arguments apply in the case of glacier avalanches. Avalanches may occur at any time, but are most likely during periods of warmer climate when glaciers on steep slopes thin and lose strength. Increased discharge of meltwater at the base of such glaciers may trigger many failures. Again, however, the timescale of climate change is important; depending on local slope and other conditions, glaciers that terminate on steep slopes and that are near the threshold of stability may fail during either advance or retreat.

The events of the last century indicate that any long-term decay in the frequency of natural processes conditioned by Pleistocene glaciation (Church and Ryder, 1972; Church and Slaymaker, 1989; Cruden and Hu,

1993) may have slowed or reversed in mountain areas during and following the Little Ice Age. In many parts of the world, glaciers achieved their maximum Holocene extent during the last few centuries (Grove, 1988); consequently the effects on mountain geomorphic systems of the rapid and substantial retreat from these maximum positions during the twentieth century have perhaps been the greatest of the last several thousand years. Even so, these effects, at least in western Canada, are likely at least an order of magnitude smaller than those associated with late Pleistocene deglaciation ca. 15,000 to 10,000 years ago.

8. Conclusion

Catastrophic events triggered by recent deglaciation include ice avalanches, landslides and slope instability caused by debuttressing, and outburst floods from moraine- and glacier-dammed lakes. Conditions favourable for ice avalanching are created when the terminus of a glacier retreats up a steep rock slope. A large number of alpine glaciers now terminate on such slopes in a state of gravitational disequilibrium.

Retreat and thinning of glaciers has left high oversteepened slopes. Many such slopes, previously buttressed by ice, have failed, giving rise to large rock and debris avalanches. Glacial debuttressing is partly responsible for two large rock avalanches in Peru in 1962 and 1970 which claimed 24,000 lives.

Some lakes in high mountains are impounded by lateral and end moraines built during the Little Ice Age. These natural dams are steep-sided, consist of loose, poorly sorted sediments which may be ice-cored, and can fail rapidly and without warning. Failure commonly occurs when waves generated by a rockfall or ice avalanche overtop and rapidly incise the dam. Melting of ice cores, piping, and earthquakes are other possible failure mechanisms. Moraine-dam failures have caused many destructive floods and debris flows in the twentieth century, including one in 1941 that killed more than 6,000 people in Huaraz, Peru.

Lakes impounded by glaciers may drain suddenly to produce large downstream floods, termed jökulhlaups. This occurs when the glacier dam collapses or is overtopped and incised, or, more commonly, when the lake empties through subglacial and englacial tunnels. Some formerly stable glacier-dammed lakes have gone through a cycle of destructive jökulhlaup activity during this century as glaciers have retreated from maximum positions achieved during the Little Ice Age. The initiation of a jökulhlaup cycle occurs when a threshold of retreat and thinning is reached. Jökulhlaups then take place with decreasing magnitude and frequency until the glacier dam ceases to exist or until the glacier readvances and forms a stronger seal. Floods from some glacially dammed lakes are enormous and can cause damage far from their source.

Post-Little Ice Age climatic warming has perturbed mountain geomorphic systems. The length of this perturbation is different for the various phenomena discussed in this paper; it is shortest (perhaps several tens of decades) for moraine-dam failures and longest (hundreds of years or more) for landslides. The recognition that climatic change as modest as that of the last century can perturb natural alpine processes has important implications for hazard assessment and future development in the mountains (Reimer, 1992).

References

Alean, J., 1985. Ice avalanches: some empirical information about their formation and reach. J. Glaciol., 31: 324–333.

Blachut, S.P. and Ballantyne, C.K., 1976. Ice-dammed lakes — A critical review of their nature and behavior. McMaster University, Department of Geography, Discussion Paper 6, 90 pp.

Blown, I. and Church, M., 1985. Catastrophic lake drainage within the Homathko River basin, British Columbia. Can. Geotech. J., 22: 551–563.

Bovis, M.J., 1982. Uphill-facing (antislope) scarps in the Coast Mountains, southwest British Columbia. Geol. Soc. Am. Bull., 93: 804–812.

Bovis, M.J., 1990. Rock-slope deformation at Affliction Creek, southern Coast Mountains, British Columbia. Can. J. Earth Sci., 27: 243–254.

British Columbia Hydro and Power Authority, 1983. Homathko development; memorandum on glacial lakes and debris flows. B.C. Hydro and Power Authority, Hydroelectric Generation Projects Division, Report Geo 18/83, 22 pp.

Broggi, J.A., 1943. La desglaciación actual de los Andes del Perú. Bol. Soc. Geol. Perú, 14/15: 59–90.

Browning, J.M., 1973. Catastrophic rock slide, Mount Huascaran, north-central Peru, May 31, 1970. Am. Assoc. Petrol. Geol. Bull., 57: 1335–1341.

Church, M. and Ryder, J.M., 1972. Paraglacial sedimentation: a consideration of fluvial processes conditioned by glaciation. Geol. Soc. Am. Bull., 83: 3059–3072.

Church, M. and Slaymaker, O., 1989. Disequilibrium of Holocene sediment yield in glaciated British Columbia. Nature, 337: 452–454.

Clague, J.J., 1982. The role of geomorphology in the identification and evaluation of natural hazards. In: R.G. Craig and J.L. Craft (Editors), Applied Geomorphology. Binghamton Symposia in Geomorphology: International Series, 11. George Allen and Unwin, London, pp. 17–43.

Clague, J.J. and Evans, S.G., 1992. A self-arresting moraine dam failure, St. Elias Mountains, British Columbia. Geol. Surv. Can. Pap., 92-1A: 185–188.

Clague, J.J. and Evans, S.G., 1993. Historic catastrophic retreat of Grand Pacific and Melbern glaciers, St. Elias Mountains: an analogue for late Pleistocene decay of the Cordilleran Ice Sheet? J. Glaciol., 39: 619–624.

Clague, J.J. and Evans, S.G., 1994. Formation and failure of natural dams in the Canadian Cordillera. Geol. Surv. Can. Bull., 464: 35 pp.

Clague, J.J. and Mathews, W.H., 1992. The sedimentary record and Neoglacial history of Tide Lake, northwestern British Columbia. Can. J. Earth Sci., 29: 2383–2396.

Clague, J.J. and Rampton, V.N., 1982. Neoglacial Lake Alsek. Can. J. Earth Sci., 22: 1492–1502.

Clague, J.J., Evans, S.G. and Blown, I.G., 1985. A debris flow triggered by the breaching of a moraine-dammed lake, Klattasine Creek, British Columbia. Can. J. Earth Sci., 22: 1492–1502.

Clapperton, C.M., 1983. The glaciation of the Andes. Quat. Sci. Rev., 2: 83–155.

Clarke, G.K.C., 1982. Glacier outburst floods from "Hazard Lake", Yukon Territory, and the problem of flood magnitude prediction. J. Glaciol., 28: 3–21.

Clarke, G.K.C., 1989. Paleohydraulic modelling of outburst floods from Neoglacial Lake Alsek, Yukon Territory, Canada [abstract]. Ann. Glaciol., 13: 295.

Clarke, G.K.C. and Mathews, W.H., 1981. Estimates of the magnitude of glacier outburst floods from Lake Donjek, Yukon Territory, Canada. Can. J. Earth Sci., 18: 1452–1463.

Clarke, G.K.C., Mathews, W.H. and Pack, R.T., 1984. Outburst floods from glacial Lake Missoula. Quat. Res., 22: 289–299.

Costa, J.E. and Schuster, R.L., 1988. The formation and failure of natural dams. Geol. Soc. Am. Bull., 100: 1054–1068.

Cruden, D.M. and Hu, X-Q., 1993. Exhaustion and steady state models for predicting landslide hazards in the Canadian Rocky Mountains. Geomorphology, 8: 279–285.

Desloges, J.R. and Church, M., 1992. Geomorphic implications of glacier outburst flooding: Noeick River valley, British Columbia. Can. J. Earth Sci., 29: 551–564.

Desloges, J.R., Jones, D.P. and Ricker, K.E., 1989. Estimates of peak discharge from the drainage of ice-dammed Ape Lake, British Columbia, Canada. J. Glaciol., 35: 349–354.

Ding, Y. and Liu, J., 1992. Glacier lake outburst flood disasters in China. Ann. Glaciol., 16: 180–184.

Eisbacher, G.H. and Clague, J.J., 1984. Destructive mass movements in high mountains: Hazard and management. Geol. Surv. Can. Pap., 84-16: 230 pp.

Ericksen, G.E., Plafker, G. and Fernández Concha, J., 1970. Preliminary report on the geologic events associated with the May 31, 1970, Peru earthquake. U.S. Geol. Surv. Circ., 639: 25 pp.

Evans, S.G., 1987. The breaching of moraine-dammed lakes in the southern Canadian Cordillera. Int. Symp. on Engineering Geological Environment in Mountainous Areas, Beijing, Proc., Vol. 2, pp. 141–150.

Evans, S.G. and Clague, J.J., 1988. Catastrophic rock avalanches in glacial environments. Proc. Fifth Int. Symp. on Landslides, Vol. 2, pp. 1153–1158.

Evans, S.G. and Clague, J.J., 1990. Reconnaissance observations on the Tim Williams Glacier rock avalanche, near Stewart, B.C. Geol. Surv. Can. Pap., 90-1E: 351–354.

Evans, S.G. and Clague, J.J., 1993. Glacier-related hazards and climate change. In: R. Bras (Editor), The World at Risk: Natural Hazards and Climate Change. Am. Inst. Phys. Conf. Proc., 277: 48–60.

Evans, S.G., Clague, J.J., Woodsworth, G.J. and Hungr, O., 1989. The Pandemonium Creek rock avalanche, British Columbia. Can. Geotech. J., 26: 427–446.

Fisher, D., 1973. Subglacial leakage of Summit Lake, British Columbia, by dye determinations. In: J.W. Glen, R.J. Adie and D.M. Johnson (Editors), Symp. on the Hydrology of Glaciers. Int. Assoc. Sci. Hydrol. Publ., 95: 111–116.

Gansser, A., 1983. Geology of the Bhutan Himalaya. Denkschr. Schweiz. Naturforsch. Ges., 96: 181 pp.

Gilbert, R., 1969. Some aspects of the hydrology of ice-dammed lakes: observations on Summit Lake, British Columbia. M.A. thesis, Univ. of British Columbia, Vancouver, B.C., 93 pp.

Gilbert, R., 1971. Observations on ice-dammed Summit Lake, British Columbia, Canada. J. Glaciol., 10: 351–356.

Gilbert, R., 1972. Drainings of ice-dammed Summit Lake, British Columbia. Environment Canada, Inland Waters Directorate, Water Resources Branch, Scientific Series, No. 20, 17 pp.

Grove, J.M., 1988. Little Ice Age. Methuen, London, 498 pp.

Haeberli, W. and Naef, F., 1988. Murgänge im hochgebirge. Die Alpen, 64: 331–343.

Hanke, H., 1966. Gletscherkatastrophen. Die Bergsteiger, 33: 433–556.

Hansen, J. and Lebedeff, S., 1987. Global trends of measured surface air temperature. J. Geophys. Res., 92: 13345–13372.

Hauser, A., 1993. Remociones en masa en Chile. Serv. Nacional Geol. Min. Chile Bol., 45: 75 pp.

Heim, A., 1948. Wunderland Peru. Verlag Hans Huber, Bern.

Hewitt, K.J., 1982. Natural dams and outburst floods of the Karakoram Himalaya. Int. Assoc. Hydrol. Sci. Publ., 138: 259–269.

Intergovernmental Panel on Climatic Change, 1990. Climate Change; The IPCC Assessment. Cambridge Univ. Press, Cambridge, 238 pp.

Jackson, L.E., Jr., 1979. A catastrophic glacial outburst flood (jökulhlaup) mechanism for debris flow generation at the Spiral Tunnels, Kicking Horse River basin, British Columbia. Can. Geotech. J., 16: 806–813.

Jackson, L.E., Jr., Hungr, O., Gardner, J.S. and Mackay, C., 1989. Cathedral Mountain debris flows, Canada. Bull. Int. Assoc. Eng. Geol., 40 35–54.

Jones, D.P., Ricker, K.E., Desloges, J.R. and Maxwell, M., 1985. Glacier outburst flood on the Noeick River: the draining of Ape Lake, British Columbia, October 20, 1984. Geol. Surv. Can. Open File Rep. 1139: 81 pp.

Jordan, P., 1987. Impacts of mass movement events on rivers in the southern Coast Mountains, British Columbia: summary report.

Environment Canada, Inland Water Directorate, Report IWD-HQ-WRB-SS-87-3, 62 pp.

Lliboutry, L., Morales Arnao, B., Pautre, A. and Schneider, B., 1977. Glaciological problems set by the control of dangerous lakes in Cordillera Blanca, Peru. I. Historic failures of morainic dams, their causes and prevention. J. Glaciol., 18: 239–254.

Marcus, M.G., 1960. Periodic drainage of glacier-dammed Tulsequah Lake, British Columbia. Geogr. Rev., 50: 89–106.

Mason, K., 1930. The Shyok flood 1929; III — The Shyok flood: a commentary. Himalayan J., 2: 40–47.

Mathews, W.H., 1951. Historic and prehistoric fluctuations of alpine glaciers in the Mount Garibaldi map-area, southwestern British Columbia. J. Geol., 59: 357–380.

Mathews, W.H., 1959. Vertical distribution of velocity in Salmon Glacier, British Columbia. J. Glaciol., 3: 448–454.

Mathews, W.H., 1964. Water pressure under a glacier. J. Glaciol., 5: 235–240.

Mathews, W.H., 1965. Two self-dumping ice-dammed lakes in British Columbia. Geogr. Rev., 55: 46–52.

Mathews, W.H., 1973. Record of two jökullhlaups. Int. Assoc. Sci. Hydrol. Publ., 95: 99–110.

Mathews, W.H. and Clague, J.J., 1993. The record of jokulhlaups from Summit Lake, northwestern British Columbia. Can. J. Earth Sci., 30: 499–508.

Matthes, F.E., 1939. Report of Committee on Glaciers. Am. Geophys. Union Trans., 20: 518–523.

McSaveney, M.J., 1992. Immediate report; the Mount Fletcher rock avalanche of 2 May 1992. New Zealand Department of Scientific and Industrial Research, Geology and Geopysics, Christchurch, New Zealand, 22 pp.

McSaveney, M.J., 1993. Rock avalanches of 2 May and 16 September 1992, Mount Fletcher, New Zealand. Landslide News, 7: 2–4.

Morales, B., 1966. The Huascaran avalanche in the Santa Valley, Peru. Int. Assoc. Sci. Hydrol. Publ., 69: 304–315.

Niyazov, B.S. and Degovets, A.S., 1975. Estimation of the parameters of catasrophic mudflows in the basins of the lesser and greater Almatinka Rivers. Sov. Hydrol., 2: 75–80.

O'Connor, J.E. and Costa, J.E., 1992. Geologic hazards in glacial basins accompanying 19th and 20th century global warming. In: The World at Risk: Natural Hazards and Climate Change; Symposium Summary. Massachusetts Institute of Technology, Center for Global Change Science, Cambridge, Massachusetts, pp. 28–29.

Oerlemans, J. and Fortuin, J.P.F., 1992. Sensitivity of glaciers and small ice caps to greenhouse warming. Science, 258: 115–117.

Plafker G. and Ericksen, G.E., 1978. Nevados Huascarán avalanches, Peru. In: B. Voight (Editor), Rockslides and Avalanches, 1. Elsevier, Amsterdam, pp. 277–314.

Popov, N., 1990. Debris flows and their control in Alma Ata, Kazakh SSR, USSR. Landslide News, 4: 25–27.

Post, A. and Mayo, L.R., 1971. Glacier dammed lakes and outburst floods in Alaska; United States Geological Survey, Hydrologic Investigations Atlas HA-455, 10 pp. and map, 1:1,000,000 scale.

Radbruch-Hall, D.H., 1978. Gravitational creep of rock masses on slopes. In: B. Voight (Editor), Rockslides and Avalanches, 1. Elsevier, Amsterdam, pp. 607–657.

Reimer, W., 1992. Landslides and reservoirs. Preprint, 6th Int. Symp. on Landslides, Christchurch, 32 pp.

Röthlisberger, H., 1978. Eislawinen und ausbruche von gletscherseen. Sonderuck aus dem Jahrbuch der Schweizerischen Naturforschenden Gesellschaft, Wissenschaftlicher Teil 1978, pp. 170–212.

Ryder, J.M., 1991. Geomorphological processes associated with an ice-marginal lake at Bridge Glacier, British Columbia. Géogr. Phys. Quat., 45: 35–44.

Schmok, J.P. and Clarke, G.K.C., 1989. Lacustrine sedimentary record of ice-dammed Neoglacial Lake Alsek. Can. J. Earth Sci., 26: 2092–2105.

Thorarinsson, S., 1939. The ice-dammed lakes of Iceland with particular reference to their value as indicators of glacial oscillations. Geogr. Ann., 21: 216–242.

Vuichard, D. and Zimmerman, M., 1987. The 1985 catastrophic drainage of a moraine-dammed lake, Khumbu Himal, Nepal: cause and consequences. Mountain Res. Development, 7: 91–110.

Yesenov, U.Y. and Degovets, A.S., 1979. Catastrophic mudflow on the Bol'shaya Almatinka River in 1977. Sov. Hydrol. Select. Pap., 18: 158–160.

Zimmerman, M. and Haeberli, W., 1992. Climatic change and debris flow activity in high mountain areas — A case study in the Swiss Alps. Catena Suppl., 22: 59–72.

ELSEVIER

Geomorphology 10 (1994) 129–138

GEOMORPHOLOGY

Erroneous perceptions of fluvial hazards

S.A. Schumm

Department of Earth Resources, Colorado State University and Resource Consultants and Engineers, Inc., P.O. Box 270460, Fort Collins, CO 80527, USA

Received January 20, 1994; revised March 21, 1994; accepted March 30, 1994

Abstract

Many times human perceptions of geomorphic hazards are fallacious. Three types of misperceptions are (1) a perception of stability, which leads to the conclusion that any change is not natural, (2) a perception of instability, which leads to the conclusion that change will not cease, and (3) a perception of excessive response, which leads to the conclusion that changes will always be major. Examples, which include the Snake, Ohio, Colorado, and Nile Rivers, incised channels, and Rocky Mountain Streams demonstrate how these perceptions can lead to litigation and unnecessary engineering works.

1. Introduction

The word hazard refers to a potential danger or risk, which is usually assumed to be a potential catastrophe or disaster that involves great damage and loss of life. In this paper, only geomorphic hazards that are characteristic of the fluvial system will be considered. The difference between geomorphic hazards and others is that geomorphic hazards may involve a slow progressive change that, although in no sense catastrophic, can be a significant hazard that will eventually involve costly preventive and corrective measures. Therefore, geomorphic hazards can be defined as any landform change, natural or otherwise, that adversely affects the geomorphic stability of a place (Schumm, 1988).

Most books on natural hazards concentrate on the spectacular events such as coastal erosion during hurricanes, volcanic eruptions, earthquakes, avalanches, landslides, subsidence, etc. (White, 1974; Bolt et al., 1975; Hewitt and Burton, 1975; Asimov, 1979; Blong and Johnson, 1986). However, there are at least three types of geomorphic hazards that involve different spans of time, different degrees of damage and different energy expenditures as follows: (1) an abrupt change that produces a catastrophic event, e.g. a landslide that occurs rapidly as a result of an equally catastrophic meteorological event, earthquake or human activity (removal of toe support); (2) a progressive change that leads to an abrupt change, e.g., weathering that leads to slope failure, gullying of a steepening alluvial fan, meander growth to cutoff, and channel avulsion; (3) a progressive change that has slow, but progressive results, e.g., meander shift, hillslope erosion, channel incision, and channel enlargement.

A type (3) hazard may not appear to represent a hazard, but the potential protective or remedial costs can be excessive. For example, the slow shift of a meander toward a bridge can cause bank erosion on one side of the approach to the bridge. Extensive bank protection can be emplaced there only to have the focus of erosion shift to the opposite bank, as the meander continues to migrate downstream. The final result can be outflanking of the bridge or its collapse.

The type (3) hazard also can lead to expensive litigation because the perception of it can be fallacious. Frequently, when a change occurs such as bank erosion,

SSDI 0169-555X(94)00018-M

the landowner believes that this is caused not by natural processes, but by human actions. The result is a law suit. Hence, another factor can be added to the costs and benefits associated with living in or having property in a hazard zone as follows (Alexander, 1992):

$$B_N = B_T - C_{Td} - C_{Ta} \pm C_{Tl}$$

where B_N is the net benefits derived from occupying hazard zone, B_T is the total benefits derived from occupying the risk area, C_{Td} is the total cost of damage caused by the hazard, C_{Ta} is the total cost of adjustment to the hazard, and C_{Tl} is the total cost of litigation.

Litigation is shown as either contributing to an increase or a decrease of net benefits depending upon the success of the litigation.

Instead of discussing specific fluvial hazards, the purpose of this paper is to consider how perceptions of hazards can greatly increase expenditures for unnecessary engineering work and for litigation. Examples will be provided from the authors experience, and they will illustrate three types of erroneous perceptions as follows:

(1) a perception of stability, which is, that any change is unnatural,

(2) a perception of instability, which is that once change occurs it will be continuous,

(3) a perception of excessive response, which is that change will always be major.

In short, faulty perceptions lead to faulty predictions, misassignment of blame, restoration or management failures, and unnecessary litigation. Most of us probably are at times victims of faulty perceptions because of inadequate experience and insufficient training (Schumm, 1991).

2. Perceptions of stability

When rivers form boundaries between states, or properties, the boundary remains fixed in position when the river changes position abruptly by avulsion, but the boundary shifts with the channel, when the change is slow as a result of bank erosion and lateral accretion (Fig. 1). However, if the original survey of the boundary, which can be the bank line of the river, is in gross error or is fraudulent, the boundary remains fixed because it is not possible to determine the exact conditions at the time of the original survey, which may have occurred a century ago. Under these conditions, any land formed by lateral accretion would not accrue to the riparian landowner, but it would become public land, which in the following example would be administered by the Bureau of Land Management (BLM). This is the situation that led to the litigation described as the first example, the Snake River in Jackson Hole, Wyoming. The second example, the Ohio River liti-

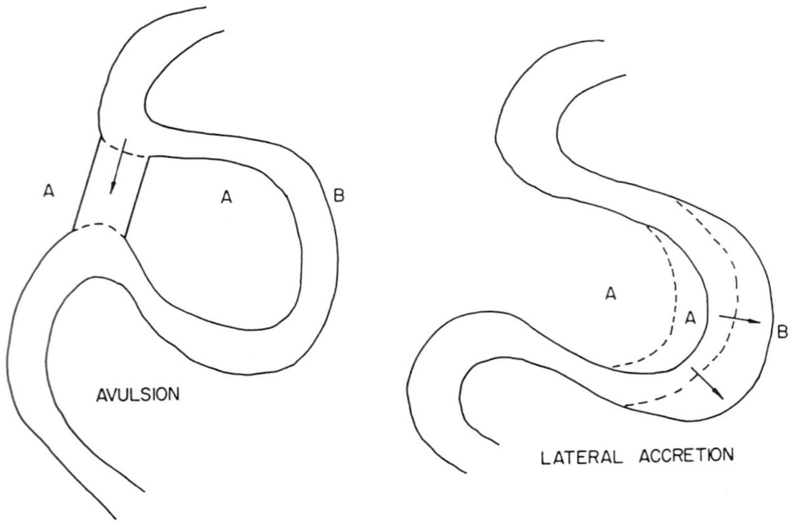

Fig. 1. River pattern change showing avulsion by meander cutoff, when the boundary between properties *A* and *B* remains fixed, and lateral accretion when property *A* gains and property *B* loses, as the boundary shifts with the river.

gation occurred because of perceptions of stability, but it did not involve boundary law.

2.1. Snake River

Three trials in the U.S. District Court in Cheyenne, Wyoming and the appointment of a Special Master to determine the ownership of 108 disputed parcels of riparian land and islands in the Snake River involved teams of experts on both sides. The dispute occurred because personnel of the BLM determined that the banks of the Snake River, as surveyed by a Government Land Office Surveyor in 1893 did not conform to the present-day banks of the river. The conclusion was that the surveys were in "gross error or fraudulent." Therefore, parcels of land along the river would revert to the Federal government for the reasons described earlier (Fig. 2). The position of the BLM, as represented by the Department of Justice (DJ), was that there had been no change in the river banks since the surveys and that late Pleistocene Pinedale loess covered all surfaces adjacent to the river. The expert team representing the owners of the Snake River Ranch (C74-91), the Hansen Ranch and the Crescent H Ranch (C74-113) demonstrated that on the areas of dispute, Pinedale loess and soils were absent. The oldest trees growing within the lines of the 1893 survey were about 60 years old, indicating that they commenced to grow after the survey, which was proof that these portions of floodplain formed after the survey. Finally, it was demonstrated by using a series of aerial photographs that the steep-gradient braided Snake River eroded its banks, eroded islands, formed bars and islands, and shifted position after the earliest aerial photographs were taken in 1945. The evidence was convincing that the steep-gradient cobble-bed Snake River could not be expected to have

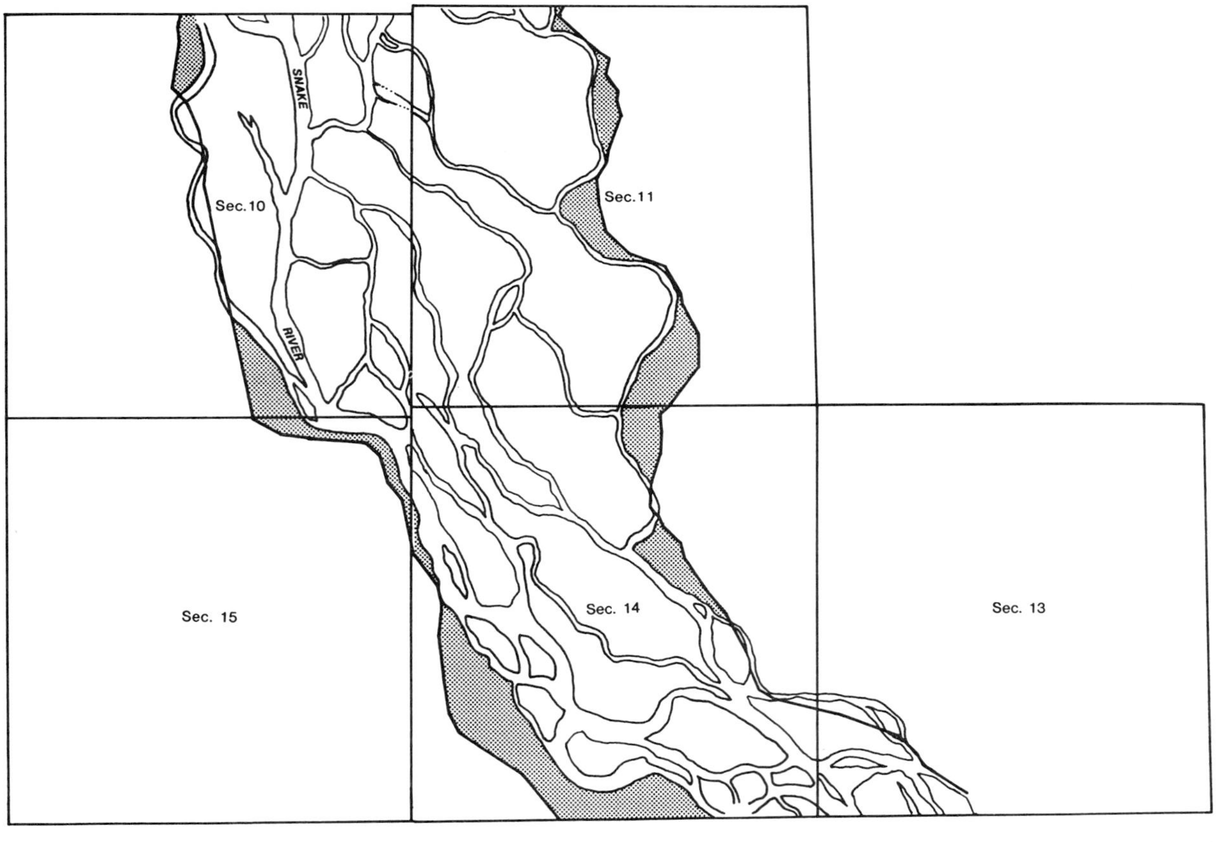

Fig. 2. Braided Snake River channel in Township 40 N, Range 117 W, Jackson Hole, Wyoming. Outer heavy lines are bank lines as surveyed in 19th century. Stippled areas are lands claimed by the U.S. Government because modern channel banks no longer match surveyed banks.

the same bank lines in the 1980s as it did in the 1890s, and the judge so ruled.

Finally, the BLM and DJ agreed that the surveys were accurate, but that islands in the river had been omitted from the survey. A Special Master was appointed by the Federal judge to evaluate the claims for 108 properties along the Snake River. Using the previous testimony and evidence for island formation and erosion, he concluded that islands were not surveyed in the 1880s because prior to flow regulation upstream, islands were absent, and the surveyors did not survey gravel bars. The BLM's perception of stability was found to be incorrect, but only at great expense to the taxpayers of the United States and to the riparian landowners.

2.2. Ohio River

The Ohio River is a major waterway formed by the junction of the Monongahela and Allegheny Rivers at Pittsburgh. It has always been a major route of waterborne commerce to the Mississippi River. Water levels fluctuate in response to meteorological conditions, and years of low water greatly hindered traffic on the river. In fact, during low water of 1911–1912, much of the bed of the river was exposed, and it was mapped by the U.S. Army Corps of Engineers (COE). In order to improve navigation and to ensure navigability during low water, the COE constructed a series of locks and dams (L&D), which would maintain a sufficient depth of water to maintain navigation, and 49 L&D were constructed by 1929. In order to enlarge the lock chambers and to provide greater water depths 19 "high-lift" dams were constructed beginning in the 1930s (Fig. 3). The dams consisted of a series of gates that could be opened and closed to regulate flow. In the 1970s, bank erosion was described in local newspapers as being a problem from Pittsburgh to Cairo, Illinois. In three trials in the United States Claims Court in Lou-

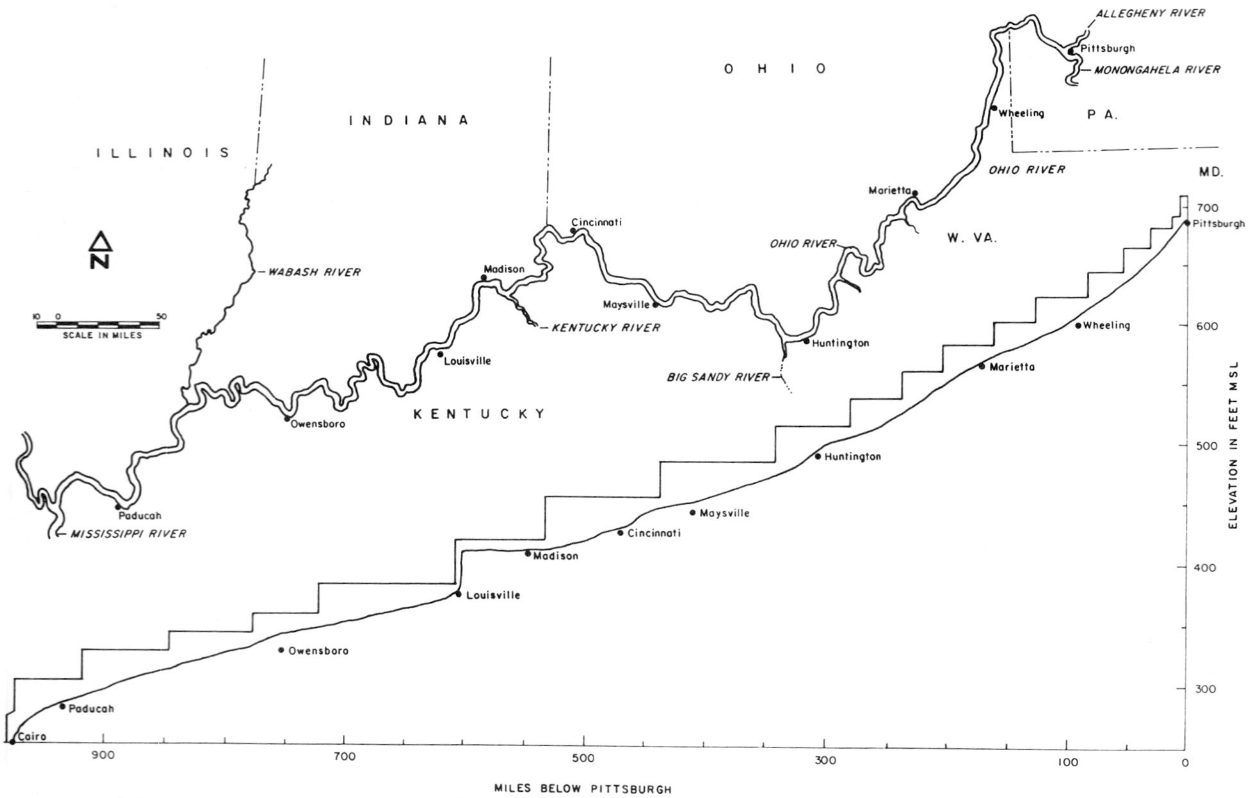

Fig. 3. Map and longitudinal thalweg profile of Ohio River from Pittsburgh to Mississippi Rivers. Stepped profile shows locations of locks and dams and the water-surface elevations of the pools at low water. At flood stage the water surface would be above the highest pool level and the profile would be smooth.

isville, Kentucky, more than 200 riparian landowners sued the Corps of Engineers for causing this bank erosion. They claimed that the ''high-lift'' dams caused water-level fluctuations that caused bank erosion and that river traffic at low-water, when water levels are maintained by the L&D, caused wave erosion of the banks. These claims were very serious because if true, the United States could be required to compensate the landowners, and the COE would be required to stabilize the banks and to maintain them permanently at great cost.

The Department of Justice assembled a team of experts which consisted of a hydraulic engineer, a geomorphologist, hydrologists, geotechnical engineers, and COE personnel who ''knew the river.'' First, a reconnaissance of the river was undertaken which eventually involved observations from Pittsburgh, Pennsylvania to Owensboro, Indiana (Fig. 3). The initial reconnaissance revealed that the condition of the banks of the river was variable and that eroding, healing, and stable banks were present. In addition, the distribution of eroding banks along the river was the same everywhere, although the new dams had raised the water level the most near the locks and dams and not at all in the farthest upstream parts of the pools.

Field studies determined that the greatest erosion occurred during high water and during the waning of the flood waters (Hagerty et al., 1983; Ullrich et al., 1986). At these times, the locks and all of the gates that formed the dams were open, and the navigation L&D acted only as roughness elements in the river. The river, in fact, was open, and the stage depended on hydrology, not on the L&D. The eroding banks were located at points of maximum attack by flowing water or they were located where landowner activities rendered the banks unstable by increasing seepage through the banks. The field investigations and measurements of bank erosion showed that erosion was related to periods of high water when the L&D did not affect the river and river traffic had ceased.

In addition to the three trials, there were several hearings to determine if further litigation was warranted. Potential litigation regarding bank erosion along the Monongahela River was not pursued when it was discovered that Monongahela means ''river with banks falling down.''

The perception of great stability held by riparian landowners led to great expense. Many of the landowners were sincere in their beliefs that erosion was caused by the COE, but others clearly were involved in the hope of monetary gain. After the first trial, a convincing argument that the erosion was essentially natural could be made, when it was observed that some of the eroding banks of the previous litigants were healing. In all of the cases, the Federal judge ruled that the bank erosion was not caused by COE activities. Vast amounts of money, that would have been spent to ''stabilize'' the banks of the Ohio River, were saved, but nevertheless, at considerable financial cost.

3. Perceptions of instability

In contrast to perceptions of stability, there can also be erroneous perceptions of instability. When a landform is perturbed, its response may appear to be persistent. That is, the period of adjustment may be long, and during one lifetime it may appear to be unending. There has been testimony supporting this conclusion regarding arroyos in southeastern Utah. Nevertheless, earth scientists understand that adjustments toward renewed stability takes place naturally. For example, although channels incise or aggrade, they eventually reach a new equilibrium. Also, as the adjustment occurs, there are times when initiating control measures will be effective and other times when they will be very costly and less effective. This can be demonstrated by using examples of incised channels and changes of sediment yield.

3.1. Incised channels

If a series of cross-sections are surveyed along a channel that is incising, as a result of natural or human-induced changes (e.g., channelization), an evolutionary model of channel adjustment can be developed (Fig. 4). Fig. 4 shows the evolution of an incised channel from initial incision (Fig. 4a, b) and widening (Fig. 4b, c, d) until widening and deposition (Fig. 4d) lead to renewed relative stability (Fig. 4e). The model was developed for incised channels in northern Mississippi, but it applies equally to arroyos of the southwestern United States and it has practical value because it permits estimation of sediment production and agricultural-land loss (Schumm et al., 1984). Fig. 4 shows clearly that even the most drastic channel behavior,

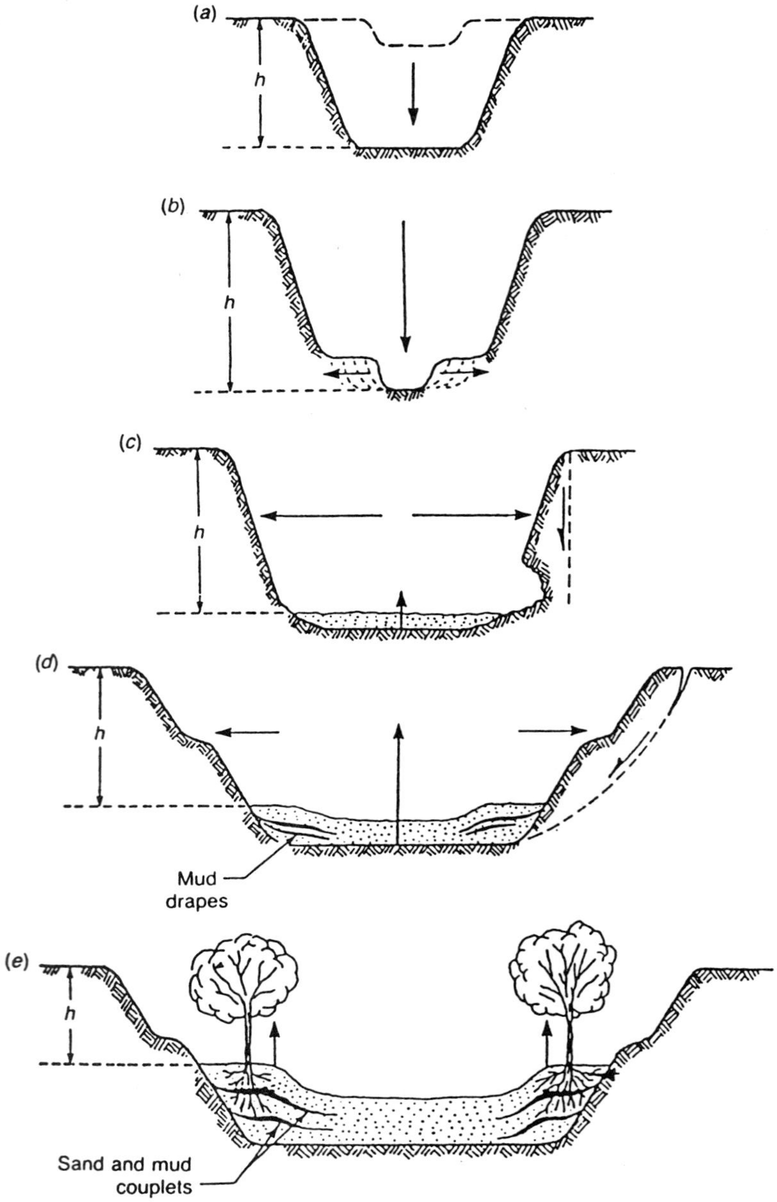

Fig. 4. Evolution of incised channels from initial incision (a, b) and widening (c, d) to aggradation (d, e) and eventual relative stability (e).

which in this case is deep incision, will moderate, and the channel will evolve to a new condition of relative stability. This is a very important concept not fully understood by many conservation engineers.

As demonstrated, time is an important variable in the development of an incised channel and, therefore, it should be an important variable in any scheme to curtail gully erosion and to reduce sediment loads. Fig. 5 is a conceptual diagram that shows the change in sediment yield and incised channel (gully) drainage density (length of gullies per unit area) with time. In a drainage basin that has been rejuvenated and in which gullies are developing, sediment production will increase as the length of incising channels increases (Fig. 5, times 1 to 4). However, at time 4 maximum headward growth of the channels has occurred, and

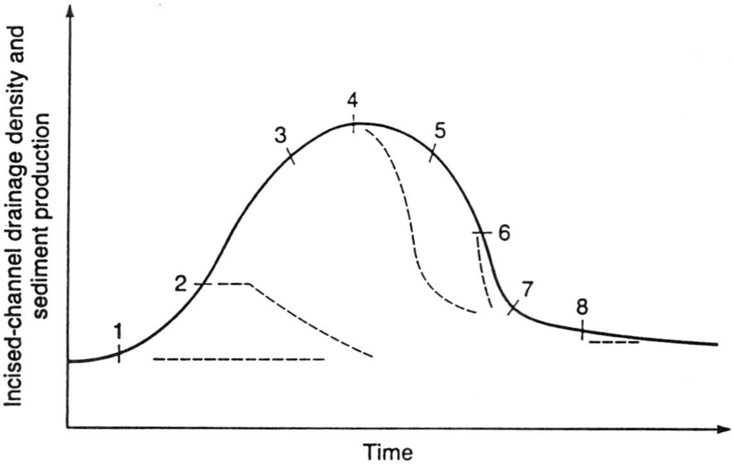

Fig. 5. Curve showing changes of incised-channel (gully) drainage density and sediment production with time. Dashed lines indicate effect of channel-control structures at various times during incised-channel evolution. Numbers indicate time.

they begin to stabilize between times 4 and 7, when there is an increase in the length of relatively stable reaches, and the length of active reaches decreases. By understanding this cycle of channel incision and gullying from initial stability (time 1) to renewed stability (time 8), it is possible to select spans of time in the cycle when land management and incised-channel control practices will be most effective. For example, gullies just initiating (times 1 or 2) and gullies almost stabilized (times 6, 7, or 8) will be the most easily controlled by artificial means. The efforts at times 6 and 8 will have little effect, as the channels are stabilizing naturally. At time 4, control will be difficult and expensive. Obviously, consideration of this complex evolving system for only short periods of record and short time spans can yield the erroneous perception that the erosion will never cease, but the development of Fig. 4 led to major modifications of channel stabilization efforts by the Soil Conservation Service in Mississippi and considerable cost reductions (Schumm et al., 1984).

3.2. Sediment delivery

The sequence of events shown in Fig. 4 can also have wider application. Throughout the arid and semi-arid regions of the southwestern United States in the 19th century, channels incised to form arroyos, that were notorious suppliers of sediment to the Colorado, Green, Rio Grande, and San Juan Rivers. Their incision also lowered water tables, and as a result, former grazing and farmlands were abandoned, as well as some small agricultural communities. Projections of the life of reservoirs on these rivers were based on the assumption that the high sedimentation rates generated by arroyo incision and widening (Fig. 5, times 3–5) would continue. However, if the sequence of incised-channel evolution as shown in Fig. 4 is generally applicable, then the arroyos would begin to stabilize (Fig. 5, times 5–7), erosion would be less and sediment would be stored in newly forming floodplains. Indeed, sediment moving through the Grand Canyon of the Colorado River has decreased significantly since the later 1930s for this reason (Gellis et al., 1991), although discharge has not. In addition, significantly less sediment is deposited in reservoirs. For example, based upon data collected from 1914 to 1957, it was estimated that 85,400 acre-feet of sediment would be deposited in Lake Powell each year. This estimate was adjusted for future upstream dam construction, on the Green, San Juan, and Gunnison Rivers. In 1963, the Glen Canyon Dam was closed, and 409 ranges were surveyed across the reservoir, which provided a means of measuring sediment accumulation in Lake Powell. In 1986, the ranges were resurveyed, and it was determined that only 36,946 acre-feet of sediment was being deposited each year (Ferrari, 1988), which is only 43% of the previous calculation. During this time, flow into the reservoir was 91% of the 1914–1957 average. Hence, an understanding of incised-channel evolution (Fig. 4) would have permitted a significant increase in the estimated reservoir life from 700 to 1600 years. Obviously, the same principle can be applied to other smaller reservoirs and to sediment delivery to lakes and bays.

Just as erroneous perceptions of stability can result in costly litigation, perceptions of instability can lead to costly and often unnecessary attempts to control an already stabilizing situation. However, in each case an understanding of system behavior is essential, and reduced sediment loads can be the result of a variety of upstream conservation measures.

4. Perceptions of adjustment

The perception that river adjustments will be major can be considered as a problem of extrapolation from one location to another with all the accompanying problems of scale, etc. (Schumm, 1991). Fisk's (1944) great monograph shows such dramatic river changes through time that an impression is conveyed that all rivers are susceptible to major changes. The fallacy of this perception can be readily demonstrated. The documented effects of meander cutoffs on the Mississippi River (Winkley, 1976) and the observation of the effects of cutoffs in sand-bed flumes convinced me that any cutoff would have a major impact upon the channel at the cutoff site and both upstream and downstream. Therefore, great efforts would be required for bank stabilization. However, Brice's (1981) study of 103 rivers, that had been channelized or subject to cutoffs, showed that only about 25% had a major response. Obviously, resistant bed and bank materials and low-energy streams would not respond to the same degree as the Mississippi River. Nevertheless, my perception of major adjustment and instability could lead to expensive stabilization efforts, that would have been unnecessary. Further examples involve the Nile River and Rocky Mountain streams.

4.1. Degradation of the River Nile

It may be that the experience of U.S. engineers with the Mississippi River has affected perceptions internationally. For example, when the High Aswan Dam (HAD) was constructed on the Nile River (between 1963 and 1968) its anticipated impact was a matter of great concern for engineers concerned with bank stability and potential channel degradation. Perhaps of greatest concern was the potential for major degradation of the Nile following construction of the HAD. Because the potential for degradation was controversial, it was studied by many researchers prior to and after construction of the HAD (Hammad, 1972; Shalash, 1980, 1983). Numerous experts estimated the amount of degradation below the HAD and downstream of irrigation diversions (Esna, Naga Hammadi and Assiut Barrages). Pre-HAD estimates of degradation ranged from 2.0 to 8.5 m, but 18 years after the dam was in operation, maximum degradation below the barrages was 0.70 m.

Several factors account for the fact that degradation has been minimal after closure of HAD. For example, it is important to recognize that, during past humid periods in Egypt tributary, tributary wadis delivered very coarse sediments to the valley, which could form an armor and act as a control of degradation. At present, many wadis appear to contribute only relatively fine sediment, which can be readily transported downstream. In fact, all of the wadis contain much stored sediment of gravel, cobble and boulder size. Archaeologists and other investigators describe cobbles and gravel at wadi mouths (Wendorf et al., 1970) and in older terrace deposits (Butzer, 1959).

Based upon an analysis of sediments obtained from borings in the Nile valley, Attia (1954, p. 105) concluded that within the valley "coarse deposits composed of coarse sand, sand and gravel or gravel — which lie beneath the fine alluvial deposits — have an irregular upper surface. The depth from ground-surface to the top of these gravels, etc., varies from 8 to 26 meters. The average depth — as worked out from the section — is 15.7 meters." This means that at many locations the bed of the Nile is in contact with sediment that is coarser than sand. Additional support for this hypothesis is provided by gravels encountered in the bores in the channel at the site of the barrages. At one there was 25% gravel at a depth of one meter in two bores. At other barrages gravels were encountered, but at greater depths. Nevertheless, it is well known that an armor can develop from only a small percentage of coarse bed material. Obviously, the engineers who estimated great degradation below the HAD were thinking of dams worldwide that had such an effect, but they neglected an important variable, coarse sediment beneath the sand-bed channel.

4.2. Mountain streams

A final example is another court case that has been estimated to have cost the State of Colorado and the Department of Justice about 7 to 8 million dollars. The litigation developed because the Forest Service (FS) claimed reserved water rights in the National Forests, specifically in Water District 1 of northern Colorado. This could mean that all diversions after the establishment of the National Forest in the late 19th century could be stopped. The FS justified its position by claiming that the phrase "maintaining favorable conditions of flow," which was one of the reasons for establishing the National Forests, meant that any diversion that caused a change of channel dimensions violated that standard. Based upon hydraulic-geometry relations, it was claimed that any diversion of flow would adversely affect the channel.

Unfortunately for the FS position, about 70% of the 244 sites at which it claimed reserved rights were on channels draining less than 10 square miles. In these small watersheds, the streams followed what was referred to as "courses of convenience" through bedrock outcrops, glacial and landslide materials, log jams, and beaver dams. There were no hydrologic data at each site, and, therefore, bankfull discharge was calculated using channel dimensions gradient and roughness. If the hydraulic geometry relations among width, depth, and discharge applied to these mountain streams, then good correlations among the variables should emerge. However, because channel dimensions were used to calculate bankfull discharge, the good relation between width and discharge ($r^2=0.75$) was obviously meaningless. Using drainage area as a substitute for calculated discharge, the relations between discharge and width ($r^2=0.50$) and depth ($r^2=0.24$) were weak with scatter over an entire log cycle. Another relation that might be expected to be strong was between drainage area and gradient, a relation established by Leopold and Bull (1979) for ephemeral streams in the Southwest, but the correlation coefficient for this relation was only 0.01. The relation between channel width and gradient was also very weak ($r^2=0.03$) in spite of a reasonable relation established by Leopold and Miller (1956). Obviously, the "laws" of hydraulic geometry that were developed for alluvial streams do not necessarily apply to steep-gradient mountain streams with other controls. Another fact of significance presented by the State's expert testimony was that where diversions had been in place for many years, a downstream effect of the diversion could not be clearly demonstrated. After a lengthy trial in 1992 and 1993, the judge ruled in favor of the State of Colorado.

It is obvious that many variables can influence a landform, and unless all factors are considered, serious errors can be made that are very expensive. In this case, if the FS had used a valid sampling procedure, they would have determined that their claims were fallacious and the millions of dollars could have been used for other purposes. The presumptions of major changes along the Nile and in the mountain streams, although apparently reasonable, were incorrect because of the significance of other factors.

5. Discussion

The three types of erroneous assumptions discussed here could have been avoided by more detailed investigations in each case. However, more detailed investigations frequently do not change closed minds. Nevertheless, if in all of the court cases used as examples, the experts from both sides had been ordered by the judge to reach a conclusion jointly, as based upon available evidence, the costly trials probably could have been avoided. Indeed, in other situations, presentation of results in depositions have led to settlement or abandonment of a case. Judges and attorneys should be encouraged to permit communication among experts in an attempt to resolve the problem before trial. Finally, the confrontation of experts in court indicates that it is not only lay people who are susceptible to erroneous assumptions concerning geomorphic hazards. It is important to recognize that these perceptions exist, and they strongly influence people who are confronted with geomorphic hazards.

References

Alexander, D., 1992. The causes of landslides: Human activities, perception, and natural processes. Environ. Geol. Water Sci., 20: 165–179.

Asimov, I., 1979. A Choice of Catastrophes. Simon and Schuster, New York, 377 pp.

Attia, M.I., 1954. Deposits in the Nile Valley and the Delta. Geological Survey of Egypt, Government Press, Cairo, 356 pp.

Blong, R.J. and Johnson, R.W., 1986. Geological hazards in the Southwest Pacific and Southeast Asian regions. Identification, assessment and impact. Aust. Geol. Geophys. J., 10: 1–15.

Bolt, B.A., Horn, W.L., MacDonald, G.A. and Scott, R.F., 1975. Geological Hazards. Springer, New York, 328 pp.

Brice, J.C., 1981. Stability of relocated stream channels. Federal Highway Administration Report FHWA/RD-80/158, 177 pp.

Butzer, K.W., 1959. Contributions to the Pleistocene geology of the Nile valley. Erdkunde, 13: 46–67.

Ferrari, R.L., 1988. 1986 Lake Powell Survey. Bureau of Reclamation, Denver, CO., 67 pp.

Fisk, H.N., 1944. Geological investigation of the alluvial valley of the lower Mississippi River. U.S. Army Corps of Engineers, Mississippi River Commission, 78 pp.

Gellis, A., Hereford, R., Schumm, S.A. and Hayes, B.R., 1991. Channel evolution and hydrologic variations in the Colorado River basin: Factors influencing sediment and salt loads. J. Hydrol., 124: 317–344.

Hagerty, D.J., Sharifounnasab, M. and Spoor, M.F., 1983. Riverbank erosion — A case study. Assoc. Eng. Geol. Bull., 20: 411–437.

Hammad, H.Y., 1972. River bed degradation after closure of large dams. Am. Soc. Civil Eng. J. Hydrol. Div., 98: 591–608.

Hewitt, K. and Burton, I., 1975. The Hazardousness of a Place. Univ. Toronto Press, Toronto, 154 pp.

Leopold, L.B. and Bull, W.B., 1979. Baselevel, aggradation, and grade. Am. Philos. Soc. Proc., 123: 168–202.

Leopold, L.B. and Miller, J.P., 1956. Ephemeral streams, hydraulic factors and their relation to the drainage net. U.S. Geol. Surv. Prof. Pap., 282-A: 37 pp.

Schumm, S.A., 1988. Geomorphic hazards: problems of prediction. Z. Geomorphol. Suppl., 67: 17–24.

Schumm, S.A., 1991. To Interpret the Earth: Ten Ways to be Wrong. Cambridge Univ. Press, Cambridge, 133 pp.

Schumm, S.A., Harvey, M.D. and Watson, C.C., 1984. Incised channels, morphology, dynamics and controls. Water Resources Publ., Littleton, CO, 200 pp.

Shalash, S., 1980. The Effects of the High Aswan Dam on the Hydrological Regime of the River Nile. Proc. Symp. Helsinki Int. Assoc. Hydrologic Science, Publ. 130, pp. 244–250.

Shalash, S., 1983. Degradation of the River Nile. Int. Water Power Dam Constr., 35: 56–58.

Ullrich, C.R., Hagerty, D.J. and Holmberg, R.W., 1986. Surficial failures of alluvial stream banks. Can. Geotech. J., 23: 304–316.

Wendorf, F., Said, R. and Schild, R., 1970. Egyptian prehistory: Some new concepts. Science, 169: 1161–1171.

White, G.F. (Editor), 1974. Natural Hazards: Local, National, Global. Oxford Univ. Press, New York, 265 pp.

Winkley, B.R., 1976. Response of the Mississippi River to the cutoffs. In: Rivers '76. Vol. 2. Am. Soc. Civ. Engineers, New York, pp. 1267–1284.

ELSEVIER

Geomorphology 10 (1994) 139–156

GEOMORPHOLOGY

Geomorphological understanding of floods

Victor R. Baker

Department of Geosciences, University of Arizona, Tucson, AZ 85721, USA

Received February 21, 1994; revised April 8, 1994; accepted April 10, 1994

Abstract

The summer 1993 flooding of the upper Mississippi River valley reminds us that floods are the most globally pervasive, environmentally diverse and continually destructive of all natural hazards. The fact that flood damages continue to rise raises commonsense questions about conventional flood science. Like much modern environmental science, conventional flood science has followed the model of theoretical physics. It advanced from early emphasis on streamflow measurement to the use of simple formulae, and finally to the abstract theoretical sophistication of modern modeling studies. Two approaches are now used to "predict" flood phenomena: (1) beginning with the conventional database of measured properties of small common floods, a conceptual generalization is made to the idealized properties of the large, rare floods from which society is assumed to be at risk, and (2) explanation of detailed, specific flood phenomena is achieved through theoretical generalization (models) based on "first principles", which are assumed to apply to the entire class of phenomena. Unfortunately, both approaches devote almost all their attention to methodology, increasingly mathematical, without questioning basic underlying assumptions. Increasingly it is the assumptions, often unstated, that serve to embody the understanding of floods as real-world particular phenomena, rather than as conceptual generalities. Such trends lead to an unease that it is not floods that are being researched by much of conventional flood science. Rather, such flood "science" is increasingly becoming the mathematical manipulation of idealized parameters that are assumed to have flood-like properties. These idealizations of flood attributes are generalized, and the resulting predicted consequences are imposed upon society through engineering designs, flood-hazard zonations, and the like.

Geomorphological understanding of floods derives from a long geological tradition of studying indices of real processes operating in the past. In contrast to the conceptual, theoretical treatment of floods as classes or generalizations, geomorphologists study particular floods revealed as a natural experience that is recorded in the sediments, landforms, and erosional scars of past floods. The strength of this approach is in its affinity to the commonsense perceptional basis that underpins human action. Geomorphological flood studies, including recent advances in paleoflood hydrology, are needed as a complement to conventional hydrological approaches. The resulting complementarity will allow the predictions of the conventional approach to be grounded in the concrete particulars of experience. Without such grounding, flood science risks continuing as an empty quest for universal ideals while humanity, paralyzed by inaction, continues to suffer from the reality of particular floods.

1. Introduction

The term "flood" has a range of meanings (Baker, 1990; Matthai, 1990), including: (1) technical hydrological definitions specified in terms of magnitude and frequency of flow, (2) commonsense definitions focused on practical consequences such as damage to property, and (3) environmental science definitions expressed as consequences for biota, the landscape, or other elements of the environment. This range of meanings reflects a range of viewpoints concerning flood phenomena, not just between science and the general public, but within science itself. Moreover, the very notion that viewpoints can influence a scientific study

SSDI 0169-555X(94)00032-M

is controversial among scientists, many of whom hold their enterprise to be completely objective, such that its facts and theories are untainted by any relativism introduced by a particular point of view. A related position is that, by focusing on general phenomena subject to fundamental scientific laws, science can reduce any subjective bias that might compromise its factual information. The reliability of this information is considered to be absolutely essential to the application of science to societal problems, which include flooding as a natural hazard.

Floods are natural, sometimes humanly perturbed, processes operating on Earth's surface. They both affect and are affected by that surface, qualities that they share with many other geomorphological processes. To understand floods means to have knowledge of them leading to comprehension, interpretation, explanation, etc. Geomorphological understanding is to have this knowledge (1) relative to Earth's surface form (''geomorph''), and (2) in terms of clear, well-reasoned and orderly thought patterns (''logical''). Thoughts or ideas come into clarity through logical processes of thought that derive one statement from another. These logical processes are known as ''inference'', and they are essential both to mathematics and to science. Different forms of inference result in very different styles of science. Styles of science are chosen by its practitioners as matters of philosophy. The point here is that understanding is not limited to received knowledge of facts applicable to the natural world. It must also include the dynamic patterns of thought that relate the scientist to the natural world.

The viewpoint of this essay will be that of an Earth scientist concerned with landforms on the surface of that planet and with the processes that shape them. Does this viewpoint convey anything different from that of the hydrologist, who approaches the problem of flooding as an engineer or as a geophysicist? What does geomorphological understanding contribute to flood science that is not already provided by hydrology? Haines-Young and Petch (1986), for example, argue that some modern process geomorphology is indeed developing into a minor branch of engineering. They claim that this trend can be reversed by carefully posing hypotheses, by following the logical implications of these hypotheses, and, above all, by critically testing and falsifying those hypotheses in the manner described by Sir Karl Popper (1959). In this spirit, consider the following hypothesis and various consequences that follow therefrom: The greatest impediment to the mitigation of flood hazards is the lack of understanding as defined above. Understanding is neither the goal of most current ''flood'' research, nor is it effectively conveyed by most flood ''scientists'' to those at risk. The goal of this essay will be to explore this hypothesis and its unfortunate consequences for flood science.

2. Understanding versus explaining floods

The term ''flood'' must be used guardedly here, for it is generally not real floods, but rather various analysts' idealizations of flood attributes that are computed and arbitrarily imposed upon society. Real floods are not matters of conception, but of perception. This perception can arise either (1) through direct experience (Fig. 1), or (2), more commonly in the case of rare phenomena, through the indirect, but real consequences of past floods. The real consequences can include flood erosional features, flood deposits, and other signs (indices) that are used to infer the nature of the causative floods. In conventional flood science, however, it is not these reality-based signs of floods that are studied, but rather idealized symbols that stand for flood concepts. The study of real floods as processes interactive with human beings is in general decline. It is being replaced by a pseudoscientific manipulation of mathematical symbols masquerading as ''pure'' science and by an irresponsible cookbookery masquerading as ''applied'' science and engineering (Klemeš, 1986, 1987).

Reasons for the decline of the science of real floods are many. Pilgrim (1986) ascribes the decline to the difficulties in conducting field research in contrast to the ease of analytical studies with the computer. The higher prestige arbitrarily accorded mathematical sophistication in science (Alvarez, 1991) may also be a factor. The scientific paper emphasizes explanation, which is the expression of knowledge, and mathematics, as discussed below, is ideal for that expression. Moreover, the overemphasis of modern philosophy of science on explanation has led to a neglect of the importance of understanding, which is the faculty of mind that allows reality to be grasped. This faculty provides meaning to science, which is essential before expla-

Fig. 1. Santa Cruz River at St. Mary's Road, Tucson, Arizona. (a) The river at flood peak, 1490 $m^3 s^{-1}$, on October 1, 1983. (b) The channel immediately after flood recession. Note the confinement by soil-cement bank protection. Photographs by P. Kresan (see Kresan, 1988).

nation, through its models and predictions, provides guidance. For guidance to serve as the product of science to the society that sustains it that guidance must be meaningful.

The late Nobel laureate immunologist Sir Peter Medawar (1990) considered the explanatory scientific paper to be a fraud. He supported this outrageous statement by pointing out that scientific papers typically argue results through a logic that reveals nothing of the science by which those results were obtained. Scientific papers employ such rhetoric in order to get published and in order to convince colleagues of results, not to enlighten them on the process by which the science was done. This is why papers and explanations are so closely tied to logical thought patterns that are analytical and mathematical. The resulting neglect of synthetic thought patterns, essential to the reality of the scientific process, is, I believe, a major factor in the decline of the science of real floods. As the science of real floods has declined, global flood damages have accelerated. We are presently in a period of marked global increases in damage to life and property from floods and from other naturally catastrophic phenomena. The problem may be partly exacerbated by anthropogenic climatic change (''global warming''), but it is certainly tied to the human propensity to build in hazard-prone areas.

Most so-called ''flood science'' in the United States now concerns neither floods nor science. Rather than a quest for understanding the reality of floods, much of flood hydrology has become the development and application of methodologies for idealizing flood attributes followed by the arbitrary imposition of those idealized attributes upon human society through regulations and structural designs. Equally disturbing is the arbitrary separation of ''pure'' (academic) flood science from ''applied'' science and engineering. The former purports intellectual rigor through the mathematical manipulation of idealized parameters that are assumed to have flood-like properties. Flood ''engineering'' has increasingly been transformed to a cookbookery that seeks replicable, consistent accounting procedures to defend design decisions against accusation and litigation. Increasingly, modern flood ''science'' fails its responsibility of testing ''flood'' idealizations against reality in nature (Klemeš, 1982). Likewise, modern flood ''engineering'' fails its responsibility of applying the best science to achieve the highest quality designs consistent with societal values.

Readers who object to this hypothesis are challenged to falsify it. If it proves not completely falsified, then the challenge is to take action to reverse the unfortunate trends that it portents. Lest this somewhat outrageous hypothesis not be taken seriously, the following elaboration will explore its dimensions with special regard to flood geomorphology.

3. Engineering, mathematics, and science

Modern science has derived considerable prestige from its ability to predict with lawlike precision phenomena as remarkable as the structure of the atom, the motions of planets, and the origin of the universe. However, science has been decidedly unsuccessful in predicting either the real occurrences or the real sizes of catastrophic floods. Conventional flood science follows two standard approaches: (1) generalization from experience with small common floods to idealized properties of the large, rare floods from which society is assumed to be at risk, or (2) explanation of detailed, specific flood phenomena through theoretical generalizations that are assumed to apply to the entire class of phenomena. The relevant mathematics applied to each approach obscures their logical basis from most of the society that is protected by the flood-mitigation measures designed through implementation of these approaches. Does continuing and increasing flood damage indicate the failure of these approaches? Perhaps these two paths suffer a common problem: lack of commonsense grounding in reality. This point will be developed in subsequent sections.

While it is correctly assumed that all flood scientists need to understand appropriate branches of mathematics, it is more rarely recognized that they also need to understand certain mathematical attitudes. A brief anecdote serves as an example. A few years ago I was asked to serve on a National Research Council committee charged with producing a report on ''Estimating Probabilities of Extreme Floods''. The committee was composed mainly of statistical hydrologists interested in flood-frequency analysis and rainfall–runoff modelers interested in quantifying hydrological processes. The action of the committee centered around reconciling these two idealizations and in criticizing the nature of available flood data, including the paleoflood analyses that I had been placed on the committee to describe. One of the other committee members was a well-known mathematical hydrologist. During one point of the deliberations he observed, ''Scientific analysis implies that all the necessary equations can be written down and programmed for a digital computer''. The reality indicated by paleoflood data was obviously but an impediment to achieving this ideal.

The mathematization of nature is one of the great ideas that led to modern science. Copernicus's 1543 book *De revolutionibus* was aptly named, since it publicized a revolutionary transformation in thinking. Newton's *Principia* (1687) completed the transformation from Aristotelian natural philosophy to mechanical philosophy. As Edwin Burtt and Andre Koyré argued in the 1920s and 1930s (Lindberg and Westman, 1990), this Scientific Revolution arose, not from fundamental empirical discoveries, but rather from a series of conceptual transformations. Whether one ascribes metaphysical foundations to these transformations (Burtt, 1952) or simply ascribes a belief in the efficacy and certainty of mathematical knowledge (Shapiro, 1990), the fact remains that the mathematical character of nature rests upon an assumption, one with exceptionally productive consequences in certain applications, but an assumption nonetheless.

Mathematics is the most certain form of knowledge. Its truth lies not in correspondence to nature, but in rigor of proof and logical succession of reason. One of the marvels of modern theoretical physics has been the rather amazing correspondence that has been found between certain mathematical expressions and the real world. This has been termed "the unreasonable effectiveness of mathematics" (Wigner, 1969). However, this effectiveness is not predestined; it appears not by design, but serendipitously, as a matter of discovery. The continuing power of mathematics in relation to science does not derive from some metaphysical truth or from perfect accord with nature. Rather, its use has to do with what Ziman (1978) considers to be an essential goal of science: to achieve maximum *consensuality*. By this he refers to the following ideal (Ziman, 1978, p. 6): "... the general body of scientific knowledge should consist of facts and principles that are firmly established and accepted without serious doubt, by an overwhelming majority of competent, well-informed scientists". This goal requires appropriate communication in an unambiguous language, for which no better means has been found than mathematics.

Logic, and ideally mathematics, function to make scientific ideas clear and unambiguous. However, it is a trivial principle of logic that this formality provides no guarantee for success in science (Ziman, 1978). To judge the quality of science by the rigor of its mathematics is analogous to judging the quality of poetry by the precision of its grammar. There is no question as to the effectiveness and desirability of mathematics and grammar in their respective domains of communication. However, neither provides a sufficient or superior criterion on which to judge overall quality for the human enterprises of which they are a part. Indeed, both those enterprises share that most precious human quality, creativity, with all its inherent subjectivity.

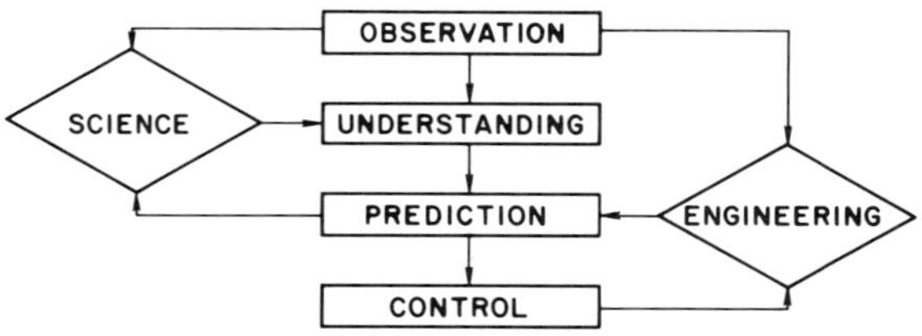

Fig. 2. Methodological distinction between science and engineering illustrated with diagrams (from Dooge, 1986, discussed by Baker, 1988).

As with science and mathematics, a kind of complementarity exists between science and engineering. Despairing of past engineering emphases in hydrology, the National Research Council (1991) proposed a new program for hydrology as a geophysical science. Yet, the same report distinguishes engineering as "problem solving" from science as "puzzle solving". This distinction fails to recognize the importance of discovering the relevant or interesting puzzle/problem to be solved. The nature of the solution involves observation leading to prediction (Fig. 2), the latter achieved through theory (models). Note that understanding precedes prediction. Scientists have the luxury of endlessly pursuing understanding while engineers must get on with the job given various constraints of time and resources. Engineers make predictions to solve some problem, such as designing a dam, while scientists make predictions to test hypotheses.

4. Flood frequency analysis

A fundamental problem that illustrates the concerns of modern flood hydrology is the estimation of the T-year flood flow for water-resources management. This problem is almost universally treated as a matter of applied statistics (Gries, 1983), rather than one of applied geology (Baker, 1982a). Statistics is a science focused on specific numerical facts or data. It treats these facts through objective collection, classification, and generalization. Its great contribution to other sciences is its quantification of uncertainty. Given adequate large numbers of facts/data on the phenomena of interest, modern statistics provides impressive authoritative statements about generalizations of the objectified phenomena.

There are two important caveats for treating flooding as a problem in applied statistics. The first is that in nature real phenomena like floods are far more diverse and complex than the simplifications objectified for statistical analysis. The latter requires the selective extraction of data or facts assumed to be representative of the phenomenon so that quantification can be achieved. This loss of individuality is especially a problem for the largest and rarest of the phenomena. Indeed the small numbers of catastrophic floods imply a violation of the probabilistic assumptions upon which statistics is based. The following statement, attributed to the social reformer, Arnold T. Toynbee, is especially eloquent on this issue:

> Science deliberately ignores those properties of phenomena that are not common to all the specimens of a species and that therefore cannot be expressed in quantitative terms. The price of quantification is the ignoring of uniqueness. This is a high price, because, in truth, uniqueness as well as uniformity is an intrinsic and essential property of phenomena.

Of course, the selection of representative data is done by excluding those aspects of reality considered to be less important. Readers familiar with the principle known as "Murphy's Law" will recognize that ignored aspects of reality have an uncanny ability to assume prominence when actions are predicated upon their unimportance.

The second caveat is more fundamental in that statistics itself has its own foundations. About these Savage (1954, p. 1) writes, "As for statistics, the foundations include, on any interpretation of which I have ever heard, the foundations of probability, as controversial a subject as one could name". Probability theory has its origin in the study of gambling, a human enterprise in which the rules are well understood, at least when the game is being fairly played. In nature the rules are for us to discover. Nature may or may not load the dice. Assuming that flood phenomena obey certain rules for games, following various idealized probability distributions, allows elegant applications of the probability calculus. But this elegance in no way assures success. This point is made especially well by the late Nobel laureate physicist Richard Feynman (1971, lecture 13, p. 8):

> There is no way of showing mathematically that a physical conclusion is wrong or inconsistent. All that can be shown is that the mathematical assumptions are wrong. If we find that certain mathematical assumptions lead to a logically inconsistent description of nature, we change the assumptions, not nature.

Appropriately, Feynman figured prominently in perhaps the most famous recent incident of exposing faulty probabilistic reasoning. This was the tragic explosion of the space shuttle Challenger in 1986, which had profound implications for public support of U.S. science and technology. Feynman served on the presidential commission investigating the disaster (Gleick, 1992). Prior to the mission, official estimates by the National Aeronautics and Space Administration (NASA) placed the chances of disaster on a given flight at approximately 1 in 10,000. Although the numbers added up, the assumptions behind them did not. Feynman made a dramatic point at a public news conference when he dunked a small rubber seal from one of the shuttle booster rockets into a glass of ice water. The cold temperatures on the morning of the launch would cause the seal to lose resiliency. The probability of failure at the time of launch was thus much higher than predicted by NASA.

Now here is the question that has relevance to the flood problem. Which probability does society care about when concerned with a possible disaster? Is it interested in theoretical probabilities in which numbers add up, but the assumptions behind these numbers are at best naive or at worst prejudiced by unstated biases? Though claimed to be objective, such probabilities are, in fact, subjective. Would it not be better to use subjective probabilities that incorporate some understanding of the reality of the situation? The probabilities at a given moment of launch, or during a season of known flood occurrence, are what matter, not the probabilities that apply to shuttle launches, or to floods, in general.

NASA's Space Shuttle experts in 1986 were making a very dangerous assumption: that the probability of launch failures in general was equal to the probability of failure for a specific launch. This assumption has no basis in reality unless it can be tested for validity. Unfortunately, the one test had tragic consequences.

The resulting inquiry showed even the assumptions about the general probabilities to be wrong. NASA now estimates probabilities of shuttle launch failures as closer to 1 in 100.

In flood-frequency analysis (FFA) an equally dangerous assumption is made. Klemeš (1989) states it as follows, "... that the historic record of a hydrological phenomenon can be regarded as one of an infinite number of equally likely realizations of a stochastic process in which ensemble averages are equal to time averages". The consequences of his assumption, if believed without challenge, are even more devastating (Klemeš, 1989, p. 46):

> The leap of logic by which the instantaneous probabilities are equated with the historic frequencies of occurrence is nothing else but a dismissal of any meaning of the historic process: if anything that happened in the past can happen at any instant with the same likelihood, then history provides no meaningful information.

This assumption manifests itself in the doubly fallacious term "hundred-year flood". These words, of course, have essentially nothing to do with real years or with real floods. Instead, the words represent an idealization based on (often fallacious) assumptions. They are an example, not of science, but of an insidious form of philosophical nominalism.

A fundamental problem with FFA is that it is a particularly weak form of reasoning, induction, which generalizes from specific cases. Induction deals with some characteristic of interest from nature; in FFA this is usually flood discharge. From finite examples of this characteristic one draws inferences concerning the larger class of phenomena. For some philosophers, such as John Stuart Mill, such inferences are possible because of the uniformity of laws of nature. There are regularities to be discovered by our objective sampling because such regularities are inherent for nature. The reply to this, of course, is to ask who are we to say that such regularities exist. Their assertion is a matter of metascience, not of science. Far from discovering an answer about nature, that floods follow some relationship, for example, the FFA poses productive scientific questions. Do all floods, even those much larger than those sampled, follow the same relationship? Why do the floods sampled in a particular region follow the relationship? Do floods from another region follow the same or different relationships? Why do the floods follow or not follow such relationships? It is not so much answers, as it is the quality of questions that drives science.

J. Hoover Mackin's (1963, p. 148) point about empirical geomorphology applies nicely to FFA: "A statement of a problem can be very valuable, but if it is mistaken for a conclusion, it is worse than useless because it implies that the study is finished when in fact it is only begun". But there is an added confusion here. The results of FFA, and of similar quantitative flood studies, may be accepted as engineering solutions for certain applications. This must be done with great care.

Based solely on statistical arguments and despite many warnings to the contrary, there is all too often an unfortunate tendency for some flood engineering to become a kind of "cookbookery" in which standard theories are applied without regard to their limitations in the real world. The following, slightly exaggerated scheme illustrates the chain of logic that might apply in an extreme case of FFA cookbookery:

(1) Assume that all floods are random samples drawn from phenomena that neatly obey certain mathematical probability distributions.

(2) Treat all the floods as simple numbers, ignoring all the other complex attributes associated with them.

(3) Concentrate the analysis on small, common floods for which the largest number of data points can be obtained.

(4) Dismiss most information on big, rare floods as pertaining to outliers or subject to measurement error.

(5) Use elegant mathematical theory, unintelligible to decision-makers or the general public, to make impressive predictions about all flood phenomena.

(6) Justify steps 1 to 5 as standard procedure, proven by their useful application in numerous other FFA studies.

Responsible engineering incorporates professional experience into the solution. Tolerances are included because of possible inapplicability of general theory to individual cases. Nevertheless, a kind of unscientific circularity is becoming rampant in the whole field of FFA. Klemeš (1987, p. 2) describes the kinds of argument that make the field insensitive to common sense:

> (a) When a hydrologist criticizes FFA for its assumptions trivializing the complex and inadequately understood processes resulting in extreme floods, and for the consequent lack of realism of its results, the FF analyst will assume the posture of a down-to-earth engineering

practitioner and say that the engineer cannot wait until everything is known and needs a result, however crude, now because he must make decisions on flood-related measures now.

(b) When an engineer summons his courage and asks why then, given the simplifications and distortions of reality implied in the basic assumptions, should he employ all the sophisticated mathematic machinery to get that crude result, and why not use some simple procedure instead, the FF analyst will immediately change into a meticulous theoretician and extol the merits of mathematical rigor in the quest for extraction of the maximum amount of information from the data. (c) And should a statistician ask him how does he know that his rigorously fitted distribution models give information relevant to flood frequencies, the FF analyst would transform into a seasoned hydrologist and put him to shame for not being aware of the empirically established time-honored ability of those models to represent hydrological data and of the good service they have rendered generations of hydrologists in such important tasks as flood frequency analysis, etc. At this point the argument returns to item (a) above.

The above contains a form of argument masquerading as science that is even more insidious than cookbookery masquerading as engineering. This is the ''mathematistry'' defined by statistician George E.P. Box (1976, p. 792):

... Mathematistry is characterized by development of theory for theory's sake, which, since it seldom touches down with practice, has a tendency to redefine the problem rather than to solve it... In such areas as sociology, psychology, education, and even, I sadly say, engineering, investigators who are not themselves statisticians sometimes take mathematistry seriously. Overawed by what they do not understand, they mistakenly distrust their own common sense and adopt inappropriate procedures derived by mathematicians with no scientific experience.

5. Modeling and predicting floods

For a number of years various colleagues and I have been interested in the interpretation of geomorphological evidence of ancient floods (Baker, 1973, 1975, 1977, 1987, 1991; Patton and Baker, 1977; Kochel and Baker, 1982; Kochel et al., 1982; Ely and Baker, 1985; Partridge and Baker, 1987; Baker et al., 1988, Webb et al., 1988; Ely et al., 1993; Enzel et al., 1993; Martinez-Goytre et al., 1994; O'Connor et al., 1994; Wohl et al., 1994). Our studies of geological evidence for ancient floods (paleofloods) have been stimulated by the excellent preservation of that evidence in the southwestern United States. We have found remarkably complete records of floods extending back as much as 10,000 years. The resulting data show the natural trends or habits of flood behavior in the region. Rather than stipulating how idealized flood concepts predict particular facts, assumed to be real, we have instead documented how particular floods have behaved. Meaningful general concepts have emerged from the synthesis of real particulars, rather than ideal particulars following logically by analysis from accepted generals. In other words, the floods have told us their properties rather than we, through our theories, telling them. This work has interesting implications regarding attitudes toward nature and the management procedures that derive therefrom, but it runs contrary to prevailing attitudes among conventional flood scientists.

The latter point, involving attitudes, is illustrated by anonymous reviewer comments for a paleoflood hydrology review paper that was subsequently published (Stedinger and Baker, 1987). The unpublished review comments were as follows:

Imagine a solid-state physicist, organic chemist or other practitioner of ''hard science'' reading this manuscript... Shaking his head with sad amusement as he muses about how far the geologists still have to go before their field of study can properly be called science, he leafs through the final section of the manuscript and references. This wipes the smile off his face, for he discovers that these paleohydrologic methods are being advocated, in all seriousness, for use in assessing the safety of dams and choosing sites for hazardous waste disposal.

Of course, since secrecy has no part in ethical science, these comments by a secret reviewer are not scientific, but they do illustrate a kind of righteousness about what constitutes good science. The underlying myth seems to be that good science requires an orderly progression from qualitative description to quantitative rigor. The history of hydrology has been invoked in support of this myth. While hydraulic engineering monopolized early quantitative approaches to analyzing the hydrological cycle, scientific hydrology derived from more qualitative attempts by geographers and meteorologists to understand the cycle. Sixty years ago Robert E. Horton (1931) could write that hydrology, newly established as a section of the American Geophysical Union, was on its way to being a quantitative geophysical science. In writing a history of hydrology, the great engineering hydrologist V.T. Chow (1964) was able to note that the field had passed through an early empirical or observational phase that occurred prior to Horton's period. This was followed by a rational phase in which understanding was achieved for hydrological processes. As this was increasingly conveyed by mathematical formulations, hydrology at last became predictive. The predictive phase of hydrology, accelerated by advances in computer modeling, has dominated the field ever since.

How can one criticize the obvious progress that provides confidence in hydrological applications? The criticism is that methodological elegance and precise prediction do not insure understanding. Science uses prediction as a tool to test explanations built upon its temporary state of understanding (Fig. 2). Engineering assumes that temporary state of understanding in proceeding directly to a useful prediction. The engineer must always be concerned that the basis in understanding is inadequate, while the scientist must always strive to elucidate the inadequacy of that understanding. All too often modern "scientists" assert their expertise on how nature works, and "engineers" make predictions as though that expertise is justified. The result is what Klemeš (1986) has termed "dilettantism". Just as ancient alchemists and astrologers utilized elaborate methodologies to make the predictions that medieval societies required of them, so modern dilettantism emphasizes useful predictions made by the most advanced physical and mathematical theory. The appearances of sophisticated explanatory power are emphasized over any reasoned testing of the powerful procedures against the reality of nature.

I wish to be very clear on this point, because it has been misinterpreted rather badly in the past. The criticism of the predictive-theoretical mindset is not meant to disparage the ideal role that prediction and theory play in science. It is through these tools that science achieves answers to questions. Moreover, it is a rather trivial matter to assert the value judgment that answered questions are superior to unanswered questions, all other things being equal. My concern is that in the real world of real phenomena, as opposed to the abstractions of the theoretical world, all other things are not equal. Theoretical predictions of that world are invariably fallible. In the real world unanswered questions are certainly no less an impediment to science as are unquestioned answers. Indeed, the case is strong that the latter are far more insidious. While the irritation of doubt associated with the unanswered question can be a great stimulus for research, leading to discovery, the smug confidence in the unquestioned answer is the greatest roadblock to scientific progress. Of course, these concerns would not apply if research were actually conducted in the ideal fashion, with slavish attention to every detail of scientific method. But this is a theoretical ideal, which is exactly the point. Science is conducted by human beings, wonderfully dedicated to their craft, to be sure, but human in spite of their aspirations.

The analytical–conceptual approaches of modern predictive physical science provide extremely powerful methodologies for extrapolation. The universality of certain physical principles allows the calculation of properties of the largest freshwater floods to have occurred on Earth (O'Connor and Baker, 1992; Baker et al., 1993; O'Connor, 1993) and even to extend this knowledge to remarkable ancient floods that occurred on Mars (Baker, 1982b; Baker et al., 1991, 1992). While such results reveal an almost magical ability of analytical logic to extend productive scientific thought, it does not logically follow that such success will compel, or even justify, belief that the same tools will provide what society desires in regard to a specific flood problem: certainty that the problem will be satisfactorily resolved.

There is a long tradition to the idea that certainty is the ultimate and most important quest of science. Dewey (1929) traces the quest for certainty to the ancient Greeks, who invented the western versions of both science and philosophy (Russell, 1945). The Greeks sought certainty for the same reasons that we have come to the subject in the present context: certainty is thought to provide security from hazards of the temporal world (Ratner, 1939). The problem with this quest is that certainty is not to be found in the material and practical world of action; instead, it is found in the eternal, immaterial realm of thought (Dewey, 1929). Although the Greeks retained a vision of a purposeful heterogeneous real world of change, the scientific revolution of the 17th and 18th century (Cohen, 1985) contributed further to denigration of the qualitative and purposeful world of action. As Dewey (1929, p. 87) describes, "All that counted for science became mechanical properties formulated in mathematical terms..."

The National Research Council (1978, p. 2) stated, "... prediction of future events is a major aim of all geophysical sciences... in no other field of science is prediction so intimately related to human safety and welfare''. Prediction is stated to be the goal of science precisely because, ''... geophysical events are among the greatest threats to people. Predictions can help in assessing... risks... They can forestall casualties and mitigate disasters''. There is a *non sequitur* here, one that unfortunately seems all too prevalent among many scientists. It is certainly true that floods and other hazards are immense threats to society. It is also true that accurate preknowledge of their size and occurrence would contribute to hazard mitigation. However, it does not follow that a prediction is the same as accurate preknowledge. The difference has a lot to do with the two roles of prediction, roles that are too rarely distinguished by scientists themselves.

Geophysical predictions can only rarely be responsibly posed in categorical terms, stating, for example, that a given size flood will or will not occur at a specific place and time. Rather they must be expressed with uncertainties. This either (1) returns us to the statistical problems discussed in the previous section, or (2) it requires progressive reduction of the uncertainties. The latter approach is well illustrated by the climatic modeling community (Mahlman, 1989, p. 71):

> Through careful, long-term research on observation, modeling, and analysis, our scientific uncertainties will decrease and our confidence for predicting details of the climate system and its changes will gradually improve. A final, very confident prediction is that the societal need for accurate and detailed climate predictions will increase as fast or faster than the scientific community can provide them. The effort to meet these societal challenges will require the combined forces of the world scientific community in a sustained effort spanning decades.

The assumption here seems to be that if we scientists will only predict the future with some appropriately low uncertainty we will have done our duty for society, which will then in its wisdom act upon this knowledge. However, common sense says that society wants a fortunate future, not a predicted one. Unlike the past, the future is available for humankind to modify. To the degree that we predict that future it is only from the assumption of present conditions or in terms of anticipated action. The whole point is to allow unanticipated action, which will then violate the premises of the prediction.

The second role of prediction is not to tell the future in the manner of ancient oracles or soothsayers. Rather, it is to express the consequences of an hypothesis, these days usually expressed through a scientific model. Moreover, the requisite attitude is not to see the prediction work but to see it fail, as Sir Karl Poper has championed since 1919, based upon the famous British eclipse expeditions to test Einstein's prediction of light bending. Popper (1976, p. 38) writes in his biography, ''Thus I arrived, by the end of 1919, at the conclusion that the scientific attitude was the critical attitude, which did not look for verifications but for critical tests; tests which could *refute* the theory tested, though they never could establish it''. This role of prediction, and the attitude that sustains it, can be considered the only proper role of prediction in science.

It is our paradoxical duty as scientists both to exalt our conceptual frameworks for their power of suggestion, simplification, or prediction, and also to criticize the same concepts for limitation of perspective, exclusion of experience, or stifling of innovation. For all its utility in scientific theorizing, predictive systems thinking restricts itself to an ideal world that cannot be the sole world of scientific inquiry. The pragmatist philosopher G.H. Mead addresses this point with particular clarity (Mead, 1938, p. 31): ''The world which is the test of all observations and hypothetical reconstructions has in itself no system that can be isolated as a structure of laws or uniformities, though all laws and formulations of uniformities must be brought to its court for its *imprimatur.*''

All scientific knowledge is fallible. To claim certainty is to run counter to the spirit of science itself, so well expressed by the mathematician William Clifford, as quoted by Bronowski (1979, p. 128):

> Remember, then, that scientific thought is the guide of action; that the truth at which it arrives is not that which we can ideally contemplate without error, but which we may act upon without fear...

The scientific bandwagons of theoretical modeling seem to be dominated by those who would argue a guide to action based on ''reduced uncertainties'', approaching a truth with our models that, ''... we can ideally contemplate without error''. Yet, if Clifford's pragmatic view is correct, what society's decision makers really need is that truth, ''... which we may act upon without fear''. Perhaps the promise of reduced uncertainty is an empty one if we expect to find this basic truth solely or even predominantly in predictions derived from our idealizations, our ''models'', of reality. Might not at least an equal portion of the necessary ''guide of action'' lie in some measured experience of reality?

6. Societal action and flood geomorphology

The summer 1993 flooding of the upper Mississippi River valley has heightened American public interest in flood hazards. Noting that damages may exceed $15 billion, Frank Press (1994) cites the pragmatist philosopher John Dewey in regard to what should follow: ''The intellectual function of trouble is to lead men to think''. As one of the founders of the International Decade for Natural Disaster Reduction (IDNRD), Press expresses hope that this and other disasters will help inform a public who will then influence their political leaders. But it is not necessary to await future floods to awaken perception to the reality of flooding. The real-world signs or indices of past floods are manifest upon the landscape, awaiting geomorphological translation into common language. Moreover, the story is not one made up according to current intellectual fads. It is read directly from Earth's own experience of its global environmental changes. The vast repository of Earth's landforms and sediments contains the entire continuum of reality that must be understood in order to interpret properly the isolated systems described by theoretical models. Moreover, we now possess the tools with which to open up that repository. Just as the computer revolution has enhanced the scientific ability to deduce future consequences of antecedent conditions, a less heralded revolution has occurred in our ability to retroduce the antecedents of those consequences that are preserved in the geological/geomorphological record. Advances in dating past events, and advances in quantifying the characteristic parameters of those events, mean that we have phenomenal opportunities to analyze multiple components of the real interactions that have occurred during past environmental changes. Regrettably, too little attention has been paid by the flood science/engineering community to these opportunities, owing, at least in part, to factors described above.

It is well to remember that the various discharges, probabilities, velocities, etc. that are calculated for floods have no practical meaning in and of themselves. Their practical meaning lies solely in their consequences. Floods produce damage, loss of life. They also produce effects on the landscape. It is solely through such effects that the word ''flood'' and concepts associated with it assume meaning for humankind. The importance of flood consequences is well illustrated by the recent history of flooding in Tucson, Arizona (Baker, 1984; Saarinen et al., 1984; Slezak-Pearthree and Baker, 1987; Webb and Betancourt, 1992; Parker, 1993). Idealizations of flood sizes, probabilities, and inundation zonation failed to be realized in the great floods of 1983 and 1993. Instead, buildings collapsed as banks were undercut (Fig. 3), and protective works designed to solve those problems induced new problems that had not arisen heretofore (Fig. 4).

Floods are part of a very broad range of water-related phenomena that have immense impacts upon human beings. These phenomena exert their influence on human affairs mainly through concerns about water supply and hazard reduction. It is out of such concerns, rather than inherent curiosity, that society provides resources for research on hydrological processes. The designs that serve to bring research results into practice are the products of various water engineers, and engineering philosophies have until recently dominated hydrological research. Due in part to heightened recognition of global environmental problems, however, there is a current initiative to emphasize ''fundamental hydrologic science... as a geoscience alongside the atmospheric, ocean, and solid earth sciences'' (National Research Council, 1991). Critical to the rationale for this new emphasis is the relationship of water geoscience to the world of human affairs: water resources management and decision making. The role of geomorphology in this conceptualization (National Research Council, 1991, p. xi) is to be one of many disciplines concerned with understanding water phenomena (''hydrologic science''). When combined with statistics, fluid mechanics, meteorology, and other fundamentals, these sciences provide the basis in understanding for forecasting, which is stated to be the proper role for engineering hydrology. The forecasting role can then be combined with economics, law, political science, and resource sciences to justify decisions by water-resource managers.

Of course, this is business as usual. In this regard, let us consider a possible model forecast, expressing a high flood risk for a certain site. Suppose that the model

Fig. 3. Destruction of apartment buildings by bank erosion of Rillito Creek, Tucson, Arizona, October 2, 1983. Photograph by S. Reynolds, Arizona Geological Survey.

has been verified for many similar situations, but, of course, the application to the specific site leaves some uncertainty. Using statistical hydrology, this uncertainty is appropriately quantified. Now suppose new developments are planned for this site, perhaps at considerable public expense. As responsible citizens, the knowledgeable scientists should, at a minimum, make these results known. It might even be suggested that some advocacy is called for. Public money might more wisely be spent on projects where risks are lower. At this point the advocates of development begin a public campaign in support of their plans. They note that some scientists have "theories" that suggest high risk, but that these theories are by no means certain. They produce other experts who explain their view that the risks are overstated, and that their calculations reveal much lower hazard probabilities. Given the great potential benefit of the project, it should not be abandoned because of the pet theories of some scientists. The uncertainty thus poses a considerable dilemma. Proper scientific methodology requires it to be acknowledged. However, the decision makers and the public they serve have little understanding of uncertainty. They want a certain basis for their actions, indeed that is what they expect from scientific facts. Given uncertainty and disagreement among experts, the decision makers can rather easily justify a position contrary to the original model prediction.

One way to avoid the above scenario is to specify a standard procedure for the risk estimation. This procedure must be sanctified in such a way that all are forced to adhere to it. Indeed this was exactly the reason for development of the concept of the "hundred-year flood", mentioned so disparagingly above. When determined by a prescribed procedure, such a concept has a very precise meaning. Moreover, when that procedure can be applied by any knowledgeable practitioner, the result is a uniform standard. Increasingly, however, such standards are being developed to afford protection, not against real floods, but against the accusation that the practitioner has failed in his or her official duties. Any responsibility for injustice to reality lies

Fig. 4. Aerial view of Rillito Creek bend shown in Fig. 3. (a) Photograph made by V.R. Baker on October 8, 1983, after flood recession. Note the prominent meander bend at lower left which migrated into vacant land immediately upstream of the apartment complex, which was partly protected by a soil-cement embankment. The erosion occurred at the unprotected bank. (b) Photograph taken on January 20, 1993, by V.R. Baker, shows new soil-cement bank protection upstream of the same apartment complex, which was undamaged during the 1993 flooding. Note, however, that extensive erosion occurred on the unprotected bank of the downstream meander bend (top center). Arrows point to an electric power line tower that is common to both photographs. The destructive effects of such partial bank protection are described by Baker (1984), Saarinen et al. (1984), and by Slezak-Pearthree and Baker (1987).

Fig. 5. Town of Winkelman, Arizona, on January 20, 1993. Discharge of the Gila River was approximately 850 $m^3 s^{-1}$ at this time. Photograph by V.R. Baker.

not solely with the practitioner; it must be shared by a litigious society thirsting after scapegoats for its own failures of perception and action.

The oft-cited reactions of the general public to the "hundred-year flood" designation reflect naiveté for hydrological idealizations and common sense for flood realities. One reaction to these responses is to attempt education and enlightenment on probability theory. Efforts in this regard have not proven particularly successful, nor is there reasonable prospect that they ever will be.

Behavioral social scientists have long argued that many, perhaps most people take action with regard to risks (and have beliefs) not because of objective measures of the threat, including model predictions or idealizations like "hundred-year" floods. Instead, they act upon personal perceptions of the risk. The residents of Winkelman, Arizona (Fig. 5), for example, acted upon their perception of a "flood-control" dam (Fig. 6) upstream of their floodplain location. They built and/or maintained their residences in a flood zone. The perception they had was that dams control floods. The residents were not provided the important concept that really big floods are not controlled by dams. Rather, it is the dams that must be protected from such events, by passing water over spillways, since failure of a dam produces a much greater flood than nature could ever generate. It would seem that many such scientific flood concepts are not properly conveyed to the public. Indeed, it is an irony of modern science that its conceptual emphasis, born of a need to bring rigor to explanations, has served to distance science from the great societal goal toward which many believe it aspires: to serve as the basis for societal action.

Herein lies a fundamental fallacy. Flood concepts from science can best help to guide action, not serve as its basis. The basis for action lies in percepts, not concepts. Perceptions derive from reality, and it is geomorphology that can provide evidence of that reality. Showing where floods once coursed across the land-

Fig. 6. Coolidge Dam on Gila River approximately 30 km northwest and upstream of Winkelman (Fig. 5) on January 20, 1993. Discharge through the spillways was approximately 850 $m^3 s^{-1}$ because of hazard posed to the dam by inflow to the nearly full reservoir that peaked at about 3000 $m^3 s^{-1}$. Fortunately the flood inflow was short-lived so that spillway flows and resulting downstream flooding (Fig. 5) would be reduced.

scape shows the long-term extent of the river. Those who construct valuable entities within the paleoflood limits are not doing so in some idealized regulatory zone defined by scientific experts. They are constructing in the river as nature has defined it. People who insist on building in rivers should both know and deserve the consequences of doing so.

Is there a way out of this dilemma: on the one hand, arbitrarily specifying how floods behave by standard procedures universally applied and conceptually justified, but meaningless to all but a few specialists; on the other hand, allowing primitive perceptions to dictate societal action, blind to the guidance of modern science? I think there is a way out of the dilemma, and that the way is the same as that which has sustained the process of science in its path to discovery as well as in its explanations of what was discovered. The way involves communicating to the public not merely the detached analytical science that so typifies the conventional approach, but also the vibrant understanding of real floods as explored by those who interpret the signs of those floods. Here then is a central role for flood geomorphologists.

Geomorphologists have generally been hesitant to bring their knowledge directly to the public, perhaps out of the recognition of its inherent uncertainty. This humility is misplaced. The reality of geomorphological flood knowledge far outweighs any uncertainty, especially in comparison to the "artificial" knowledge most often conveyed by conventional hydrology. People's perceptions, on which they are prepared to act, are open to what they can see and touch. Geomorphological flood evidence is as accessible to their senses, including their common sense, as it is to those of the hydrologist.

This cannot be said of the latter's concepts. Perhaps this model for science/society interaction deserves to be tried. Then it will have the opportunity to achieve reality through its consequences.

7. Conclusions

In the legalistic traditions of the American political system, scientific knowledge enters decision making in an authority mode. No role of scientific knowledge could be more inimical to the spirit of science itself. It is one of the great fallacies of popular modern thought that science can somehow provide fundamental truths or concrete facts that provide an infallible, certain basis for societal action. Quite the contrary! The whole process of science itself is possible because of continual questioning of the received wisdom, an infinite voyage of discovery, hypothesizing, determining consequences, and testing against reality. To impose upon society that which science would not impose upon itself is hypocritical in the extreme.

Scientific knowledge, predictions, and the rest are not sacred ideals which we should expect to be slavishly followed because of their proclaimed superiority as bases for human belief. Rather, these concepts are valuable as reliable guides to action, consistent with but immensely extending the commonsense reasoning processes of all humankind (Bronowski, 1979; Ziman, 1978). The key to enlightening humankind with this guidance is not to replace their assumed ignorance with the conceptual "truths" of science. Rather, the human perception that is the basis for belief needs to be stimulated with real understanding. In the perception of floods, geomorphologists have more to contribute than most other sciences, including much of conventional hydrology.

Acknowledgments

I thank S.A. Schumm for critical comments contributing to the revision of this essay. Flood geomorphological research by the author has been supported in recent years by the National Science Foundation Grants EAR-88-05321, BCS-8901430, EAR-9104489, and EAR-9305252.

This publication is contribution No. 30 of the Arizona Laboratory for Paleohydrological and Hydroclimatological Analysis (ALPHA), University of Arizona.

References

Alvarez, W., 1991. The gentle art of scientific trespassing. GSA Today, 1: 29–31, 34.

Baker, V.R., 1973. Paleohydrology and sedimentology of Lake Missoula flooding in eastern Washington. Geol. Soc. Am. Spec. Pap., 144: 79 pp.

Baker, V.R., 1975. Flood hazards along the Balcones Escarpment in central Texas: Alternative approaches to their recognition, mapping and management. Univ. Texas, Bureau of Economic Geology Circ. 75-5, 22 pp.

Baker, V.R., 1977. Stream channel response to floods with examples from central Texas. Geol. Soc. Am. Bull., 88: 1057–1070.

Baker, V.R., 1982a. Geology, determinism, and risk assessment. In: Scientific Basis of Water-Resource Management. National Academy Press, Washington, D.C., pp. 109–117.

Baker, V.R., 1982b. The Channels of Mars. Univ. Texas Press, Austin, 198 pp.

Baker, V.R., 1984. Questions raised by the Tucson flood of 1983. In: Hydrology and Water Resources in Arizona and the Southwest. American Water Resources Association and Arizona–Nevada Academy of Sciences, Proc. of the 1984 Meetings, 14, pp. 211–219.

Baker, V.R., 1987. Paleoflood hydrology of extraordinary flood events. J. Hydrol., 95: 79–99.

Baker, V.R., 1988. Geological fluvial geomorphology. Geol. Soc. Am. Bull., 100: 1157–1167.

Baker, V.R., 1990. Floods. In: Magill's Survey of Science: Earth Science Series. Salem Press, Pasadena, CA., pp. 719–725.

Baker, V.R., 1991. A bright future for old flows. In: L. Starkel, J. Thornes and K.J. Gregory (Editors), Fluvial Processes in the Temperate Zone during the Last 15,000 Years: Temperate Palaeohydrology. Wiley, New York, pp. 497–520.

Baker, V.R., Kochel, R.C. and Patton, P.C., 1979. Long-term flood frequency analysis using geological data. Internat. Assoc. Hydrological Sciences Publ. No. 128, pp. 3–9.

Baker, V.R., Kochel, R.C. and Patton, P.C. (Editors), 1988. Flood Geomorphology. Wiley, New York, 503 pp.

Baker, V.R., Strom, R.G., Gulick, V.C., Kargel, J.S., Komatsu, G. and Kale, V.S., 1991. Ancient oceans, ice sheets and the hydrological cycle on Mars. Nature, 352: 589–594.

Baker, V.R., Carr, M.H., Gulick, V.C., Williams, C.R. and Marley, M.S., 1992. Channels and valley networks. In: H.H. Kieffer, B. Jakosky and C. Snyder (Editors), Mars. Univ. Arizona Press, Tucson, pp. 493–522.

Baker, V.R., Benito, G. and Rudoy, A.N., 1993. Paleohydrology of late Pleistocene superflooding: Altay Mountains, Siberia. Science, 259: 348–350.

Box, G.E.P., 1976. Science and statistics. J. Am. Stat. Assoc., 71: 791–799.

Bronowski, J., 1979. The Common Sense of Science. Harvard Univ. Press, Cambridge, MA, 154 pp.

Burtt, E.A., 1952. The Metaphysical Foundations of Modern Science. Humanities Press, Atlantic Highlands, NJ, 352 pp.

Chow, V.T., 1964. Hydrology and its development. In: V.T. Chow (Editor), Handbook of Applied Hydrology. McGraw-Hill, New York, pp. 1-1 to 1-22.

Cohen, A.B., 1985. Revolution in Science. Harvard Univ. Press, Cambridge, MA, 711 pp.

Dewey, J., 1929. The Quest for Certainty. Minton, Balch and Co., New York, 318 pp.

Dooge, J.C.I., 1986. Looking for hydrologic laws. Water Resour. Res., 22: 46S–58S.

Ely, L.L. and Baker, V.R., 1985. Reconstructing paleoflood hydrology with slackwater deposits, Verde River, Arizona. Phys. Geogr., 6: 103–126.

Ely, L.L., Enzel, Y., Baker, V.R. and Cayan, D.R., 1993. A 5000-year record of extreme floods and climate change in the southwestern United States. Science, 262: 410–412.

Enzel, Y., Ely, L.L., House, P.K., Baker, V.R. and Webb, R.H., 1993. Paleoflood evidence for a natural upper bound to flood magnitudes in the Colorado River Basin: Water Resour. Res., 29: 2287–2297.

Feynman, R., 1971. Lectures on Gravitation. California Inst. of Technology, Pasadena, 78 pp.

Gleick, J., 1992. Genius: The Life and Science of Richard Feynman. Pantheon Books, New York, 532 pp.

Gries, N.P., 1983. Flood frequency analysis: A review of 1979–1982. Rev. Geophys. Space Phys., 21: 699–706.

Haines-Young, P.H. and Petch, J.R., 1986. Physical Geography: Its Nature and Methods. Harper and Row, London, 230 pp.

Horton, R.E., 1931. The field, scope, and the status of the science of hydrology. EOS Trans. Am. Geophys. Union, 12: 189–202.

Klemeš, V., 1982. Empirical and causal models in hydrology. In: Scientific Basis of Water Resource Management. National Academy Press, Washington, DC, pp. 95–104.

Klemeš, V., 1986. Dilettantism in hydrology: Transition or destiny? Water Resour. Res., 22: 177S–188S.

Klemeš, V., 1987. Hydrological and engineering relevance of flood frequency analysis. In: V.P. Singh (Editor), Hydrologic Frequency Modeling. Reidel, Dordrecht, pp. 1–18.

Klemeš, V., 1989. The improbable probabilities of extreme floods and droughts. In: Ö. Starosolszky and O.M. Melder (Editors), Hydrology of Disasters. James and James, London, pp. 43–51.

Kochel, R.C. and Baker, V.R., 1982. Paleoflood hydrology. Science, 215: 353–361.

Kochel, R.C., Baker, V.R. and Patton, P.C., 1982. Paleohydrology of southwestern Texas. Water Resour. Res., 18: 1165–1183.

Kresan, P.L., 1988. The Tucson, Arizona, flood of October 1983. In: V.R. Baker, R.C. Kochel and P.C. Patton (Editors), Flood Geomorphology. Wiley, New York, pp. 465–489.

Lindberg, D.C. and Westman, R.S. (Editors), 1990. Reappraisals of the Scientific Revolution. Cambridge Univ. Press, New York, 551 pp.

Mackin, J.H., 1963. Rational and empirical methods of investigation in geology. In: C.C. Albritton (Editor), The Fabric of Geology. Addison-Wesley, Reading, MA, pp. 135–163.

Mahlman, J.D., 1989. Mathematical modeling of greenhouse warming: How much do we know? In: R.S. DeFries and T.F. Malone (Editors), Global Change and Our Common Future: Papers from a Forum. National Academy Press, Washington, DC, pp. 62–72.

Martínez-Goytre, J., House, P.K. and Baker, V.R., 1994. Spatial variability of small-basin paleoflood magnitudes for a southeastern Arizona mountain range. Water Resour. Res., 30: 1491–1501.

Matthai, H.F., 1990. Floods. In: M.G. Wolman and H.C. Riggs (Editors), Surface Water Hydrology. Geol. Soc. America, The Geology of North America, 0-1, pp. 97–120.

Mead, G.H., 1938. The Philosophy of the Act. Univ. Chicago Press, Chicago, 696 pp.

Medawar, P., 1990. The Threat and the Glory. Oxford Univ. Press, Oxford, 291 pp.

National Research Council, 1978. Geophysical Predictions. National Academy of Sciences, Washington, DC, 215 pp.

National Research Council, 1991. Opportunities in the Hydrological Sciences. National Academy Press, Washington, DC, 348 pp.

O'Connor, J.E., 1993. Hydrology, hydraulics, and sediment transport of Pleistocene Lake Bonneville Floods on the Snake River, Idaho. Geol. Soc. Am. Spec. Pap., 274: 83 pp.

O'Connor, J.E. and Baker, V.R., 1992. Magnitudes and implications of peak discharges from glacial Lake Missoula. Geol. Soc. Am. Bull., 104: 267–279.

O'Connor, J.E., Ely, L.L., Wohl, E.E., Stevens, L.E., Melis, T.S., Kale, V.S. and Baker, V.R., 1994. A 4500-year record of large floods on the Colorado River in the Grand Canyon, Arizona. J. Geol., 102: 1–9.

Parker, J.T.C., 1993. Channel Change on the Santa Cruz River, Pima county, Arizona, 1936–1986. U.S. Geol. Surv. Open-File Rep., 93-41: 65 p.

Partridge, J. and Baker, V.R., 1987. Palaeoflood hydrology of the Salt River, Arizona. Earth Surf. Process. Landforms, 12: 109–125.

Patton, P.C. and Baker, V.R., 1977. Geomorphic response of central Texas stream channels to catastrophic rainfall and runoff. In: D. Doehring (Editor), Geomorphology of Arid and Semi-arid Regions. Allen and Unwin, London, pp. 189–217.

Pilgrim, D.H., 1986. Bridging the gap between flood research and design practice. Water Resour. Res., 22: 165S–176S.

Popper, K.R., 1959. The Logic of Scientific Discovery. Basic Books, New York, 480 pp.

Popper, K., 1976. Unended Quest: An Intellectual Autobiography. Open Court, LaSalle, IL, 258 pp.

Press, F., 1994. Human kind and Earth's natural system. Geotimes, 39: 4.

Ratner, J. (Editor), 1939. Intelligence in the Modern World: John Dewey's Philosophy. Random House, New York, 1077 pp.

Russell, B., 1945. A History of Western Philosophy. Simon and Schuster, New York, 895 pp.

Saarinen, T., Baker, V.R., Durenberger, R. and Maddock, T., Jr., 1984. The Tucson, Arizona Flood of 1983. National Academy Press, Washington, DC, 112 pp.

Savage, L.J., 1954. The Foundations of Statistics. Wiley, New York, 310 pp.

Shapiro, A.E., 1990. New turnings. Science, 250: 1600–1601.

Slezak-Pearthree, M.S. and Baker, V.R., 1987. Channel changes along the Rillito Creek system of southeastern Arizona 1941 through 1983. Arizona Bureau of Geology and Mineral Technology Geological Survey Spec. Pap. 6, 58 pp.

Stedinger, J.R. and Baker, V.R., 1987. Surface water hydrology: Historical and paleoflood information. Rev. Geophys., 25: 119–124.

Webb, R.H. and Betancourt, J.L., 1992. Climatic variability and flood frequency of the Santa Cruz River, Pima County, Arizona. U.S. Geol. Surv. Water-Supply Pap., 2379: 40 pp.

Webb, R.H., O'Connor, J.E. and Baker, V.R., 1988. Paleohydrology reconstruction of flood frequency on the Escalante River. In: V.R. Baker, R.C. Kochel and P.C. Patton (Editors), Flood Geomorphology. Wiley, New York, pp. 403–418.

Wigner, E.P., 1969. The unreasonable effectiveness of mathematics in the natural sciences. In: T.L. Saaty and F.J. Weyl (Editors), The Spirit and Uses of the Mathematical Sciences. McGraw-Hill, New York, pp. 123–131.

Wohl, E.E., Fuertsch, S.J. and Baker, V.R., 1994. Sedimentary records of late Holocene floods along the Fitzroy and Margaret Rivers, Western Australia. Aust. J. Earth Sci., in press.

Ziman, J., 1978. Reliable Knowledge: An exploration of the Grounds for Belief in Science. Cambridge Univ. Press, Cambridge, 197 pp.

ELSEVIER

Geomorphology 10 (1994) 157–168

GEOMORPHOLOGY

Geomorphic and hydrologic aspects of monsoon floods on the Narmada and Tapi Rivers in central India

Vishwas S. Kale[a], Lisa L. Ely[b], Yehouda Enzel[c], Victor R. Baker[d]

[a]*Department of Geography, University of Poona, Pune 411007, India*
[b]*Earth System Science Center, Pennsylvania State University, University Park, PA 16802, USA*
[c]*Department of Physical Geography, Institute of Earth Sciences, Hebrew University, 91904 Jerusalem, Israel*
[d]*Department of Geosciences, The University of Arizona, Tucson, AZ 85721, USA*

Received January 27, 1994; revised March 20, 1994; accepted March 30, 1994

Abstract

High-magnitude floods during the monsoon season are considered to be India's recurring and leading natural disaster. Such large floods are extremely important events, not only in terms of human impact, but also from the standpoint of geomorphic effectiveness and geomorphic work. The Narmada and Tapi Rivers in central India are characterized by one of the most intense flood regimes in the seasonal tropics. Modern, historical and palaeoflood records indicate that large floods are relatively common hydrologic and geomorphic events on the Narmada and Tapi Rivers. A catastrophic flood on Tapi in July 1991, that occurred from heavy rains and a dam break, represents one of the highest unit discharges yet recorded in any part of central India. Flood competence studies indicate that the flood power of the two rivers is comparable with some of the largest historic and prehistoric floods reported from other parts of the world. The rivers have preserved geomorphic and sedimentologic records of palaeofloods for the last two millennium, and in Choral River a 5000 year old flood record has been located. Examination of the synoptic conditions associated with the flood generating low-pressure systems reveals that all but a few of them are the result of Bay of Bengal depressions. The flooding in such large rivers profoundly challenges flood-hazard management, because of limited instrumental records and large spatio-temporal variation. It is therefore imperative that the design engineering community in India incorporates the geomorphic, sedimentologic and botanic information in the planning and design of water resource development projects.

1. Introduction

The Indian monsoon is one of the most impressive, but at the same time most enigmatic, features of the planetary atmospheric circulation. Central India lies at the heart of the classic monsoon region and is very vulnerable to repeated hazards of floods and heavy downpours that accompany low pressure systems (LPS) during the monsoon season. Such large floods are extremely important events, not only from the standpoint of human impact, but also in terms of geomorphic effectiveness and work. This paper describes the flood geomorphology and hydrology of two large and extremely flood-prone rivers namely the Narmada and Tapi, in an environment typical of the monsoonal tropics, with periodic high-magnitude rainfall. The large floods in these two rivers are an example of the interplay between meteorology and drainage basin characteristics that provides central India with the most intense flood phenomena in seasonal tropics.

SSDI 0169-555X(94)00021-I

2. Narmada and Tapi Rivers: An introduction

The Tapi and Narmada are the two largest west flowing rivers in India (Fig. 1). Both basins are elongated in shape and the Narmada basin is relatively hilly. The rivers reflect suitable basin physiography for the generation of high-magnitude floods. The Narmada rises near Amarkantak in the state of Madhya Pradesh at an elevation of 1057 m and has a length of 1312 km and a total drainage area of 98,796 km^2. The river flows through three narrow picturesque gorges — Marble Canyon, Punasa Gorge and Dhadgoan Gorge (Fig. 1). The Tapi rises near Multai in Madhya Pradesh at an elevation of 730 m. The total length of the river is 724 km and its catchment area is 65,145 km^2. Except in the source region, for a major part of its course the Tapi flows over thick alluvial deposits. About 35 km from its source, the Tapi enters the 60 km long Bhainsadehi Gorge (Fig. 1). Chandora Dam is located 26 km upstream of the gorge.

The average rainfall in the Narmada and Tapi basins is 1250 mm and 830 mm, respectively. About 90% of the annual rainfall falls during the monsoon months, i.e., mid-June to early October. Some of the southern tributaries of Tapi, namely Girna and Panzara, head in the Western Ghats, whereas other tributaries head in central India (Fig. 1). These diverse source areas are responsible for the peculiar nature of floods on Tapi. Heavy rainfall either in the Western Ghats or central India can cause floods on the Tapi River. The Narmada, on the other hand, receives a very low proportion of its total run-off from heavy rainfall in the western part of the basin.

Tectonic and flood processes constitute the most important controls on the channel morphology of the Narmada (Rajaguru et al., in prep.) and Tapi Rivers. The channels are box-shaped in nature and the form ratio (width/depth) varies between 11:1 and 59:1. For most of their length the channel gradient of the rivers is relatively low, between 0.0008 and 0.001. However,

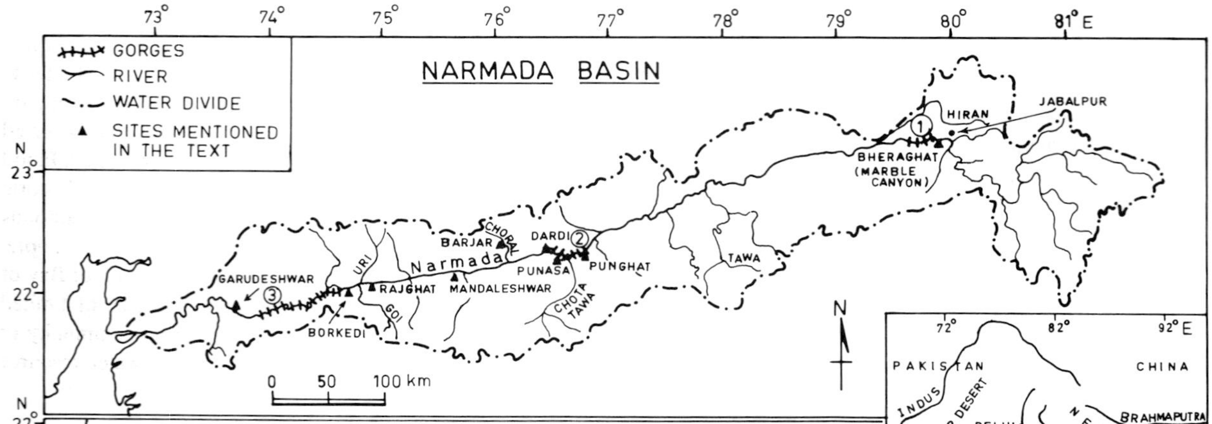

Fig. 1. Map showing the location of sites mentioned in the text. 1 = Marble Canyon, 2 = Punasa Gorge, 3 = Dhadgoan Gorge, 4 = Bhainsadehi Gorge.

there are numerous waterfalls and rapids where the bed elevation drops between 0.5 m and > 10 m giving an average slope of > 0.02. The channels are deeply incised either in bedrock or alluvium, and even high flows are insufficient to fill the whole channel (Gupta, 1993). However, there are reports that at least three or four times in this century the Narmada rose slightly beyond its high banks (about 11–22 m high). On the Tapi, bank failures and gullying in the alluvial tract is common during high floods (Patil, 1993). In comparison, the Narmada is confined within relatively stable banks, even in the alluvial sections. In both the rivers, floodplain formation and point bar development is restricted within the channel banks and most of the geomorphic work of the rivers is accomplished during a few high-magnitude events (Gupta, 1993; Rajaguru et al., in prep.).

3. Flood hydrology

High-magnitude rainfall occurs on a regular basis during the monsoon season (mid-June to early October) in central India. The rivers rise in floods filling the deep channels at least once every 10–20 years. The unit discharges on the Narmada and the Tapi Rivers range between 0.49 and 1.25 cm/km^{-2}, and 0.50 and 3.14 cm/km^{-2}, respectively. The high unit discharges imply the dominant control of flood processes on the morphological properties of the rivers. Systematic records of such catastrophic floods are generally short or discontinuous. Gauge data for several sites are, however, available for the last 20 to 40 years. The available data indicate that the rivers have experienced devastating floods with peak discharges ranging from 10,000 to 60,000 m^3 s^{-1} (Table 1). Most floods occur during August and September. Nevertheless early and late monsoon floods cannot be considered as rare events. At Mortakka, the 100, 500 and 1000 year return period discharges are 66,487 m^3 s^{-1}, 83,562 m^3 s^{-1} and 90,903 m^3 s^{-1}, respectively (Goel et al., 1987). The magnitude of the historical Narmada flood lies at the global extreme for floods of this drainage area (Baker et al., 1994).

Table 1
Highest discharges recorded on Narmada, Tapi and Mahanadi Rivers

River	Station	Date	Discharge ($m^3 s^{-1}$)
Tapi	Kathore	14 Jul. 1941	13,600
		6 Aug. 1942	21,500
		24 Aug. 1944	25,500
		17 Sept. 1959	37,300
		6 Aug. 1968	41,700
	Savakheda	11 Aug. 1979	20,845
		30 Aug. 1987	24,200
		31 Jul. 1991	11,181
Narmada	Garudeshwar	24 Sept. 1954	29,900
		15 Sept. 1959	43,200
		17 Sept. 1961	47,700
		6 Aug. 1968	58,000
		6 Sept. 1970	59,400
		31 Aug. 1973	40,428
		30 Aug. 1978	40,744
		20 Aug. 1984	49,500
		20 Aug. 1987	49,500
Mahanadi[a]	Barmul (Naraj)	1955	35,092
		1961	35,856
		1973	26,234
		30 Aug. 1982	44,827

Source: Central Water Commission, Water Year Books and UNESCO World Catalogue of large floods.
[a]After Puri and Ravishanker (1983).

The upper reaches of the Tapi River receive annual rainfall of more than 800 mm, and extremely high rainfall during a few days may lead to catastrophic floods. Such was the case on 30 July, 1991 (Kale et al., in prep.). On 28 July an intense low pressure system moved landward onto the eastern coast of India in Orissa and heavy to very heavy rainfall occurred in several parts of the Narmada, Tapi and Wardha basins. The heavy rains (1 day rainfall depths between 20 and 45 cm) in the headwaters were followed by breaching of the Chandora Dam (gross storage capacity 18.2 MCM), upstream of the Bhainsadehi Gorge. There are no accurate records of the flood peak, as none of the gauging stations in the Bhainsadehi Gorge could measure the discharge. The HEC2 step-backwater hydraulic model (HEC-2, 1987) and surveyed channel cross sections yielded a peak flood discharge of 9500–10,000 m^3 s^{-1} at Dhanora, and Manning's formula yielded values of 12,000 m^3 s^{-1} ($n = 0.03$) to 19,000 m^3 s^{-1} ($n = 0.02$) at Teska, associated with a scour line carved by this catastrophic flood in weathered basalt (Kale et al., in prep.). These figures indicate an extremely high unit discharge (> 12 cm/km^{-2}) and suggest that this

Table 2
Dates of historic floods

River	Large flood years
Central India	1291, 1345, 1630, 1770, 1783, 1818–19, 1823–27, 1832, 1854–55, 1866, 1868, 1878, 1891, 1893–94, 1894–95, 1898
Tapi	1554, 1727, 1776, 1782, 1810, 1822, 1835, 1837, 1849, 1872, 1873, 1875, 1876, 1882, 1884, 1894, 1904, 1927, 1937, 1941, 1942, 1944, 1970
Girna	1872, 1946, 1947, 1970
Narmada	1818, 1823, 1832, 1854, 1855, 1868, 1878, 1891, 1893, 1894, 1895, 1898, 1905, 1907, 1923, 1926, 1937, 1944, 1945
Mahanadi	1834, 1855, 1866, 1872, 1896, 1926, 1933, 1937, 1940

Based on historical books, Imperial Gazetteer, and other reports.

flood was perhaps the largest yet recorded in central India.

Long and well documented historic flood records for the Narmada and Tapi Rivers are not available. However, there are several references to heavy rains and/or floods on these rivers in historical books and letters, the Imperial Gazetteer and other government reports. In addition, flood marks on buildings, temples and bathing ghats indirectly document the magnitude of large floods in the last hundred years or so. Relatively, the Tapi has a longer historical record extending back to the early sixteenth century (Table 2). This is primarily because of the greater number of large historical towns located on the banks of Tapi. The occurrence of large floods in central India has also been indicated by reference to high-magnitude floods in the adjoining basins. For example, on the Mahanadi River, exceptionally large floods were recorded in 1834, 1855, 1866, 1926, 1937, 1961, 1973 and 1982 (Ahuja and Majumdar, 1959; Puri and Ravishanker, 1983; Satpathy, 1987). Most of these years also represent severe flood years in the Narmada basin (Table 2). The records given in Tables 1 and 2 indicate that large floods are not rare hydrologic and geomorphic events on the Narmada and Tapi Rivers. Such powerful floods are likely to exceed the competence threshold for channel erosion and transportation of coarse sediments, more frequently than on other tropical rivers.

4. Flood hydraulics and dynamics

Geomorphic erosion can be analyzed as the dynamic action of forces on resistant geological materials (Baker, 1988). Flood-prone rivers like Narmada and Tapi, have probably experienced significant geomorphic activity in recent times. Calculations for hydraulic geometry equations using annual maximum series data for an alluvial, but stable, reach on the Narmada at Rajghat yield the following relationships:

$$w = 496.3Q_p^{0.037} \quad (r^2 = 0.91) \tag{1}$$

$$\bar{v} = 0.03Q_p^{0.46} \quad (r^2 = 0.78) \tag{2}$$

$$\bar{d} = 0.07Q_p^{0.5} \quad (r^2 = 0.93) \tag{3}$$

Similarly, the corresponding equations for Tapi at Burhanpur are as follows:

$$w = 31.0Q_p^{0.24} \quad (r^2 = 0.88) \tag{4}$$

$$\bar{v} = 0.32Q_p^{0.26} \quad (r^2 = 0.68) \tag{5}$$

$$\bar{d} = 0.10Q_p^{0.51} \quad (r^2 = 0.91) \tag{6}$$

where Q_p is the annual peak discharge in cms, w is the width in m, $\bar{d}$ is the mean depth in m, and $\bar{v}$ is the mean velocity in m/s.

Although the equations are based on limited and discontinuous data, they reflect the general hydraulic characteristics of the Narmada and Tapi Rivers during large floods. The rate of increase in mean depth ($f = 0.5$) and velocity ($m = 0.46$) with discharge (Q) is high for the Narmada. For Tapi, the rate of increase in velocity ($m = 0.26$) is relatively lower than that of the mean depth ($f = 0.51$). Concomitantly, due to the box-shaped nature of the channel, the rate of change in width is low for the Narmada ($b = 0.037$) and Tapi ($b = 0.24$) Rivers. The bedrock reaches undoubtedly constrain the width of the flow and therefore the increase in discharge is primarily compensated by a marked increase in the velocity and depth. It can therefore be inferred that the bedrock reaches feature higher velocities and consequently extremely high flood power. Velocity measurements available for some gauging stations indicate that during high-magnitude floods the maximum velocities increase relatively rapidly with discharge and the values range between 4 and >7 m/s (Fig. 2). Similarly, HEC-2 modeling for the Punasa Gorge and Bhainsadehi Gorge indicates that during large floods the velocities can be as high 10–15

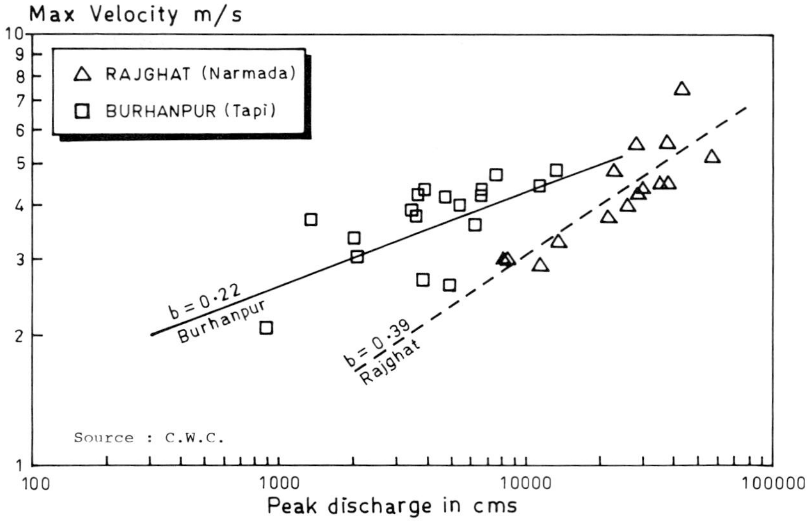

Fig. 2. Plot showing the relationship between annual peak discharge and maximum velocities for two stations (see Fig. 1 for location).

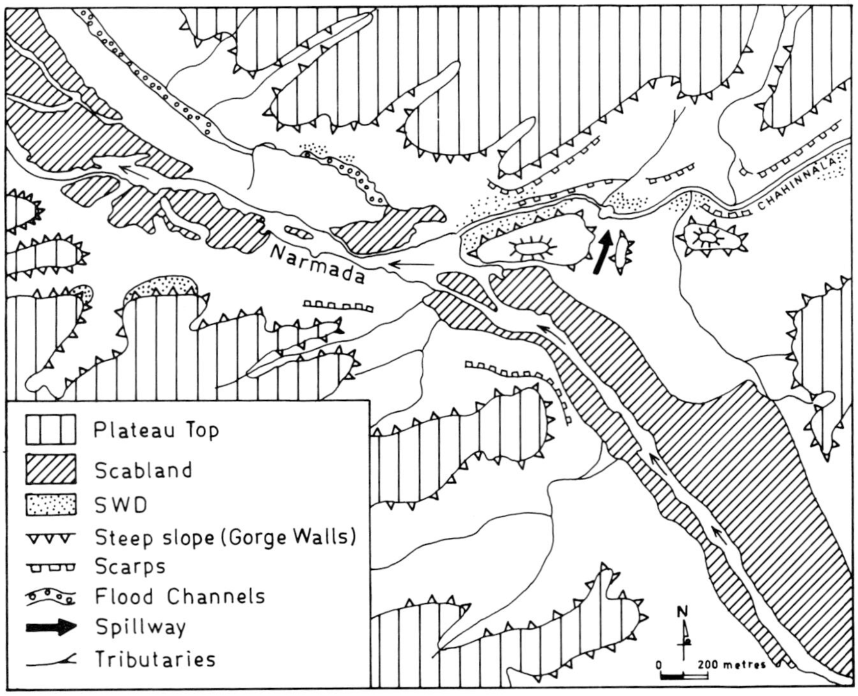

Fig. 3. Geomorphic map of the Punasa Gorge, based on aerial photo and IRS image interpretation.

m/s in the constricted and steep portions of the bedrock gorge (Kale et al., 1992; Kale et al., in prep.).

The ability of a stream to erode and transport material is related to channel boundary shear stress (τ), expressed in newtons per square metre, and is the product of specific weight of water (9800 N m^{-2}), hydraulic radius and the slope:

$$\tau = \gamma RS \tag{7}$$

Another useful concept for sediment transport is unit stream power (ω) that expresses the energy developed in watts per square metre and is calculated as the product of unit weight of water, discharge, and slope divided by width:

$$\omega = \gamma QS/w \tag{8}$$

By applying Eqs. (7) and (8), average values for unit stream power and shear stress were computed, which range between 193 and 800 W m^{-2}, and 200 and 500 N m^{-2}, respectively. However, as mentioned earlier, the rivers feature numerous rapids and waterfalls, where the local slopes exceed 0.02. For such reaches, even conservative estimates yield maximum bed shear stress of >3000 N m^{-2} and unit stream power of 6000 to 13,000 W m^{-2} (Kale et al., 1992; Rajaguru et al., in prep.). These values reflect the unusually high ability of the rivers to erode and transport sediments. These estimates and the hydraulic characteristics of the Narmada further suggest that high flows can easily move cobbles in suspension, and large boulders as bedload (Baker and Costa, 1987).

In addition to the above estimates, impressive scablands at Bheraghat, Punghat, Punasa (Fig. 3), Dardi, and downstream of Borkedi on the Narmada, and between Dhanora and Burhanpur on the Tapi indicate that Froude numbers close to 1.0 have been reached several times for a sustained length of time in the recent past. Intense bedrock scouring, resulting from cavitating flow conditions is also reflected by erosional features such as flute marks, polished surfaces and pot holes (Baker, 1988; Kale et al., 1993a).

The record-breaking discharge on the Tapi River on 30 July, 1991 was accompanied by perhaps one of the highest transportation and erosion capacities. Downstream of the Chandora Dam, a 2 m thick black soil was stripped from a very large area, exposing late Pleistocene sediments rich in fossils and palaeolithic artifacts. Rock and concrete slabs, measuring 1–3 m long and 0.5 to 1.5 m wide, were torn from a bridge and a temple, carried a minimum of 3 to 133 m downstream, and imbricated on top of boulder bars (Kale et al., in prep.). Several big trees were uprooted and washed away in the high flow. HEC-2 modeling of the discharge associated with the scour line indicates that the channel velocities during this event ranged between 5 and 11 m/s and the unit stream power in steeper reaches exceeded 3000 Wm^{-2} (Kale et al., in prep.).

Table 3
Boulder and flow competence data

Site	River	Intermediate axis (cm, max.)	$\bar{v}$ (m/s)	τ (N m^{-2})	ω (W m^{-2})
Dardi	Narmada	240	3.18	408	1550
Punghat	Narmada	210	2.97	357	1309
Gwarighat	Narmada	193	2.86	328	1176
Borkedi	Narmada	231	3.12	393	1477
Barjar	Choral	42	1.33	71	170
Dhanora	Tapi	160	2.60	272	926
Hanumanghat	Tapi	72	1.74	122	336
Ghuttigarh	Tapi	94	1.99	160	472

The presence of large boulders along the gorges provides some clue to the competence of the flows. Using the empirical relationships developed by Williams (1983), the threshold values of bed shear, unit stream power, and mean velocity necessary to transport the boulders were calculated (Table 3). The equations used in the boulder transport calculations are as follows:

$$\bar{v} = 0.065d^{0.5} \tag{9}$$

$$\tau = 0.17d \tag{10}$$

$$\omega = 0.079d^{1.27} \tag{11}$$

where $\bar{v}$ is the mean velocity in m/s, τ is the shear stress (N m^{-2}), ω is the unit stream power (W m^{-2}) and d is the intermediate diameter of the particle (mm).

The estimated values, when compared with the measured maximum velocities or the values generated by the HEC-2 modeling program, indicate that for the Narmada and Tapi, flows are higher than the threshold values for the entrainment of boulders. The calculated figures further suggest that most of the annual peak floods are competent to transport the largest boulders present on the bed. This is particularly true for the bedrock reaches of the Narmada and Tapi Rivers.

5. Palaeoflood geomorphology

In recent years, the scientific study of slackwater flood deposits and palaeostage indicators (SWD–PSI) in bedrock channels, has made it possible to reconstruct chronologies of large historic and prehistoric floods (Kochel and Baker, 1982). Slack-water deposits

Table 4
Sedimentological properties of SWD (phi scale)

Site	Mean (Mz)	Sorting (σ_1)	Skewness (SK_1)	Kurtosis (K_G)	Colour (Munsell)
Borkedi	3.73	0.66	−0.57	0.52	10YR 5/3
Goi Nadi	2.91	0.74	0.37	0.49	10YR 4/3
Morkatta	2.90	0.59	0.63	0.74	10YR 5/4
Punasa	2.60	0.67	0.21	1.22	10YR 5/6
C. Tawa	2.78	0.69	0.36	0.90	10YR 4/4
Barjar	2.83	0.61	0.51	0.55	10YR 5/4
Satrasen	3.28	0.89	0.13	0.48	10YR 3/3
Satrasen	2.98	0.65	0.19	0.34	10YR 4/2

(SWD) consist of silt or sand that generally settle rapidly in backwater locations during large floods (Kochel and Baker, 1982). These flood sediments selectively preserve records of large floods spanning 10^2 to 10^3 years (Ely and Baker, 1985). Such palaeoflood records were investigated in the Narmada and Tapi basins and some of their tributaries.

The Narmada and Tapi Rivers are characterized by spectacular bedrock chasms and thus are ideal for palaeoflood hydrology. The gorges are developed on resistant Proterozoic rocks of the Vindhyan Supergroup or Cretaceous–Eocene basalts. The narrow bedrock gorges have preserved evidence of high-magnitude floods in the form of slack-water deposits (SWD), scour lines, and flood debris. The SWD occur as benches parallel to the channel or at the mouth of tributaries. The thickness of the deposits varies from 3 to > 10 m. The majority of SWD reveal multiple flood units, delineated by abrupt changes in texture, colour and compositions, representing successive floods of increasing magnitude. The slackwater deposits are characterized by strongly fine-skewed, moderately well sorted fine sands (0.08 to 0.18 mm) and coarse silts (Table 4). Charcoal occasionally occurs at the top or within the individual flood units. At some places, the fine-grained flood deposits are separated by coarser and poorly sorted colluvium or incipient soils, representing a distinct break between two successive floods. At most sites, the flood deposits contain 5 to > 30 flood units (Kale et al., 1992). A few C-14 dates (Ely et al., 1993) and archaeological material denote that the deposits are

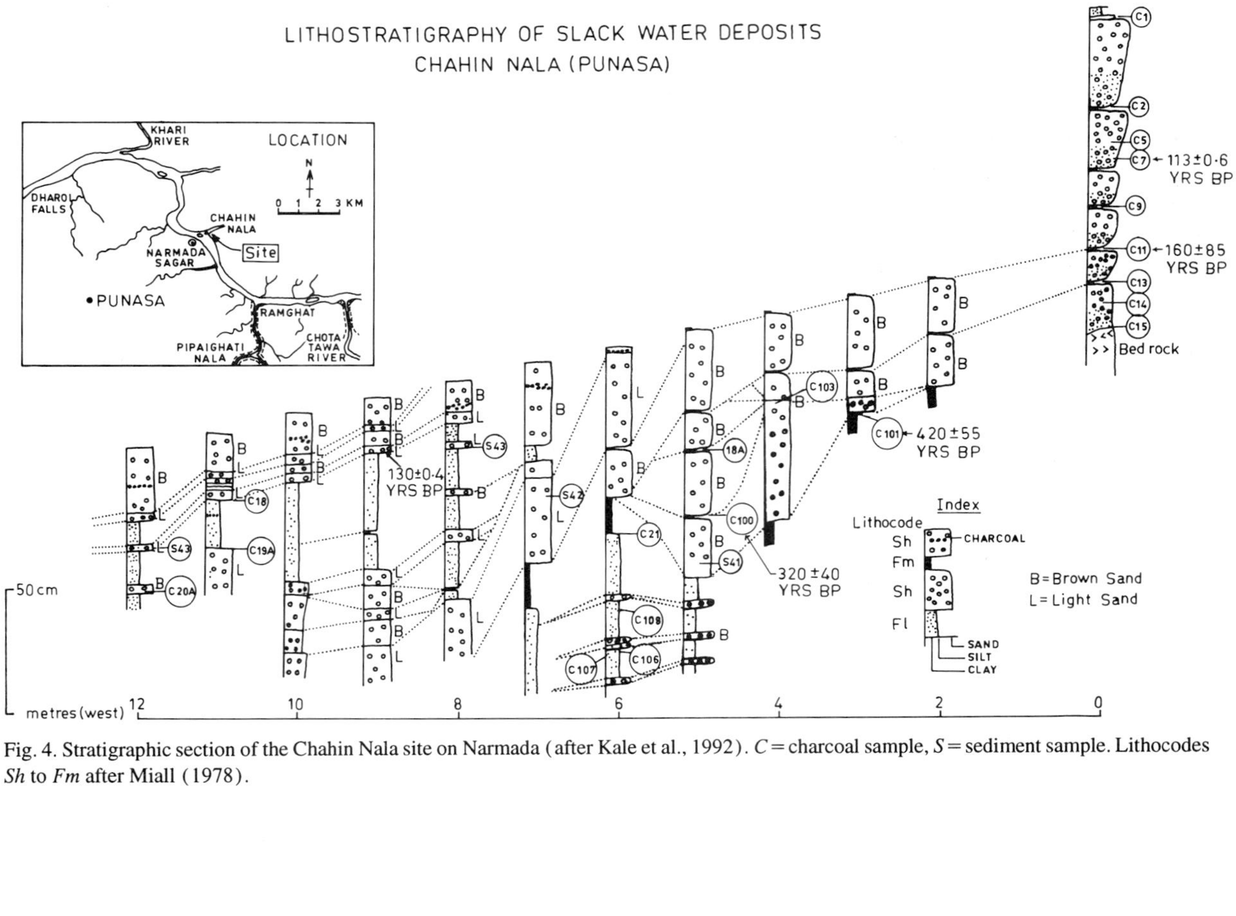

Fig. 4. Stratigraphic section of the Chahin Nala site on Narmada (after Kale et al., 1992). C = charcoal sample, S = sediment sample. Lithocodes *Sh* to *Fm* after Miall (1978).

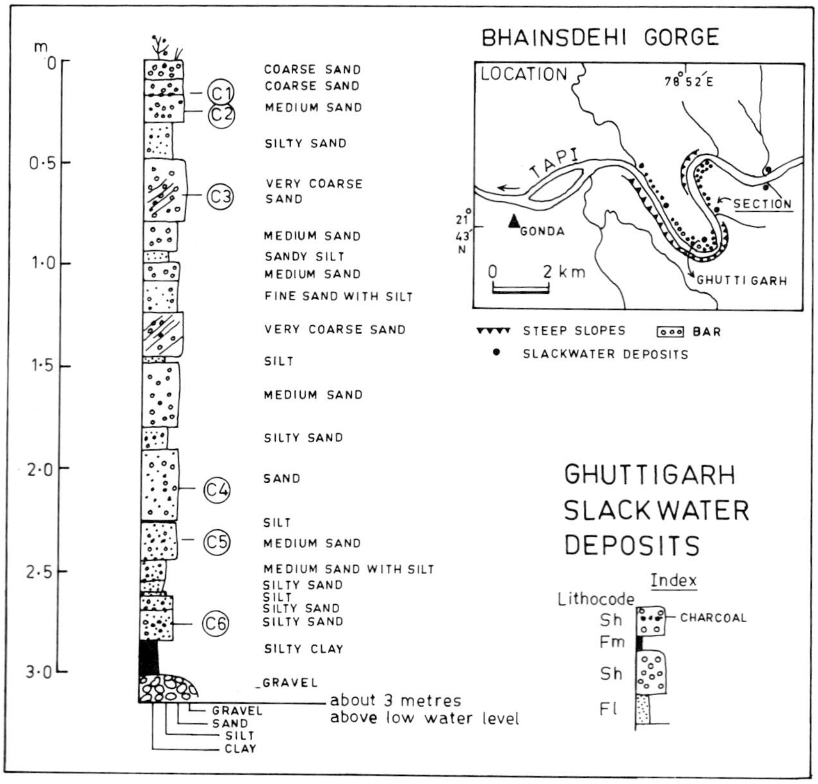

Fig. 5. Stratigraphic section of the Ghuttigarh site on Tapi (after Kale et al., in prep.).

ultramodern to <2000 yr old. A radiocarbon date on charcoal collected from the lowest flood unit of SWD at Barjar on Choral River is 5170 ± 135 yr BP implying that the overlying flood deposits were emplaced in the last five thousand years (Kale et al., 1993b). The sections of the Punasa and Ghuttigarh site (Figs. 4 and 5), illustrate the nature of SWD stratigraphy. At most places on Narmada and Tapi the recent deposits are coarsest indicating higher velocities (Baker et al., 1994).

Calculations of the discharges associated with highest occurring SWD in the Punasa Gorge were performed using the step-backwater computer modeling procedures. Modeling of the discharges, using the HEC-2 computer program (HEC-2, 1987) indicates that the SWD–PSI evidence is associated with flood discharges close to 55,000 $m^3 s^{-1}$ (Kale et al., 1992), giving a unit discharge of 0.89 cm km^{-2}. If one considers that the top of the SWDs may be lower than the actual water level by approximately 10% (O'Connor and Webb, 1988), then in the Punasa Gorge the SWDs were emplaced by floods that were not less than 60,000 $m^3 s^{-1}$. Similarly, discharge estimates using the step-backwater computer program have yielded values of >4500 $m^3 s^{-1}$ for a tributary of Narmada, the Choral River at Barjar (Kale et al., 1993b). The estimated palaeoflood discharges and the modern gauge records suggest that high-magnitude floods are not a very uncommon hydrologic feature on the Narmada River.

The palaeoflood record from the Choral River is the longest yet discovered in central India (Kale et al., 1993b). Flood deposits on the main Narmada River cover at least the last 1500–2000 years, but forthcoming dates from SWD will probably increase the length of flood records in central India (Kale et al., 1993b).

6. Flood hydrometeorology

Floods are caused by extraordinary synoptic situations that deliver more precipitation to a drainage basin than can be readily absorbed or stored within the basin

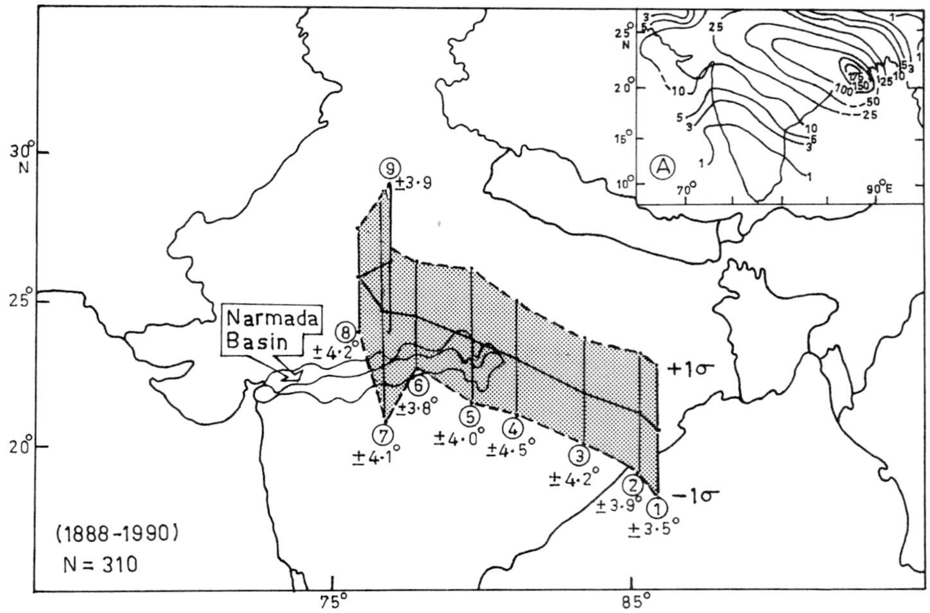

Fig. 6. Mean track of the LPS (continuous line) from 1st to 9th day (encircled numbers) over central India. The shaded area enclosed within dashed lines represents one latitudinal standard deviation and the figures at the bottom of the shaded area represent one longitudinal standard deviation for each day. (A) Frequency of depressions and cyclonic storms over Indian land area during monsoon (1891–1970). (After Dhar et al., 1980.)

Table 5
Highest rainfall depths experienced by the Narmada basin (1891 to 1985). Values in cm

Reach of Narmada	1-day	2-day	3-day	4-day
Source to Jamatara	20.4 (1926)	39.4 (1926)	53.0 (1926)	59.2 (1926)
Jamatara to Hoshangabad	16.6 (1913)	22.6 (1891)	28.4 (1891)	33.9 (1891)
Hoshangabad to Mortakka	23.7 (1905)	29.4 (1905)	35.4 (1944)	37.4 (1944)
Mortakka to Garudeshwar	14.9 (1970)	24.1 (1905)	29.6 (1945)	27.0 (1970)

After Dhar and Nandargi (1989).

(Hirschboeck, 1991). In central India, intense precipitation from low pressure systems (LPS) during the monsoon season is the major cause of high-magnitude floods, and these floods represent the largest recorded peaks in a flood time series. Examination of the synoptic conditions associated with the flood-generating LPS reveals that all but a few of them are the result of Bay depressions and that the remaining are associated with active monsoon conditions over central India or land depressions. The LPS form in the Bay of Bengal and thereafter move in a west-northwest or northwest direction across the catchment areas of the Mahanadi, Narmada, Wardha and Son Rivers. More than 95% of the flood events on the Narmada River are associated with the LPS that pass through the source or near-source region of the river in a period of 6–10 days. Fig. 6 shows the mean latitudinal and longitudinal locations

Table 6
Maximum rainfall depths experienced by the Tapi basin (1891 to 1968). Values in cm

Storm period	1-day	2-day	3-day	4-day
20–21 Jul., 1894	8.7	14.6	–	–
13–16 Sep., 1904	8.2	15.5	19.2	22.0
6– 8 Sep., 1937	8.1	11.3	14.9	–
1– 2 Jul., 1941	8.1	14.9	–	–
4– 5 Aug., 1942	7.4	14.4	–	–
14–16 Jul., 1944	6.3	10.3	13.9	–
21–22 Aug., 1944	6.9	11.7	–	–
14–15 Sep., 1959	7.7	15.0	–	–
15–16 Sep., 1961	4.4	6.6	–	–
4– 6 Aug., 1968	11.1	19.0	21.1	–

After Abbi and Jain (1971).

and standard deviation of the LPS that arrive from the Bay of Bengal and travel towards central India. Analysis shows that the duration of the LPS over the basin, antecedent precipitation and the pressure deficiency determine the rainfall depths and consequently the intensity of floods (Kale et al., 1992). The floods are higher when the LPS travel more or less parallel to the basin axis, as in case of August 1968 and September 1970 floods. The rate of movement of depressions also plays an important role in generating floods. If the LPS moves slowly or remains stationary for a few days (as in the case of the 1926 flood), it causes heavy rain for a longer spell resulting in severe floods (Dhar, 1958).

Due to the peculiar orientation of the southern tributaries of the Tapi River, floods on this river are also associated with heavy orographic rainfall in the Western Ghats. The Tapi watershed receives heavy rainfall from Bay depressions as well as heavy rainfall spells over Western Ghats. However, only those Bay depressions which move in a west/northwest direction close to or through the Narmada basin have been found to cause major rainstorms and floods in the Tapi basin (Abbi and Jain, 1971).

Analyses of the heavy rainfall spells by Dhar and Nandargi (1989) for the Narmada basin and by Abbi and Jain (1971) for the Tapi basin indicated that heavy rainspells are mostly of 1 to 2 days duration and rarely of more than 4 days duration (Tables 5 and 6). For the Tapi basin, the value of the maximum probable storm for one day has been estimated to be 17.0 cm (Abbi and Jain, 1971). Similarly, for the Narmada basin, the total rainfall depth for a 5-day storm has been estimated to be 40.6 cm (Ramasastri and Seth, 1986).

7. Flood hazard assessment

The foregoing discussion has brought out six interesting aspects of the flood geomorphology of the Narmada and Tapi Rivers that have significant implications for flood hydrology and flood hazard assessment. First, the rivers are characterized by some of the largest unit discharges observed in seasonal tropics. Second, large floods are not rare geomorphic and hydrologic events on the Narmada and Tapi Rivers. Third, due to the box-shaped nature of the channels, any increase in discharge creates a rapid increase in the depth and velocity. Fourth, the flood power estimated for large modern and palaeofloods in the two rivers is comparable with some of the largest historic and prehistoric floods reported from other parts of the world. Fifth, the rare, extreme floods are invariably associated with low pressure systems from the Bay of Bengal that traverse through the source or near-source region. Sixth, a fairly long record of late Holocene floods has been preserved in central India.

The flooding in the Narmada and Tapi Rivers profoundly challenges flood-hazard management, because of the limited instrumental records and the high spatio-temporal variability of floods. In recent years palaeoflood geomorphology has emerged as an effective engineering tool for securing information on the extreme floods. The conventional techniques used to estimate design floods and design storms in India seriously suffer from inadequacy of data for model verification. The geomorphic, sedimentologic and botanic evidence of large floods cannot be overlooked as the floods are critically important in the hydrological planning and design of water resource development projects.

The geomorphic and sedimentologic evidence of large palaeofloods in the Narmada and Tapi basins suggest that the palaeoflood information is essential for improving risk analyses on these rivers. Palaeoflood records provide an independent means for testing the reliability of flood-frequency analyses based on conventional data as well as flood parameters derived from physically-based watershed models, such as those used to estimate the probable maximum flood (PMF) (Baker et al., 1990). The palaeoflood information can also considerably improve the understanding of the meteorological processes associated with monsoon depressions and cyclones. The climatic implications of extremes in the monsoon extend far beyond the Indian subcontinent due to the broad global teleconnections between the monsoon and other climatic phenomena such as the ENSO (Webster, 1987).

It is evident from the variety of proxy records, that the large floods observed in the present century were as large as the palaeofloods. This implies that some of the largest modern floods, may approach the upper physical or hydroclimatological limit of the magnitude of floods that can be produced in the basins. This is particularly true in case of the catastrophic flood on Tapi in July 1991. Furthermore, a comparison of the design floods estimated by hydrologists for the rivers and those inferred by employing palaeoflood and geomorphic techniques indicates that the two do not even remotely concur. For example, the design flood peaks for Narmada Sagar and Sardar Sarovar Dams on the Narmada River are 142,466 m^3 s^{-1} and 210,137 m^3 s^{-1} (Ramasastri and Seth, 1986). These values are 170–200% higher than their corresponding observed peak flood magnitude on Narmada (Ramasastri and Seth, 1986). In comparison, the largest floods for which there is any evidence in the palaeoflood records or modern geomorphic records are ca 60,000 m^3 s^{-1} for the Narmada River near Narmada Sagar. In the study of 10^2 or 10^3 year flood events for hazards to settlements and man-made structures, palaeoflood and historical-flood information can make a key contribution to hazard analysis. The importance of such data in central India are emphasized by its use in defining the spillway design discharge for several dams contemplated on the Narmada River. Failure of these dams (as in the case of the Chandora Dam on Tapi in 1991), would put a large population at risk. It is therefore of utmost importance that the benefits of these techniques in hazard assessment are recognised by the design engineering community in India and that all of the available geomorphic and palaeohydrological information is incorporated into estimates of design floods and PMF values on rivers such as these in highly flood-prone regions.

Acknowledgements

This research was supported by the Department of Science and Technology, New Delhi, Grant No. ESS/CA/A3-04/90 to VSK. Additional support to LLE, YE and VRB was provided by U.S. National Science Foundation Grant No. EAR-9104489. The authors are grateful to Sheila Mishra, S.N. Rajaguru, Avijit Gupta, Hema Deodhar and Pramod Hire for their support in the field, and for their critical comments and helpful suggestions. VSK thanks the officials of C.W.C. Western Circle, Nagpur for supplying the discharge data. This publication is contribution No. 30 of ALPHA, University of Arizona, Tucson, USA.

References

Abbi, S.D.S. and Jain, B.C., 1971. A study of major rainstorms of Tapi basin for evaluation of design storm. Ind. J. Met. Geophys., 22: 203–212.

Ahuja, P.R. and Majumdar, K.C., 1959. A sampling approach to the estimation of floods. Ind. J. Power River Valley Develop., 9: 23–49.

Baker, V.R., 1988. Flood erosion. In: V.R. Baker, R.C. Kochel and P.C. Patton (Editors), Flood Geomorphology. Wiley, New York, pp. 81–95.

Baker, V.R. and Costa, J.E., 1987. Flood power. In: L. Mayer and D. Nash (Editors), Catastrophic Flooding. Allen and Unwin, London, pp. 1–21.

Baker, V.R., Ely, L.L. and O'Connor, J.E., 1990. Palaeoflood hydrology and design decision for high risk projects. In: H.H. Chang and J.C. Hill (Editors), Hydraulic Engineering Proc. of the 1990 National Conference. Am. Soc. Civ. Eng., New York, pp. 433–438.

Baker, V.R., Ely, L.L., Enzel, Y. and Kale, V.S., 1994. Understanding India's rivers: late Quaternary palaeofloods, hazard assessment and global change. In: S. Wadia, R. Korisettar, V.S. Kale and R.K. Ganjoo (Editors). Geol. Soc. India Mem. Ser., in press.

Dhar, O.N., 1958. Meteorological situations associated with actual major floods in different parts of India. Proc. Meteorological and Hydrological Aspects of Floods in India, New Delhi, pp. 57–62.

Dhar, O.N. and Nandargi, S., 1989. Distribution of rainfall over different sections of the Narmada basin associated with the highest floods in recent times. Ind. J. Power and River Valley Develop., 39: 210–214.

Dhar, O.N., Ghose, G.C. and Mandal, N., 1980. A brief appraisal of India's worst recorded floods. J. Ind. Assoc. Hydrol., 4: 1–11.

Ely, L.L. and Baker, V.R., 1985. Reconstructing palaeoflood hydrology with slackwater deposits, Verde River Arizona. Phys. Geogr., 6: 103–126.

Ely, L.L., Enzel, Y., Baker, V.R. and Kale, V.S., 1993. A recent increase in the magnitude and frequency of severe floods on the Narmada River, central India. Geol. Soc. Am. Abstr. Programs, 25: A-396.

Goel, N.K., Seth, S.M. and Chandra, S., 1987. Design flood estimation for Narmadasagar Project using PDS. In: V.P. Singh (Editor), Application of Frequency and Risk in Water resources. Reidel, pp. 113–122.

Gupta, A., 1988. Large floods as geomorphic events in the humid tropics. In: V.R. Baker, R.C. Kochel and P.C. Patton (Editors), Flood Geomorphology. Wiley, New York, pp. 301–315.

Gupta, A., 1993. The changing geomorphology of the humid tropics. Geomorphology, 7: 165–186.

HEC-2, 1987. HEC2 water surface profile program user's manual. Hydrologic Engineering Centre, Davis, CA.

Hirschboeck, K.K., 1991. Hydrology of floods and droughts. National water summary 1988–1989 — Floods and droughts: hydrology. U.S. Geol. Surv. Water-Supply Pap., 2375: 67–88.

Kale, V.S., Mishra, S., Enzel, Y, Ely, L., Baker, V.R. and Rajaguru, S.N., 1992. Geomorphic investigations of floods in Central Narmada Basin, India. First Nat. Symp. on Environmental Hydraulics. C.W.P.R.S., Pune, pp. 211–220.

Kale, V.S., Mishra, S. Enzel, Y., Ely, L.L., Rajaguru, S.N. and Baker, V.R., 1993a. Flood geomorphology of the Indian peninsular rivers. Int. Seminar on Hydrology Abstracts, Kathmandu, Nepal, pp. 16–18.

Kale, V.S., Mishra, S. Baker, V.R., Rajaguru, S.N., Enzel, Y. and Ely, L., 1993b. Prehistoric flood deposits on the Choral River, central Narmada Basin, India. Current Sci., 65: 877–878.

Kale, V.S., Hire, P. and Baker, V.R. Geomorphological effects of catastrophic flooding on Tapi River, India, in prep.

Kochel, R.C. and Baker, V.R., 1982. Palaeoflood hydrology. Science, 215: 353–361.

Miall, A.D., 1978. Lithofacies types and vertical profile models in river deposits, a summary. In: A.D. Miall (Editor), Fluvial Sedimentology. Can. Soc. Petrol. Geol. Mem., 5: 597–604

O'Connor, J.E. and Webb, R.H., 1988. Hydraulic modeling for Palaeoflood analysis. In: V.R. Baker, R.C. Kochel and P.C. Patton (Editors), Flood Geomorphology. Wiley, New York, pp. 393–402.

Patil, Y.V., 1993. Development of Badlands along Tapi channel. Ph.D. dissertation submitted to University of Poona, Pune, India, unpubl.

Puri, S.R. and Ravishanker, A.D., 1983. Uncommon floods of Orissa in 1982. In: V.V.J. Sarma (Editor), Proc. Seminar on Hydrology, Hyderabad, pp. 57–64.

Rajaguru, S.N., Gupta, A., Kale, V.S., Mishra, S., Ganjoo, R.K., Ely, L.L., Enzel, Y. and Baker, V.R., Channel forms and processes of the flood-dominated Narmada River, India, in prep.

Ramasawamy, C., 1985. Meteorological aspects of severe floods in India (1923–1979). MMH, No. 10, India Meteorological Department, New Delhi.

Ramasastri, K.S. and Seth, S.M., 1986. Estimation of design flood for Narmada Sagar and Sardar Sarovar. Seminar on Flood Estimation and Control, Roorkee, pp. II1–17.

Satpathy S., 1987. Floods and flood control policies: analysis with reference to the Mahanadi Delta in Orissa. M. Phil. dissertation, Jahawarlal Nehru University, New Delhi, unpubl.

Webster, P.J., 1987. The variable and interactive monsoon. In: J.S. Fein and P.L. Stephens (Editors), Monsoons. Wiley, New York, pp. 269–330.

Williams, G.P., 1983. Paleohydrological methods and some examples from Swedish fluvial environments. I. Cobble and boulder deposits. Geogr. Ann., 65A: 227–243.

ELSEVIER

Geomorphology 10 (1994) 169–182

GEOMORPHOLOGY

Spatial variations in the magnitude of the 1993 floods, Raccoon River basin, Iowa

Karen L. Prestegaard[a], Anne M. Matherne[b], Brendan Shane[a], Kevin Houghton[a], Michael O'Connell[a], Nancy Katyl[a]

[a]*Department of Geology, University of Maryland, College Park, MD 29742, USA*
[b]*Conservation and Survey Division, University of Nebraska, 113 Nebraska Hall, Lincoln, NE, USA*

Received March 25, 1994; revised April 21, 1994; accepted April 22, 1994

Abstract

The persistent position of a mid-level circulation pattern in the summer of 1993 supported the formation of frequent storms in the upper Midwest (Bell et al., 1993). Storm events that occurred during this period caused multiple episodes of flooding in the region (Wahl et al., 1993). The high floods along the upper Mississippi River generated debates about the effectiveness of flood control measures and the effects of land use on flood peak discharges.

In order to examine spatial variations in flood peak discharges, we surveyed flood channels and flood profiles on the Raccoon River and its tributaries in west-central Iowa. The Raccoon River basin has variable topography and land use, including some of the most intensely agricultural land in the United States. Extensive ditch networks and subsurface tile drain systems have been installed to enhance runoff and accelerate drainage. We found that sites within and downstream of modified channels had higher magnitude floods than than other comparably sized basins in the Raccoon River basin and the upper Midwest for which data were available. Erosion patterns also followed land use patterns; the upper portions of channelized Raccoon River tributaries experienced short times of flooding and had less erosion than downstream channelized reaches that were severely eroded and had significant damage to bridges and other structures.

1. Introduction

The maintenance of an abnormal mid-level circulation pattern over the upper Midwest from May to September, 1993 resulted in storms falling upon successively more saturated soils (Bell et al., 1993; Kroczynski et al., 1993). These rainstorm events created large magnitude floods over the region, but flood magnitude was spatially variable, even within individual river basins. Flood magnitude at many locations was determined and reported by the U.S. Geological Survey (Boning et al., 1993; Parrett et al., 1993). Although the effects of the 1993 floods were most dramatic in the main channels of the Mississsippi and Missouri River systems, flood discharges were primarily generated in upstream drainage basins that contain most of the contributing basin area. The downstream increase in flood magnitude in the Mississippi River basin appears to be largely due to the north to south movement of heavy precipitation patterns during the summer of 1993, resulting in convergence of flood discharges in the main portion of the Mississippi River valley (Parrett et al., 1993).

The 1993 floods were among the highest on record in many parts of the upper Mississippi basin, but there were major differences in flood magnitude throughout the region. Parrett et al. (1993) reported that flood

SSDI 0169-555X(94)00033-N

discharges exceeded estimates of the 100-year R.I. flood at 46 gaging stations.

The purpose of this paper is to examine spatial variations in flood discharges in one major tributary to the Mississippi River drainage system. We examined the Raccoon River basin of west-central Iowa. It was chosen because of the high magnitude of flooding in the basin and the damage that the floods caused to bridges and to the water treatment plant in Des Moines. We will examine whether spatial variations in rainfall, land use, topography, or a combination of these factors can best explain the distribution of flood magnitudes in the Raccoon River basin.

The Raccoon River Basin has a watershed area of 9399 km^2 and joins the Des Moines River at Des Moines. We examined the southern tributaries which have a total area of 2650 km^2 (Fig. 1). During the summer of 1993, flooding on the Des Moines River preceded flooding on the smaller Raccoon River. The convergence of flood flows from the Raccoon River into an already swollen Des Moines River caused flooding of the city of Des Moines water treatment plant (Parrett et al., 1993). Therefore, study of this basin may provide valuable information on factors affecting flood magnitude in the Raccoon River Basin.

Large magnitude floods have been recorded in the Raccoon River basin in 1947, 1973, 1979, 1986, and 1990 (Heinitz, 1980; Baebenroth and Schaap, 1992). Therefore, peak discharges measured for this study can be compared with recent flood discharges (1986 and 1990) in the Raccoon River basin, with other 1993 flood peaks in the upper Mississippi River basin, and with extreme floods on record for the United States (Costa, 1987a,b).

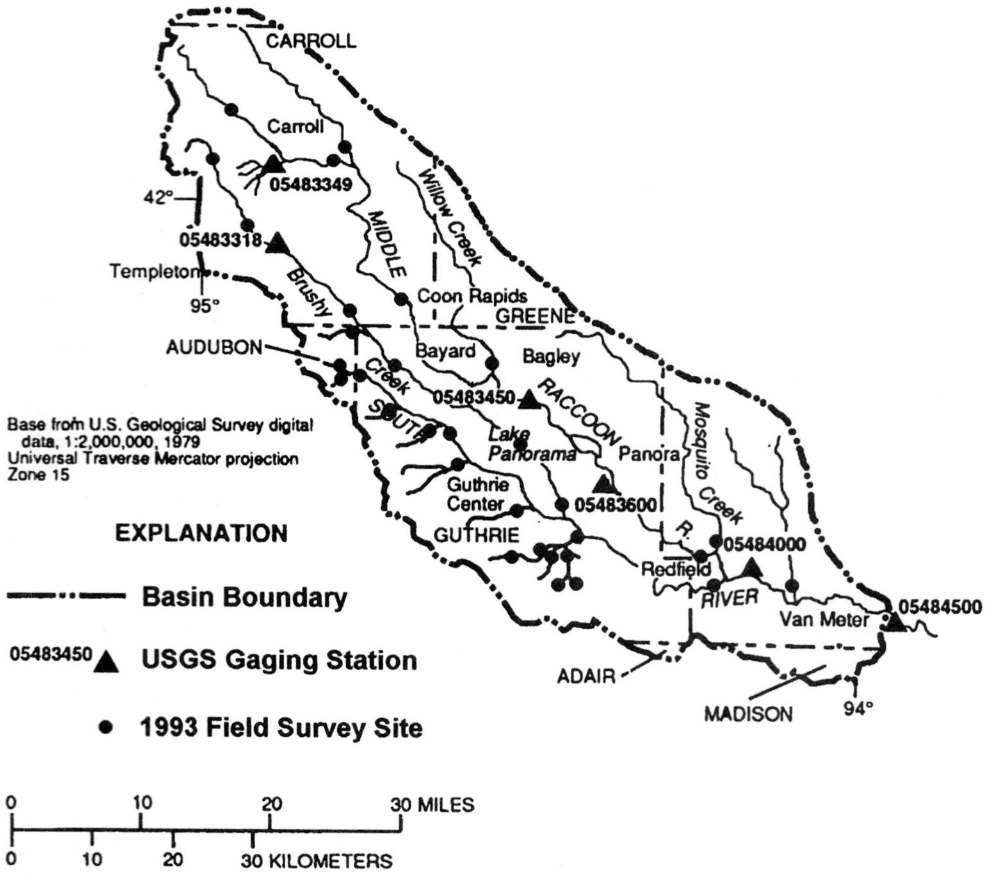

Fig. 1. Study sites within the southwestern tributaries of the Raccoon River basin. Dots indicate 1993 survey sites (this study); triangles and station numbers identify U.S. Geological Survey gaging stations. Dashed lines are county boundaries. Figure adapted from Babenroth and Schaap (1992).

2. Study site and methods

Our study focused on three of four main tributaries of the Raccoon River. From a northeast to southwest direction the three tributaries are: the Middle Raccoon River, Brushy Creek, and the South Raccoon river. These rivers drain successively smaller basin areas from north to south (Fig. 1). The Middle Raccoon River is the approximate position of the southernmost advance of Wisconsinan glaciation that ended approximately 14,000 years ago (Ruhe, 1969). Therefore, the study basin encompasses two distinct geomorphic areas. The larger North Raccoon River was not examined because its watershed did not receive significant rainfall at the same time as the southern tributaries.

Heavy rains that generated high magnitude floods occurred in the basin twice during the summer of 1993. These rains occurred on July 8–9 and August 30. During the second of these two events, we went to Iowa to survey the extent of the flooded areas. Field work, conducted between September 3 and October 15, 1993, consisted of surveying the extent of inundated areas and sampling flood materials in the study basins.

We chose 30 channel sites, located along the three southern tributaries of the Raccoon River, for the flood surveys (Fig. 1). At each of these sites, we surveyed channel cross sections of the flooded area and flood water surface profiles. Bed and water surface profiles at low flow were also surveyed. We sampled or conducted pebble counts of bed material for grain size determinations and we collected samples of deposited overbank sediment. In some cases, both cross-bedded sand and laminated clay were deposited on the floodplain; these sediments were sampled separately if both were present. Each site was photographed to document erosion and sedimentation features and to document channel characteristics.

Survey sites were chosen that were representative of the channel character in the reach. Sites had to be upstream of bridge backwater or downstream of bridge scour effects; visible evidence of high flows had to be located within the study reach. We used several different indicators to determine maximum extent of flooding in the river valleys. These indicators included the highest level of overbank deposits (least reliable indicator), debris and scars on or in trees, and goosenecked soybean and other plants that were originally flattened by the floods and then grew upwards from a horizontal position. An example of the survey data collected at each site is shown in Fig. 2. These data were used to determine cross sectional channel area for the flood flows and flood water surface slopes.

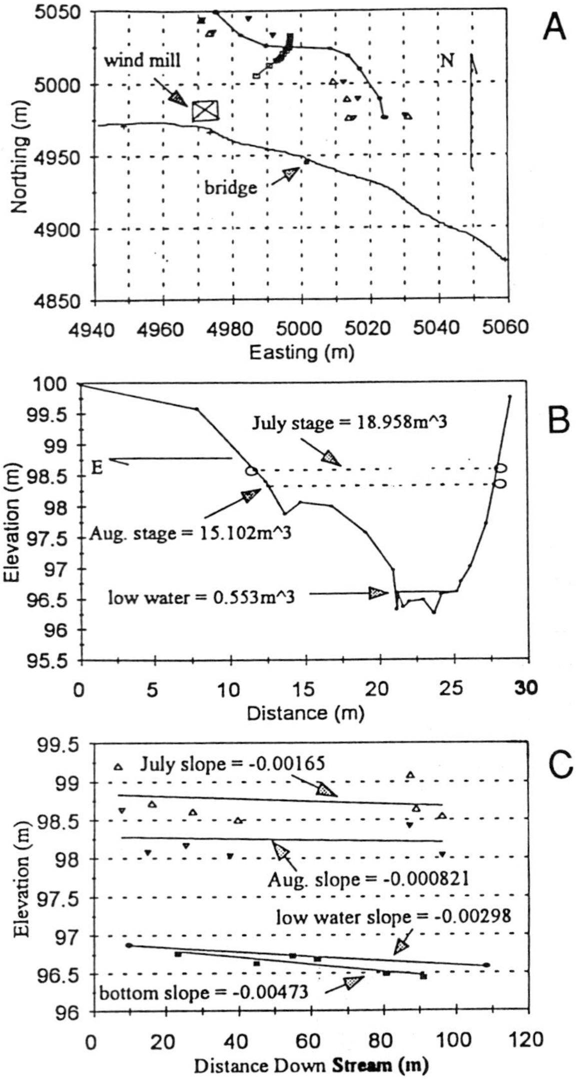

Fig. 2. Examples of channel surveys in the Raccoon River basin (A) Plan view map showing locations of high flow indicators, (B) example of surveyed cross section, and (C) example of water surface elevation profile.

Flood discharges calculated from flood surveys are estimates of actual flood discharges; errors are introduced in the initial survey and in estimates of flood flow velocities. The amount of error in the calculation of flood cross sectional area and water surface slope varied from site to site depending upon the number and reliability of high flow indicators. The variance in channel cross sectional area between the uppermost indicator and lowermost indicator was generally between 2 and 15% of the total cross sectional area. The errors in the estimation of flood water surface slopes were more variable; regression coefficients for flood water surface elevation data averaged 0.85, but ranged from 0.13 to 0.99; for poorly constrained water surface slope profiles bed slope was used instead (this was done for 3 sites). Land use for each site was evaluated from topographic maps and air photos. The drainage area for each site was determined from topographic maps.

Other hydraulic information was calculated from the survey data. The average width of the flooded area at each site was determined directly from flood survey data. Shear stresses for both channel sites and floodplains were calculated with the survey data for water surface slopes, flood depth on the floodplain, and hydraulic radius of channel flows (channel area/wetted perimeter). The accuracy of these calculations are dependent upon the accuracy of the survey data.

Flood discharges for one or both high flow peaks were calculated from the slope–area method (Rantz et al., 1982). The velocities of flood flows were estimated by several methods. Low flow discharges, velocities, and channel slopes were measured at the time of flood surveys. These velocity measurements were used to calculate low flow channel roughness. The variation of roughness and, therefore, velocity with increasing discharge was estimated by comparison with dimensionless discharge versus flow resistance curves (Q/Q_{bf} versus Mannings' n) that we derived for Raccoon River channels at gaging station locations. Roughness for channels with different characteristics than gaging station sites was calculated from relative roughness (Limerinos, 1970) or estimated from channel characteristics (Barnes, 1967).

The velocities of flows over the floodplain were constrained by different methods than the channel velocities. Dimensionless rating curve data indicated that many of the channels have higher flow resistance once they have left their banks. The size and presence of bedforms, and the grain size of the sand in which they were formed, were used to estimate flow velocities on the floodplains. An example of these overbank deposits is shown in Fig. 3. Grain size and bedform dimensions were compared with experimental data on flow depth–velocity–grain size diagrams (Middleton and Southard, 1984). These data on flow velocities were used to calculate flow resistance on the floodplains that could

Fig. 3. Floodplain deposits on Brushy Creek near mouth. Cross-bedded sand is 0.45 m thick.

be extended to similar sites for which data were not available. Total discharge at each site was determined by adding the discharge of channel flows and overbank flows.

From this brief description, it is clear that flood flow resistances and, therefore, flood velocities and discharges are estimates of actual values. Calculations of flood velocities and discharges are not as accurate as the data for flooded area widths, flood water surface slopes, flood depths, and shear stresses, which are determined directly from field measurements or data derived from field measurements.

In addition to the survey of the extent of flood damage, we collected baseline data on water quality in stream flow after the storm. We measured dissolved load and suspended sediment load for baseflow conditions. The suspended load data were used to estimate minimum suspended load concentration and water density during floods.

3. Land use and land management in the basin

Land management in the Raccoon River basin is of interest for two reasons. First, most of the Raccoon River basin is intensely used for agricultural purposes. The upper reaches of Middle Raccoon, Brushy Creek, and South Raccoon are all primarily used for row crops; some headwater tributaries of these creeks are almost 100% row crops. Second, both the main channels and tributaries of upper Middle Raccoon Creek, Brushy Creek, and upper South Raccoon have been straightened and deepened. The northern tributaries of Middle Raccoon Creek are networks of drainage ditches known by their assigned ditch numbers. Land modification in this part of Iowa began in the last century with the ditching and draining of poorly drained lands (Daniels, 1960). This was followed by additional channel straightening efforts that followed the 1947 flood (Renze, pers. comun.). Tile drains and other drainage methods were installed in the 1960's to the present (pers. commun. with Soil Conservation Service and individual landowners).

The subwatersheds that are the least affected by agricultural land uses are southwestern tributaries of South Raccoon Creek. Agricultural land in these basins is used primarily for pasture, although row crops occupy 20% or more of some of the basins. These tributary basins also have forested regions, primarily in riparian areas and along steeper hillslopes. This southwestern portion of the Raccoon River basin is south of the glacial deposits left by the Des Moines Lobe of Wisconsinan age. The southwestern region is more highly incised and is characteristic of landforms developed on older glacial regions in Iowa (Ruhe, 1969).

Potter (1991) has examined decreases in flood magnitude for the Pecatonica River in southwestern Wisconsin that are associated with increases in conservation tillage and an increase in the percentage of the basin that is forested. These changes in land use have not been observed in the Raccoon River Basin, with the exception of the southwestern tributaries. For example, the Pecatonica River in Wisconsin has less than 20% of the land used for row crops, while tributary basins of the upper Brushy and Middle Raccoon Rivers are 80–100% in row crops (soybeans and corn). Both topography and land use can have important effects on flood runoff generation (Dunne, 1978; Beven, 1986). The Raccoon River basin illustrates spatial variations in both variables that resemble the covariations in land use and topography that exist throughout the upper Midwest. Watersheds with significant relief and thin soils usually have small proportions of the land area used for row crops.

4. 1993 Rainfall and floods in the upper Midwest

The U.S. Geological Survey has published several circulars that describe the storms and floods of the upper Midwest (Parrett et al., 1993; Wahl et al., 1993). Heavy rainstorms first fell in Minnesota and Wisconsin in June, 1993. Over the next several months, heavy rains moved towards the south affecting Iowa, Illinois, and Missouri. Flood discharges moved down the upper Mississippi River to meet flood flows coming from more southern tributaries along the main stem of the Mississippi River. Therefore, the record floods on the main channel of the central Mississippi River were not due to heavy rains during one storm, but from an unfortunate aspect of the southern movement of major storms in a southern flowing river system (Parrett et al., 1993; Potter et al., 1993).

The effects of the converging flood flows can be seen in a diagram of flood peak discharge versus drainage basin area for the Upper Mississippi River drainage

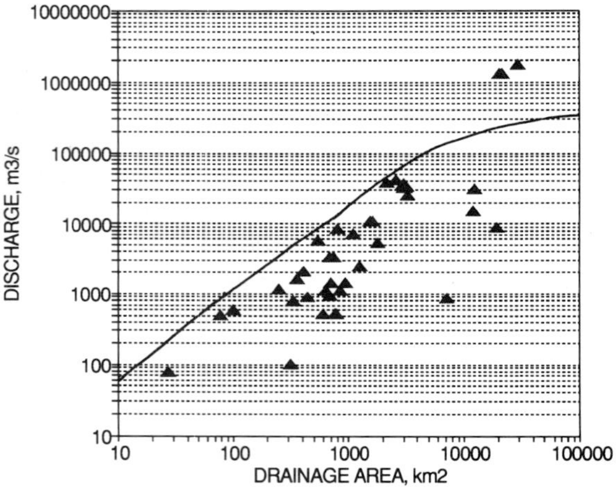

Fig. 4. The relationship between 1993 peak discharge and drainage basin area for gaging station sites in the upper Mississippi River basin that had reported recurrence intervals greater than 50 years (Parrett et al., 1993). Solid line is envelope curve for largest rainfall–runoff floods in the United States (from Costa, 1987a,b).

(Fig. 4). An envelope curve for the largest U.S. floods on record is also shown (Costa, 1987a,b). The 1993 Midwest floods are smaller than record floods for drainage basins less than 1000 km^2, but meet or surpass the largest floods on U.S. record for the larger basins. The largest U.S. floods on record for small basins have occurred in the semi-arid west and southwest (Costa, 1987a) where high intensity small storm centers are common. The 1993 Midwest floods surpass the maximum recorded floods for basins around 10,000 km^2. Basins of this size are less numerous than small basins and relatively small intense precipitation cells rarely can cover the whole basin. Extremely high magnitude floods on large basins typically result from uniquely timed contributions from large portions of the basin; this only occurs under more rare hydrometeorologic conditions, such as during the 1993 floods.

5. Spatial distribution of rainfall in the Raccoon River Basin

The Raccoon River basin contains fairly few official National Weather Service rain gage sites (NOAA, 1993). Therefore, we supplemented the NOAA precipitation data with data from sites monitored for radio stations, monitored by individuals and reported to us, or published in county papers (Fig. 5A). These additional sources of data agreed well with the data from the NOAA gage sites, and provided additional data that we used to derive isohyetal patterns for of the two storm events.

5.1. Storm of July 8–9, 1993

The rainfall data for this storm indicate that the highest rainfall totals (28 cm) were located in the upper and middle reaches of Brushy Creek, the Middle Coon River, and South Raccoon River (Fig. 5B). The North Raccoon drainage basin received significantly less rainfall over most of its drainage area than the other three basins. The isohyetal map of rainfall distribution was used to determine the amount of rain that fell in the basin area upstream of each stream site on which we conducted surveys.

5.2. Storm of August 30, 1993

Rainfall data for the storm of August 30, 1993 were compiled using the same techniques as for the previous storm. The second major storm was smaller in both area and maximum rainfall. The center of highest rainfall for this storm was located in the southeast portion of the drainage basin, near Des Moines. Maximum precipitation in the Raccoon River basin was 15 cm. The July storm was the largest magnitude storm in the basin for all but three of our surveyed sites in the basin.

6. Floods in the Raccoon River basin

6.1. Spatial distribution of floods

The peak discharges for each survey site are shown in Fig. 6. The spatial scaling of the Raccoon River floods follows a different pattern than for high magnitude floods recorded in the summer of 1993 in the upper Midwest (Fig. 4). The relationship between peak discharge and drainage basin area shows significant scatter. Most of the channelized reaches of river have higher flood peaks than unchannelized reaches, most of which are steeper southwestern tributaries to the S. Raccoon River.

Flood discharges in the Raccoon River basin were higher than recorded discharges for basins less than 1000 km^2 in the upper Mississippi River basin (Figs.

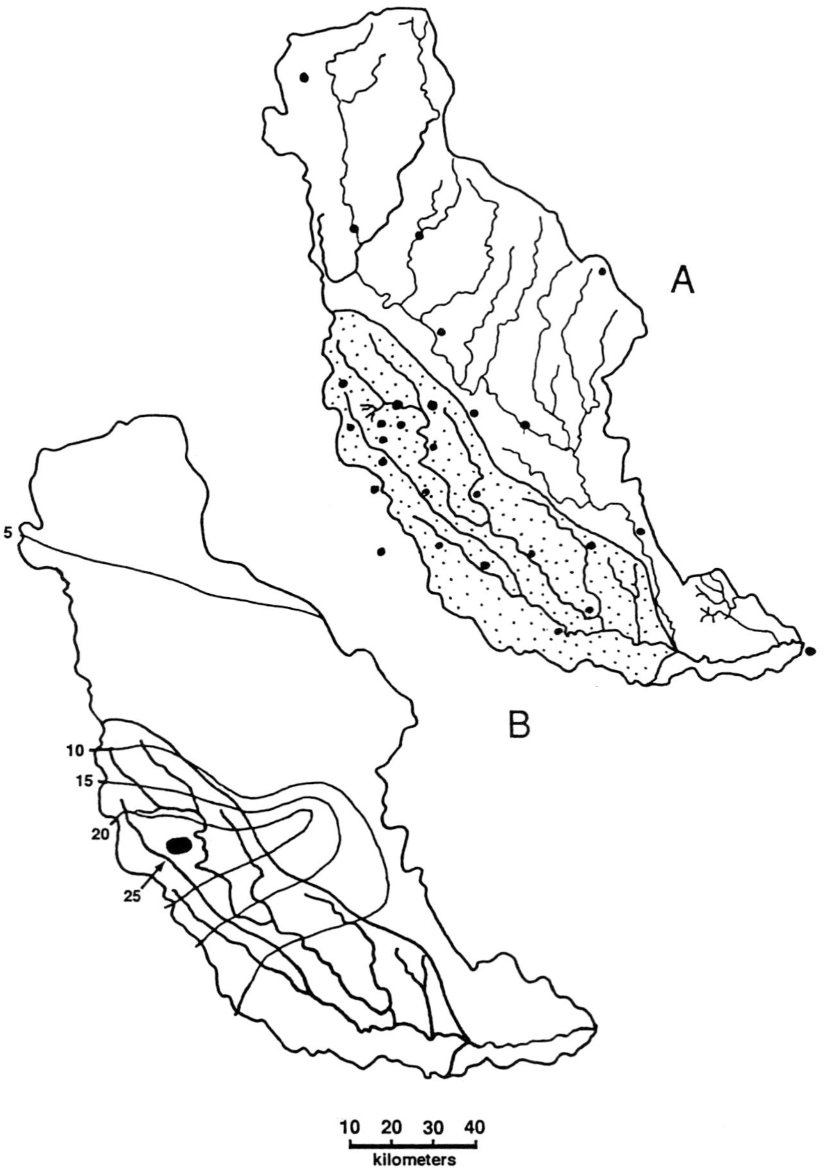

Fig. 5. (A) Stations for which rainfall data were available, July 8–9, 1993; (B) isopach map of rainfall totals, July 8–9, 1993.

4 and 6). This difference could be due to the lack of gaging station data for small streams, errors in the analyses or due to real differences in land use, precipitation patterns, or other causes. The Raccoon River basin is one of the most intensely agricultural basins in the United States; the percentage of land that is used for row crops is greater than 60% in the Raccoon River basin, this is a higher percentage than in much of the upper Midwest that also experienced flooding (Goolsby et al., 1993). The timing of the peak floods in the Raccoon River basin was later than in more northern parts of the upper Midwest. Therefore, antecendent moisture conditions could have been higher than in northern areas. Conclusive arguments as to the causes of differences in behavior between Raccoon River sites and other sites in the upper Midwest will not be resolved without further research.

There are, however, significant differences in land

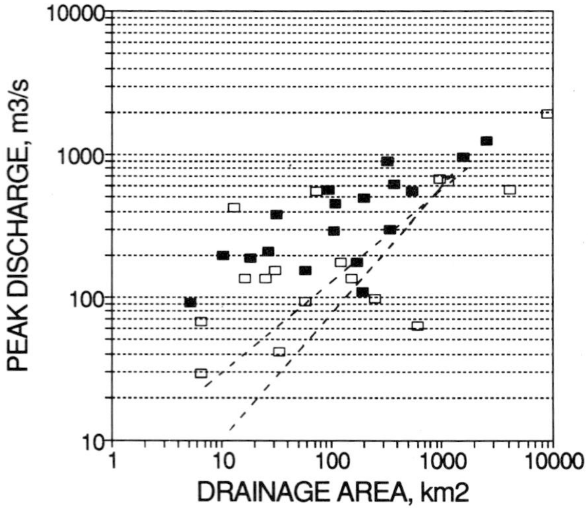

Fig. 6. Peak discharge versus drainage basin area for Raccoon River Basin. Ninety percent of the data are for the July flood. Dashed lower line is the 100-year flood estimates for region 4 from Lara, 1987. Dashed upper line is the 100-year flood estimate for region 3 from Lara, 1987. Solid squares are channelized reaches, open squares unchannelized.

use in the Raccoon Basin itself. These differences primarily follow the pattern of Pleistocene glaciation. Most of the basin, with an exception of the southwestern tributaries to the South Raccoon River, is underlain by portions of the Des Moines Lobe of the Wisconsin ice sheet (Ruhe, 1969; Prior, 1976). This is a poorly drained area in its natural state and much of the land has been ditched or drained with subsurface pipes to increase the rate of water runoff during and after storms (Fig. 7A). Most of the upper tributaries of the Middle Raccoon River, Brushy Creek, and the South Raccoon River been channelized in this manner. Peak flood discharge from these areas was compared with other sites draining less agricultural basin without channelized reaches of streams.

6.2. Relationship between flood magnitude and rainfall amounts

A comparison between flooding in different parts of the Raccoon River basin cannot be made without first evaluating the effects of rainfall on flood magnitude. Data from Fig. 5B and the corresponding map for the August storm were used to evaluate average rainfall in the sub-basins above each flood discharge evaluation point. Although rainfall intensity is known to affect runoff generation amounts and processes, it was not

Fig. 7. Channel and land character in the basin. (A) Upper channelized reach of the Middle Raccoon River. Note percentage of the basin in row crops; watershed divide is on the skyline. (B) Intermediate channelized reach, Brushy Creek at Highway 171, where the bridge failed during the floods. (C) Brushy Creek at mouth.

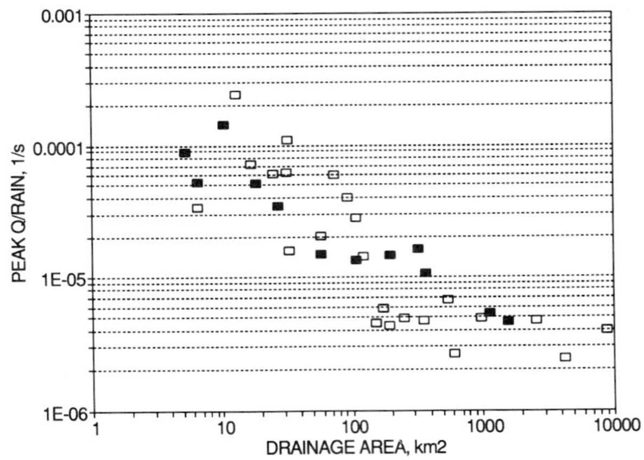

Fig. 8. Relationship between peak discharge (m^3/s) normalized for volume (m^3) of rainfall in the basin (1/s) and basin area (km^2). Solid squares are straightened and ditched areas; open squares are unchannelized areas.

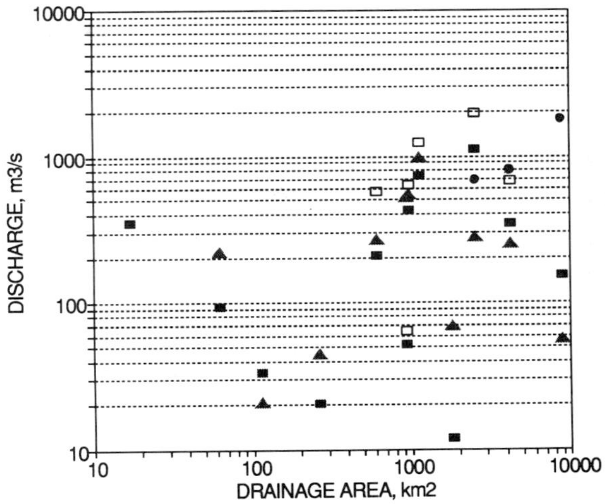

Fig. 9. USGS gaging station and post-flood survey data for peak floods of 1947 (○), 1986 (■), 1990 (▵), and 1993 (□) versus drainage basin area in the Raccoon River Basin.

available at most sites. The effects of spatial rainfall heterogeneity on peak discharges were evaluated by examining the relationship between flood discharge normalized by average rainfall volume in the basin and basin area (Fig. 8).

The peak discharge normalized to rainfall volume shows a distinct inverse relationship to drainage area (Fig. 8). Many of the channelized areas in Brushy Creek and the upper basins were also in areas of highest rainfall totals. Peak discharge normalized for rainfall volume was higher for the unchannelized southwestern tributaries of S. Raccoon River for basins smaller than 100 km^2. Channelized reaches had larger normalized discharges for basins larger than 100 km^2 (Fig. 8). The major loss to bridge structures on Brushy Creek were at these basin scales (Fig. 7B). This spatial pattern of flood discharges is consistent with other studies in channelized rivers that show increases in peak discharges and flooding downstream of channelized reaches (Lane, 1947; Emerson, 1971; Campbell et al., 1972).

6.3. Previous high magnitude floods in the Raccoon River basin

High magnitude floods in the Raccoon River basin have been recorded in previous years (e.g. 1947, 1986 and 1990, among others). Previous studies of the spatial variation in the magnitude of floods within the basin were conducted for the 1986 and 1990 floods (Baebenroth and Schaap, 1992). The 1993 floods were recorded at four gaging stations in the Middle and South Raccoon Rivers and at the downstream confluence of these rivers. The 1993 floods on the South Raccoon River were the highest in the period of record (Fig. 9). The estimates of flood peak discharges described in this study suggest that the floods were generally higher in magnitude than previous floods for almost all drainage basin sizes (see Figs. 6 and 9). The June 29–July 1, 1986 storm caused similar amounts of precipitation over the basin (20–30 cm), but the precipitation occurred primarily in the upper Middle Raccoon basin and rainfall occurred over a longer time interval (Baebenroth and Schaap, 1992).

6.4. Relationship of basin morphological characteristics to flood peaks

Drainage basin morphology develops from channel and hillslope sediment movement during floods. Drainage basin networks, therefore may reflect the prevailing flood magnitude and frequency relationships (Patton and Baker, 1976; Baker, 1977). Drainage basin characteristics are usually significantly different among climatic regions with different runoff processes (Dunne, 1978). Patton and Baker (1976) examined the relationship between morphometric characteristics of small watersheds and flood history. They found that basin magnitude, basin ruggedness, and first order channel frequency correlated with flash flood frequency indices. Newson (1978) found that drainage basin size, slope, and channel network characteristics were valuable in predicting flood characteristics. In a study of the largest rainfall–runoff floods in small basin in the United States, Costa (1987a) found that basin morphologic characteristics had little relationship with flood magnitude. These results are similar to those of Benson (1965) who found that drainage basin size and rainfall intensity were most closely associated with peak discharge. Benson hypothesized that flooding during extreme events may not be constrained by the characteristics of the drainage net. Research on runoff processes has shown that the drainage network can expand into extended networks of saturated areas during storm events (Dunne, 1978; Dunne and Black, 1970).

Previous studies of flood discharges in the Raccoon River basin have shown relationships between surficial geology, drainage character and the magnitude of floods for a given recurrence interval event (Lara, 1987; Baebenroth and Schaap, 1992). Lara found that for basins less than 1000 km^2, estimated values of the 100-year flood were higher for stream sites in the southwestern tributaries where the drainage density is higher than in the Wisconsinan-age glacial terraine to the north. The 100-year flood equation for the southwestern regions (Lara's region 3) is:

$$Q_{100} = 3.30A^{0.75}$$

where Q is the 100-year flood in m^3/s and A is the drainage area in km^2. Regions underlain by the Wisconsinan-aged glacial deposits had lower peak discharges for basins less than 5000 km^2 (Lara's region 4):

$$Q_{100} = 1.03A^{0.92}$$

These equations are shown in Fig. 6.

The peak discharges that we estimated from survey data and the reported USGS gaging station data for the Raccoon River basin are higher than estimates of the 100-year flood (Fig. 6). The best-fit line through the 1993 data follows the relationship:

$$Q_{max} = 29A^{0.5}$$

indicating both a different spatial pattern and significantly higher peak discharges than the 100-year flood estimates for the basin.

We found that the magnitude of the 1993 floods in the Raccoon River basin were closely associated with both drainage basin area and rainfall over the basin (Figs. 6 and 8). We found that peak discharge was higher for channelized reaches of rivers in the originally poorly drained areas than for the steeper southwestern, integrated drainage basins of the S. Raccoon River. When peak discharge is normalized by rainfall volume, however, the channelized reaches have higher peak discharges only for basins larger than 100 km^2. A comparison with Lara's analysis suggests that the effects of channelization on peak flood discharges become more pronounced with larger flood magnitudes.

The most surprising result, however, is that the flood discharge relationship to basin was so similar for two distinctly different geomorphic areas. This could be interpreted as either supporting Benson's (1965) view that basin characteristics are not important in the flood responses to extreme events or that the draining and channelization efforts of the past century have made the two dissimilar drainage areas behave similarly. Data from this study suggests that stream channelization and drainage of lands can influences the magnitude of extreme events.

The inundation time of floods in upstream drainage basins may be a fraction of the original flood time prior to ditching and channelization. During the 1993 floods, flood inundation was measured in hours rather than in days or weeks as may have occurred under natural conditions (Shankman and Samson, 1991). The ditching of unchanneled lands and straightening of the rivers was done to prevent damage to crops by flooding (Daniels, 1960). In this study, we noted little flood damage to fields in upstream tributaries of the three streams that we studied (Fig. 7A). Possible causes of the pattern of flood damage will be explored in the next section.

7. Hydraulics of flood flows

7.1. Width of flooded areas

The width of the flooded area during the storm generally increases with drainage basin area (Fig. 10). There is a rapid increase in flooded area width with

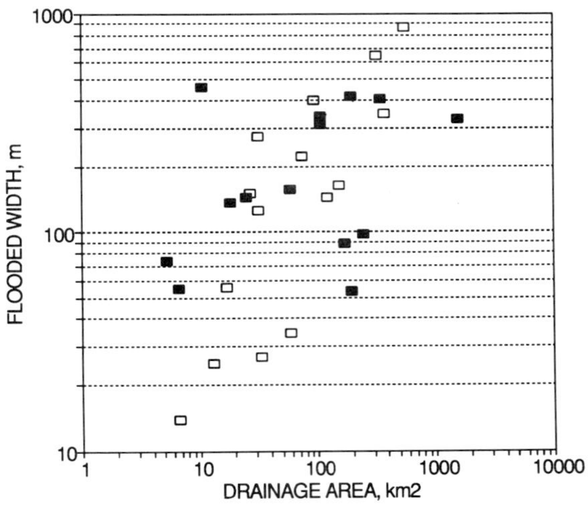

Fig. 10. Relationship between flooded area width and drainage basin area. Solid squares are channelized sites, open squares are unchannelized sites.

increasing drainage area for the unchannelized southwestern tributaries of the S. Raccoon River. This relationship reflects the morphology of the valleys that become wider downstream in this region of well-developed drainage basins. There is not a good relationship between flooded areas and drainage basin area for the channelized reaches of river, indicating that large basins are not temporarily storing more water than upstream, smaller basins (Fig. 10)

Floodplain velocities are lower than channel velocities due to the shallow depths of overbank flows and high floodplain roughness producing high relative roughness of the floodplain (Bhowmik and Demissie, 1982). Therefore, flooded area width is one measure of potential temporary storage of both sediment and water on the floodplain during and after the storm. These data on flooded width of channel indicates that overbank flows in the unchannelized reaches of river are a significant factor in the downstream movement of the flood waves.

7.2. Spatial distribution of flood shear stresses

Although crop damage was significant during the 1993 floods in the Raccoon River basin, the most dramatic damage was to bridges on Brushy Creek (Fig. 7B). The spatial pattern of channel erosion and bridge damage suggested to us that erosive power for the floods may have been concentrated in the central portions of Brushy Creek and other rivers. We calculated both channel shear stress and floodplain shear stress for each site. Shear stress was calculated by use of the DuBoy's equation:

$$\tau = \rho g R S$$

where τ is shear stress, ρ is the density of water, R is hydraulic radius, and S is the water surface slope. For the flow over the floodplain, the average floodplain depth was used in place of hydraulic radius, R. For channel flow, the hydraulic radius was used in the calculation of shear stress.

The distribution of channel shear stresses is shown in Fig. 11A. Channel shear stresses were high in both headwater streams and in the larger streams, with an average shear stress of 70 N/m^2, which is capable of moving cobbles on the stream bed. Most of the stream beds were composed of sand, which could easily be moved by the channel flows. The floodplain shear stresses are considerably lower (Fig. 11B), for both channelized and unchannelized river reaches. The distribution of shear stress does not clearly explain the dramatic failure of bridges along Brushy Creek and the S. Raccoon River. The most likely explanation is both shear stress and velocities in the channels are high, and considerable bed scour can occur around bridge pilings and around the lateral bridge supports. The rectangular shape and smooth banks of the channelized reaches should result in high velocities near the edges of the channels where bridge pilings and supports are located.

Previous researchers have indicated that the smaller the basin, the larger the percentage of total sediment that is likely to be carried by large floods (Wolman and Miller, 1960). Research on swales above first order streams indicates that erosion of these features is often episodic (Dietrich et al., 1986; McHugh, 1989). These data support other flood data that indicate that erosion might be most extreme in headwater regions (Baker and Costa, 1987). Field examination of field sites, however, did not find this to be true throughout the Raccoon River basin. Many headwater sites in agricutural areas have been deepened and straightened by channelization and previous high flood events (Daniels, 1960). Therefore, the channel was sufficiently large to contain much of the flood discharge, resulting in temporary high channel shear stresses in grass-lined channels. The short time of flooding and the roughness provided by

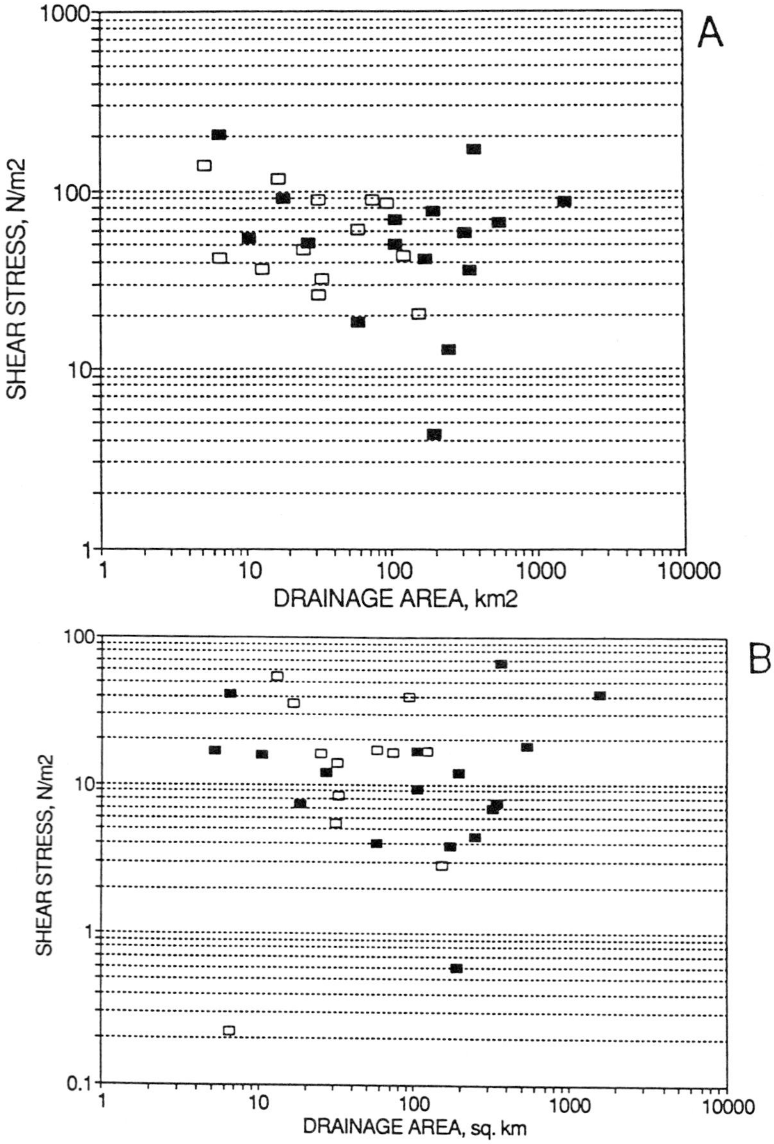

Fig. 11. (A) Relationship between channel shear stress and drainage basin area, and (B) between floodplain shear stress and basin area. The channelized reaches are indicated with solid squares.

the low width to depth ratio of these grass-lined channels may explain the lack of channel erosion in upstream tributaries. Low overbank shear stresses and short time of flooding also likely explain the lack of floodplain erosion in most headwater locations, where the erosive power of the floodplain flow was not sufficient to remove juvenile soybean plants, which were our most valued flood level indicators.

8. Discussion

There has been considerable discussion as to whether land use had a major effect on the magnitude of the 1993 Midwest floods. Many people have made the point that the major wetlands of the Mississippi River bottomlands were under water and had little effect on mitigating flood magnitude. An assessment of this issue is beyond the scope of our investigation. We did find, however, that the landscape of west-central Iowa has been distinctly changed by human processes and the modification of these upstream drainage basins may have affected flood magnitudes. In particular, poorly drained, recently glaciated terraine has been modified to enhance land drainage. Other researchers (Emerson, 1971; Campbell et al., 1972; Shankman and Samson, 1991; Shankman and Pugh, 1992) have reported on the effects of land drainage on flood magnitude. We found that the most modified parts of the landscape contributed the largest magnitude floods when scaled to rainfall volumes for basin areas larger than 100 km^2. Bridge failure was not associated with sites of particularly high shear stresses, but channel velocities were high and bank flow resistance was low, which affected velocity distributions in the channel. Land use practices did not cause flooding in the midwest, nor are they primarily responsible for the convergence of floods in the mainstem of the Mississippi River, but land use may have affected the magnitude of floods and the timing of floods within small and large watersheds.

To put the effects of land use on flood magnitude in perspective, the water treatment plant on the Raccoon River in Des Moines was inundated by water that overtopped the levee by one foot. The channelization of the Raccoon River tributaries influenced both the amount and timing of runoff. The increase in downstream flood magnitude associated with channelization may have contributed to the flooding problems in Des Moines. The main cause of the flood, however, was due to the persistent hydrometerological conditions and the southern movement of storm systems throughout the summer.

Acknowledgements

This work was done with the help of many people in the Raccoon River Basin. In particular, we benefitted from conversations with people at the USGS in Iowa City, Iowa and with Soil Conservation Service employees in every county in our study area, especially Carroll county. We received rainfall data from individual farmers, from the radio station in Carroll, Iowa and from county newspapers.

This study was funded by a supplement to NSF grant EAR-9220170 (K.L. Prestegaard).

References

Baebenroth, R.W. and Schaap, B.D., 1992. Floods of 1986 and 1990 in the Raccoon River Basin, West-Central Iowa. U.S. Geol. Surv. Open-File Rep., 92-94.

Baker, V.R., 1977. Stream-channel response to floods with examples from central Texas. Geol. Soc. Am. Bull., 88: 1057–1071.

Baker, V.R. and Costa, J.E., 1987. Flood power. In: L. Mayer and D. Nash (Editors), Catastrophic Flooding. Allen and Unwin, Boston, pp. 1–21.

Barnes, H.H., Jr., 1967. Roughness characteristics of natural channels. U.S. Geol. Surv. Water Supply Pap., 1849: 213 pp.

Bell, G.D., Rodenhuis, D.R. and Miskus, D., 1993. The great midwestern flood of 1993. EOS, 74(43).

Benson, M.A., 1965. Factors affecting the occurrence of floods in the southwest. U.S. Geol. Surv. Water Supply Pap., 1580-D: 72 pp.

Beven, K.J., 1986. Hillslope runoff processes and flood frequency characteristics. In: A.D. Abrahams (Editor), Hillslope Processes, Allen and Unwin, Boston, pp. 187–202.

Bhowmik, N.G. and Demissie, M., 1982. Carrying capacity of flood plains. J. Hydraul. Div. ASCE, 108: 443–452.

Boning, C.W., 1993. U.S. Geological Survey data collection and analysis in response to the 1993 Mississippi River Floods, EOS, 74(43): 61.

Campbell, K.L., Kumar, S. and Johnson, H.P., 1972. Stream straightening effects on flood–runoff characteristics. Trans. Am. Soc. Agric. Eng., 15: 94–98.

Costa, J.E., 1987a. Hydraulics and basin morphometry of the largest flash floods in the conterminous United States. J. Hydrol., 93: 313–338.

Costa, J.E., 1987b. A comparison of the largest rainfall–runoff floods in the United States with those of the Peopole's Republic of China and the world. J. Hydrol., 96: 101–115.

Daniels, R.B., 1960. Entrenchment of the willow drainage ditch, Harrison County, Iowa. Am. J. Sci., 258: 161–176.

Dietrich, W.E., Wilson, C.J. and Reneau, S.L., 1986, Hollows, colluvium and landslides in soil-mantled landscapes. In: A.D. Abrahams (Editor), Hillslope Processes. Allen and Unwin, Winchester, MA, pp. 361–388.

Dunne, T., 1978. Field studies of hillslope flow processes. In: M.J. Kirkby (Editor), Hillslope Hydrology. Wiley, London, Ch. 7.

Dunne, T. and Black, R.D., 1970. An experimental investigation of runoff production in permeable soils, Water Resour. Res., 6: 478–1311.

Emerson, J.W., 1971. Channelization; A case study. Science, 173: 325–326.

Goolsby, D.A., Battaglin, W.A. and Thurman, E.M., 1993. Occurrence and transport of agricultural chemicals in the Mississippi River Basin. July through August, 1993. U.S. Geol. Surv. Circ., 1120-C: 22 pp.

Heinitz, A.J., 1980. Floods in the Raccoon River basin, Iowa. U.S. Geol. Surv. Open-File Rep., 80-162: 110 pp.

Kroczynski, S., Richards, F., Miskus, D., 1993. Hydrometeorological conditions preceding and resulting in the ''Great flood of 1993''. EOS, 74(43): 62.

Lane, E.W., 1947. The effects of cutting off bends in rivers. Proc. 3rd Hydraulics Conf. Bull., 31. Univ. of Iowa, Iowa City, pp. 230–240.

Lara, O.G., 1987. Method for estimating the magnitude and frequency of floods at ungaged sites on unregulated rural streams in Iowa. U.S. Geol. Surv. Water Resour. Invest. Rep., 87-4132: 34 pp.

Limerinos, J.T., 1970. Determination of the Manning coefficient from measured bed roughness in natural channels. U.S. Geol. Surv. Water Supply Pap., 1898-B: 47 pp.

McHugh, J.P., 1989, Hydrologic and erosional processes in a gullied headwater basins, SW Wisconsin. M.S. Thesis, Univ. of Illinois at Chicago, 114 pp.

Middleton, G.V. and Southard, J.B., 1984. Mechanics of Sediment Movement, SEPM short course #3, 2nd ed. SEPM, Tulsa, OK, 401 pp.

NOAA (National Oceanic and Atmospheric Administration), 1993. Climatologic data, Iowa: Asheville, N.D. monthly summaries, Vol. 104, No. 1–9.

Newson, M.D., 1978. Drainage basin characteristics, their selection, derivation and analysis for a flood study of the British Isles. Earth Surf. Process., 3: 277–293.

Parrett, C., Melcher, N.B. and James, R.W., Jr., 1993, Flood discharges in the Upper Mississippi River Basin, 1993. U.S. Geol. Surv. Circ., 1120-A: 14 pp.

Patton, P.C. and Baker, V.R., 1976. Morphometry and floods in small drainage basins subject to diverse hydrogeomorphic controls. Water Resour. Res., 12: 941–952.

Potter, K.W., 1991. Hydrological impacts of changing land management practicies in a moderate-sized agricultural catchment. Water Resour. Res., 27: 845–855.

Potter, K.W., Rogers, C.A. and Kunkel, K., 1993. How rare were the 1993 floods in the upper Mississippi River valley. EOS, 74(43): 61.

Prior, J.C., 1976. A regional guide to Iowa landforms. Iowa Geol. Surv. Educ. Ser., 3: 72 p.

Rantz, S.E. and others, 1982. Measurement and computation of streamflow. Vol. 2. Computation of discharge. U.S. Geol. Surv. Water Supply Pap., 2175: 285–631.

Ruhe, R.V., 1969. Quaternary Landscapes in Iowa. Iowa State Univ. Press, Ames, 255 pp.

Shankman, D. and Pugh, T.B., 1992. Discharge response to channelization of a coastal plain stream. Wetlands, 12: 157–162.

Shankman, D. and Samson, S.A., 1991. Channelization effects on Obion River flooding, Western Tennessee. Water Resour. Bull., 27: 247–254.

Wahl, K.L., Vining, K.C. and Wiche, G.J., 1993. Precipitation in the Upper Missisippi River Basin, January 1 through July 31, 1993. U.S. Geol. Surv. Circ., 1120-B: 13 pp.

Wolman, M.G. and Miller, J.P., 1960. Magnitude and frequency of forces in geomorphic processes. J. Geol., 68: 54–74.

ELSEVIER

Geomorphology 10 (1994) 183–196

GEOMORPHOLOGY

Flooding of property by runoff from agricultural land in northwestern Europe

John Boardman[a], Laurence Ligneau[b], Ad de Roo[c], Karel Vandaele[d]

[a]*School of Geography and Environmental Change Unit, University of Oxford, Mansfield Road, Oxford OX1 3TB, UK*
[b]*A.R.E.A.S., 25 Rue de Dieppe, 76460 St Valery en Caux, France*
[c]*Department of Physical Geography, University of Utrecht, P.O. Box 80115, 3508 TC Utrecht, The Netherlands*
[d]*Laboratory for Experimental Geomorphology, KU Leuven, Redingenstraat 16bis, B-3000 Leuven, Belgium*

Received February 10, 1994; revised April 7, 1994; accepted April 8, 1994

Abstract

In the last twenty years there has been an increase in the incidence of flooding of property by runoff from agricultural land in many areas of northwestern Europe. These events take the form of inundations by soil-laden water associated with erosion and the formation of ephemeral or talweg gullies developed in normally dry valley bottoms. Costs of such events may be considerable e.g. almost US$2M at Rottingdean, southern England, in 1987. These costs are largely borne by individual house occupants, insurance companies and local councils. The distribution of flooding is widespread but areas of high risk can be identified: the hilly area of central Belgium, parts of northern France, the South Downs in southern England and South-Limburg (the Netherlands). All these areas have silty, more or less loessial soils. Two types of flooding may be distinguished: winter flooding associated with wet soils and the cultivation of winter cereals, and summer flooding due to thunderstorm activity and runoff particularly from sugar beet, maize and potato crops. The distribution of these types of erosion varies in relation to the interaction between physical characteristics (soils and topography), climatic conditions and land use across the region. The reason for the recent increase in flooding events appears to be changes in land use, in the area of arable cropping, and the continued intensification of farming such as the use of chemical fertilizers, the decline in aggregate stability, the increase in the size of fields and compaction by farm vehicles. In some regions the risk of flooding has also increased because of expansion of urban areas in valley bottom locations. Communities have responded to the flooding hazard with emergency or protective measures usually involving engineered structures rather than land use change. The policy response to the increased risk of flooding has been very limited especially at the national and provincial level, the exception being plans developed with farming organisations in South-Limburg and the Pays de Caux. In southern England initiatives have been few and largely consist of protective measures undertaken by urban municipalities.

1. Introduction

In this study we have selected four areas of northwestern Europe where property damage by runoff from agricultural land has become more frequent in recent years (Fig. 1). The flooding is linked to erosion since runoff is generally from arable areas and carries with it soil. Damage is therefore the result of "muddy floods". The areas examined are South-Limburg (The Netherlands), the Pays de Caux (France), the area between Brussels and Leuven (Belgium) and the South Downs (England). They have features in common: dissected terrain with dry valley systems of periglacial origin; soils derived from loess; recent intensification and specialisation of agriculture, and housing areas proximal

SSDI 0169-555X(94)00031-L

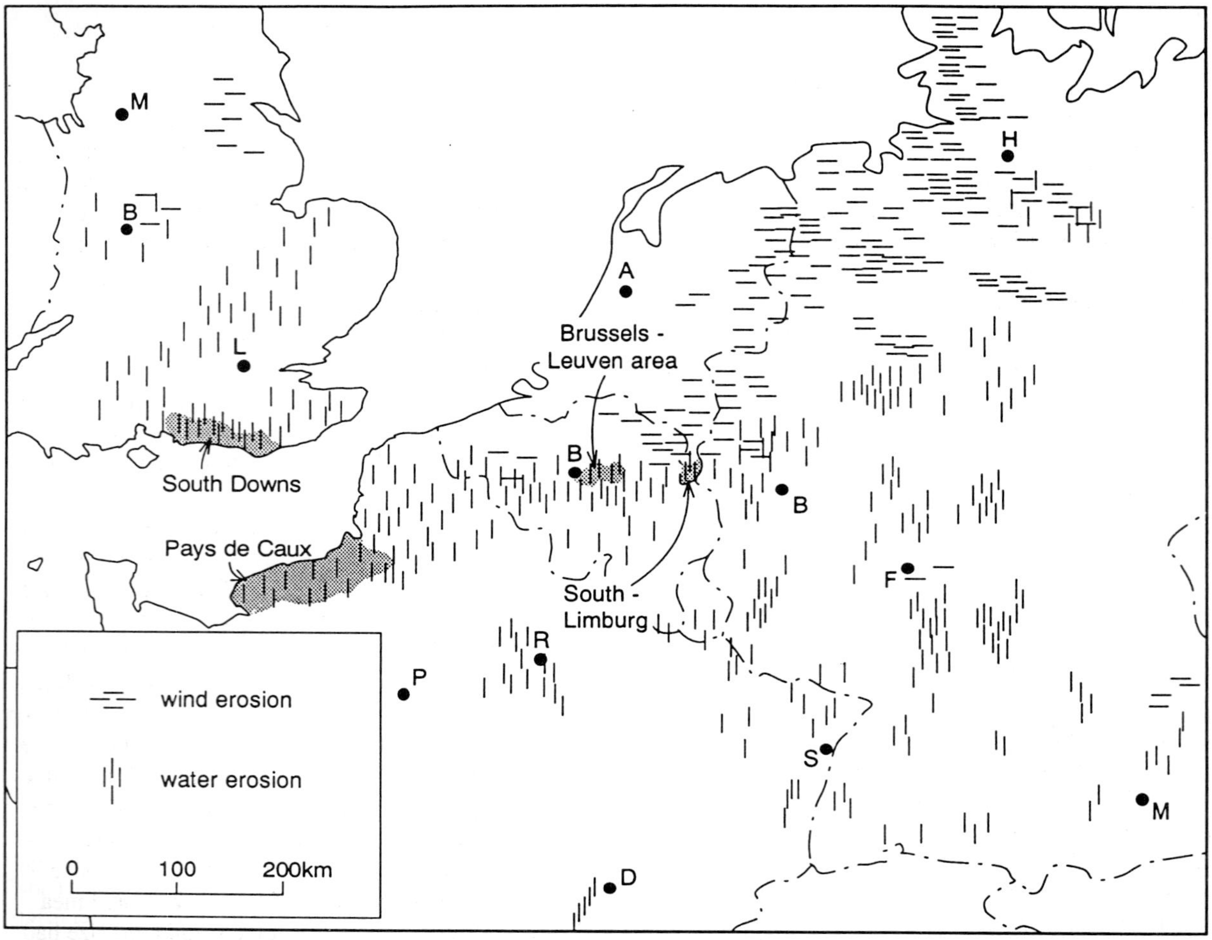

Fig. 1. Location of study areas and parts of the loess belt subject to wind and all forms of water erosion (after De Ploey, 1989b).

to farmland. There are, however, differences between the study areas.

This paper addresses the reasons for the increase in flood damage to property, the scale of that damage and the responses to the flood hazard.

1.1. South-Limburg

The soils of South-Limburg are developed on loess and are especially vulnerable to surface runoff and soil erosion. The characteristic landforms of the area are dry valleys, incised (''hollow'') roads and lynchets (man made prehistoric/medieval cultivation terraces, usually parallel to the contours). Re-allotment schemes have resulted in larger fields, causing surface runoff to be more erosive. Changes in land use also contribute to increasing erosion, total runoff and peak runoff. The area of grassland has decreased in favour of urban areas and, after 1975, in favour of arable land (De Roo, 1993). Between 1960 and 1986, the kinds of crops which give rise to a higher erosion risk, such as maize and sugarbeet, have increased in area, replacing cereals such as winter wheat. The alternating pattern of small arable fields, grassland, and orchards has been replaced by single large fields. Another and related consequence of re-allotment is the removal of many hollow (sunken) roads and lynchets which can lead to a changing drainage pattern and the formation of rills and gullies at different locations (De Roo, 1993). Lynchets were constructed to prevent soil erosion on steep slopes. Nowadays they slow down surface runoff and therefore play an important role in soil conservation but have to be maintained regularly. In the re-allotment area of Ransdaal (Ransdalerveld) between 1913 and 1975, 40 to 50% of the lynchets disappeared (Van der Helm, 1987). Diemont and Van de Westeringh (1978) reported that between 1910 and 1950 the total length of lynchets in South-Limburg decreased from 200 to 120 km (40%). Since 1950 there has been a further decrease of 30% (Saris, 1984). There is no reason to believe that this process has stopped. Even now, at some locations farmers remove lynchets very gradually, a small part during every tillage operation leading to instability and gully formation. The removal of lynchets increases the pressure of soil macro-fauna (e.g. mice, moles, rabbits) on the remaining lynchets (De Ploey, 1986, p. 87; Vleeshouwer and Damoiseaux, 1990). If farmers maintain a buffer zone of 1–2 m at the sides of hollow roads and lynchets, they act as a useful soil conservation measure.

1.2. The Pays de Caux

The Pays de Caux is an extensive plateau underlain by chalk; overlying the chalk is 5–6 m of clay with flints of Tertiary age. The soils are developed in loess up to 6 m depth. Brown soils contain 10 to 14% clay, and about 60% silt with organic matter percentages of 1.4 to 1.6 on cultivated soils. The soils are particularly sensitive to crusting. Asymmetric dry valleys dissect the plateau with long, gentle slopes of less than 10%. Mean annual rainfall ranges from 700 to 1000 mm with a peak in November; daily rainfalls of 40 mm have a return period of 5 to 10 years.

Intensive agriculture began in the 1960s and is carried out in open fields (Fruit, 1991). The principal crops are cereals (wheat and barley), grass leys, sugar beet and oil seed rape. In the last ten years there has been an increase in peas, potatoes, oil seed rape and maize cultivation and a decrease in permanent grass and animals. The consequence of these changes has been an increase in the total arable area and in spring crops; this has resulted in more bare unused soils in the winter between successive spring cultivations.

There has been a decrease in the number of farms and farmers and an increase in average field size, partly as a result of re-allotment (Fruit, 1991). The average farm size is 35 ha (1993).

1.3. The loess area between Brussels and Leuven, central Belgium

The Belgian loess belt is situated in the central part of the country and is composed of an incised plateau at about 100 m above sea level. Incision of the valleys can reach 60 m. The region is characterised by a semi-continuous loess cover and the topsoils have a very high silt content (70–80%), a clay content of 10–20%, and a sand content below 15% (Govers, 1991). The loess overlies Tertiary sands which outcrop locally. About one third of the area is covered with colluvial deposits mainly in topographically defined drainage lines (Bollinne, 1978). The depth of loess cover varies but is of the order of a few decimetres to several metres (Goossens, 1993).

The area is characterised by a dense network of dry valleys, lynchets and sunken roads. Slope angles are low with 69% of a sample area of 40 km^2 having slopes of less than 5%, 27% with slopes of 5–10% and 4% of the area with slopes of more than 10%. Between 1950 and 1990 the area of arable land decreased from 46 to 36% reflecting the growth of urban areas and afforestation. There was also an increase in the area under winter cereals. In 1990 about 60% of the fields grew winter cereals with 30% under spring-sown crops (potatoes, maize, sugar beet and chicory). The size of fields is small but increasing, with an average size in 1979 of 0.7 ha (Gulinck, 1985). Based on field measurements, Vandaele found that the mean field size had increased to almost 1 ha by 1990.

1.4. The South Downs

The South Downs are a range of low rolling hills composed of chalk, a soft Cretaceous limestone. They stretch for a distance of 120 km and rise to just over 200 m edged with a steep north-facing scarp. The dip slope is deeply dissected by dry valleys with relief up to 150 m and slopes in the range of 10 to 48%. Soils are typically shallow stony rendzinas containing about 70% silt of loessic origin and 20–50% stones of chalk or flint. The thin A horizon of 150–250 mm overlies chalk or chalk rubble. In the early Holocene the Downs were wooded and soils were considerably thicker than now. Erosion began as a result of Neolithic and Bronze Age clearing and farming and continued intermittently until the present. Valley floor deposits are thicker than those on the slopes and contain archaeological evidence of the erosional history of the Downs (Bell, 1983). Mean annual rainfall varies from 750 to 1000 mm with a peak in the period October to January. For much of the Downs, a daily rainfall of 40 mm has a return period of between 1 and 2 years.

During the 1920s and 1930s the Downs were largely grassland for sheep grazing but in the Second World War many parts were ploughed. After the war the arable area continued to expand. Since the mid-1970s there has been a dramatic change in emphasis with the dominant crops being winter cereals as opposed to spring cereals. The farms have become very specialised, sometimes without any livestock, and with occasional break crops of oil seed rape and, latterly, linseed. In the early 1980s about 55% of the cultivated area of the eastern Downs was under winter cereals (Boardman, 1990); in some areas the figure was considerably higher. During the last forty years many field boundaries, ancient field systems and lynchets have been removed, farm machinery has become heavier and more powerful, and chemical fertilizers are now the norm. Cereals are grown on slopes of up to 40% whereas in the 1960s soils on slopes of greater than 20% were classified as "ill suited to regular cultivation" (Hodgson, 1967).

In comparison to loessic soils of the other study areas, those of the South Downs are much less erodible because of their stony character. However, this to some extent is offset by steep slopes, large fields and the dominance of a single crop.

2. Geomorphological processes

Erosion due to runoff on agricultural land is frequent and widespread in the study areas. Three types have been identified:

- sheet (or interrill) erosion on slopes of less than 3%. In northern France rates are less than 2 t ha^{-1} yr^{-1} (Ludwig et al., 1994). On the South Downs this type of erosion is unimportant compared with rilling and gullying (Boardman, 1993).
- rill–interrill erosion. Auzet et al. (1994) report rates of 5 to 25 t ha^{-1} yr^{-1} on rilled fields. On the South Downs rates of 100 t ha^{-1} yr^{-1} (Boardman and Robinson, 1985) and, exceptionally, over 200 t ha^{-1} yr^{-1} have been recorded (Boardman, 1988).
- erosion by concentrated flow in valley bottoms and depressions has been termed "talweg" or "ephemeral gullying". Auzet et al. (1994) suggest that 80% of erosion in studied catchments in northern France is due to this process.

In a review of the relative proportion of erosion due to interrill compared to rill processes, Govers and Poesen (1988) point out that the ratio varies in space and time and generalisations cannot be made. Even within the loess belt of western Europe different systems can be identified depending on topographic and soil conditions where splash, rill, sheetwash or ephemeral gully erosion are dominant (De Ploey, 1989a).

For the study of the flooding of properties, the most important erosional system is that where ephemeral gullying occurs. On the South Downs (1976–93) all major sites of damage to property were associated with ephemeral gullies (Boardman, 1994). In central Belgium, in several intensively studied catchments, ephemeral gully erosion was the dominant sediment source and contributed about 41% of total soil loss. Over a three year period almost 90% of eroded material left the catchment and as a result downstream villages can suffer severe damage (Vandaele, 1993b). Frequent, low-intensity winter rain leads to the development of narrow (0.3 m) but deep (more than the depth of the plough layer) gullies; in constrast, high intensity summer storms produce wide shallow gullies (Auzet et al., 1994).

Experimental work in South-Limburg has led to the suggestion that overland flow generation in summer is controlled by rainfall intensity and conforms to the Horton model; whereas that in winter is primarily controlled by rainfall amount and conforms to the saturation or storage control model (Kwaad, 1991). Many workers in western Europe have emphasised the importance of soil crusting and under these conditions Hortonian overland flow occurs in winter also. The seasonality of erosion is primarily controlled by crop cover. In areas where maize, sugar beet and vegetables are cultivated erosion and flooding occur in summer. On the South Downs there is very little bare ground in spring or summer and flooding is exclusively a winter phenomenon.

3. History of the flooding problem

For the purpose of this study a flood event is defined as runoff from a catchment, usually via an ephemeral

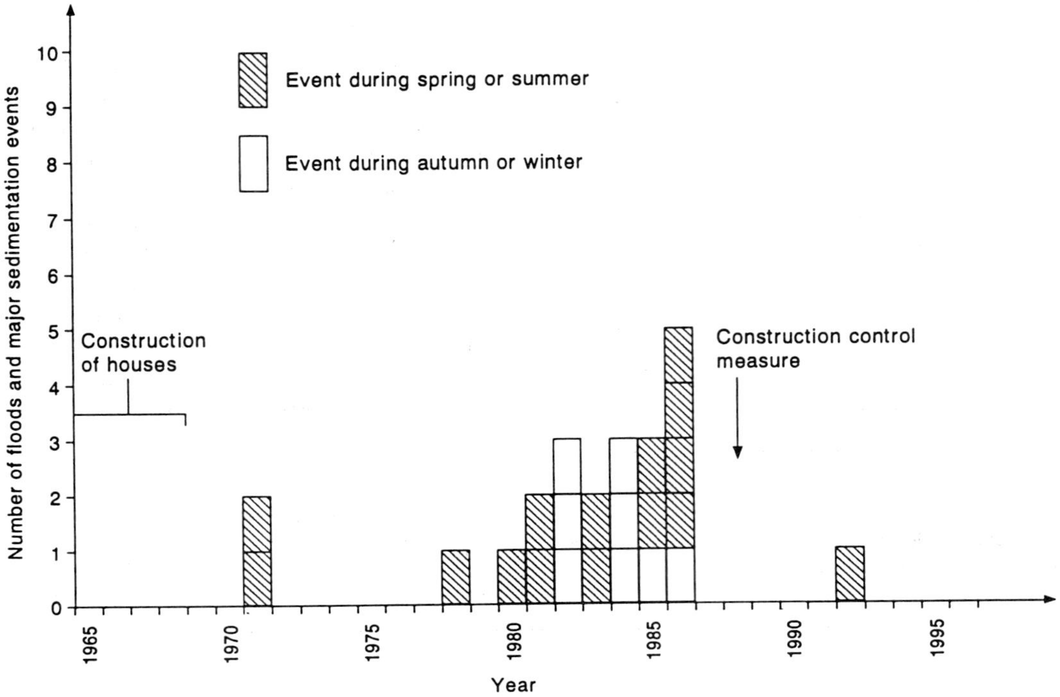

Fig. 2. Occurrence of floods and major sedimentation events in the Hammeveld 2 catchment, southwest of Leuven, 1965–1993 (Vandaele, 1993b).

gully, which affects property, most commonly, houses. Flood events may be of any magnitude and, depending on timing, discharge and land use, will vary in the amount of sediment transported and therefore in their impact.

Information about the incidence of flooding events varies from region to region. In central Belgium, the Hammeveld catchment (ca. 25 ha), suffered 22 flood events between 1971 and 1986 with an increasing frequency of events in the 1980s (Fig. 2) (Vandaele, 1993b). In South-Limburg, Schouten et al. (1985) mention increased problems of erosion and flooding in recent years. In particular they discuss extreme events in the early 1980s, e.g. at Valkenburg (August 1982), Ransdalerveld (May 1987), Etzenrade (May 1988), the Valkenburg area (May 1988) and Oirsbeek (June 1993). The first serious damage occurred on the South Downs in 1976, followed by 1982, 1987 1991 and 1993, all of which were autumn and early winter events (Boardman, 1994).

4. Causes of flooding

4.1. Land use

The increase in arable land and in the area under erosion-sensitive crops varies from region to region. In South-Limburg, the total area of grassland has decreased each year and since 1960 there has been a substantial increase in row crops (maize and sugar beet). These crops are 15 to 20 times more sensitive to erosion than cereals (Van der Helm, 1987). Here, as in the Pays de Caux, land planted with successive spring crops is left fallow in winter, either harrowed (with a tendency to crust) or unploughed with stubble/crop debris, the latter may be compacted and may therefore generate runoff (e.g. Ouvry, 1987). More rarely a cover crop is sown with the aim of protecting the soil in winter.

In the loess belt of Belgium erosion mainly occurs during (1) autumn and winter on fields left bare after

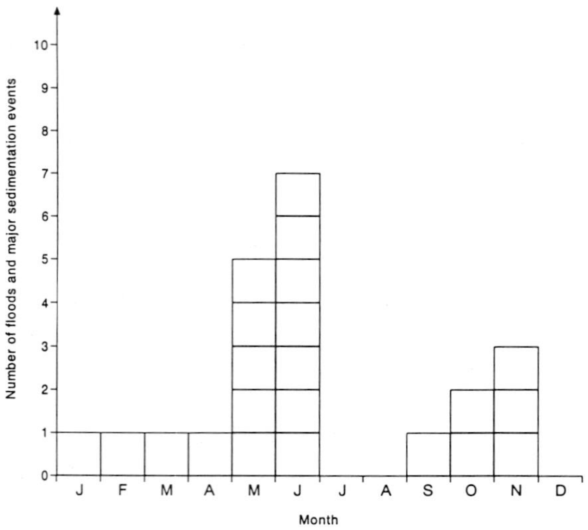

Fig. 3. Distribution of floods (1965–1986), Hammeveld 2 catchment, southwest of Leuven (Vandaele, 1993b).

harvesting or planted with winter cereals, and (2) in spring and early summer on fields planted with summer crops (Govers, 1991; Vandaele, 1993a). In some cultivated catchments between 60 and 70% of total soil loss occurs during high-intensity rain in spring and summer (Vandaele, 1993a,b). Most of the problems in the Hammeveld residental area occur in May and June with a small peak in October–November (Fig. 3). The area of the catchment under spring-sown crops appears to control the length of ephemeral gullies (Fig. 4); the relationship is less good for the winter period.

On the South Downs bare ground associated with cultivation of winter cereals coincides with the autumn and early winter peak in annual rainfall. In some catchments the majority of fields are under cereals and runoff is able to pass from one to another along valley bottom routes.

4.2. Topography and soil

The risk of flooding is high in northwestern Europe because of the presence of dry valley systems where

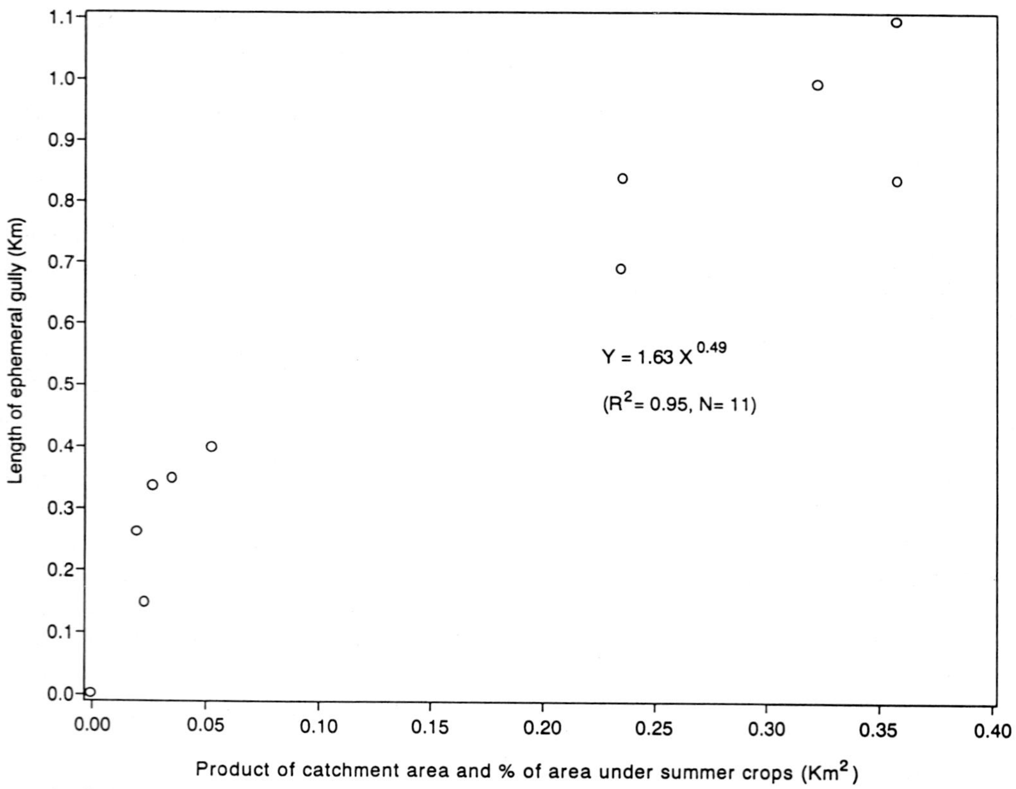

Fig. 4. Ephemeral gully length vs. catchment area (based on aerial photographs). Data from several catchments in Hammeveld area, for summer crops, 1963 (Vandaele, 1993b).

Table 1
Flooding and major sedimentation events for the Hammeveld residential area, 1971–1993. Only events with known rainfall parameters were included. Data from newspapers, local authorities, inhabitants and the Royal Institute for Meteorology

Date	Rainfall in 24 h (mm)	Peak rainfall intensity in 15 min ($mm\ h^{-1}$)	Rainfall in previous 5 days (mm)
15/05/71	18.5	35.0	12.1
16/05/71	33.5	35.0	30.6
17/05/71	22.2	35.0	64.1
01/05/78	43.1	60.0	–
07/10/82	41.2	10.0	11.3
14/10/82	12.6	7.0	36.3
09/11/82	11.0	7.0	–
27/04/83	34.5	10.0	18.7
23/06/83	52.0	110.0	–
07/02/84	10.9	16.0	39.9
22/11/84	24.5	17.0	17.6
26/01/85	21.6	10.0	19.1
14/05/85	29.0	56.0	10.8
06/06/85	45.0	100.0	–
25/03/86	10.0	4.0	10.5
20/05/86	38.5	68.0	–
07/05/86	65.0	11.0	–
17/06/86	39.0	40.0	–
23/06/86	29.0	75.0	–
15/09/86	41.6	17.0	31.3
09/08/86	45.0	160.0	12.0

runoff may be concentrated in valley bottoms. Due to field enlargement, monoculture of certain crops (cereals on the Downs), and the large arable area, valley-side slopes are able to generate runoff which reaches the valley floor. In many cases of flooding of property, recent removal of field boundaries has created long uninterrupted slopes which connect interfluves, valley-sides and valley floor into one potentially efficient runoff-producing and sediment-transporting system (e.g. Boardman and Robinson, 1985).

The soils of the region are potentially erodible because of their high silt and often low organic matter content. In South-Limburg, *K*-values ranged from 0.040 to 0.042 in SI units or 0.30 to 0.32 in American units (Römkens et al., 1986). The loessic soils are particularly sensitive to crusting. Low values for aggregate stability have been reported with significant seasonal changes (Van Eijsden and Imeson, 1985; Huigen, 1986; Blackman, 1992). Crusting is a particularly important factor in runoff generation. On winter cereal fields in the Pays de Caux, 45–60 mm of rain is sufficient to form a crust if there is a delay before the first rain, and only 20–30 mm if there is no delay (Boiffin et al., 1988). Infiltration capacity on recently tilled soils may be of the order of 30–50 $mm\ h^{-1}$ but this is reduced on a well crusted surface to 1–2 $mm\ h^{-1}$ (Auzet et al., 1990).

4.3. Farm management techniques

Certain farm management techniques increase the risk of runoff generation. The selection of erosion susceptible crops and the creation of larger fields have already been referred to. Other practices include:

- The preparation of fine seedbeds, often by use of powered harrows, is widespread and increases the risk of aggregate breakdown by rainfall.
- Reliance on chemical fertilizers rather than farmyard manure, and lack of grass in rotations, reduces the organic matter content of soils and their aggregate stability.
- The rolling of fields after drilling (e.g. cereals on the South Downs, peas in the Pays de Caux) produces more compact surfaces of low micro-topography.
- Vehicle wheel tracks act as sites of localised runoff and erosion due to low infiltration rates; on the Downs these are associated with the spreading of herbicides and nitrates on winter cereal fields.
- On moderate and steep slopes fields are frequently drilled down the line of maximum slope rather than along the contour. In South-Limburg, 67% of erosion cases on maize, sugar beet and potatoes were on fields with a ploughing direction downslope (Van der Helm, 1987).

4.4. Rainfall and runoff

Little attention has been paid to the threshold conditions for the onset of ephemeral gullying. However, on the basis of a physical–mathematical model, De Ploey (1990) found that thalweg incision in central Belgium and the Paris Basin should start at rainfall intensities between 8.5 and less than 1 $mm\ h^{-1}$, provided that the topsoils are already saturated or that rainfall amounts exceed storage. According to Papy and Douyer (1991) most flooding in the Pay de Caux occurs during heavy thunderstorms with rainfall intensities above 10 $mm\ h^{-1}$. In the Hammeveld catchment,

	S	O	N	D	J	F	M	A	M	J	J	A
1976-77	x	⊗⊗	xxx									
1977-78												
1978-79												
1979-80												
1980-81		xx										
1981-82												
1982-83		x⊗	⊗x									
1983-84			xxx									
1984-85	x	x										
1985-86												
1986-87												
1987-88		xxxxxx xxxxxx xx⊗⊗⊗	xx⊗⊗			x						
1988-89		x										
1989-90												
1990-91		xxxxx			xx	x					x	
1991-92												
1992-93												
1993-94		xxxxxx		xxxxx								

⊗ Autumn : date unknown

Fig. 5. Timing of damage to property on the South Downs, 1976–1993.

central Belgium, spring and summer flooding is associated with rainfall intensities above 35 mm h^{-1} (for 15 minutes) (Table 1). The return period of these events is about once every two years. In autumn and winter the problem can start on saturated soils at intensities of 5 to 10 mm h^{-1}. Laurant and Alexandre (1977) and Demaree (1990) found evidence for an increase in the maximum precipitation intensity over the period 1934–1976/1983 at Ukkel, Belgium. The number of thunderstorm days increases from the late 1960s. This may partly explain the observed increase in flooding incidents in that area.

On the South Downs 30 mm in two days has been suggested as a threshold at which rilling is initiated on typical winter cereal fields (Boardman, 1990). During the erosion-risk period (1 September–30 November) of the years 1982–91, there were 22 such events. Exceedence of this threshold does not always lead to damaging floods but they are relatively common in autumn months (Fig. 5). If rill systems on slopes and ephemeral gullies in valley floors are already established, then flooding will take place at much lower rainfall threshold values than that necessary to initiate the system. For example, severe flooding on the Downs in October 1987 resulted from a daily rainfall of 63 mm, flooding continued at Breaky Bottom throughout October with rainfall amounts as low as 7 mm (Boardman, 1988).

4.5. Location of property

In all of the study areas expansion of housing onto sites at risk of flooding has created the problem. On the South Downs most of this expansion occurred in the

1920s and 1930s when the dominant land use was grass; the risk of the generation of runoff was therefore low. Any flooding that took place was by clean water and was not perceived as a major hazard. Housing was located on valley floor and valley-side slopes in order to protect the skyline from visual intrusion (Robinson and Blackman, 1990a). In other areas, housing has been constructed more recently e.g. in the Hammeveld catchment (Fig. 2).

In summary, it should be emphasised that in three out of four of the study areas there is a risk of flooding in both autumn and summer associated with autumn and spring-sown crops. On the South Downs there is minimal risk in summer due to the lack of spring crops.

In all areas, changes in land use and farm practices have increased the risk of flooding at sites where previously the risk was regarded as low or non-existent.

The consensus view in northwestern Europe is that increases in erosion and flooding in recent years are primarily the result of changes in land use, farm practices and location of property and that changes in climate (for which there is little evidence) are unimportant.

5. The impact of flooding on property

The impact of soil erosion on crop yields in this region is, in general, insufficient over the short term to convince farmers that erosion is a problem. Attention therefore turns to off-farm impacts. Parts of the region are devoid of watercourses or lakes and therefore the impact of runoff carrying polluting agricultural chemicals is minimal. This is true of the South Downs. Water quality and sedimentation problems occur locally in the three other study areas. However, because of the relatively high population density, and the proximity of arable land to housing areas, the major off-farm impact is the flooding of property.

In the area between Brussels and Leuven, a recent survey revealed that about 70% of municipalities had experienced flooding and major sedimentation events (Ballings et al., 1993). For the small cultivated catchment of Hammeveld peak discharges during flooding events can range between 0.1 and 3 $m^3 s^{-1}$ in summer with suspended sediment loads at the outlet of the catchment of about 23 $g l^{-1}$ (Vandaele, 1993b). Measurements in a retention structure at the outlet of this 25 ha catchment indicate that about 350 m^3 of material was stored during a four year period (Vandaele, 1993b). This means that during the study period 5 to 7 $m^3 ha^{-1} yr^{-1}$ left the catchment. Flood events between 1965 and 1993 are shown in Figs. 2 and 3. A preliminary conservative estimate of costs of flood damage in the municipality of Bertem during a 16 year period amounted to about US$1,250,000. A cloudburst in the region between Brussels and Leuven in June 1986 caused damage of several million USD. A large amount of eroded material reaches streams and rivers: the city of Leuven dredges about 10,000 m^3 of silt out of the river Dijle each year (Poesen et al., 1988).

In South-Limburg, people living in the valleys are frequently troubled by muddy floods in their houses and on public roads, which cause serious financial loss for householders, the local authorities, and insurance companies. Cars are damaged, cellars have to be pumped dry and cleaned, private gardens are damaged. In August 1982, in the village of Valkenburg, water and mud caused damage which was estimated at several million Dutch guilders. Provincial and local governments have labour and equipment costs of repairing flooded roads and removing sediment. They have to repair and maintain infrastructural works such as waterways, bridges, culverts, storage ponds, and the sewerage system (Schouten et al., 1985). Estimated annual expenses for the maintenance and repair of damage to private properties by erosion and flooding to seven communities in South-Limburg and to the Waterboard "Roer en Overmaas" are almost US$800,000 (1985). In 1993, the costs of two events in Valkenburg (May 1) and Oirsbeek (May 12) have been thoroughly investigated. The costs of the removal of sediment by the authorities are estimated at US$113,300 and US$52,700 respectively. The costs of the damage to agricultural fields and private properties are still being assessed (J. Duijsings, pers. commun., 1993). Boardman (1990) reported that in the UK these off-farm costs probably exceed on-farm costs by an order of magnitude. This is likely to be the same in Limburg.

Ecologically valuable areas on slopes and in valleys are affected by the penetration of partly polluted water and sediment carrying nutrients and pesticides. The increased amount of surface runoff in South-Limburg is reflected in a flashier discharge regime, e.g. the River Geul. Leenaers and Schouten (1989) found that, under

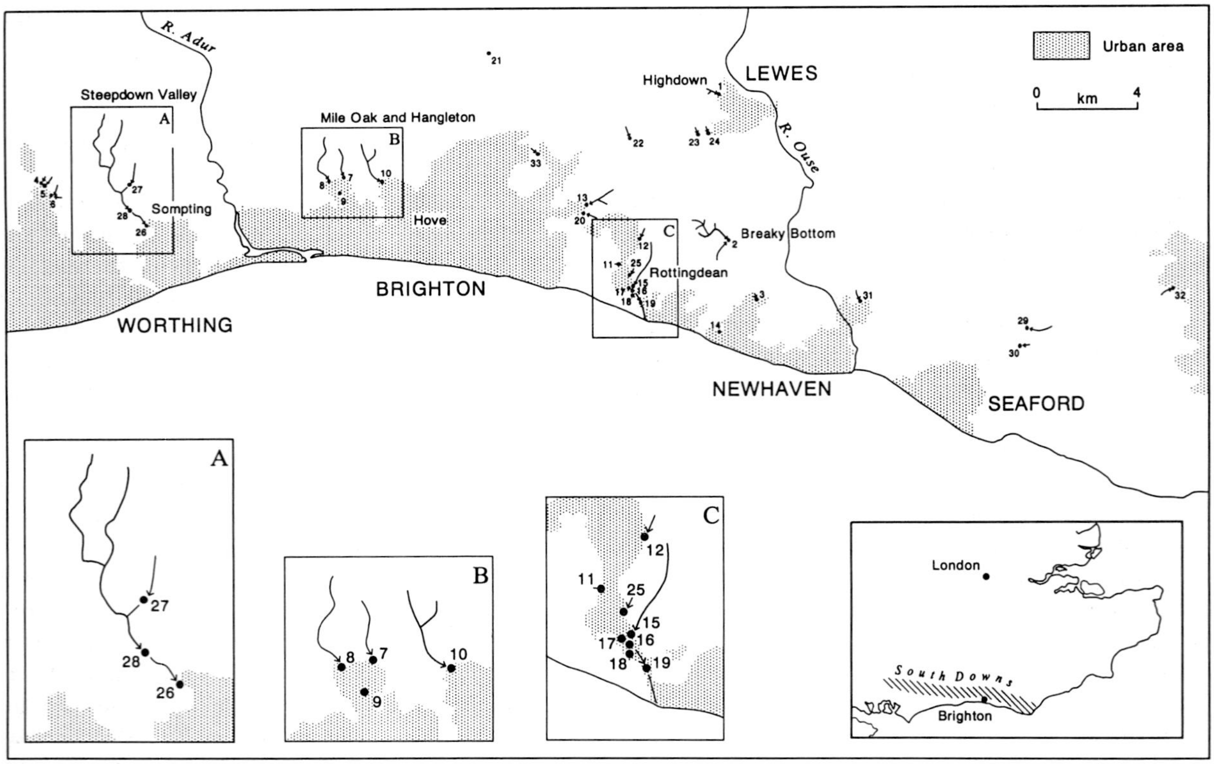

Fig. 6. Location of damage to property on the South Downs, 1976–1993.

similar rainfall conditions, a higher peak discharge is produced in the present situation, as compared to the situation in the 1950s. Also, due to increased peak discharges and high sediment loads, streams and rivers of a high ecological value suffer from bank erosion and increased bed migration (Leenaers, 1989).

On the South Downs damage to property has occurred regularly since 1976 (Fig. 5) with a concentration of incidents in the months of October–December and in the wetter years of 1976, 1982, 1987, 1990 and 1993 (Boardman, 1994). One reason for property damage is that peak discharges from cultivated catchments when they are bare in the autumn can be high. At Rottingdean (sites 15–19, Fig. 6), a catchment with 68 ha of recently drilled winter cereals, a flow of over 70 m width and 1 m depth was generated by a rainfall event of about 63 mm. At Breaky Bottom (site 2), also in October 1987, sediment concentrations of 2.3, 1.6, 10.1 and 3.5 $g\,l^{-1}$ were measured in rills and ephemeral gullies (Boardman, 1988). These amounts were responsible for an estimated transport of 18 tonnes per hour of soil into the flooded farmhouse and vineyard.

Fig. 6 shows 33 sites where damage has occurred; several sites have been flooded more than once. Breaky Bottom farmhouse (site 2) was flooded in 1976, 1982 and on about ten occasions in October 1987 (Boardman, 1988). Flooding of houses is recorded at Shepherds Mead (sites 4–6) in five separate years since 1976. At site 20, a single house owned by Brighton Borough Council, flooding of the house or garden has occurred on eight occasions since 1983. Houses in Sompting (site 26) have been flooded five times since 1980. Flooding of roads often goes unrecorded since it is regarded as of minor significance compared to that of houses. Several roads have been temporarily closed due to inundation by muddy water or soil.

Some detailed information is available on costs of damage on the Downs. At Highdown (site 1), in 1982, local council costs were estimated at US$17,500. At Sompting (site 26) costs to date for Adur District Council are in excess of US$67,000. Costs to flooded householders in both these instances are unknown. The flooding of Breaky Bottom farmhouse and vineyard in 1987 are estimated to have caused over US$116,800 worth of damage. The most detailed survey of costs of flooding was done by Robinson and Blackman (1990b) in relation to damage in the Brighton area in October 1987. They estimated costs for the Mile Oak and Hangleton area (sites 7–10) of over US$154,700 to householders and about US$223,380 to local councils; at Rottingdean (sites 15–19) costs were about US$438,000 and US$175,200 respectively. In the latter case, half of the total off-farm costs were borne by household insurance and US$146,000 by the householders.

There is no detailed information on the impact of flooding in the Pays de Caux. Papy and Douyer (1991) list the catastrophic floods that have occurred over a thirty year period based on newspaper reports. In December 1993 many French regions suffered flooding; costs to public services in the commune of Saint Valery en Caux (Pays de Caux) were estimated at US$52,400.

6. Responses

Boardman (1994) categorises the responses to flooding on the South Downs as:

1. Do nothing and hope for the best.
2. Engineering approaches.
3. Land use change by agreement.
4. Integrated plans including engineered and land use change elements.
5. Relocation of at-risk properties or people inhabiting the properties.
6. Recourse to the law.

Many incidents result in nothing being done: inhabitants of damaged properties believe that there is little chance of it happening again. This is partly to avoid spending money or engaging in legal proceedings but also because of a myth perpetuated by the farming community, supported by the media, that flood events are "Acts of God" and have little to do with human activities such as land use change.

Engineering approaches include the construction of ditches, banks, dams and piping water away from threatened areas. A distinction may be made between emergency measures, where renewed flooding is expected, and long-term protective measures. The problems and disadvantages of reliance on engineered structures include capital cost, maintenance cost, safety, loss of land, and the problem of how large to build structures (protection against the ten year flood, or the hundred year flood?) (Stammers and Boardman, 1984; Boardman, 1994).

Land use change (generally from arable to grass) has been negotiated where farmers are tenants of local councils. In other instances land use change has taken place where farmers have feared legal action if flooding were to recur.

A successful integrated plan of land use change together with small dams has been carried out at Rottingdean (sites 15–19) following severe flooding in 1987 (Robinson and Blackman, 1990a,b). Relocation of property has not occurred except for the rebuilding of a golf course green (site 21) away from the flood-prone area at a cost of about US$36,500.

Several individuals and local councils have taken legal advice following flooding. At Breaky Bottom damages of over US$116,800 were claimed in an action against the neighbouring farmer for negligently allowing runoff from winter cereal fields to cause damage. The case was terminated with an out-of-court settlement in 1993. Under English law, to successfully sue for negligence the plaintiff must show that the defendant, or a reasonable farmer in his position, ought to have foreseen the risk to his neighbour. Clearly, the case is stronger if flooding has previously occurred at the site (Boardman, 1994).

The response in other areas of western Europe has been rather different. In the Leuven area no-till experiments were set up with farmers and were shown to maintain yields, reduce erosion rates and also energy, labour and time (De Ploey, 1988). However, farmers are not really concerned about on-site costs of soil erosion and therefore are not willing to fund anti-erosion measures. The need for control measures to reduce the risk of flooding mainly concerns people living downstream. The most common approach has been the construction of deviation channels and retention basins with the costs being met by municipalities and government (Table 2). Only 15% of municipalities that are subject to flooding take control measures (Ballings et al., 1993).

In South-Limburg, emergency responses vary from putting bales of hay in gullies and using tyres to prevent runoff entering a stable, to raising private drives and

Table 2
Costs and method of financing structures to prevent damage to property due to muddy floods, Leuven–Brussels area (Ballings et al., 1993)

Municipality	Cost (USD)	Percent share of financing by:		
		Municipality	Government	Householders
Bertem	321,429	40	60	–
Hoegaarden	28,929	10	60	30
Hoegaarden	32,145	?	?	?
Hoeilaart	146,570	35	65	–
Kortenberg	?	–	100	–
Kortenberg	?	–	100	–
Kortenberg	160,715	100	–	–
Overijse	32,145	100	–	–
Overijse	257,145	100	–	–
St. Gen. Rode	?	100	–	–
St. Gen. Rode	?	50	–	50

constructing small dams on them. Long-term responses consist of the construction of reservoirs to buffer direct storm runoff. These reservoirs are built, financed and maintained by the Water Board. The location and size of a reservoir is based on an assessment of local discharge as a consequence of a storm with a 25 year return period, taking into account that some fields will have a changed land use and/or management due to conservation rulings which are presently under development. Recently, the sediments in these reservoirs, unless they had been removed to maintain the buffer capacity, have formed the basis of scientific research into the soil erosion history of their catchments (Laronne and De Roo, 1994).

Ongoing research, which will lead to renewed conservation rulings at the end of 1994, deals with the introduction of permanent grass-strips, cover crops, direct drilling, mulch sowing, and reduction in field sizes. Kwaad and Van Mulligen (1991) investigated different cropping systems to decrease soil erosion amounts.

Since 1991, a soil erosion project called the ''Erosienormeringsprojekt'', has been carried out in three catchments in the loess area of South-Limburg, funded by 14 municipalities (almost all villages involved), the Province of Limburg, the Water Board ''Roer en Overmaas'' and the Ministry of Agriculture. Three research institutes (the Departments of Physical Geography of the Universities of Utrecht and Amsterdam, and the Soil Physics Division of the Winand Staring Centre in Wageningen) cooperate to investigate soil conservation measures, monitor the present situation, and to develop and validate a new physically-based hydrological and soil erosion model, which can be used for planning and conservation purposes: the LImburg Soil Erosion Model (LISEM) (De Roo et al., 1994). All parties involved, including the Dutch Agricultural Board ('Landbouwschap') have regular formal meetings and talks on soil conservation issues.

Since 1992, a ruling from the Dutch Agricultural Board, obliges farmers to carry out conservation practices based on the slope steepness and size of an individual field e.g. on a field with an average slope greater than 18% no arable agriculture is allowed. Also, all farmers have a reporting obligation whenever soil erosion phenomena occur on their property. The effectiveness of this ruling is being evaluated together with possibilities for improvement.

In Limburg, there is an ongoing ''demonstration'' project showing farmers the effects of different tillage and management systems on soil erosion amounts and the consequences for their farming practices. Farmers voluntarily contributed to this project. Also, the experimental farm at Wijnandsrade, which has performed soil erosion experiments for several years, informs farmers of the progress of its work.

In the Pays de Caux a pilot scheme planned to last five years was started in 1985. Its aim was to investigate erosion and runoff processes and experiment with agronomical solutions to control concentrated flow at the watershed scale (Ouvry and Ligneau, 1992; Ouvry 1989–90). The scheme was sponsored by national and local government agencies. At the end of the trial period the local representatives decided to continue the scheme and the association they created in 1985 (AREAS) set up an office providing information, coordinating research and action on erosion and runoff problems. AREAS is particularly active in providing advice during re-allotment of fields. It also advises Water Boards on water quality problems. It works at the watershed scale with groups of municipalities using grassed waterways and water detention basins to prevent erosion and flooding. Communes can obtain 50% grant aid from Departements to fund anti-flooding work. Recent introduction of Set Aside requiring farmers receiving grant aid to set aside 15% of their land, cannot yet be evaluated in terms of providing flood protection and erosion control.

7. Conclusion

While the basic problem of flooding is similar across northwestern Europe there are some striking differences between areas. The timing of flood risk is similar in three areas being associated both with spring and autumn-sown crops; on the Downs it is confined to the autumn because of the dominance of autumn-sown cereals. Throughout the regions the causes are similar and there is a good understanding of the reasons for flooding. The information available on the extent, locations and costs of floods is variable. In the Pays de Caux there is little information, on the the South Downs there is a quite detailed record for 1976–93. There are some impacts on watercourses in terms of quality and sedimentation in the continental areas. The physical response to flooding is most impressive in South-Limburg and central Belgium with emphasis on engineered structures. Policy initiatives involving land use change, advice to farmers and integrated catchment management are best developed in the Pays de Caux and South-Limburg. It is ironic that on the South Downs with its detailed record of flood damage there is only one example of successful integrated planning with protection provided by small dams and substantial land use change; the British approach is, ''unimaginative and dilatory'' (Boardman, 1994).

References

Auzet, A.V., Boiffin, J., Papy, F., Maucorps, J. and Ouvry, J.F., 1990. An approach to the assessment of erosion forms and erosion risk on agricultural land in the northern Paris basin, France. In: J. Boardman, I.D.L. Foster and J.A. Dearing (Editors), Soil Erosion on Agricultural Land. Wiley, Chichester, pp. 383–400.

Auzet, V., Boiffin, J. and Ludwig, B., 1994. Concentrated flow erosion in cultivated catchments: influence of soil surface state. First Int. ESSC Congr., Silsoe, in press.

Ballings, P., Jansen, A. and Torfs, S. 1993. Studie van wachtbekkens in Vlaams-Brabant. Unpubl. MSc thesis.

Bell, M., 1983. Valley sediments as evidence of prehistoric land-use on the South Downs. Proc. Prehistoric Soc., 49: 118–150.

Blackman, J.D., 1992. Seasonal variation in the aggregate stability of downland soil. Soil Use Manage., 8(4): 142–150.

Boardman, J., 1988. Severe erosion on agricultural land in East Sussex, UK, October 1987. Soil Technol., 1: 333–348.

Boardman, J., 1990. Soil erosion on the South Downs: a review. In: J. Boardman, I.D.L. Foster and J.A. Dearing (Editors), Soil Erosion on Agricultural Land. Wiley, Chichester, pp. 87–105.

Boardman, J., 1993. The sensitivity of Downland arable land to erosion by water. In: D.S.G. Thomas and R.J. Allison (Editors), Landscape Sensitivity. Wiley, Chichester, pp. 211–228.

Boardman, J., 1994. Damage to property by runoff from agricultural land, South Downs, southern England, 1976–93. Geogr. J., submitted.

Boardman, J. and Robinson, D.A., 1985. Soil erosion, climatic vagary and agricultural change on the Downs around Lewes and Brighton, Autumn 1982. Appl. Geogr., 5: 243–258.

Boiffin, J., Papy, F. and Eimberck, J., 1988. Influence des systèmes de culture sur les risques d'érosion par ruissellement concentré. I. Analyse des conditions de déclenchement de l'érosion. Agronomie, 8(8): 663–673.

Bollinne, A., 1978. Study of the importance of splash and wash on cultivated loamy soils of Hesbaye (Belgium). Earth Surface Process., 3: 71–84.

Demaree, G., 1990. Climatological and hydrological aspects of the intensity of the precipitation at Uccle/Ukkel and the theory of the extreme values distributions. Unpublished PhD thesis, Vrije Universiteit, Brussel.

De Ploey, J., 1986. Bodemerosie in de Lage Landen, een Europees milieuprobleem. ACCO, Leuven/Amersfoort, 108 pp.

De Ploey, J., 1988. No-tillage experiments in the central Belgian loess belt. Soil Technol., 1: 181–184.

De Ploey, J., 1989a. Erosional systems and perspectives for erosion control in European loess areas. Soil Technol., 1: 93–102.

De Ploey, J. 1989b. Soil Erosion Map of Western Europe. Catena, Cremlingen, Germany.

De Ploey, J., 1990. Threshold conditions for thalweg gullying with special reference to loess areas. Catena Suppl., 17: 147–151.

De Roo, A.P.J., 1993. Modelling surface runoff and soil erosion in catchments using Geographical Information Systems; Validity and applicability of the 'ANSWERS' model in two catchments in the loess area of South Limburg (The Netherlands) and one in Devon (UK). Ph.D. Thesis. Netherlands Geographical Studies, No. 157, Utrecht University, 304 pp.

De Roo, A.P.J., Wesseling, C.G., Cremers, N.H.D.T., Verzandvoort, M.A., Ritsema C.J. and Van Oostindie, K., 1994. LISEM: a physically-based model to simulate runoff and soil erosion in catchments: model structure. Proc. Third Int. Conf. on Geomorphology, Hamilton, Canada, 1993, in press.

Diemont, W.H. and Van De Westeringh, W., 1978. Graften in Zuid-Limburg: groen of kaal, levend of verdwijnend. Tijdschr. Kon. Ned. Heidemij.

Fruit, J.P., 1991. Etudes Normandes No. 1: Agriculture et espace en Normandie. Problèmes et perspective.

Goossens, D., 1993. The Belgian loess deposits: an overview. Geomorphological processes in the Belgian loess belt. Excursion Guide: Memorial Symposium Prof. J. De Ploey. Experimental Geomorphology and Landscape Ecosystem Changes, 24 March 1993, Leven, pp. 5–10.

Govers, G., 1991. Rill erosion on arable land in central Belgium: rates, controls and predictability. Catena, 18: 133–155.

Govers, G. and Poesen, J., 1988. Assessment of the interrill and rill contributions to total soil loss from an upland field plot. Geomorphology, 1: 343–354.

Gulinck, H., 1985. Landbouw, bodembesscherming en landschapszorg. Landbouwtijdschr., 1: 37–47.

Hodgson, J.M., 1967. Soils of the West Sussex Coastal Plain. Soil Survey of Great Britain, Bull. No. 3, Harpenden.

Huigen, P., 1986. Bodemstructuur en erosie en wateroverlast in Zuid Limburg. MSc Thesis, Department of Physical Geography, University of Utrecht.

Kwaad, F.J.P.M., 1991. Summer and winter regimes of runoff generation and soil erosion on cultivated loess soils (The Netherlands). Earth Surface Process. Landforms, 16: 653–662.

Kwaad, F.J.P.M. and Van Mulligen, E.J., 1991. Cropping system effects of maize on infiltration, runoff and erosion on loess soils in South-Limburg (the Netherlands): A comparison of two rainfall events. Soil Technol., 4: 281–295.

Laronne, J. and De Roo, A.P.J., 1994. Magnitude and frequency of erosion based on reservoir event stratification — South Limburg, The Netherlands. Earth Surf. Process. Landforms, in press.

Laurant, A. and Alexandre, J., 1977. Acccrossement recent de l'intensité des fortes averses dans la banlieu bruxelloise. Bull. Soc. Geogr. Liège, 13: 127–132.

Leenaers, H., 1989. The dispersal of metal mining wastes in the catchment of the River Geul (Belgium – The Netherlands). Dissertation, University of Utrecht, 200 pp.

Leenaers, H. and Schouten, C.J., 1989. Soil erosion and floodplain soil pollution: related problems in the geographical context of a river basin. IAHS Publ. 184, pp. 75–83.

Ludwig, B., Boiffin, J. and Masclet, A., 1994. Spatial distribution of sediment sources and the relative contribution of erosion forms to soil losses in a cultivated catchment. Proc. First Int. ESSC Congr., Silsoe, 1992, in press.

Ouvry, J.F., 1987. Bilan des travaux. Campagne 1986–1987. Rapport A.R.E.A.S., 153 pp.

Ouvry, J.F., 1989–1990. Effet des techniques culturales sur la susceptibilité des terrains a l'érosion par ruissellement concentré: expérience du Pays de Caux. Cah. ORSTOM Ser. Pedol., XXV(1–2): 157–169.

Ouvry, J.F. and Ligneau, L., 1992. A specific strategy set up with farmers to succeed erosion control: the experience of a French region "Pays de Caux". In: S. Wicherek (Editor), Farm Land Erosion, in Temperate Plains Environment and Hills. Elsevier, Amsterdam, pp. 497–502.

Papy, F. and Douyer, C., 1991. Influence des états de surface du territoire agricole sur le déclenchement des inondations catastrophiques. Agronomie, 11: 201–215.

Poesen, J., De Ploey, J. and Govers, G., 1988. Bodemerosie in Midden-Belgie: oorzaken, gevolgen en bestrijding. Jaarbulletin van de vrienden van Heverleebos en Meerdaalwoud, pp. 24–38.

Robinson, D.A. and Blackman, J.D., 1990a. Soil erosion and flooding: consequences on land use policy and agricultural practice on the South Downs, East Sussex, UK. Land Use Policy, 7: 41–52.

Robinson, D.A. and Blackman, J.D., 1990b. Some costs and consequences of soil erosion and flooding around Brighton and Hove, autumn 1987. In: J., Boardman, I.D.L. Foster and J.A. Dearing (Editors), Soil Erosion on Agricultural Land. Wiley, Chichester, pp. 369–382.

Römkens, M.J.M., Prasad, S.N. and Poesen, J.W.A., 1986. Soil erodibility and properties. Trans. 13th Congr. ISSS, Hamburg, 5: 492–504.

Saris, F.J.A., 1984. Lijnvormige landschapselementen, belangrijke onderdelen van een ecologische infrastructuur. Landschap, 1: 91–108.

Schouten, C.J., Rang, M.C. and Huigen, P.M.J., 1985. Erosie en wateroverlast in Zuid-Limburg. Landschap, 2: 118–132.

Stammers, R. and Boardman, J., 1984. Soil erosion and flooding on downland areas. Surveyor, 164: 8–11.

Vandaele, K., 1993a. Assessment of factors affecting ephemeral gully erosion in cultivated catchments of the Belgian Loam belt. In: S. Wicherek (Editor), Farm Land Erosion in Temperate Plains Environment and Hills. Elsevier, Amsterdam, pp. 125–135.

Vandaele, K., 1993b. Soil erosion and floods in the Hammeveld catchments (Leefdaal). In: Excursion Guide Geomorphological Processess in the Belgian Loess Belt, March 1993. Memorial Symp. Prof. J. De Ploey: Experimental Geomorphology and Landscape Ecosystem Changes, pp. 29–43.

Van der Helm, P.P.M., 1987. Bodemerosie en wateroverlast; een milieuprobleem in het Zuid-Limburgse heuvelland. MSc Thesis, Dept. of Physical Geography, University of Utrecht.

Van Eijsden, G.G. and Imeson, A.C., 1985. De relatie tussen erosie en enkele landbouwgewassen in het Ransdalerveld, Zuid-Limburg. Landschap, 2: 133–142.

Vleeshouwer, J.J. and Damoiseaux, J.H., 1990. Bodemkaart van Nederland 1: 50000; Toelichting bij kaartblad 61–62 West en Oost Maastricht – Heerlen. Staring Centrum, Wageningen, 177 pp.

ELSEVIER

Geomorphology 10 (1994) 197–219

GEOMORPHOLOGY

Geomorphology applied to flooding problems of closed-basin lakes... specifically Great Salt Lake, Utah

Genevieve Atwood

Earth Science Education, 30 U Street, Salt Lake City, UT 84103, USA

Received March 4, 1994; accepted April 4, 1994

Abstract

Closed-basin lakes fluctuate in level and area in response to variations in the climate over their drainage basins. Rises of the level of these lakes present severe flooding hazards to developments on the lake bed. Effective planning for development of these lake beds requires a knowledge of the lake-level fluctuations that should be anticipated and also the effect of wind, earthquakes, and human activity on flooding. The historic record of fluctuations of the closed-basin lakes of the Great Basin of the western United States is too short to adequately define the recurrence intervals of high lake levels. Geomorphic evidence can supplement the historic record. Great Salt Lake is the largest closed-basin lake in the Great Basin and its lake bed has been the most extensively developed. In the 1980s the level of Great Salt Lake rose dramatically in response to a period of much above-average precipitation. The resulting flooding caused extensive damage to developments on the lake bed and posed serious dilemmas to Federal, state and local governments. Studies subsequent to the flooding reveal how geomorphic evidence can supplement the historic record and provide sufficient understanding of the lake's flooding hazard to guide development and planning. Under the climatic conditions that have existed for the past several thousand years, the still-water level of Great Salt Lake can be expected to rise to near the historic highstand, or slightly above, about once every one hundred years. Wind action extends flooding in some locations to elevations 2 m above the still-water elevation of the lake. If an understanding of flooding hazards guides the planning and engineering of developments on and around the lake, losses from future high-lake cycles can be significantly reduced.

1. Introduction

The water level and area of terminal lakes in closed topographic basins fluctuate in response to changes in climate. Transportation, recreational, industrial and residential developments on the shores of closed-basin lakes are exposed to major flooding hazards caused by rising lake levels. The flooding hazards of these lakes differ substantially from the flooding hazards of rivers. The periods of climate cycles that produce large increases in the areal extent of closed-basin lakes are long relative to historic records. An understanding of lake-level fluctuations cannot be obtained from historic records alone. Geomorphic features associated with prehistoric fluctuations provide information on prehistoric lake levels, recurrence intervals of fluctuations, and the factors that control the fluctuations. This paper reports on how geomorphic evidence has contributed to an understanding and definition of flooding hazards associated with closed-basin lakes in the Great Basin of the western United States and specifically with the largest of these lakes, Utah's Great Salt Lake. Shoreline and lake-level data for this paper were collected in inch/pound (U.S. customary) units, i.e., feet above sea level. Except for statements in quotations, this paper reports elevations in metric units generally rounded to

SSDI 0169-555X(94)00022-J

the nearest tenth of a meter and also reports elevations in feet above sea level reported to the significant figure appropriate to the original data.

1.1. Closed-basin lakes

Lakes differ markedly worldwide. Most lakes in open basins impound surface water as wide, low-gradient portions of rivers. Surface water discharges from these lakes via drainages which regulate the elevation of the lakes. These lakes are relatively short-lived geologic phenomena that rapidly fill with sediments and cease to function as lakes.

Closed basins are physiographic features with topographic closure caused by such diverse geologic fac-

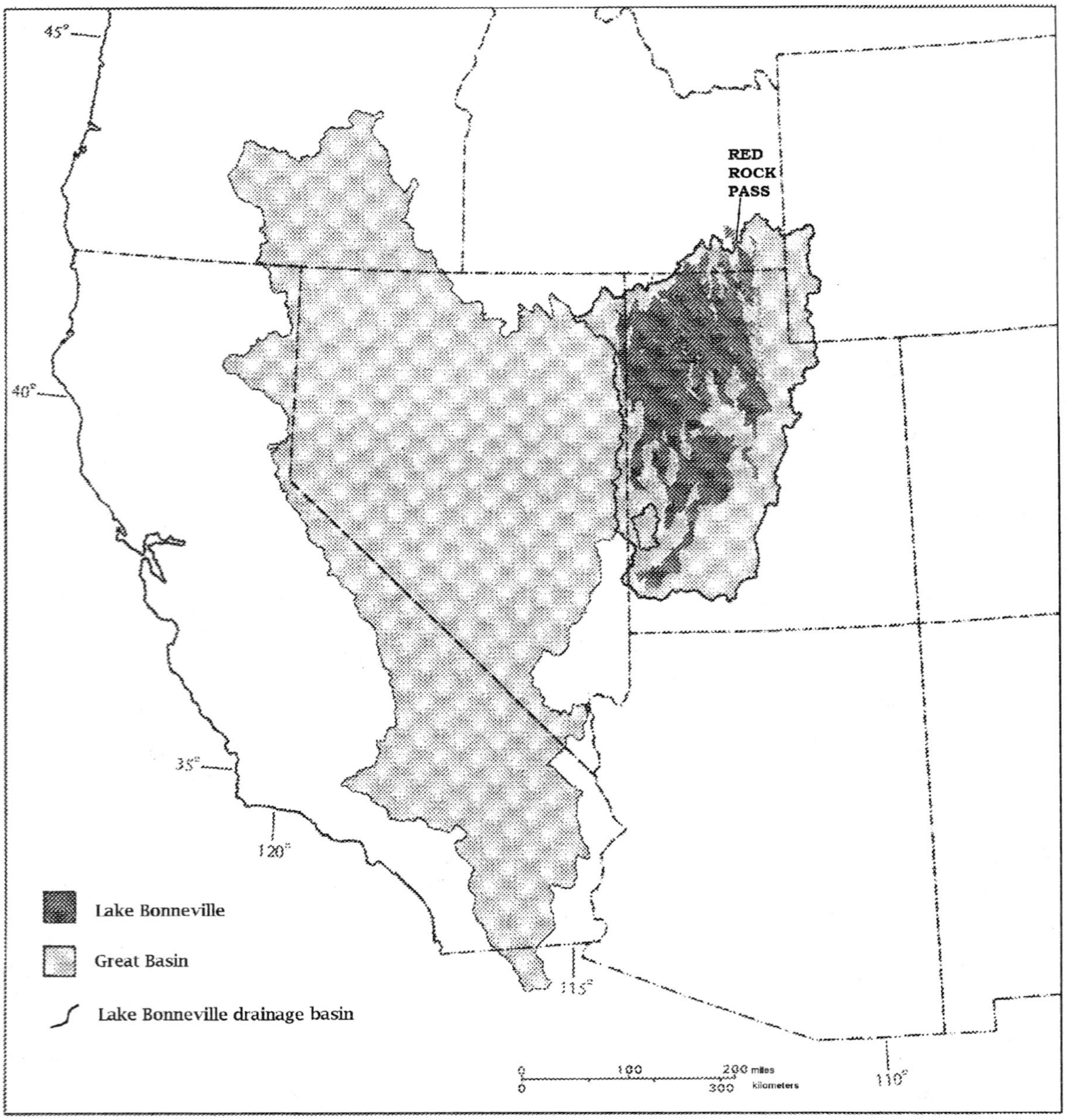

Fig. 1. Map showing the extent of the Great Basin, the Bonneville drainage basin, and the maximum extent of Lake Bonneville (adapted from Currey, 1980).

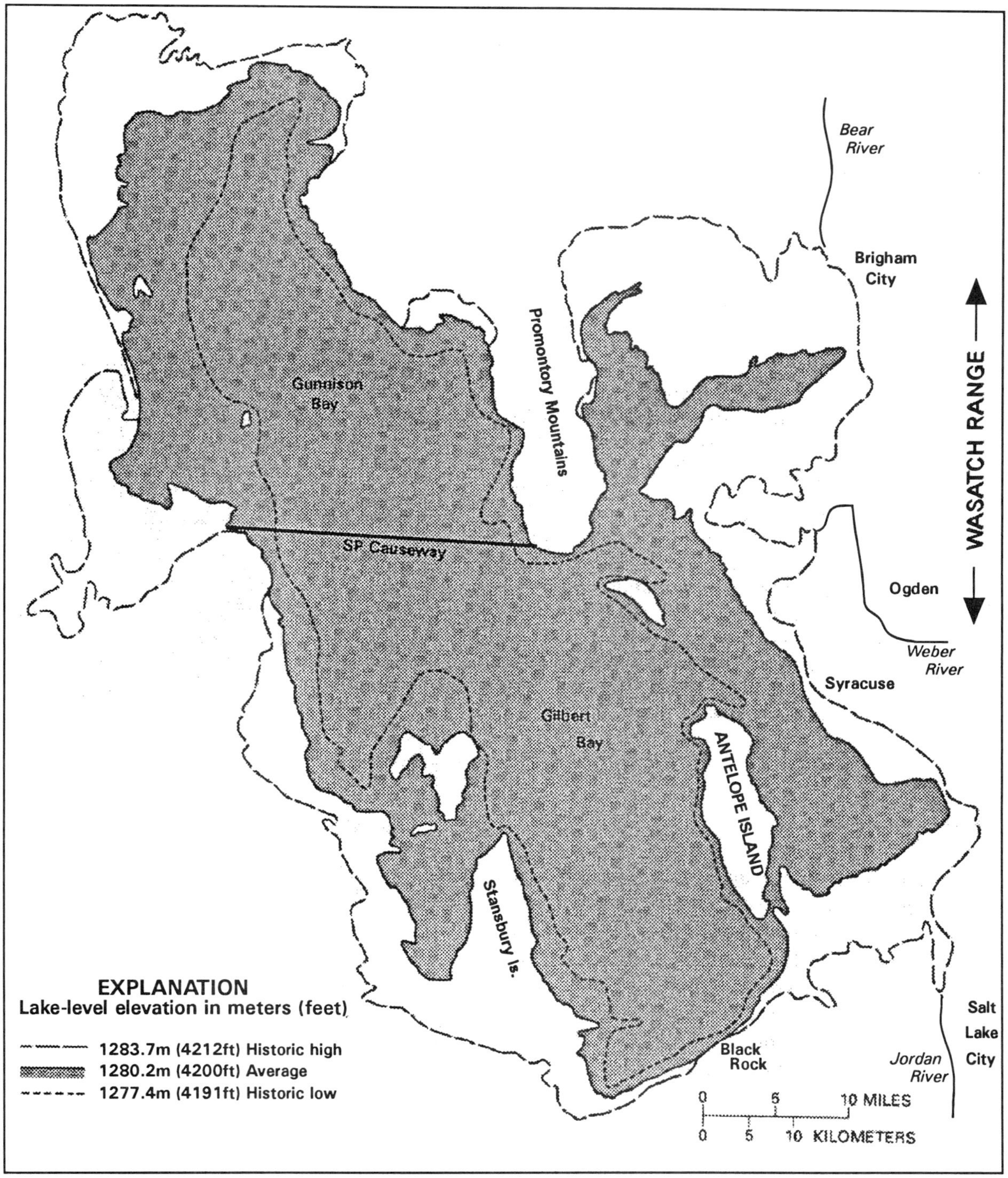

Fig. 2. Map of Great Salt Lake and vicinity showing historic maximum, minimum and average extent of Great Salt Lake (adapted from Currey, 1980).

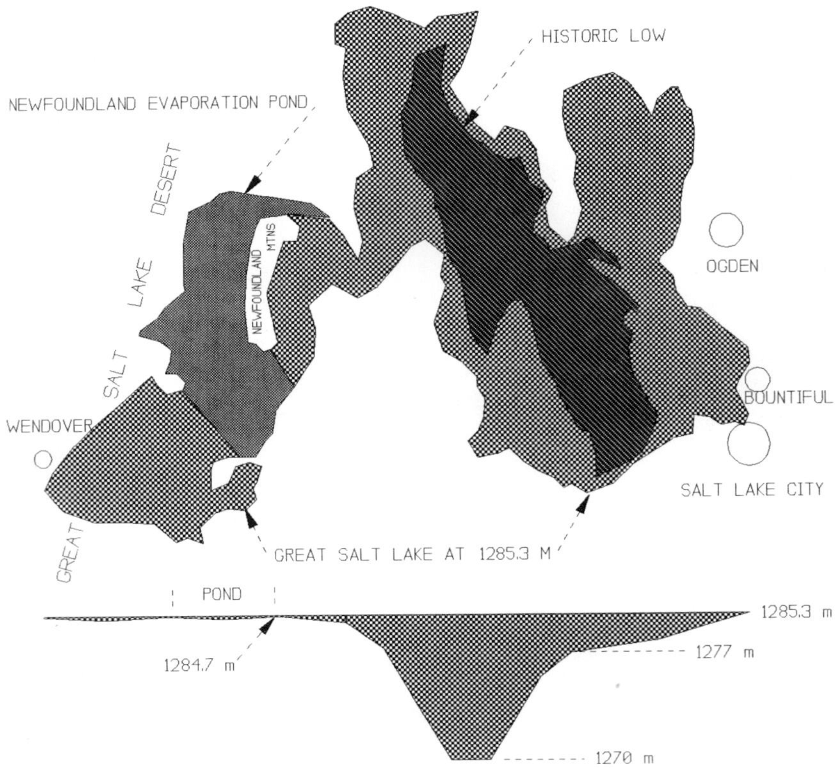

Fig. 3. Sketch section through the Great Salt Lake closed-basin and the Great Salt Lake Desert showing the profile of the lake bed.

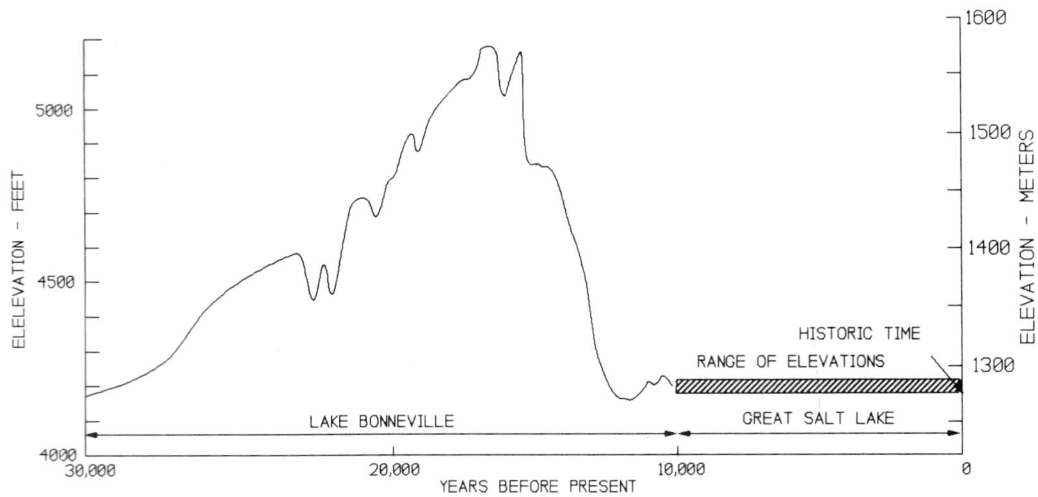

Fig. 4. Hydrograph of Lake Bonneville and Great Salt Lake (adapted from Currey et al., 1984).

tors as glaciation, tectonism, sedimentation and volcanism. Closed-basin lakes have no surface discharge until the lake-level rises to and spills over the low point of the basin's rim, its topographic divide to the neighboring basin. The height of the topographic divide of the closed basin determines the deepest lake the closed basin can contain but many basins show no evidence of having filled to overflowing.

The Great Basin of the western United States is defined as an area of no external drainage (Grayson, 1993). It is characterized by generally north-trending fault-controlled mountain ranges separated by basins. The closed basins of the region contain "about two dozen large perennial lakes and more than 100 dry ones" (Hunt, 1974). The Great Basin can be further divided into sub-basins based on drainage systems and their terminal lakes such as the Bonneville basin to the east (Fig. 1). Major basins consist of smaller sub-basins, some of which are nested one inside the next, and others which eventually coalesce across topographic divides. Although tectonic activity is the primary cause of the development of closed basins in the Great Basin, depositional and erosional processes have modified the divides that control the elevation of the topographic closures of the basins.

The present topographic closure of the closed-basin area that contains Great Salt Lake (hereafter called the Great Salt Lake closed-basin) consists of the 7000 km^2 lowest area of the Bonneville basin (Fig. 2). Contained within the Great Salt Lake closed-basin and flooded by the Great Salt Lake is a 520 km^2 closed basin to the west of the Promontory Mountains in the area of the lake's Gunnison Bay. The lowest point of the Great Salt Lake closed-basin is about 1270.1 m (4167 ft) above sea level (a.s.l.). The topographic divide between the closed-basin area west of the Promontory Mountains and the Great Salt Lake closed-basin is approximately 1271.3 m (4171 ft) a.s.l. and trends southwest from the southern end of the Promontory Mountains.

The Great Salt Lake closed-basin is separated from closed basins in the Great Salt Lake Desert to the west by a threshold at an elevation of about 1284.7 m (4215 ft) a.s.l. south of the Newfoundland Mountains (Fig. 3) (Chapman and Sappington, 1986). The closed basin that includes Great Salt Lake closed-basin and the Great Salt Lake Desert in turn is included in larger and larger coalescing sub-basins of the Bonneville basin. The total extent of the Bonneville basin is 140,000 km^2. Great Salt Lake is one of several terminal lakes of the drainage basin. In historic time these lakes have covered only part of their lake beds and, for much of the time, most of the lakes other than Great Salt Lake have disappeared.

During the Pleistocene epoch, wetter and colder climates caused the terminal lakes in the Bonneville basin to rise and coalesce into a single lake. The most recent of these high-level lakes called Lake Bonneville began to develop approximately 30,000 years ago (Fig. 4). The lake reached a maximum depth of over 300 m (1000 ft) about 15,000 years ago and flooded 52,000 km^2 of the 140,000 km^2 Bonneville basin drainage area. The lake level rose until water spilled over the approximately 1551 m (5090 ft) a.s.l. topographic divide that separated the Bonneville basin from the Columbia River basin. From the basin divide just north of present Red Rock Pass in Idaho, water flowed down the present Marsh Creek drainage, to the Portneuf River, to the Snake River near present-day Pocatello, to the Columbia River and on to the Pacific Ocean. The significantly eroded threshold divide between the Bonneville basin and the Columbia River basin now at Red Rock Pass controls the maximum possible rise of Great Salt Lake at an elevation of 1457 m (4780 ft) a.s.l.

2. Evidence of lake level fluctuations

2.1. Measurements of levels of Great Salt Lake

Observations by John C. Fremont in 1843 recorded first measurements of the level of Great Salt Lake. He wrote, "From a discussion of the barometrical observations made during our stay on the shores of the lake, we have adopted 4200 feet for its elevation above the Gulf of Mexico" (Jackson and Spencer, 1970, p. 510). Fremont's barometric observations cover a considerable range and cannot be used to determine the exact elevation of the lake surface at that time. In 1846 he made a more useful observation when he recorded that as he rode his horse across the ford on the bar that connects Antelope Island to the mainland, the water nowhere reached above the saddle girth. The elevation of the low point in the ford to the island is about 1280.2 m (4200 ft) a.s.l. Thus Fremont's observation indicates a lake level of about 1280.8 m (4202 ft) a.s.l. When

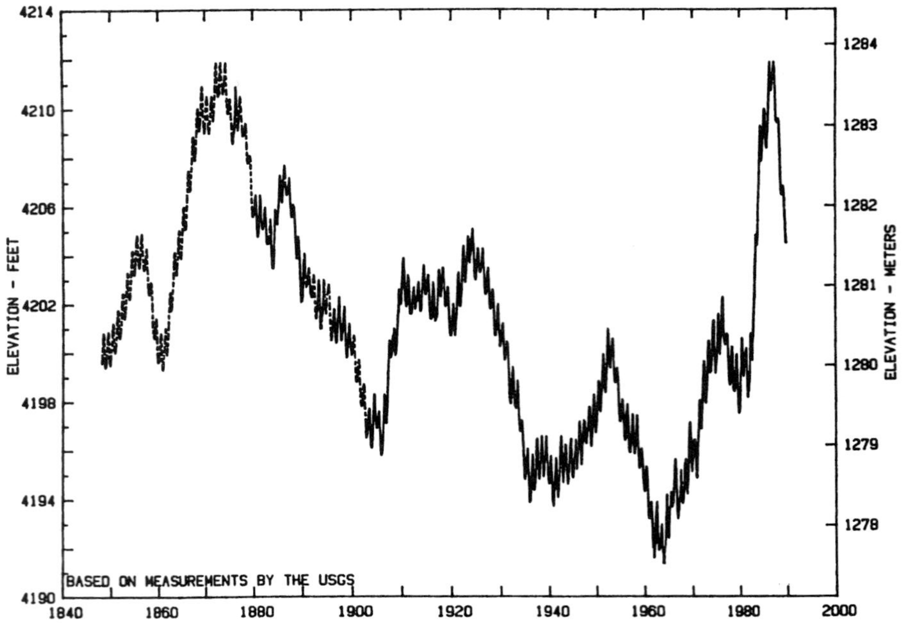

Fig. 5. Hydrograph of the historic fluctuations of the main body of Great Salt Lake.

G.K. Gilbert made his study of Lake Bonneville in 1875–78, he interviewed stockmen, who had moved animals between Stansbury and Antelope Islands and the mainland, to determine their recollections of the water depth on the bars to islands for the period from 1847 to 1875. Based on the information they supplied, Gilbert constructed a hydrograph for this period (Gilbert, 1890).

Gilbert reported that the lake reached its highest level in the summers of 1872, 1873 and 1874. The exact year and exact elevation of the highest lake level in the 1870s is not known (Mabey, 1986). Recorded lake-level measurements began in 1875 with the installation of a gauge at Black Rock. Since then the lake level has fluctuated over a range of 6.2 m (20.5 ft) requiring several relocations of the lake gauges. Gilbert acquired from Brigham Young a block of the granite that was being used in the construction of the Salt Lake Temple and set it at Black Rock as a benchmark. By leveling and simultaneous measurements he determined the elevation of the early gauges relative to this benchmark. Later, these elevations were tied to a sea-level reference.

In 1877, Gilbert measured the elevation of the debris line formed on the east side of Antelope Island at the highest level of the 1860–1870s cycle and related it to the gauged lake level. Using the sea level datum for this level provides an elevation of 1284.3 m (4213.5 ft) a.s.l. for the debris line. This was the only direct measurement he made of the highest lake levels of the 1870s. Today the U.S. Geological Survey and most others use hydrographs based on Gilbert's early data. These hydrographs generally show the highest lake level in 1873 at about 1283.7 m (4211.5 ft) a.s.l.

A nearly complete set of records for the level of Great Salt Lake exists from 1875 to the present (Fig. 5). The U.S. Geological Survey presently monitors two lake-level gauges on Great Salt Lake and reports the measurements for the 1st and 15th day of each month. The gauges are located in Gilbert and Gunnison Bays and record still-water lake levels south and north of the Southern Pacific Railroad causeway.

In historic time the lake level has ranged from 1277.5 m (4191.3 ft) a.s.l. to 1283.8 m (4211.8 ft) a.s.l. This translates to a maximum lake depth of about 7.3 m (24 ft) when the lake was at its 1963 low and covered 2300 km^2 and to a maximum lake depth of about 13.7 m (45 ft) when the lake was at its historic high in the 1870s and 1980s and covered an area of 6000 km^2. During historic time the divide to the lowest closed basin has not been exposed nor has the lake level risen to the elevation of the divide to the west.

2.2. Geomorphic evidence of levels of Great Salt Lake

In some closed basins the highest extent of the flooding hazard is defined by the topographic closure of the basin. However, in most major closed basins the topographic closure is so great that it is of little use in flood–hazard analysis. For example the topographic closure on the present-day Bonneville basin is over 180 m (600 ft).

As closed-basin lakes fluctuate, they create suites of geologic and geomorphic evidence that document physical, chemical and biological characteristics of the lake. The sediments captured in closed-basin lakes can provide nearly continuous records of the erosional history of the drainage basin. Closed-basin lacustrine sediments have attracted considerable attention as indicators of Pleistocene and Holocene climate (Street-Perrott and Harrison, 1985; Benson and Paillet, 1989; Forman, 1989; Sack, 1993). The chemical, biological and physical properties of these lacustrine sediments provide evidence of recurrence of lake-level rises (McKenzie and Eberli, 1985; Oviatt et al., 1994) but are difficult to calibrate to specific highstand elevations.

Geomorphic features, especially depositional shorelines, provide evidence of specific highstand elevations. Geologists from G.K. Gilbert to the present have used shoreline features to unravel Great Basin closed-basin lake histories (see Sack, 1989 for a case history of geomorphic studies of Lake Bonneville). Differences in elevations of known shorelines have been used to measure isostatic deformation associated with Pleistocene Lakes Bonneville and Lahonton (Crittenden, 1963; Mifflin and Wheat, 1971; Bills and May, 1987) and to document rates of regional deformation (Currey, 1988). Shoreline features have been used to define the recurrence intervals of highstands of Great Salt Lake (Atwood and Mabey, 1994) and this paper shows how shorelines can be used to define flooding hazards associated with closed-basin lakes.

3. Flood hazards and closed basins

The nearly flat basin floors of the Great Basin attract development often near lakes. Most of the people in the region live and work in the basins. Native Americans of prehistoric time, early explorers and trappers, emigrants bound for the west coast, and engineers of transcontinental railroads, highways and interstate freeways avoided mountain ranges by crossing basins. Gas lines, power lines, and other utilities take advantage of easy construction of straight routes across sparsely vegetated and easily excavated basin floors.

The highly populous metropolitan area neighboring Salt Lake City, hemmed in on the east by the Wasatch Range, has been built across the basin floor and encroaches on the flood plain of Great Salt Lake. Numerous lifelines including two transcontinental railroads and an interstate freeway cross the lake bed. Recreational facilities and brine extraction industries are located on the lake shores to take advantage of lake waters. Communities build sewage treatment plants near the lake to allow gravity to carry wastes to the lowest point of the collection system.

Flooding along the shores of Great Salt Lake can occur in several ways including climate-caused lake highstands, storm surges, earthquake-induced surges, seiches, tectonic tilt of the lakebed, and failure of dikes. Each type of lake flooding has occurred in historic time except tectonic tilt. Any of these types of flooding can cause damage at any lake level, and all are likely to be more destructive the higher the still-water level of the lake. Assessing the hazard associated with any type of Great Salt Lake flooding involves consideration of fluctuations in the lake level.

Great Salt Lake, the largest lake in the Great Basin and the closest to a major population center, presents the greatest flooding hazard. However other terminal lakes in the Great Basin present significant flooding hazards, and as development increases, the flooding hazards from terminal lakes will increase. As noted by FEMA (Federal Emergency Management Agency Region VIII, 1986 p. iii, 2–3):

> Lake flooding is widespread — not a problem unique to the west. Without a concerted management effort, losses attributable to it will likely exceed US$1 billion by the year 2000.
>
> * * *
>
> The magnitude of the property damages in the United States due to floods resulting from fluctuations in lake levels has exceeded a quarter billion dollars in the past five years alone. Lakeshore industries, rail lines, highways, residential, commercial and agricultural, wildlife refuges and recreation facilities have been damaged or destroyed as a result of rising lake levels encroaching upon developed land. Economic data suggest that lake-rise flooding is a significant hazard in the United States. Between 1983 and January 1985, dam-

ages resulting from flooding around Malheur Lake, Oregon, had reached US$13.5 million. Around the Great Salt Lake damages have exceeded US$200 million since 1983.

3.1. Case history: responses to fluctuations of Great Salt Lake

Pre-1970

When the level of Great Salt Lake rose rapidly in the late 1860s, two decades after Mormon settlements were established near its shores, concern developed for the flooding that was occurring and that which might occur. However, in the 1870s the lake began a long-term, gradual and fluctuating decline that continued until 1963 when the lake reached its historic low. By the 1960s bird sanctuaries, recreationists, and salt industries faced problems associated with very low lake levels. High lake levels seemed a phenomenon of the past. During the 1960s' construction boom, residential, industrial, and transportation-related development encroached on the lake's historic flood plain.

The 1970s

As lake levels rose during the 1970s, brine-extraction industries operating in and around the lake raised legal and economic concerns to Utah's governor and legislature about dilution of the lake's brines and the economic consequences of lake flooding to the state's economy. In 1975 the legislature created the Great Salt Lake Division within Utah's Department of Natural Resources and a Great Salt Lake Board to develop a comprehensive plan and manage the Great Salt Lake below a lake-level elevation of 1280.8 m (4202 ft) a.s.l. (Utah Department of Natural Resources, 1976). In 1978 the lake receded and, as the lake level lowered, so did legislators' level of concern. For the next five years the lake level fluctuated at about the 1280.5 m (4201 ft) level.

The 1980s

A major wet cycle began in September 1982 heralded by the wettest month in the history of observations at the Salt Lake City International Airport. From September 1982 through June 1983 the lake rose a record 1.58 m (5.2 ft) and increased in volume by 7.4 billion cubic meters (6 million acre-ft) eliminating the volume deficit accumulated over 136 years of consumptive use. During the succeeding four years, much-above average precipitation occurred over most of the Great Salt Lake drainage basin with the Salt Lake International Airport recording about 80 cm of excess precipitation over average, or about 50 percent above average annual precipitation (Fig. 6). The level of Great Salt Lake rose rapidly in response to this precipitation excess and in the Spring of 1986 reached the highest level since the first gauge was established in 1875. The lake level again peaked at the historic high in 1987 and then declined rapidly in response to lower-than-average precipitation and pumping of Great Salt Lake waters into the Great Salt Lake Desert.

The rise of Great Salt Lake coincided with the debris flows and ground failures of 1983–85. State agencies mobilized for the range of wet-cycle events. In 1983 the Recommendations for a Great Salt Lake Contingency Plan of the Department of Natural Resources and Energy concluded: "It is anticipated that lake levels will peak at approximately 4203 feet a.s.l. in 1983 with potential resultant damages of US$20 to US$30 million. Unfortunately there are presently insufficient data on which to base firm action recommendations for long term solutions." (Utah Department of Natural Resources, 1983, p. 44).

In October 1983, the University of Utah Bureau of Economic Business Research reported already-sustained capital damages associated with the rising level of Great Salt Lake of US$60.3 million at a lake level of 1281.7 m (4205 ft) a.s.l. The report calculated capital damages, annual revenue reductions of private firms, annual household income losses, and annual reductions in state and local tax revenues for lake levels ranging from 1280.8 m (4202 ft) to 1283.8 m (4212 ft) a.s.l. (Steffen, 1983).

In September 1984, Utah's Division of Comprehensive Emergency Management estimated that "Even with normal precipitation patterns, the National Weather Service (NWS) predicts that in 1985 the level will increase to 4210.5 feet a.s.l. as a result of above normal groundwater conditions in the basin. These abnormal conditions may continue for several years." (Interagency Task Force, 1984, p. 3). Already-damaged facilities included flooded waterfowl management areas, resort areas, recreation areas, buildings, and extensive damage to evaporation ponds and mineral extraction industries. Flood waters had breached the causeway to Antelope Island isolating the State Park.

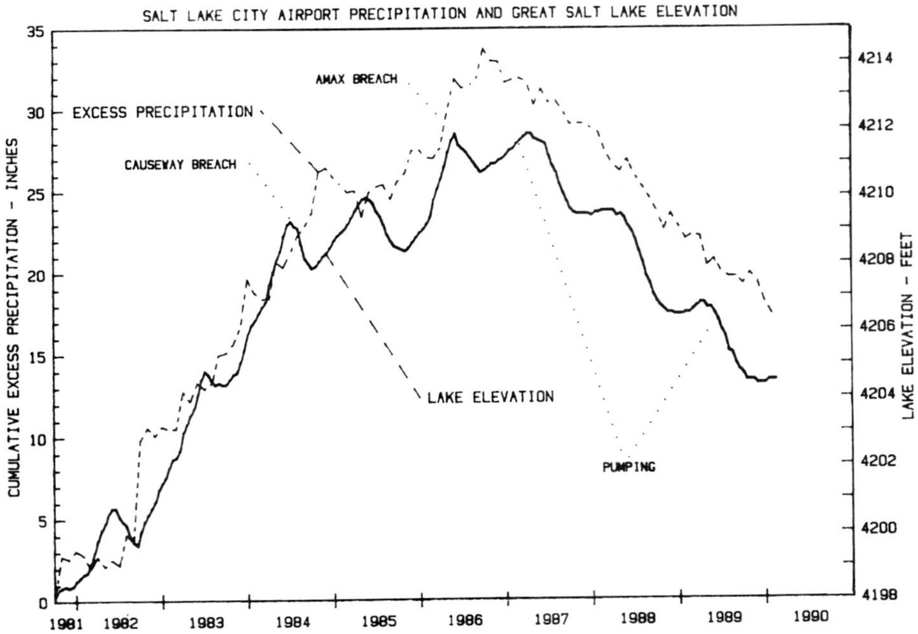

Fig. 6. Hydrograph of the fluctuations of Great Salt Lake during the 1980s wet cycle plotted with a graph showing cumulative excess precipitation over average as measured at the Salt Lake City International Airport.

Great Salt Lake State Park facilities on the south shore of the lake suffered severe damages and the park had been closed. The road grade of the transcontinental railroad near the south edge of Great Salt Lake had been raised to 1285.0 m (4216 ft) a.s.l. A stretch of Interstate 80 had been flooded and sections of the freeway were being raised to an elevation of 1284.4 m (4214 ft) a.s.l. at a cost of US$15 million, not including indirect costs of delays of interstate traffic.

The State of Utah explored a wide variety of actions to control the lake or decrease losses should the lake continue to rise. Possible actions studied included:

- diverting part of the Bear River which provides about half of the surface inflow into the lake, from the Bonneville basin into the Snake River drainage;
- pumping water from the lake into the Great Salt Lake Desert;
- diking selected facilities;
- dyeing the lake to increase evaporation; and
- breaching the Southern Pacific Railroad causeway across the lake which impounded water south of it.

The Southern Pacific Railroad causeway, completed in 1960, restricted circulation in the lake to such an extent that the lake had become two different water bodies. The lake north of the causeway and west of the Promontory Mountains (Gunnison Bay) receives little direct surface inflow and had became much more saline and dense than the lake to the south (Gilbert Bay). Because of the density difference in the brines separated by the causeway, two-directional flow developed in the culverts and through the causeway with dense brine from the north flowing south under the north-flowing less-concentrated brine from the south. Gilbert Bay south of the causeway became stratified as the dense brine flowed into the deepest areas of the bay. Ice rafts developed on the low salinity brines and damaged structures in the lake. The causeway maintained the lake level of Gilbert Bay about 1 m (3 ft) above the lake level of Gunnison Bay. Virtually all the facilities at risk bordered parts of the lake south of the causeway. In August 1984, the State of Utah cut a 100-meter gap in the causeway and the resulting equilibration of water levels lowered the level of Gilbert Bay about 0.3 m. Communities diked critical facilities such as sewage treatment plants and residential areas constructed on the historic floodplain of the lake.

When the lake continued to rise the State developed a pumping plan to accomplish artificially what the lake would accomplish naturally at about 1284.7 m (4215 ft) a.s.l., move water across the closed-basin divide into the Great Salt Lake Desert. As originally designed, pumps would lift low salinity water from Gilbert Bay into a canal to circulate through a large shallow pond west of the Newfoundland Range where part of the water would be evaporated. The concentrated brine would return to the lake's Gunnison Bay. With this system the pond would not fill with salt and the Great Salt Lake's salt resources would not be depleted.

West Desert Pumping

In 1986 the Governor selected a scaled down version of the plan that would pump more concentrated brine from Gunnison Bay into an evaporation pond somewhat smaller than that planned originally. The Utah State legislature approved the West Desert Pumping Project with an appropriation of US$60 million recognizing that flooding of Great Salt Lake "posed a threat to life, health and property, and in particular may result in extensive damage to public lands, major transportation routes and other public facilities" (Utah House of Representatives, 1986, p. 2).

The massive West Desert Pumping engineering and construction project consisted of a 16-kilometer access road along the Southern Pacific Railroad right of way, a pumping station, canals, trestles, dikes, a 60-kilometer natural gas pipeline, and a 1300-square-kilometer evaporation pond in the desert area west of the divide between the Great Salt Lake closed-basin and the Great Salt Lake Desert.

> Three large pumps and a pumping station lift water diverted from the north arm of the lake to a 4.1 mile-long outlet canal. The first pump began operating [on a trial basis] in April, 1987. All three pumps were scheduled to be operating in June, 1987. The canal begins at an elevation of 4224 feet a.s.l. and discharges water into the large evaporation pond located at 4217 feet a.s.l. west of the Newfoundland Mountains. The pumping system will be able to divert approximately 2 million acre-feet of water a year from the Great Salt Lake into the pond from which approximately 825 thousand net acre-feet of water can be evaporated each year (Utah Department of Natural Resources, 1987, p. 1, parenthetical remark added).

Project construction began in July 1986. In December 1986 the U.S. Army Corps of Engineers forecast: "if precipitation amounts for 1987 and 1988 continue along the pattern of 1986 (about 140% of normal), the lake is expected to rise to about 4213 feet a.s.l. even with the pumping project in operation. Potential damages to public and private developments have been estimated at $315 million for a lake stage of 4214 feet a.s.l. and $792 for a stage of 4217 feet." (U.S. Army Corps of Engineers, 1986).

But precipitation patterns had changed. The lake peaked shortly after the pumps began operation. Pumping helped lower the lake level faster than it would have lowered without pumping but had no appreciable effect on the maximum level reached. The wet cycle was over and the lake declined rapidly. The 1986–87 peak of Great Salt Lake had caused US$240 million of economic loss to Federal, state, and private interests (Utah Department of Natural Resources, 1989).

3.2. *Case history: availability of technical information about fluctuations of Great Salt Lake*

The Governor had based his decisions primarily on information provided by industry on the economic and political consequences of losing the Southern Pacific Railroad which crossed the lake. The legislature based its decisions on the Governor's recommendations, and on high-lake-level projections from a variety of sources.

We can learn from examining the reasons that government, industry, and the Great Salt Lake's neighbors were so surprised and unprepared for the dramatic rise of the lake level in the 1970s and 1980s. Although prior to the 1970s, the lake's flooding patterns had not been studied in detail for scientific curiosity or defined as a hazard, the physical evidence was sufficient to warn of the problem.

Recorded information of fluctuations of the level of the lake began with observations of the Mormon settlers who arrived on its shores in 1847. Nearly complete records of lake-level measurements existed since 1875. Weather records, although far from complete, defined the general nature of the precipitation events that produced the high lake levels a century before (Karl and Young, 1986). The effect of consumptive use of water was known (Whitaker, 1971). These three sets of information, properly analyzed, indicated that the lake would at some time return to near the levels reached in the 1870s.

Pre-1970

G.K. Gilbert reported his studies of Great Salt Lake and Lake Bonneville in a series of publications including U.S. Geological Survey Monograph I published in

1890. By 1890 the lake had declined about 3 m (10 ft) from the 1870s historic highstand. Gilbert concluded that consumptive use of water taken from streams flowing into Great Salt Lake would cause the lake level to continue to decline. Gilbert's impressive studies had answered most of the questions that had been raised about Great Salt Lake. That he failed to understand that the cumulative effect of over 100 years of consumptive use could be overcome in one very wet year and that future cycles involving several wet years would produce important lake flooding was one of the few weaknesses of his studies.

From 1890 to 1963 during the lake's decline most government agencies had other priorities. No well-circulated or professionally recognized studies advised industries and communities to consider flooding a major hazard to development around the lake. Government officials who had approved questionable development on the lake bed could truthfully say that they had not been warned that the lake would rise again to the high levels of the 1870s. The cyclic nature of lake-level fluctuations had not been understood. The downward trend of the hydrograph of the lake was interpreted during the 1960s as a permanent decline in the level of the lake resulting from consumptive use of stream water primarily for irrigation.

No student of the lake pointed out that while the level of the lake was more than 1.5 m (5 ft) lower than it would have been without consumptive water use, the volume of water represented by this deficit would have a much smaller effect on the lake level when the lake level rose and that cycles of several wet years should be expected that would raise the lake to flooding levels despite the effects of consumptive use. We know now, and should have known then, that consumptive use results in lower lake levels during high lake cycles but will not prevent costly flooding of facilities built on the lake bed and has little impact on the rise of the lake. Construction of dams and the filling of reservoirs do not permanently lower the lake significantly. After the initial filling of a reservoir, its net outflow approximates its net inflow minus evaporation.

A 1961 map of the Great Salt Lake distributed by the Utah Geological and Mineralogical Survey reflects professionals' thinking about lake levels (Eardley, 1961). Great Salt Lake is shown in solid blue at the 1278.6 m (4195 ft) contour. A stippled blue pattern includes areas below 1283.2 m (4210 ft) a.s.l. and the map shows a spillway into Great Salt Lake Desert at 1283.2 m (4210 ft) a.s.l. The map's purpose is "to furnish a suitable base for hydrologic computations and engineering and recreational planning." That the lake had peaked only a century before at nearly 1283.8 m (4212 ft) a.s.l. was forgotten or dismissed.

The 1970s

When the lake rose during the 1970s, planners recognized the inadequacies of their technical information base. The Great Salt Lake Board adopted as one of its seven goals to "define and identify the Great Salt Lake flood plain and recognize it as a hazard zone for management and development... Perhaps the greatest potential threat, and hence the hazard that requires immediate attention, is flooding. Flooding may come about primarily as a result of any one or more of three major factors: (1) the rising waters of the lake, (2) wind tides, and (3) rising ground water table. There is presently insufficient information available on these factors affecting flooding." (Utah Department of Natural Resources, 1976, pp. 8–9).

In the absence of sound scientific studies to support actions to deal with the developing flood, advice based on an inadequate scientific base flourished. Trends based on the Great Salt Lake hydrograph were the most-frequent basis for projections of future lake levels. These studies analyzed the historic lake fluctuations as recorded in the 107 years of measurements supplemented by an additional 27 years inferred levels. This historic record is too short to reveal much about long-term fluctuations of the lake. Attempts to predict the performance of Great Salt Lake based on the historic record included trying to identify cycles and sometimes correlate the cycles with other phenomena, such as sunspot activity, that could to a degree be predicted. Some analyses were not based on any kind of a physical model, for example a college professor released a projection based on the same technique he used to analyze stock market trends. One of the lake industries concluded that the lake would continue to rise for seven years after the wet cycle ended. The correlation between precipitation and the response of the lake level was not well understood.

The 1980s

The Utah Water Research Laboratory of Utah State University attempted to quantify the Great Salt Lake's flooding hazard using estimates of modified stream flows to predict lake levels. The report expressed the realization that methods for riverine flooding estimates "are not appropriate for terminal lakes because they do not account for the dependence of lake levels on the levels of immediately previous years. Great Salt Lake levels vary over many feet, rising over series of years when cumulative inflows exceed cumulative evaporation and falling when the opposite conditions prevail. These cycles last decades and make the level during the previous year the major parameter in estimating the level for the following year." (James et al., 1981, p. 1). Projections for future annual lake levels grossly underestimated the next decade's events as did later updates of probabilistic predictions.

The problems of inadequate technical information and inadequate analysis persisted into the mid-1980s. "Future levels of the lake are generally unknown. The lake's elevation is a function of weather patterns, and since even a 5-day weather forecast by the U.S. Weather Service is accurate only 50 percent of the time, the reliability of any longer forecast is extremely questionable. The Water Research Laboratory of Utah State University has attempted to minimize the uncertainty of future lake levels by assigning probability to various lake levels based upon the current level, the volume of water upstream, and historic lake levels since 1873." (Steffen, 1983).

Predictions based on stream inflow to the lake were frustratingly oblivious to geomorphic processes. The concept of a 100-year flood similar to riverine flooding simply does not apply to the rise and fall of Great Salt Lake. Although some geologists recognized the importance of the morphology of the lake bed, particularly the threshold by which the lake had access into the Great Salt Lake Desert, this morphology was not known in detail. Detailed topographic maps of the area were not available until they were developed as part of the West Desert Pumping Project in 1986 (Chapman and Sappington, 1986). Previous U.S. Geological Survey topographic maps mislocated the threshold by about 30 km and the elevation was shown as about 0.6 m (2 ft) higher than in later surveys. Hydrologists and geologists attempted to explain why the lake rose but engineers needed numbers that said what would happen. Engineers tended to make key decisions.

Beneficial development area

In March 1985 the University of Utah convened a conference of meteorologists, climatologists, geologists, geographers, and hydrologists (Kay and Diaz, 1985).

> Invited speakers presented reviews of the current status of their research into climatic fluctuations, hydrologic modeling, short and long term records of lake level fluctuations, and related topics. The theme was an assessment of our scientific knowledge of the lake system; the conference was not intended to debate or formulate policy positions... It was abundantly clear that there is still a great deal to learn about the lake and the climatic system that drives it; a series of recommendations arising from the conference are summarized at the end of the volume (p. 5).
>
> * * *
>
> Although the conference participants were discouraged from enumerating the solutions to the problem of the Great Salt Lake level changes, the participants were encouraged... to define the nature of the problem itself; in particular, to define the "planning" level of the Great Salt Lake... There was considerable discussion on this topic and a universal agreement that levels below 4212 were too low to be considered as planning levels for the Great Salt Lake... Virtually all the participants... considered the threshold level the best alternative for a planning level because of the basin's morphology. Judith McKenzie stated that her data show that during the last 500 years this level has been reached perhaps as many as five times and certainly a couple times. Most participants believed the 4217–4218 is a "rational number to work with." One participant believed it is "too low" because only the Provo level (4750) could be considered a "truly safe number." The participants believed that the 4222 elevation was "not well enough defined," not rationally controlled by topography, and that the lake, although it had probably reached this level in the last 2000 years, had not reached it with the relative frequency as the threshold level.
>
> The participants expressed many uncertainties about the 4217–4218 level. Not only is the threshold poorly defined topographically, members of the conference would have felt much more comfortable with that number if we knew how often the lake had reached this level in the recent past and its persistence at that level. Many of the specific research topics suggested for further research would directly contribute to the better understanding of the threshold and its consequences to society (pp. 291–292).

Based in considerable part on the consensus of the conference, the Division of Comprehensive Emergency Management developed the concept of a Beneficial Development Area to be established for the shorelines of the Great Salt Lake at an "interim working elevation of 4217 feet."

> It is intended that 4217 feet be used in Federal planning and design activities as the best available estimate of the maximum lake level, however, it is not being suggested that this elevation be used as the design level for all new Federal, State or local projects or improvements on existing projects. The design level for each new project or improvement must depend on the specific economic conditions of

the individual project or improvement. Potential damages may suggest a maximum design level of 4217 feet or higher, whereas from an economic standpoint, a lower design level with calculated risk costs taken into effect may be preferable (Utah Department of Public Safety, 1985, work element 1).

Another important contribution of the conference was an analysis of the historic weather records. Karl and Young (1986) reported that precipitation events such as occurred in the 1980s and a century before were normal events for the area and, while their timing could not be predicted, they should be expected.

Although considerable research had been devoted to determining how and why the level of Lake Bonneville had fluctuated in late Pleistocene time, before the 1980s wet cycle, little work had been done to determine the fluctuations of Great Salt Lake over the last few thousand years. Research done as a result of interest generated by the flooding in the 1980s has provided substantial information on Great Salt Lake including studies of the detailed morphology of the lake bed, particularly the portion leading into the Great Salt Lake Desert. Geomorphic features, particularly low-level shoreline features have provided the basis for much of this information.

4. Lake processes

4.1. Processes that control the still-water level of Great Salt Lake

U.S. Geological Survey has published a summary of much of what is known about the hydrology of Great Salt Lake (Arnow and Stephens, 1990). The still-water level of Great Salt Lake fluctuates in response to variations in the amount of water added to the lake and the amount of water evaporated from the lake. Inflow of water into the lake is primarily a function of weather across the lake's drainage basin. Evaporation is controlled by the surface area of the lake, the weather over the lake, and the temperature and salinity of the lake.

Inflow to the lake

In an average year, streams contribute about 66 percent of the water entering Great Salt Lake, precipitation onto the lake contributes 31 percent, and groundwater discharge directly into the lake contributes 3 percent. The Bear River is 59 percent of the surface inflow, the Weber River 20 percent, and the Jordan River 13 percent.

The normal annual cycle of the lake is to begin to rise in the winter when evaporation is low, rise more rapidly during spring when precipitation and stream flows are highest, decline rapidly in the summer when evaporation is highest, and decline more slowly in the fall. The seasonal fluctuation averages about 0.5 m (1.5 ft) but has varied in historic time between no rise to a maximum rise of about 1.5 m (5 ft) and from no decline to a maximum decline of about 1 m (3 ft). The annual cycle is superimposed on longer term fluctuations related to long-term variations in the weather and to human activity. Across the Great Salt Lake drainage basin, annual precipitation ranges from less than 25 cm in the desert areas to more than 100 cm in the high areas of the Wasatch Range and the mountains to the east. No long-term precipitation records are available for stations on the lake. The record at the Salt Lake City International Airport, which is located a few kilometers from the lake shore and about 6 m above the average level of the lake, dates from 1929. Annual precipitation at the airport averages about 38 cm and has ranged from 20 to 63.5 cm. In the mountainous areas to the east, where most of the surface inflow into the lake originates as snowpack, the variations from average are even greater. The percentage variation in stream flow into the lake is considerably greater than that of precipitation, ranging from 40 to 300 percent of average.

The amount of ground water inflow into the lake cannot be measured directly. As the lake expands and inundates springs around the lake, groundwater inflow increases but under no conditions is the groundwater inflow likely to exceed a few percent of the total inflow to the lake.

Evaporation from the lake

The surface area of the lake, weather conditions, and the salinity of the lake waters control the volume of water evaporated from the lake. As the lake rises, its surface area increases until evaporation from the lake equals the sum of inflows to the lake. The lake stabilizes at this level. The lake level tends to stabilize at elevations where the increase in surface area is large per unit increase in lake level such as at an elevation of approx-

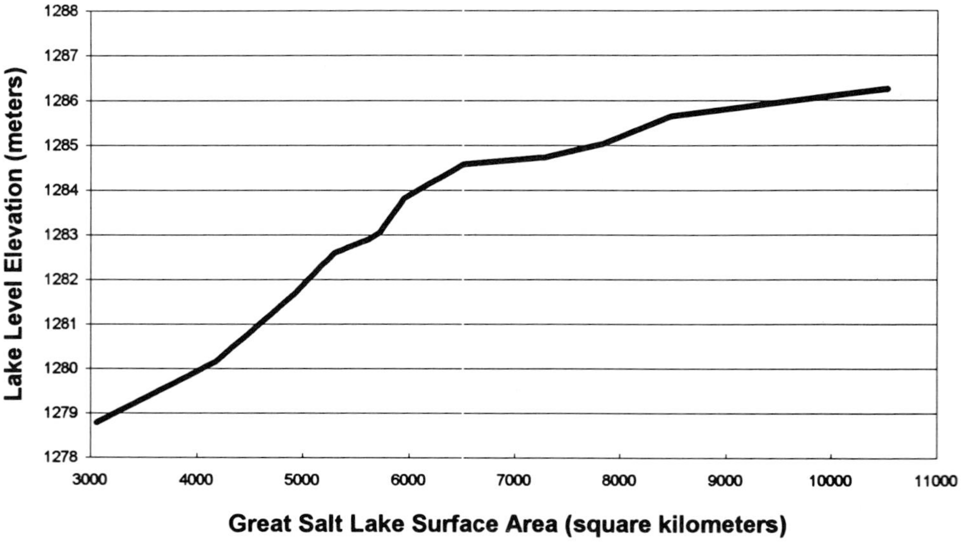

Fig. 7. Graph of lake surface area for Great Salt Lake for lake elevations above 1280.2 m (4200 ft) a.s.l.

imately 1284.7 m (4215 ft) a.s.l. where the lake expands from the Great Salt Lake closed-basin across the divide into the Great Salt Lake Desert (Currey et al., 1984). The one-meter rise between 1281 m and 1282 m a.s.l. elevation results in 450 km^2 increase in surface area in comparison to the one-meter rise between 1285 m and 1286 m a.s.l. which results in a 1880 km^2 increase (Fig. 7).

4.2. Processes that affect flooding associated with still-water levels of Great Salt Lake

Hydrographs report measurements of the still-water level of Great Salt Lake unaffected by wind setup, wave setup, wave runup, and seiches. Measurement of wind setup and resulting seiches have been made for at least two locations south of the causeway, but no accurate direct measurements of lake levels during storms have been reported. Geomorphic features such as gravel beaches and shorelines reflect the level reached by the waves that created them, not the still-water lake level that would be interpreted as the official lake level for that time period.

Wind generated lake-level changes

The height of a storm line on a beach is determined by three wind driven processes (Coastal Engineering Research Center, 1984). Wind setup (storm surge) is the super-elevation of the water level generated by a wind blowing over the water in a constant direction for several hours. Wave setup is the super-elevation of the water surface due to on-shore transport of water by wave action alone. Runup is the up-rushing of waves onto the beach above the average water level. The approximate elevation above the still-water level of a storm line expected to be produced by these three processes can be computed using estimated sustained wind velocity, wind direction, fetch, lakebed and beach configurations. For this purpose the fetch is defined as the region in which the wind velocity and direction is reasonably constant, and for Great Salt Lake is usually determined by the distance from the beach to the upwind shoreline, dike, or causeway that restricts circulation.

In the 1980s the State of Utah considered a major diking project to regulate the level of the eastern part of Great Salt Lake. To determine the design for the dikes, engineers calculated the height at which these dikes should be maintained based on combined setup and runup for storms for two wind directions and for a series of still-water levels using methods approved by the U.S. Army Corps of Engineers. The sustained winds of greatest concern on Great Salt Lake come from the northwest and southwest. The computations for dike

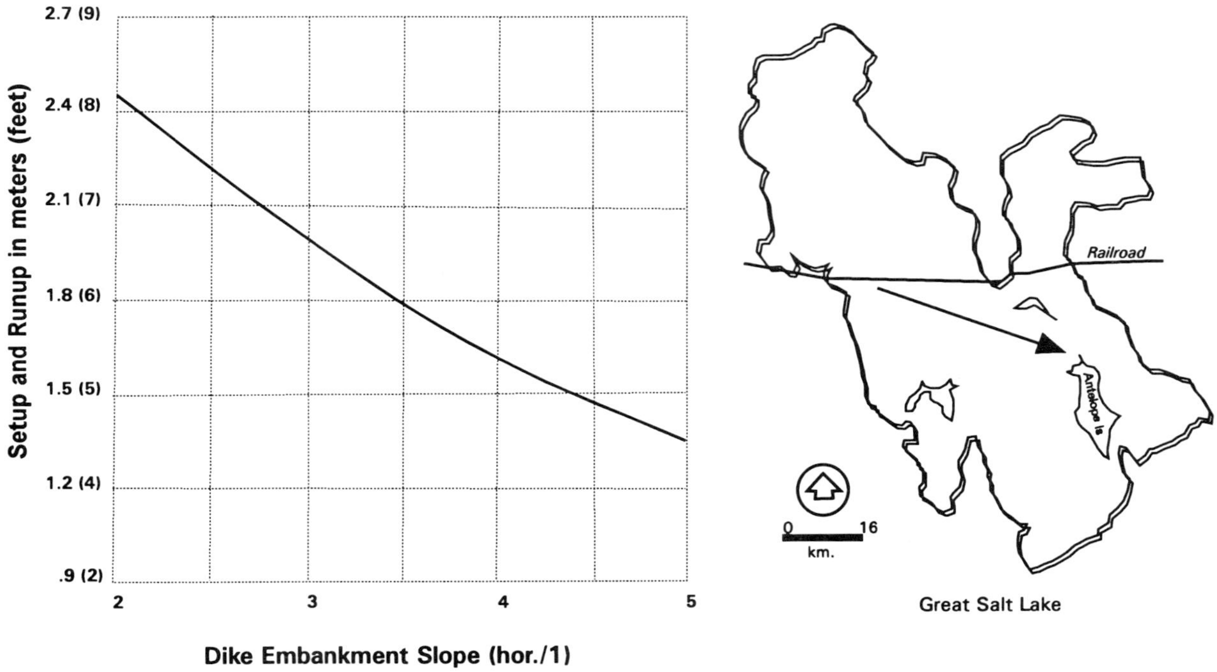

Fig. 8. Curve of computed combined setup and runup caused by northwest winds at a proposed dike location north of Antelope Island assuming a static lake-level elevation of 1283.8 m (4212 ft). Computation based on a wind speed of 71 km/h and a fetch length of 32 km (adapted from Rollins, Brown, and Gunnell Inc, 1987).

designs north and south of Antelope Island provide estimates of the height above still-water level that storm lines might be expected on the west side of the island (Fig. 8). The values computed for sustained wind velocities of from 53 to 70 km/h and lake levels of 1283.8 m (4212 ft) to 1285.0 m (4216 ft) a.s.l. and a dike slope of 1:1, ranged from 1.2 m (4 ft) to 2.4 m (8 ft) for dikes trending approximately north from locations both north and south of Antelope Island (Rollins Brown and Gunnell, Inc. and Creamer and Knoble Engineers, 1987).

Wind seiches

Seiches are standing waves of longer period in enclosed or partly enclosed bodies of water caused by an abrupt increase or decrease in the water level in one part of the body of water. In detail, seiches are complex but the primary effect is an oscillation of the water to raise the level on one shore as it is lowered on the opposite shore. The period of the seiche is determined by the width and depth of the water body, and the amplitude of the seiche by the disturbance that initiated it. Wind seiches are initiated on Great Salt Lake when a wind that has produced a significant wind setup on the lake decreases abruptly to a velocity that cannot maintain the setup.

Lin and Wang (1978) analyzed one wind tide event in January 1969, recorded at two gauging stations on the lake. For about 12 hours the wind blew from the south at an average velocity of about 25 knots (47 km/h). The water level declined at Black Rock at the south end of the lake 0.3 m (1.0 ft) and rose 0.2 m (0.8 ft) at the lake shore at the southern end of the Promontory Mountains (Promontory Point). When the velocity of the wind dropped to less than 10 knots (12 km/h), oscillation started. On the first cycle the lake level at Black Rock rose about 0.8 m (2.5 ft) to a level about 0.5 m (1.5 ft) above the still-water level. At Promontory Point the level dropped 0.5 m (1.8 ft) to 0.3 m (1.0 ft) below the still-water level. Oscillations continued with a period of six hours and decreasing amplitude for two days. Lin and Wang calculated that the construction of the Southern Pacific Railroad causeway had reduced the period of the seiches from nine hours to six hours.

Earthquake effects on lake levels

Earthquakes can produce dramatic changes in water

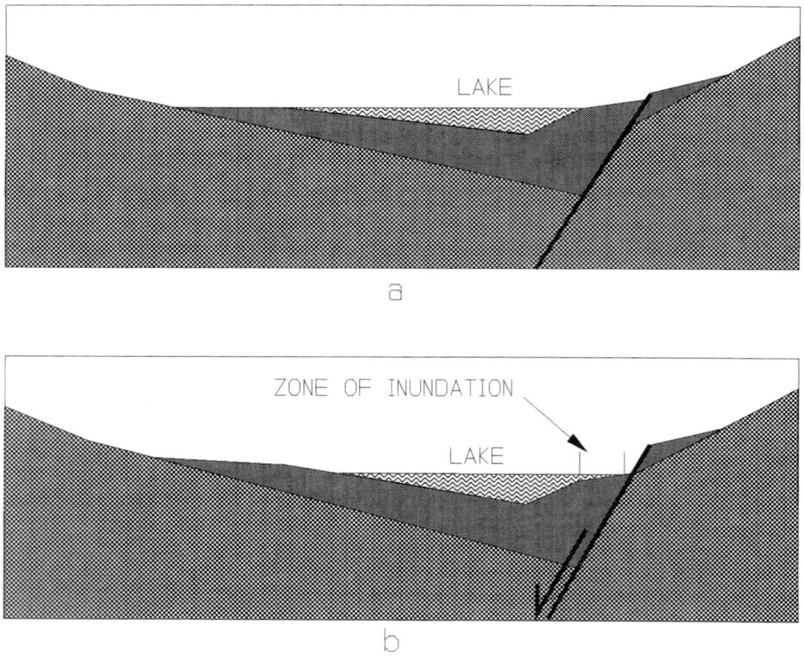

Fig. 9. Sketch showing how tectonic tilt caused by movement along a normal fault bounding a closed basin affects the lake in the basin.

level of Great Salt Lake. Earthquake-induced surges on the lake also can initiate seiches. An earthquake near the north end of the lake in 1909 produce a surge that sent water over the Southern Pacific Railroad causeway and the bath house pier at Saltair (Deseret Evening News, October 6, 1909). Movement on one of several major active faults in or near the lake could permanently tilt the lake bed causing abrupt inundation of parts of the lake shore (Fig. 9) (Smith and Richins, 1984). Such an event occurred during the Hebgen Lake earthquake in Montana in 1959 when one shore of Hebgen Lake dropped 2.4 m (8 ft) (Myers and Hamilton, 1964).

Effects of diking and other human activities on lake levels

Pumping and diking of portions of Great Salt Lake floodplain for evaporation of brines for salt extraction and diking for waterfowl refuges, flood control, recreation, and transportation artificially alter the still-water level of Great Salt Lake. Consumptive use of water in the drainage basin decreases the amount of water that reaches the lake. The U.S. Geological Survey calculated that by 1963 the consumptive use of water had lowered the level of the Great Salt Lake 1.5 m (5 ft) below what it would have otherwise been (Whitaker, 1971). Interbasin transfers of water such as from the Colorado River basin add comparatively minor volumes of water to Great Salt Lake. The West Desert Pumping project has the capacity to impact future fluctuations of Great Salt Lake. It can withdraw 2.5 billion cubic meters (2 million acre-ft) of water from the lake per year and, depending on the return flow to the lake, this could be a 0.3 m (1 ft) to 0.6 m (2 ft) impact on the lake level, depending on the lake level elevation and inflow into the lake.

Diking sections of the lake bed results in greater amplitude fluctuations of the remaining portions of the lake. Much of the lake bed provides poor foundation for construction and maintenance of dikes. The consequences of dike failure should be considered in approving any diking project. In June 1986 during the flooding events of the 1980s, storm waves of Great Salt Lake breached evaporation pond dikes west of Stansbury Island and allowed lake waters to invade a large area west of Stansbury Island which had been used for salt evaporation. As a result the still-water lake level of Gilbert Bay dropped 0.15 m (0.5 ft) and the evaporation ponds were badly damaged and put out of commission through the duration of the wet cycle. Schemes to build dikes to form a fresh water lake along the eastern shore of Great Salt Lake and maintain it at near constant level would greatly increase the amplitude of fluctuations of the level of Gilbert and Gunnison Bays. Failure of a portion a dike when the still-water level of the main body of the lake is higher than the fresh-water lake could cause sudden inundation of thousands of acres of developed land with consequences similar to a catastrophic failure of a dam.

5. Evidence of flooding levels

5.1. Measurements of elevations of shoreline features on Antelope Island

In the summer of 1986 after the lake had receded from the still-water record high lake level of 1283.8 m (4211.8 ft) a.s.l., Don R. Mabey, Donald R. Currey, and the author measured the elevation of the 1986 debris line at three locations on the eastern side of Antelope Island. The goal was to compare the 1986-recorded highstand with the 1870s highstand as determined by Gilbert's measurement of the elevation of the 1870s debris line. Gilbert had made his measurement somewhere on the eastern side of Antelope Island in 1877, a few years after the Great Salt Lake crested in the 1870s. The average elevation measured for the 1986 debris line was 1284.3 m (4213.5 ft) a.s.l., the same as Gilbert's elevation for the 1870s debris line. The agreement of the 1986 measurements with Gilbert's measurements suggests that the maximum elevation reached by the lake in the two historic high cycles was approximately the same and is the best determination that has been made of the 1870s still-water level.

In 1987–88 the Utah Geological and Mineral Survey included a study of Great Salt Lake shorelines as part of a cooperative program of geological investigations of Antelope Island for the Utah Division of Parks and Recreation (Atwood and Mabey, 1994). The purpose of the study was to document the elevation and geomorphic effects of the 1980s lake cycle on shorelines of Antelope Island and to better understand recurrence intervals of high lake-level cycles of Great Salt Lake. Antelope Island proves to be a superb field area for observing shorelines because of its undisturbed shoreline zone, its variety of beach conditions, the abundance of manufactured debris interspersed in the beach materials that clearly delineated 1980s shorelines, and its exposure to storms from all directions. Well-preserved low-elevation shorelines of Great Salt Lake on Antelope Island record fluctuations of the level of Great Salt Lake over the last 10,000 years.

5.2. Description of Antelope Island shorelines

The numerous shorelines on Antelope Island formed by Great Salt Lake are expressed in four general ways: debris lines at or occasionally partly above the shoreline, wave-constructed gravel ridges and beaches, erosional steps, and vegetation lines. Commonly two or more of these features occur together.

Debris lines

The high shoreline formed in 1986–87 is nearly everywhere marked by debris. Wood, both processed, such as railroad ties, and unprocessed, is the most common material. Also included are non-wood manufactured materials such as automobile tires and a wide array of plastic items. Because some of the lighter debris is moved by wind as well as wave action, some debris has been carried up the beach above the gravel shoreline. Most of the debris in these shorelines clearly postdate the 1870s highstand although the possibility that some of the old wood was in place earlier cannot be ruled out. No debris shorelines higher than the 1980s debris shoreline were identified. Lower debris lines formed since 1987 consisted mostly of the remains of dead insects with minor amounts of wood and other material.

Gravel ridges and beaches

The crests of the 1986–87-constructed gravel ridges were surveyed as the 1980s highstand except along one portion of the east side of the island which had no constructional beach features at all. The amount of gravel on the shorelines varies considerably around the island. In some areas the 1986–87 shoreline consists entirely of well-sorted gravel and in other areas no gravel is present. The 1986–87 shoreline gravels were always bleached with no observable surface stain. At several locations two or more gravel ridges were formed and preserved at essentially the same elevation. Gravel beaches and ridges occur both above and below the 1986–87 shorelines. Those above usually show some degree of surface staining of the gravel with all gravels significantly higher than the 1986–87 shoreline appearing to be about equally stained. Lower gravels are all bleached.

Erosional steps

Where the lake waters eroded a shoreline in relatively soft sediments rather than constructed it of gravels, the shoreline was marked by a near vertical step. Debris or gravel collect at the base of some of these steps. The steps range in height up to more than 3 m (10 ft). Damage to structures built on these steps could occur by ground failure long after the high water had receded. The vertical face of these steps begins to slump and erode as soon as it is formed and 1986–87 features were considerably modified by 1989. Step shorelines above the 1986–87 shorelines were consistently modified into gentle slopes. No step shorelines were observed below the 1986–87 shoreline in 1988 and 1989. In areas with no constructional shorelines such as along portions of the eastern shore of the island, the 1980s highstand-level was surveyed at the debris line at the base of the erosional step.

Vegetation lines

Some shorelines are clearly distinguished by their absence of vegetation, zones of distinctive vegetation, or changes in vegetation. In some areas of subtle shorelines, vegetation zones at constant elevations appear to mark shoreline elevations. The saline water of the lake kills most plants it inundates and in some areas the upper limit of the vegetation-kill zones clearly marks the 1980s highstand. Anomalous non-killed areas occurred where springs discharged fresh water and maintained plants even though they were partly or completely submerged by lake water. Gilbert a few years after the 1870s highstand noted a kill zone below which he observed no sagebrush and concluded that it would take perhaps several hundred years for sagebrush to become reestablished in inundated zones. This led him to conclude that the lake had not been above the 1870s highstand level for several hundred years. The vegetation on Antelope Island had fully recovered in the period between 1877 when Gilbert made his observations and the next highstand in 1986. Some of the gravel shorelines both at the 1986–87 level and above are marked by plants that grow well in disturbed areas, or areas with coarser sediments. Sunflowers are particularly common on some of the gravel ridges. Vegetation evidence alone did not provide sufficient information for distinguishing a shoreline.

5.3. Elevations of 1986–87 storm-line shorelines

Elevations of the 1986–87 Great Salt Lake shoreline were measured at 21 locations distributed around Antelope Island (Table 1). The shoreline elevations for the 1980s highstand (1283.8 m/4211.8 ft still-water level maximum) ranged from 1283.9 m (4212.4 ft) a.s.l. to 1285.7 m (4218.1 ft) a.s.l., a range of 0.2 m (0.6 ft) to 1.9 m (6.1 ft) above the highest still-water level.

Interpretation

The differences in elevation for the 1980s high shoreline reflect different exposures of the shores to storms and different geometries of the lake bed. Shorelines formed by 1986–87 processes are generally higher on the west side of the island than on the east (Fig. 10). Depending on exposure, the shape of the beach and its slope, wave action can be expected 1.8 m (6 ft) above the still-water lake level at some locations on Antelope Island and may occur higher in pocket beaches. These measurements are consistent with theoretical calculations based on procedures of the U.S. Army Corps of Engineers shore protection manuals (Coastal Engineering Research Center, 1984). Their calculations indicate that storm-lines could be expected at even higher elevations than those measured for the Antelope Island study for some exposures on the west side of the island.

The 1986–87 shorelines provide a datum for recognizing potential wave damage on Antelope Island associated with a still-water lake level of 1283.8 m (4211.8 ft) a.s.l. The range in elevations also indicates the care with which elevations on prehistoric shorelines must be used when inferring still-water levels of the lake that created it or when using a shoreline as an originally horizontal datum.

5.4. Elevations of pre-1980s storm-line shorelines on Antelope Island

Elevations of pre-1986–87 shorelines were also measured at 24 profiles at the 21 locations distributed around Antelope Island. Table 1 summarizes observed shoreline data and groups these data based on elevation into a history of fluctuations. Each high lake cycle

Table 1
Measured elevations of five low-level shorelines on Antelope Island and inferred elevations of the associated still-water levels of the lake

Location	Shoreline elevation (m)				
	Historic shoreline	Shoreline B	Shoreline C	Shoreline D	Shoreline H "Holocene High"
WRB	1284.7				1287.5
BUP	1284.8				1286.6
BUS	1284.8	1285.3			
LFW	1284.5				
LFE	1284.4	1284.8		1285.8	
BUB	1284.9				
TLS	1284.5	1285.0			1286.9
CAF	1284.3	1285.0			1286.9
BOL	1284.4			1286.2	
SGN	1284.5				
SGE	1283.9		1285.1	1285.4	
MUF	1284.1				
UNC	1284.2	1284.6			
UNP	1284.4	1285.0			1286.8
SOP	1285.4				1287.2
SOW	1284.9	1285.0			1287.6
18S	1284.9	1285.2	1285.9	1286.6	
DCS	1284.9	1285.5	1286.0		
RRC	1285.7	1286.3			1287.3
Still-water elevation (m)	1283.8	1284.3	1284.7	1285.3	1286.2
Still-water elevation (ft)	4212	4214	4215	4217	4220

The historic shoreline was formed by the 1870s and 1980s highstands. Shoreline B is gravel-stained and little modified by erosion. Shoreline C is somewhat modified by erosion but not cut by debris-flow channels. Shoreline D is somewhat modified by erosion and cut by debris-flow channels. Shoreline H, considerably more modified by erosional and depositional processes, correlates with the Holocene high of Murchison (1989). Still-water elevations in feet rounded to the nearest foot.

masks or destroys the surface expression of lower and older shorelines formed by lake levels that lasted only briefly. Because the lake-level highstands of the 1870s and 1980s were both approximately 1283.8 m (4212 ft) a.s.l., most shoreline features produced in the 1870s high cycle are at the same level as, and are masked by, those produced in 1986–87. At several locations modified multiple beach ridges or stained gravel appear to identify features related to the 1870s cycle where in very local areas it had not been destroyed by wave action in 1986–87.

The next shoreline (shoreline B) above that formed by the last two high cycles was identified at 10 of the survey locations. The gravel is stained but the gravel ridges are little modified by erosion. This shoreline must be older than the 143-year historic record, and the lack of major modification suggests it is no more than a few hundred years old. At three locations a shoreline (shoreline C) slightly above shoreline B was identified. Erosion has modified shoreline C more than shoreline B but, unlike shoreline D, shoreline C was not cut by debris-flow or other stream channels. The elevation of shoreline D was measured at four locations. The gravels of this shoreline are cut by erosion channels including debris-flow channels. The highest Great Salt Lake shoreline measured was identified at eight locations. This shoreline (designated shoreline H) was considerably more modified by erosional and depositional processes and thus likely considerably older than the lower shorelines. This shoreline correlates with the Holocene high of Murchison (1989) which has been radiocarbon-dated as between 3440 and 1400 B.P.

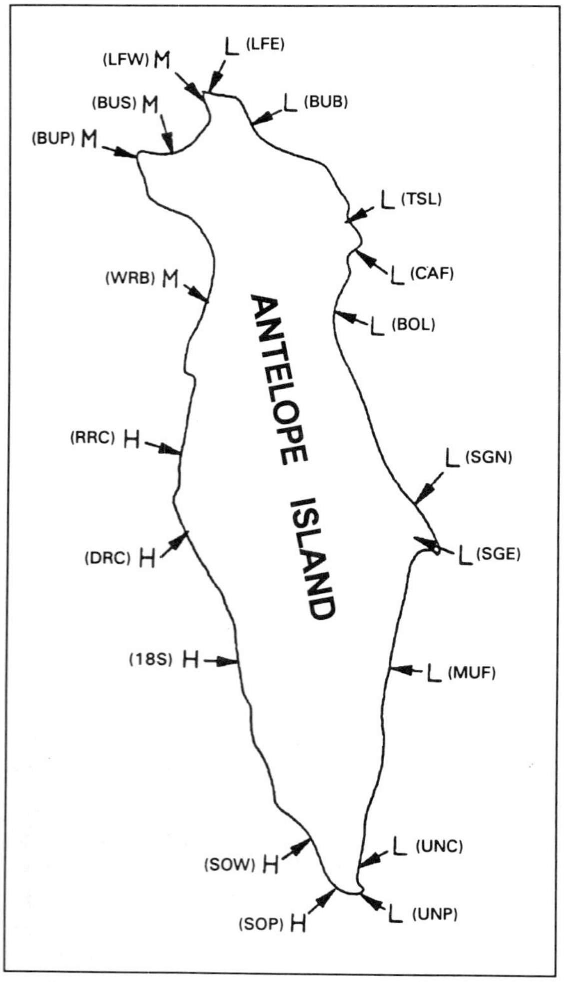

Fig. 10. Map of Antelope Island showing locations of measured shoreline profiles and relative height of the 1986–87 shoreline above the still-water level. $L < 0.7$ m, $M = 0.8$–1.0 m, $H = 1.1$–2.0 m.

Interpretation

Based on the pattern of shoreline elevations associated with the known lake-level history of 1986–87, the older shorelines have been assigned still-water lake levels (Table 1). The observed shoreline elevations indicate pre-1980s still-water levels of Great Salt Lake of 1284.3 m (4214 ft) a.s.l. (shoreline B); 1284.7 m (4215 ft) a.s.l. (shoreline C); 1285.3 m (4217 ft) a.s.l. (shoreline D) and 1286.2 m (4220 ft) a.s.l. (shoreline H).

5.5. Implications for recurrence of high lake levels

Because each high lake cycle destroys the surface expression of older shorelines, the record of lake-level fluctuations preserved in the shorelines is confined to higher and higher levels of older and older ages. Shorelines are excellent indicators of the elevation of flooding at a site during a high lake cycle. If proper corrections are made for setup and runup caused by wind, shorelines are also indicative of the high still-water lake level (highstand) of a lake cycle.

Shoreline data can be supplemented with other data to develop an understanding of lake cycles. Where lagoons are associated with high lake cycles, sediments deposited in the lagoons may survive subsequent inundations and provide a record of multiple lake cycles. If the sediments contain organic material, radiocarbon dating may provide a chronology of the lake cycles (Murchison, 1989). Analysis of the chemistry of lake sediments obtained from cores and trenches of the lake bed can provide an indication of the salinity of the lake water when the sediment was deposited, and the salinity can be used to calculate the approximate level of the lake (McKenzie and Eberli, 1985). Lake cycles reflect precipitation over the lake's drainage basin. Statistical analysis of historic precipitation data provides an indication of the frequency that above average precipitation events are likely to recur (Karl and Young, 1986). Tree-ring studies can provide information on prehistoric climate.

Using historic lake level and precipitation records, geomorphic evidence from shorelines, and the general configuration of the lake bed, conclusions can be made concerning future cycles of Great Salt Lake (Fig. 11). Within the normal climate of the region, a series of three- to five-years of much-above-average precipitation can be expected on the average of about once every 100–120 years. Such a precipitation event will cause Great Salt Lake to rise abruptly. The still-water level the lake reaches depends upon the lake level at the start of the event and the magnitude of the event. From 1982 to 1986, 16 billion cubic meters (13 million acre-ft) of water were added to the lake causing it to rise 3 m (10 ft) from 1280.8 m (4202 ft) to 1283.8 m (4212 ft) a.s.l. Without the effect of 140 years of consumptive use of water, the still-water lake level would have been significantly higher, about 1284.3 m (4214 ft) a.s.l. If the wet cycle had began 20 years earlier when the lake

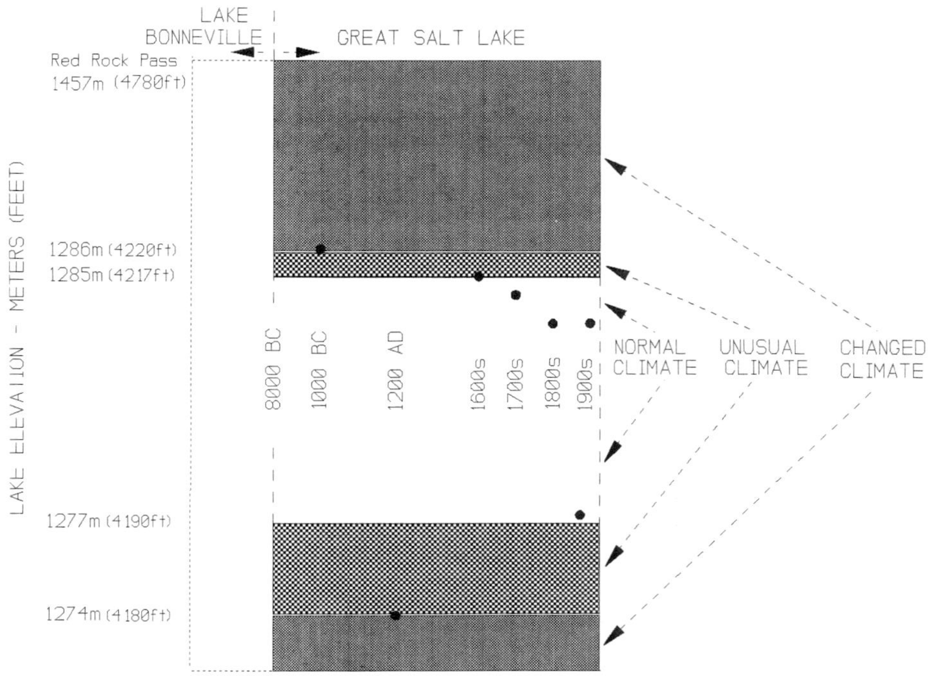

Fig. 11. Still-water elevations of Great Salt Lake. Dots refer to high stands and low stands of Great Salt Lake.

level was lower, the maximum would have been about 1282.6 m (4208 ft) a.s.l. Every few hundred years the lake has risen to the 1284.7 m (4215 ft) still-water lake level, where it has access to the Great Salt Lake Desert. Longer or more intense wet cycles have occurred in the past 10,000 years and caused the lake to rise to the 1285.3 m (4217 ft) still-water level. The recurrence interval of these cycles has not been determined but is likely to be more than several hundreds of years. Only once in the last 10,000 years has the lake reached a still-water level of 1286.2 m (4220 ft) a.s.l. This once in 10,000-year event required unusual climatic conditions. A rise of the lake to levels associated with Lake Bonneville requires a significant change in climate and would take thousands of years to develop.

6. Conclusions

Lake level rises occur slowly and are not inherently life threatening. Therefore, most decisions relating to hazards associated with lake-level rises can be based on considerations other than life safety. Two exceptions are important to consider for Great Salt Lake and other seismically active areas: inundation caused by earthquakes and hazards associated with dike failures if the dikes impound considerable volumes of water. A large earthquake with an epicenter in or near Great Salt Lake could produce a destructive surge and resulting lake seiches. Surface displacement of faults in or near the lake could produce tectonic tilt of the lake bed and cause abrupt and permanent inundation of areas of lake shore. The danger from either of these phenomena is greatest when the lake level is high. Both provide incentives for not placing developments at low elevations around the lake. Proposals to build dikes on the bed of Great Salt Lake to provide permanent control of the lake level for selected stretches of the shoreline have been advanced for many years. These proposed projects would greatly increase the hazard associated with high lake levels. If dikes maintain the water level along a developed shoreline at a level below that of the main lake, the risk is compounded by the dikes' vulnerability to groundshaking from earthquakes. Large differences in water level that would develop across the dikes when the lake level is high pose a threat to any development along the shore of the stabilized part of the lake. Lake level rises are now a threat to property: some proposed diking projects would make high lake levels a threat to lives.

Transportation corridors can be severely disrupted by high lake levels. Some transportation routes that cross flood plains of closed-basin lakes provide essential arteries for interstate travel and must be kept functioning. Other routes such as access to islands in closed-basin lakes are desirable but not essential. Deciding whether a roadway should be built to function 100% of the time versus 95% or 90% of the time is a political and economic decision. Geomorphic and historic evidence of the storm-water lake level and theoretical calculations of wave runup and setup provide the basis for predicting the probability of road-closure for different time intervals. For example, a road to provide access to Antelope Island from Syracuse, Utah on the east shore of Great Salt Lake could be designed at a level for 85%, 90%, 95% and 100% accessibility with major differences in cost of construction and problems of subsidence and groundfailure from loading the soft sediments of the lake bottom.

Developments on and adjacent to Great Salt Lake should consider the following factors relating to lake-level fluctuations:

(1) Rises of the lake such as have occurred twice in the last 130 years are "normal" and any development should plan for them.

(2) During high lake cycles the still-water lake level may rise to 1284.7 m (4215 ft) a.s.l. but is not likely to exceed 1285.3 m (4217 ft) a.s.l.

(3) Lake levels higher than the 1285.3 m (4217 ft) a.s.l. still-water level are so infrequent that they can be ignored for planning purposes.

(4) The impact of high still-water lake levels on a site is controlled by the exposure, the geometry and geology of the lake bed and land surface as well as the elevation. Planning for still-water lake levels alone is not sufficient to avoid flooding hazards. Many of the effects of storms on individual stretches of shorelines can be computed. In many shoreline areas geomorphic evidence can be used to test the computations.

(5) Construction of structures on the lake that restrict the movement of lake waters can have serious and unforeseen consequences. Some diking projects may substantially increase the flooding hazard they seek to control.

(6) Fluctuations of salinity accompany lake-level fluctuations and may create additional problems such as allowing the formation of ice on the lake with resulting ice rafting damage to structures on the lake or lake shore.

Fluctuations of the level of Great Salt Lake and similar closed-basin lakes are controlled by long-term weather which we do not presently have the ability to predict. Nevertheless our understanding of past lake fluctuations based on geomorphic and historic evidence and our knowledge of effects of wind and earthquakes on lakes provide sufficient information to be used in estimating flooding risks associated with development alternatives. Properly applied with sound engineering and planning practices, this knowledge can control losses associated with future high lake levels.

Acknowledgements

This reseach was supported by the Utah Geological and Mineral Survey and the Utah Division of Parks and Recreation as part of a cooperative geologic investigation of Antelope Island. I thank Donald R. Currey, Don R. Mabey, Marie Morisawa, and Gary Tackman for their helpful reviews of this paper.

References

Arnow, T. and Stephens, D., 1990. Hydrologic characteristics of the Great Salt Lake, Utah: 1847–1986. U.S. Geol. Surv. Water Supply Pap., 2332: 32 pp.

Atwood, G. and Mabey, D.R., 1994. Shorelines of Antelope Island as evidence of fluctuations in the level of Great Salt Lake. In: H.H. Doelling (Editor), Geology of Antelope Island. Utah Geol. Surv. Bull., in press.

Benson, L.V. and Paillet, F.L., 1989. The use of total lake-surface area as an indicator of climatic change: examples from the Lahonton basin. Quat. Res., 32: 262–275.

Benson, L.V., Currey, D.R., Dorn, R.I., Lajoie, K.R., Oviatt, C.G., Robinson, S.W., Smith, G.I. and Stine, S., 1990. Chronology of expansion and contraction of four Great Basin lake systems during the past 35,000 years. Palaeogeogr. Palaeoclimatol. Palaeoecol., 78: 241–286.

Bills, B.G. and May, G.M., 1987. Lake Bonneville: constraints on lithospheric thickness and upper mantle viscosity. J. Geophys. Res., 92: 11,493–11508.

Chapman, W. and Sappington, W., 1986. Basin contours of the northern section Great Salt Lake Desert, Utah. U.S. Geol. Surv. Open-File Rep. 86-009 Research Project, map.

Coastal Engineering Research Center, 1984. Shore Protection Manual. U.S. Army Corps of Engineers, 2 Vols., 3-88 to 3-127.

Crittenden, M.D., 1963. New data on the isostatic deformation of Lake Bonneville. U.S. Geol. Surv. Prof. Pap., 454-E.

Currey, D.R., 1980. Coastal geomorphology of Great Salt Lake and vicinity, In: J.W. Gwynn (Editor), Great Salt Lake: A Scientific, Historical and Economic Overview, Utah Geol. Miner. Surv. Bull., 116: 69–82.

Currey, D.R., 1988. Seismotectonic kinematics inferred from Quaternary paleolake datums, Salt Lake City seismopolitan region, Utah. In: National Earthquake Hazards Reduction Program, Summaries of Technical Reports. U.S. Geol. Surv. Open-File Rep., 88-673: 189–214.

Currey, D.R., Atwood, G. and Mabey, D.R., 1984. Major levels of Great Salt Lake and Lake Bonneville. Utah Geol. Mineral. Surv. Map 73.

Eardley, A.J., 1961. Great Salt Lake, Utah. Utah Geol. Mineral. Survey Map.

Federal Emergency Management Agency Region VIII, 1986. Closed-basin lake flooding: case studies and mitigation opportunities. Assoc. of State Floodplain Managers Symp., March 24–28, 1986, 31 pp.

Forman, S.L. (Editor), 1989. Dating methods applicable to Quaternary geologic studies in the western United States. Utah Geol. Miner. Surv. Misc. Publ., 89-7, 80 pp.

Gilbert, G.K., 1890. Lake Bonneville. U.S. Geol. Surv. Monogr., 1: 438 pp.

Grayson, D.K., 1993. The Desert's Past: A Natural Prehistory of the Great Basin. Smithsonian Institution Press, Washington, DC, 356 pp.

Hunt, C.B., 1974. Natural Regions of the United States and Canada. W.H. Freeman, San Francisco, 725 pp.

Interagency Task Force, 1984. Interagency hazard mitigation report. Federal Emergency Management Agency, September 14, 1984, FEMA-720-DR-Utah, 26 pp.

Jackson, D. and Spencer, M.L., 1970. The Expeditions of John Charles Fremont, 1. Travels from 1938 to 1848. Univ. of Illinois Press, Chicago, 854 pp.

James, L.D., Bowles, D.S., James, W.R. and Canfield, R.V., 1981. Update on estimation of water surface elevation probabilities for the Great Salt Lake. Water Resources Planning Series UWRL/P-81/01. Logan UT, Utah Water Research Laboratory, Utah State University.

Karl, T.R. and Young, P.J., 1986. Recent heavy precipitation in the vicinity of Great Salt Lake; just how unusual? J. Clim. Appl. Meteorol., 25: 353–363.

Kay, P.L. and H.F. Diaz (Editors), 1985. Problems of and Prospects for Predicting Great Salt Lake Levels. Center for Public Affairs and Administration, University of Utah, 309 pp.

Lin, A. and Wang, P., 1978. Wind tides of the Great Salt Lake. Utah Geol., 5(1): 17–35.

Mabey, D.R., 1986. Notes on the historic high level of Great Salt Lake. Utah Geol. Miner. Surv. Notes, 20(2): 13–15.

McKenzie, J.A. and Eberli, G.P., 1985. Late Holocene lake-level fluctuations of Great Salt Lake as deduced from oxygen-isotope and carbonate contents of cored sediments. In: P.L. Kay and H.F. Diaz (Editors), Problems of and Prospects for Predicting Great Salt Lake Levels. Center for Public Affairs and Administration, Univ. of Utah, pp. 25–39.

Mifflin, M.D. and Wheat, M.M., 1971. Isostatic rebound in the Lahontan basin, northwestern Great Basin. Geol. Soc. Am. Abstr. Programs, 3: 647.

Murchison, S.B., 1989. Fluctuation history of Great Salt Lake, Utah, during the last 13,000 years. Ph.D. dissertation, Univ. of Utah, 137 pp.

Myers, W.F. and Hamilton, W., 1964. Deformation accompanying the Hebgen Lake earthquake of August 17, 1959. U.S. Geol. Surv. Prof. Pap., 435: 55–98.

Oviatt, C.G., McCoy, W.D. and Nash, W.P., 1994. Sequence stratigraphy of lacustrine deposits: a Quaternary example from the Bonneville basin, Utah. Geol. Soc. Am. Bull., 106(1): 133–144.

Rollins Brown and Gunnell Inc. and Creamer and Knoble Engineers, 1987. Great Salt Lake Inter-Island Diking Project — Final Design Report. Vols. 1 and 2. Prepared for the Utah Division of Water Resources.

Sack, D., 1989. Reconstructing the chronology of Lake Bonneville. In: K.J. Tinkler (Editor), History of Geomorphology. Unwin, London, pp. 223–256.

Sack, D., 1993. Geomorphic evidence of climatic change from desert basin paleolakes. In: A.D. Abrahams and A.J. Parsons (Editors), Geomorphology of Desert Environments. Unwin Hyman, London, pp. 616–630.

Smith, R.B. and Richins, W.D., 1984. Seismicity and earthquake hazard of Utah and the Wasatch front: paradigm and paradox. In: Evaluation of Regional and Urban Earthquake Hazards and Risk in Utah. U.S. Geol. Surv. Open-File Rep., 84-763: 73–112.

Steffen, C.C., 1983. The Great Salt Lake: major economic impacts of high lake levels. Utah Economic and Business Review University of Utah, 43(9/10): 1–7.

Street-Perrott, F.A. and Harrison S.P., 1985. Lake levels and climate reconstruction. In: A.D. Hecht (Editor), Paleoclimate Analysis and Modeling. Wiley, New York, pp. 291–340.

U.S. Army Corps of Engineers, 1986. Great Salt Lake, Utah, Reconnaissance Report. December, 1986.

Utah Department of Natural Resources, 1976. Great Salt Lake comprehensive plan. Division of Great Salt Lake. December, 1976, 68 pp.

Utah Department of Natural Resources, 1976. Great Salt Lake environs report. Division of Great Salt Lake. June, 1976, 30 pp.

Utah Department of Natural Resources, 1983. Recommendations for a Great Salt Lake contingency plan for influencing high and low levels of Great Salt Lake. Division of State Lands and Forestry. January, 1983, 47 pp.

Utah Department of Natural Resources, 1987. West Desert Pumping Project. Division of Water Resources information handout, May 1987, 7 pp.

Utah Department of Natural Resources, 1989. Great Salt Lake pumps to be shut down. Press release. June 26, 1989, 8 pp.

Utah Department of Public Safety, 1985. State Hazard Mitigation Plan — 1985. Division of Comprehensive Emergency Management, 21 pp.

Utah House of Representatives, 1986. H.B. 6, Diking and West Desert Pumping. Second Special Session, 11 pp.

Whitaker, G.L., 1971. Changes in the elevation of Great Salt Lake caused by man's activities in the drainage basin. Geological Survey Research 1971, U.S. Geol. Surv. Prof. Pap., 750-D: D187–D189.

ELSEVIER

Geomorphology 10 (1994) 221–230

GEOMORPHOLOGY

The landslide hazard in the Himalayas: geological control and human action

John Gerrard

School of Geography, The University of Birmingham, Birmingham, UK

Received February 4, 1994; revised April 1, 1994; accepted April 4, 1994

Abstract

Mountains are high-energy environments characterised by instability and variability. Much of this instability is the result of a variety of landsliding processes; mudflows, rockfalls and debris slides/avalanches being especially important. This landsliding constitutes a major hazard in the Himalayas. Early work has suggested that certain rock types are more predisposed to failure than others. It also seems possible that specific types of failure are related to certain rock types and properties. This analysis suggests that these relationships can be employed in a hazard assessment. Phyllite rocks are the most susceptible to landsliding followed by shales, schists, poorly cemented sandstones, gneiss, granites and quartzite. This order is also appropriate for slope failures within weathered material although weathered gneiss appears to be more susceptible to rill and gully erosion. Clearly rock type exerts a major influence on landsliding, but assessing the role of rock type in comparison to the effect of human activity is difficult. Human activity is often influenced by the nature of slopes, including the type and depth of material on the slopes. As these are affected by rock type, it is difficult to differentiate the contributing factors. Because of this it would be dangerous to assume that landsliding in the Himalayas can be completely explained and predicted by the nature of the rock type.

1. Introduction

Landslides, a major world-wide hazard, are greatest in areas of weak rock and steep slopes. For these simple reasons, landsliding tends to be extensive in mountainous areas. Mountains are high-energy environments characterised by instability and variability. Casual observations are sufficient to indicate many examples of active landslides, mudflows, rockfall and debris avalanches. In the classification of mountain hazards, produced by Messerli and Ives (1984), landsliding has a prominent part, occurring in all mountain zones. This classification also emphasises that human activity can increase the frequency and nature of landsliding. This is a crucial issue as mountains world-wide are coming under increasing pressure from a variety of human activities. Many research workers have tried to establish linkages between human activities and landsliding.

In the Himalayas, this link forms a major component of what is known as the Himalayan Environmental Degradation Theory (Ives and Messerli, 1989). This relationship, however, is far from clear. Laban (1979), following an airborne reconnaissance of Nepal, concluded that geological structure and lithology accounted for more than 75% of all observed landslides. A similar conclusion was reached by Joshi (1987) after an extensive survey in part of the Garhwal Himalayas. This paper analyses relationships between landsliding and geology in the Himalayas and will use such relationships to assess the nature of the natural/human-induced landsliding conundrum. This account will draw heavily on previously published results but the assessment will also be based on the author's experience working in the Darjeeling region of northeast

SSDI 0169-555X(94)00023-K

India, the Kumaun Himalayas of northwest India, and three years involvement in a land use, soil conservation, and water resource management project in the Middle Hills, Nepal, sponsored by the British Overseas Development Administration and directed by Rita Gardner, King's College, University of London. Before the specific case of the Himalayas is examined, some general statements concerning the nature of landsliding are necessary. Landsliding throughout this account is taken to include all forms of mass movement but the slow downslope movement of material such as soil creep and solifluction is excluded.

2. The nature of landsliding

Landsliding, like most natural denudation processes, is governed by a complex set of factors. An external trigger, such as heavy rainfall, slope undercutting, or seismic activity initiates the process. But the nature of the rock governs the response of the slope to external stresses. Thus, at one level landslides are clearly a function of the nature of the rock materials. What is important is the establishment of more specific relationships between rock type and the nature and intensity of landsliding. If such relationships can be established, then they can be used in the creation of landslide hazard and risk assessment maps and in the planning of new developments, such as road construction and dam building. In this analysis the general term rock type will include lithological and structural characteristics, although some attempt will be made to differentiate the two factors.

Landsliding may be influenced by rock type in two main ways, reflecting whether the failure zone of the landslide occurs within the rock itself or in the surface layer of weathered material or regolith. The manner in which rock slopes fail depends on the geometry and mechanical properties of joints and bedding planes. The size and shape of the detached mass is determined by the nature of these rock discontinuities, the state of weathering, and slope geometry. The variety of discontinuities that can occur will influence the potential instability of any rock slope. Rock slopes can be classified as structural and astructural. Slopes parallel or nearly parallel to the structural planes are known as structural slopes and slopes not parallel to such planes are known as astructural slopes. In Nepal, most large landslides occur on structural slopes. Wagner (1981, 1983) noted that 33% of the 100 landslides he studied in the Middle Hills, Nepal, occurred on slopes formed by bedding of inclined rock and 67% occurred on structural slopes formed by a plane of fracture. Similarly, Haigh (1984) stressed that instability along roads in the Garhwal Himalayas was influenced by joints and bedding planes dipping out of slopes. Road construction techniques, especially blasting, accentuate the structural weaknesses in rock and thus enhance the relationship between landsliding and rock structure. For many reasons, great care needs to be taken when assessing landsliding along highways.

On slopes mantled with regolith, rock influences are more subtle. Properties of the regolith will influence the processes acting on the slope. Weathering and landsliding are closely related. As weathering attacks rock, the strength and elasticity of the material decreases more or less continuously, porosity and saturation moisture content increase and grain size decreases. Such changes usually lead to a decrease in shear strength which will increase the likelihood of landsliding. Variations within the weathered profile and relict rock structures may exert a great influence on the nature of any slope failure. Rock type influences the way that weathering occurs and the nature of the materials that are produced. Structural disintegration occurs more readily in materials that are brittle and/or that have the capacity for volume change than in materials that are plastic and/or that have little capacity for volume change. Weathering profiles vary considerably between different rock types and in this way rock type has a major influence on slope instability.

Large rock slides will be influenced by structural features such as joints, faults and bedding planes, whereas smaller landslides will be more influenced by the nature of the weathered mantle. Mudflows and mudslides may require an appreciable clay content for movement to occur. In many cases it will be important to differentiate the influence of lithology from that of structure. Landslides are just one element in the overall denudation of the landscape, although in high, active mountain areas such as the Himalayas, it is often a dominant process. Rock type and structure also may influence a variety of sub-aerial processes and indirectly affect landsliding. An especially strong relationship exists between river action and slope development. Drainage patterns and the intensity of dissection are

Table 1
Rock type and characteristics, Kumaun Himalaya, India (after Tandon, 1984)

Formation	Age	Lithology
Bhabar and Terai	Recent	Gravel, coarse sand
Upper Siwalik	Pliocene–Pleistocene	Conglomerates, sandstones
Middle Siwalik	Miocene–Pliocene	Arkoses, some shales
Lower Siwalik	Miocene	Greywackes, clays
Krol Series	Permo–Triassic	Limestones, shales slates, quartzites
Quartzite Series	Devonian?	Quartzites with slate, phyllites

often governed by geological factors. General relationships between landform elements and landsliding are also likely to reflect relationships between rock type and landsliding.

In an interesting study using rocks from South Africa and the United States, Cooks (1983) has correlated various measures of rock strength and elasticity with certain drainage basin parameters. Rock types studied were sandstone, granite and gneiss and rock properties measured were compressive, tensile and shear strength, Young's modulus, shear modulus and Poisson's ratio. The drainage basin properties examined were drainage density, angle of valley-side slope and basin hypsometric integral. All rock properties were significantly related to drainage basin properties except for Poisson's ratio. The nature of the relationships indicated that material of low compressive strength was more easily incised and removed from drainage basins by erosional processes than material of greater strength. High drainage densities and high slope angles were associated with low-strength material. Low drainage densities and gentler slope angles were generally associated with high-strength rock. In general, the results indicated that more channels, deeper incision and more rapid removal occur on low-strength rocks and fewer channels, shallower incision and less rapid removal occur on high-strength rocks. The link with landsliding is that it is assumed that landsliding is a major component of landscape evolution.

A similar interpretation can be made for the results of a study by Tandon (1974) in the Kumaun Himalayas. He examined a variety of drainage basin characteristics on six major rock formations (Table 1). Metamorphic rocks support the least number of streams and maintain longer stream lengths, with larger drainage basins. Relief ratios and ruggedness are high because of greater relief, in spite of lower drainage densities. Basins on the Krol Series possess smaller areas, medium to coarse texture, high relief and ruggedness. Drainage density is higher because of low porosity and low infiltration capacity. The Lower and Middle Siwalik units possess similar characteristics because of homogeneity in lithology except for a slightly higher degree of induration in the Lower Siwalik sandstones. Both formations possess medium to coarse texture and drainage networks with moderate relief and ruggedness. Drainage basin areas are greater than on rocks of the Krol Series. The Upper Siwalik formation, of conglomerates and sandstones, possesses a higher stream frequency and drainage density, medium to coarse texture with low values of relief and ruggedness. The major characteristics of the sands and clays of the Bhabar and Terai units are the large number of lesser-order streams, medium to coarse texture, gentle slopes, low values of relief and ruggedness. This study by Tandon (1974) suggests that an analysis of landsliding in the Himalayas from the point of view of rock type will be fairly fruitful.

3. Himalayan geology and structure

The Himalayan mountain chain possesses a highly complex combination of rock types and geological structure, but it can be subdivided into three reasonably distinct lateral belts running approximately west to east. The Siwalik Hills form the outermost southern unit followed by the Middle Hills or Middle Mountains and then the High or Greater Himalayas (Ives and Messerli, 1989). The Tibetan Plateau lies to the north of the Himalayas and is sometimes included in the Himalayan structural unit. A fourth morphological component in the Himalayas is the Terai, a gently sloping surface of deposition at the foot of the Siwalik Hills. The basic structural pattern has been formed by the Indian tectonic plate thrusting beneath the central Asian plate, creating a 150 km wide zone of intensely folded and faulted rocks of varying lithologies. Because this collision is continent to continent, igneous rocks are uncommon although small areas of granite occur in some areas. The thrusting has also resulted in three major thrust or fault zones: the Main Frontal Thrust, approximately at the outer limit of the Siwalik Hills; the main Boundary Fault approximately at the junction of the Siwaliks and the Middle Mountains; and the Main Central Thrust within the transition belt of the Middle Mountains and the High Himalayas. All three zones exert a major influence on landsliding and many of the larger and most catastrophic landslides are associated with movements along these thrusts (Nakata, 1982).

The basic Himalayan structure consists of large thrusts formed from the basement of the advancing of the Indian continent. The Siwalik rocks consist of folded molasse-like sedimentary rocks that are relatively soft and unconsolidated; folding occurred in Pliocene–Pleistocene times. The Middle Hills are essentially thrust sheets composed of Precambrian schists and gneisses derived from the Indian shield. The High Himalayas are formed of Precambrian gneisses intruded with Tertiary granites. Thus, the main lithological units are gneisses, schists, phyllites, altered sedimentary rocks and some igneous rocks. Metamorphic rocks dominate, but complexity makes it very difficult to establish general characteristics. Also, very few areas of the Himalayas have been mapped in detail. Major recrystallization of minerals has occurred and directed pressure has produced alternations of parallel schistose layers and light-coloured granular quartz and feldspar. The boundary between the schists and gneisses is ill-defined but gneiss consists of minerals associated with granite, such as quartz and feldspar, whereas the schists contain very small amounts of feldspar. It is often difficult to assess which metamorphic rock is influencing landsliding and which rock properties are important. Many of these issues are now examined in greater detail with reference to three specific areas of the Himalayas.

4. Nepal

A number of studies provide an insight into landsliding in Nepal. Shrestha (1980), working in the Karnali watershed, Nepal, has shown that high potential erodibility occurred on schists and broken sandstones, medium potential erodibility on limestone, sandstone and dolomite and low erodibility on granites and gneiss. One of the most detailed studies is that of Brunsden et al. (1981) in the Middle Hills area of eastern Nepal. Rocks vary in age from Precambrian to mid-Tertiary (Table 2) and have been subjected to severe tectonic disturbance. The various forms of mass movements are the dominant processes controlling hillslope form on all rock types except the gneiss of the Lower Himalaya unit. On slopes in gneiss, failures are associated with the development of gullies in the deeply weathered regolith. The gneiss has weathered to depths of up to 20 m producing a silty sand with a variable proportion of core stones. Soil erosion and piping in this material have produced numerous broad, deep, vertical-sided, flat-floored gullies with sharp, well-defined heads. It can be argued, however, that this is an indirect effect of rock type in governing the nature of the regolith. Steep slopes have developed on quartzite and schist rocks and rockfalls frequently occur especially noticeable where quartzite bands form buttresses and pinna-

Table 2
Stratigraphy of the Dharan–Dhankuta area, eastern Nepal (after Bordet, 1961; Brunsden et al., 1981)

The Border Unit	Dharan Bazar Series	Sandstones Phyllites	Lower Miocene
	Sangure Series	Quartzites Phyllites	Permo–Triassic
Lower Himalayan Unit		Upper Phyllites	Devonian–Carboniferous
		Calc-schists	Devonian
		Upper Quartzites	Silurian–Devonian
		Lower Phyllites	Silurian
		Lower Quartzites	Cambrian–Silurian
		Mica Schists	Precambrian
		Migmatic Gneiss	Precambrian

Table 3
Landslide frequencies in the Dharan–Dhankuta area, eastern Nepal, classified according to slide type and rock type (after Brunsden et al., 1981)

Lithology	Landslide type			
	Debris	Rock	Rotational	Frequency
Phyllite	26 (63%)	13 (31%)	2 (5%)	41
Shale	34 (74%)	7 (15%)	5 (11%)	46
Schist	23 (60%)	9 (24%)	6 (16%)	38

cles. Falls include wedge, planar and toppling failures governed by the pattern of rock discontinuities. Rockslides develop where major discontinuities dip out of the slope face and occur mostly on shales, phyllites and schists. Debris slides, the commonest form of mass movement, are usually shallow (1–3 m) and elongate (100–1000 m), and are found on all rock types. Rotational slides are essentially confined to those areas where deeply weathered shales and schists are undercut by rivers or gullies. Mudslides are relatively uncommon but in all cases are associated with deeply weathered phyllites and schists.

In many of these failures schistosity and bedding appears to control the shape of the failure surfaces. This partly explains why most of the landsliding is occurring on shales, schists and phyllites with few on gneiss outcrops. The paucity of failures on gneiss may be explained because the bedrock structure and deep irregular weathering does not permit the development of planar slide surfaces and unstable weathering mantles. Landslide frequencies associated with the three unstable rock types are shown in Table 3. The differences exhibited by the data are not statistically significant. This suggests that, although rock type appears to be a major factor, other factors have been involved in governing landslides. The frequency of type of landsliding on phyllites appears to be different from those on the other rock types. Even when phyllites are tested against the other rock types separately, however, no significant differences are detected. This evidence suggests that the three rock types are equally susceptible to landsliding as a whole but that phyllites are more susceptible to rock slides than other rock types and shales to debris slides.

Analysis of landslide frequency and slope angle distribution also show few differences between rock types (Brunsden et al., 1981). If only undercut slopes are examined, however, a separation occurs between shales and schists (modal slope angle group 36 to 39 degrees, mean 36.2), phyllites (39 to 42 degrees, mean 39.2) and gneiss (30 to 33 degrees, mean 30.8). This reflects, perhaps, the production of finer calibre material on gneiss and could signify a difference between shallow mass movement processes on shales, schists and phyllites and rainsplash and gully erosion on the gneiss.

One of the most detailed analysis of relations between rock type and landsliding has been undertaken in the Jikku Khola drainage basin in the Middle Hills of Nepal to the northeast of Kathmandu (Dongol, 1991). Assessment of the entire basin makes the study even more useful. Metasandstones and metasiltstones are characterised by deep weathering profiles and are more prone to rill and gully erosion rather than landsliding. Phyllites and rocks with phyllite bands were the most susceptible to landsliding with eighteen failures. Slopes on massive quartzites are steep and very stable. Slopes on mica schists can be unstable if the mica content is appreciable as a high mica content induces rapid weathering and a loss of rock strength. Slopes on coarse crystalline marble possess deeply weathered soils which may lead to instability on steep slopes. Slopes on small areas of granitic gneiss and basic intrusive rocks are generally stable and maintain steep slopes. This example has demonstrated again the general instability of slopes on phyllites and the influence of the nature and depth of weathered material.

The last point is important. The majority of the smaller landslides and the greatest number of landslides, overall, occur in weathered material, especially when landslides on cultivated land are considered. A most interesting study, that addresses many of these issues was completed by Caine and Mool (1982) in the upper part of the Kolpu Khola drainage basin, a few kilometres northwest of Kathmandu. The main rock types are phyllites, granites, biotite gneiss and augen gneiss but all the slides have occurred in untransported weathered bedrock. On the basis of an analysis of friction angles, the weathered mantle on augen gneiss is the strongest and that on phyllites the weakest (Table 4). Also 10% of the strength of phyllite weathered debris is lost on remoulding but a loss of only 5% of strength occurs with the other rock types. This demonstrates that the untransported material on phyllites is

Table 4
Critical slope angles on regolith of a variety of rocks (based on Caine and Mool, 1982)

Lithology	Critical slope angles (deg)	
	(drained)	(saturated)
Phyllite	32.8	18.5
Biotite gneiss	41.5	20.8
Granite	42.8	21.8
Augen gneiss	44.2	24.4

Table 5
Brittleness indices of weathered material in the Kolpu Khola basin, Nepal (after Caine and Mool, 1982)

Bedrock	Brittleness index
Augen gneiss	0.343
Phyllite	0.460
Granite	0.531
Biotite gneiss	0.673

Table 6
Infiltration rates (mm/min) for undisturbed and failed material in the Kolpu basin, Nepal (after Caine and Mool, 1982)

Lithology	Undisturbed	Landslide debris
Biotite gneiss	2.07	4.94
Phyllite	3.36	6.85
Augen gneiss	4.89	11.32
Granite	10.00	15.00

Table 7
Effects of wetting on penetration resistance. Values are ratios of wetted to dry (after Caine and Mool, 1982)

Lithology	Undisturbed	Landslide debris
Granite	0.285	0.415
Phyllite	0.346	0.716
Biotite gneiss	0.409	0.649
Augen gneiss	0.548	0.700

the weakest and this becomes even more pronounced once failure has occurred.

Failed slopes on phyllites are more susceptible to future failures than slopes on other rock types. This can be emphasised by calculating values of the brittleness index (Bishop, 1973). The index (I_b) is defined as:

$$I_b = \frac{S_{cu} - S_{cd}}{S_{cu}}$$

where S_{cu} is the compressive strength of the undisturbed material and S_{cd} is the compressive strength of the failed material.

A summary of the brittleness values for the various weathered lithologies is shown in Table 5. Two important points can be made. First, all the values are high with, on average, more than 50% of the material strength being lost on failure. In some cases the loss was as high as 90%. These values indicate a greater loss of strength than indicated earlier for remoulding and reworking. Second, the pattern of brittleness approximately matches that of weak and strong material indicated earlier. Statistical analysis, however, reveals no significant differences when all the lithologies are examined together. This reflects high within-lithology variability but analyzing lithologies in a pair-by-pair way might reveal some individual differences.

Somewhat similar relationships are found with respect to material–water behavior. Infiltration rates vary with lithology (Table 6). Infiltration is most rapid into the coarse-textured materials derived from the weathering of granite and augen gneiss. The data also show that infiltration rates are higher in the failed material. The comparatively low infiltration rates for the failed phyllite and biotite gneiss regolith suggest that these materials will be more susceptible to subsequent rill and gully erosion. Such failed material may take longer to recover. Penetration resistance is reduced when material becomes wet but the magnitude of the change seems to be partially related to the nature of the material. This can be seen in Table 7 where ratios of penetration resistances for wet/dry states for undisturbed and failed debris are shown. In general, the effect of wetting on undisturbed material is to lower the values on average by almost 50%. For the granite regolith, the value has been lowered by over 70%. The effect of wetting on failed material is less, presumably because the initial penetration resistance has been lowered by changes brought about by failure.

This detailed analysis by Caine and Mool (1982) indicates that relationships exist between rock type, regolith characteristics and potential for instability but that these relationships are complex and not necessarily obvious. Of the rock types in this analysis, phyllites emerge as being most susceptible to failure.

5. Eastern Himalayas

The mountain region of West Bengal, in northeast India, especially the Darjeeling area, is an area prone to major landslides (Table 8). The landslide events listed have been the most devastating; landslides occur each year but most go unrecorded. The region encompasses all of the major subdivisions of the Himalayas noted earlier. The geological framework varies from unaltered sedimentary rocks in the south to various grades of metamorphic rock and some intrusive rocks in the middle and northern zones. The lithologies have been affected by multiple folding and faulting, shearing and fracturing. This has weakened the rocks and made them especially vulnerable to water percolation. The Terai is composed of horizontal beds of unconsolidated sand, silt, gravel and cobbles of Quaternary age. The Tertiary Siwalik beds, mostly arkosic and micaceous sandstones, siltstones and conglomerates, are more compacted and cemented and have been affected by mountain building episodes. The Siwaliks dip northward under the overthrusts of Damuda Series rocks, which are generally coarse sandstones and quartzites with a few impersistent coal seams. North of the Damuda rocks the mountains are formed in low-grade metamorphic rocks, mostly quartzites, slates, schists, phyllites and gneiss. The Daling Series (phyllites and schists) occurs in the Kalimpong area to the east of the River Tista while Darjeeling gneiss dominates the western region.

Earthquake induced landslides are a complicating factor. The region suffers from persistent activity along the Main Boundary and Himalayan Front Faults. A number of severe earthquakes have occurred over the last 100 years, including the Bengal earthquake (14 July 1885), Assam earthquakes (12 June 1897, 8 July 1918, 10 July 1969), north Bihar earthquake (15 January 1934) and the Tibet/Assam earthquake (15 August 1950). The 1950 Assam earthquake denuded the Dibang valley and tributaries of forest cover (Kingdom-Ward, 1955). The severity of this destruction by landsliding was related to the angle of slope and nature of the rock.

Establishing relationships between landsliding and rock type in the Darjeeling area is very difficult. Except for the 1934 Bihar earthquake, major landsliding is the result of rainfall of above average intensities. The landslips of 24 and 25 September 1899 were caused by 60.96 cm (24 inch) of rain, and those of 1950 by 84.20 cm (33.15 inch) between 10 and 14 June. The detailed study by Starkel (1972) of the effects of the rainfall between 3 and 5 October 1968 suggest that land use patterns were a major influence on landsliding. Tea gardens were badly affected with about 20% on average of the area of every garden destroyed. Terraced slopes experienced damage of between 30 and 50 percent. In contrast, destruction in forests was one-twentieth to one-tenth of that on the tea slopes and tended to be concentrated along roads. This analysis would suggest that geology and rock type has very little influence on

Table 8
Recorded major landslides of Darjeeling area

Year	Details	Rainfall (mm)	Effect on landscape	Casualties
1899	24–25 Sep.	610	Innumerable landslips	Heavy
1934	15 Jan.	–	Associated with Bihar earthquake	–
1950	10–14 Jun.	842	Widespread landslides; communication badly affected	Total disruption of civil life
1968	3–5 Oct.	1266	Communication links badly affected	Heavy toll of human life
1980	27 Aug. 3–4 September	– –	2966 acres of cultivated land and 30 acres of plantations damaged	–
1984	1–5 Sep.	450–700	150 m stretch of main road washed off	

landsliding in the Darjeeling area. This might be misleading because land use patterns reflect geology and the nature of the soils that have developed. Moreover, in general, discontinuities within the rocks, joints and bedding planes, have an influence on slope failure.

Clearer relationships between rock type and landsliding have been established in the Lish Basin, West Bengal (Basu and Ghatowar, 1988). In the Lish basin landslides are especially common on rocks of the Daling Series and rare on Darjeeling gneiss (Table 9). The Daling Series consists of purple slates, quartzites and crumpled phyllites. The bands of phyllites make the slopes especially susceptible to landsliding. Most failures on slopes of Darjeeling gneiss were shallow debris slides in the weathered regolith. Slopes in rocks of the Siwalik and Damuda Series were of intermediate stability. Some of the landsliding in the Damuda Series rocks was the result of poor mining operations in the coal-bearing strata. Rather unusually, the intensity of landsliding is moderate on the soft sandstone, mudstones, shales and conglomerates of the Siwalik Series.

Table 9
Landslides and geology in the Lish Basin, West Bengal (after Basu and Ghatowar, 1988)

Geology	No. of slides	Percentage area of landsliding
1929–30		
Siwaliks	8	0.3
Damuda	10	3.8
Daling	115	4.2
Darjeeling Gneiss	2	0.4
1955		
Siwaliks	8	0.4
Damuda	9	4.1
Daling	115	5.4
Darjeeling Gneiss	2	0.4
1964–65		
Siwaliks	10	9.1
Damuda	4	4.1
Daling	41	8.1
Darjeeling Gneiss	2	0.4
1981–84		
Siwaliks	16	11.3
Damuda	4	4.4
Daling	41	9.4
Darjeeling Gneiss	3	0.2

6. Western Himalayas

An insight into landsliding in the northwest Himalayas is provided by the study by Bartarya and Valdiya (1989) in the Kumaun Himalayas, India, near Nainital. Three main litho-tectonic units can be distinguished. The first is the Precambrian Almora Group of medium-grade metamorphic rocks (schists and micaceous quartzites interbedded with augen gneiss). The second is the Ramgarh Group of mainly slates, phyllites, marbles and quartzites plus highly mylonitized porphyritic granite and quartz porphyry. The third group is the Bhawali quartzite associated with the Bhimtal volcanics of Precambrian age. In addition to these units slates, quartzites and conglomerates of the Upper Jaunsar Group and Krol limestones and dolomites of the Mussoorie Group are present. The whole area is affected by numerous faults trending especially NW–SE, N–S and NE–SW. The majority of streams of third, fourth and fifth orders follow such lineaments.

The steepest slopes (over 40 degrees) are related either to fault scarps or to the hard rocks of Bhawali quartzites or Ramgarh porphyries. The rock types most prone to landsliding are granites, quartzites, and basic volcanics (Table 10). Landslides are most frequent in the volcanic rocks and quartzites of Bhawali and Blaini formations (189), followed by the Amritpur granite (108), and the sandstones of the Siwaliks and schists of the Ramgarh Group (49). With respect to geological formations the Amritpur granite has the greatest density of landslides ($2.6/km^2$), followed by the Infrakrol Formation ($2.33/km^2$), the Bhawali quartzite ($1.58/km^2$) and Siwalik Group ($1.12/km^2$) (Table 11). Landslide densities on the other geological formations are low.

Table 10
Landslides in relation to lithology in the Guala Catchment area, Nainital (after Bartarya and Valdiya, 1989)

Lithology	No. of landslides
Slates	14
Quartz-porphyry	31
Gneiss and granite	49
Schists	50
Sandstone	85
Volcanics	87
Quartzites	102
Granite	108

Table 11
Landslides on different geological formations of the Gaula Catchment area, Nainital, India (after Bartarya and Valdiya, 1989)

Geological formations	No. of landslides	Frequency (No./km^2)
Almora Group	3	0.092
Ramgarh Group	104	0.368
Nagthat Formation	170	1.58
Blaini Formation	17	0.88
Amritpur Granite	139	2.66
Krol Formation	2	0.18
InfraKrol Formation	14	2.33
Siwalik Group	85	1.12

Table 12
Relationships between landslides and geology along two highways in northwest India (data adapted from Mukerji, 1984)

Geological Series	Rock type	Percentage of road with landsliding
Rishiketh–Badrinath Highway		
Krol	limestones, shales, sandstones	0.0
Nagthat	orthoquartzite, shales	0.20
Damta	greywackes	0.17
Chandpur	shales, slates quartzites	0.0
Basics	intrusive, extrusive	0.13
Almora	granite, gneiss	0.50
Tejam	phyllites, slates	0.13
Joshimath	gneiss, quartzite	0.80
Kalka–Rampur Highway		
Lower Siwaliks	greywackes, conglomerates	4.62
Subathu/Murree	shales, limestones	2.12
Krol	shales, limestones, sandstones	2.31
Blaini	siltstones, slates	0.0
Nagthat	orthoquartzite, shales	0.67
Low grade	quartzite, slate	1.69
Basantpur	black shales, basics	0.38
Basics	intrusive, extrusive	0.58

These figures compare with an average density of 1.6/km^2 recorded by Ramsay (1986) for all landslides in the Phewa Valley of Nepal. Bartarya and Valdiya (1989) believe that lithology was the most important factor governing landsliding in the study area, but that the localisation of landslides within lithologies was the result of fractures and sheared zones and the dip of the beds. Debris flows were often initiated by springs emanating from fracture zones.

Table 13
Landsliding and geology along highways in northwest India (data adapted from Mukerji, 1984)

Geological Series	Rock type	Percentage of road landslipped
Mandi–Manali Highway		
Larji Shali	limestone, dolomite	3.20
Salkhala	schist, quartzite	0.16
Manjir	conglomerate	1.05
Central Gneiss	gneiss, schist	2.50
Jammu–Leh Highway		
Middle Siwaliks	sandstone, clay conglomerate	0.00
Murree	orthoquartzite, clay conglomerate	0.80
Lower Precambrian	phyllites, slates, quartzite, limestone	1.56
Pir Panjal Trap	shale, limestone	2.11
Recent alluvium		1.09
Triassic	limestone	0.14
Cambrian	slates	0.03
Lower Tertiary	volcanic	0.00

Similar results were obtained by Mukerji (1984), also in northwest India. A reworking of his data is shown in Tables 12 and 13. High incidence of landslides occur on rocks of the Pir Panjal Trap (limestones and shales), Subathu/Murree Series (shales and limestones) and on the Krol Series (shales, limestones and sandstones). Landslides were less frequent on gneisses, intrusive rocks and quartzites.

7. General conclusions

This analysis suggests that sufficient consistent relationships exist to place the most common rock types in order of susceptibility to landsliding. With respect to the three broad divisions of the Himalayas, the Siwalik rocks, composed of essentially poorly consolidated rocks, appear to be the most susceptible to landsliding. Little difference occurs between the Middle Hills and the High Mountains in landslide susceptibility. Differences are related to variations in lithology and broad structural units. The Middle Hills are under greater pressure from human activity and this activity has increased the incidence of landslides. The High Mountains experience less human pressure but possess greater gravitational and tectonic stresses.

With respect to specific rock types, the order from most susceptible to least susceptible to landsliding appears to be phyllites, shales, schists, poorly cemented sandstones, limestones, gneiss, granites and quartzites. This order is sufficiently consistent in all the examples reviewed here to be used as an initial assessment of landslide susceptibility although in some cases specific structural features might have to be taken into consideration. This sequence could provide a useful input in hazard assessment.

The close relationship between rock type and landsliding has led several workers to suggest that landsliding is not necessarily related to human activity. Landsliding might be related to rock type but this does not mean that human activity is not involved in the landslide process. As noted earlier, many general landscape characteristics are also related to rock type and structure. These landscape characteristics also influence the nature and intensity of human activity. Slope angle and the nature and depth of soil are features that influence settlement and land use patterns: much landsliding occurs within the zone of weathered material. The evidence suggests that many small mass movements are partially influenced by human activity but conditioned by the nature of the weathered material. The larger failures may be more determined by rock type and structure.

References

Bartarya, S.K. and Valdiya, K.S., 1989. Landslides and erosion in the catchment of the Gaula River, Kumaun Lesser Himalaya, India. Mt. Res. Develop., 9: 405–19.

Basu, S.R. and Ghatowar, L., 1988. Landslides in the Lish Basin of the eastern Himalayas and their control. In: S. Singh and R.C. Tiwari (Editors), Geomorphology and Environment. Allahabad Geographical Society, Allahabad, pp. 428–443.

Bishop, A.W., 1973. The stability of tips and spoil heaps. Q. J. Eng. Geol., 6: 335–376.

Bordet, P., 1961. Recherches géologiques dans l'Himalaya du Népal, Région du Makalu. Centre National de la Recherche Scientifique, Paris.

Brunsden, D., Jones, D.K.C., Martin, R.P. and Doornkamp, J.C., 1981. The geomorphological character of part of the Low Himalaya of eastern Nepal. Z. Geomorphol. Suppl., 37: 25–72.

Caine, N. and Mool, P.K., 1982. Landslides in the Kolpu Khola drainage, Middle Mountains, Nepal. Mt. Res. Develop., 2: 157–73.

Cooks, J., 1983. Geomorphic response to rock strength and elasticity. Z. Geomorphol., 27: 483–93.

Dongol, G., 1991. Geological data base of the Jhikhu Khola watershed and the potential risk assessment study. In: Soil Fertility and Erosion Issues in the Middle Mountains of Nepal, Workshop Proc. Integrated Survey Section, Kathmandu and Dept. of Soil Science, Univ. of British Columbia, Canada, pp. 180–187.

Haigh, M.J., 1984. Landslide prediction and highway maintenance in the Lesser Himalaya, India. Z. Geomorphol. Suppl., 51: 17–37.

Ives, J.D. and Messerli, B., 1989. The Himalayan Dilemma. The United Nations University and Routledge, London and New York.

Joshi, B.C., 1987. Geo-environmental studies in parts of Ramganga catchment Kumaon Himalayas. PhD thesis, Univ. of Roorkee, Roorkee, India, unpubl.

Kingdom-Ward, F., 1955. Aftermath of the Assam earthquake of 1950. Geogr. J., 121: 290–303.

Laban, P., 1979. Landslide occurrence in Nepal, Phewa Thewa Tal Project. Rep. No. SP/13, Integrated Watershed Management project, Kathmandu.

Messerli, B. and Ives, J.D., 1984. Gongga Shan (7556 m) and Yulongxue Shan (5596 m). Geoecological observations in the Hengduan Mountains of Southwestern China. In: W. Lauer (Editor), Natural Environments and Man in Tropical Mountain Ecosystems. Franz Steiner, Wiesbaden, pp. 55–77.

Mukerji, A.B., 1984. Geomorphic damages in the western Himalaya: Regional patterns and some human dimensions revealed by newspaper reports. In: W. Lauer (Editor), Natural Environments and Man in Tropical Mountain Ecosystems. Franz Steiner, Wiesbaden, pp. 127–146.

Nakata, T., 1982. A photogrammetric study on active faults in the Nepal Himalaya. J. Nepal Geol. Soc., 2: 67–80.

Ramsay, W.J.H., 1986. Erosion problems in Nepal Himalaya: An overview. In: S.C. Joshi (Editor), Nepal Himalaya: Geoecological Perspectives. H.R. Publishers, Nainital, India.

Shrestha, B.P., 1980. Geology, soil erosion and litho-structural division of Nepal. Dept. of Soil and Water Conservation, Ministry of Forests, Kathmandu.

Starkel, L., 1972. The role of catastrophic rainfall in the shaping of the relief of the Lower Himalaya (Darjeeling Hills). Geogr. Pol., 21: 103–47.

Tandon, S.K., 1974. Litho-control of some geomorphic properties: an illustration from the Kumaun Himalaya, India. Z. Geomorphol., 18: 460–71.

Wagner, A., 1981. Rock structure and slope stability study of Walling area, central Nepal. J. Nepal Geol. Soc., 1: 37–43.

Wagner, A., 1983. The principal geological factors leading to landslides in the foothills of Nepal: a statistical study of 100 landslides; steps for mapping the risks of landslides. Helvetas/Swiss Technical Cooperation/ITECO, Kathmandu.

ELSEVIER

Geomorphology 10 (1994) 231–252

GEOMORPHOLOGY

The geomorphology of some debris flows in the southern Sierra Nevada, California

Jerome V. DeGraff

USDA Forest Service, Sierra National Forest, 1600 Tollhouse Road, Clovis, CA 93611, USA

Received March 7, 1994; accepted April 4, 1994

Abstract

Debris flows are one of the natural hazards present in the southern Sierra Nevada of California. Historic debris flow activity is documented at a USDA Forest Service research facility in the Kings River drainage related to a 1937 storm event. No subsequent study of this phenomenon was undertaken until 1982. Observations of debris flows over the succeeding 10 years offer an initial assessment of the physical geomorphology of debris flows in this area. This information provides a starting point for future efforts to avoid or limit the effect of this natural hazard.

Observations were made in the Tuolumne, Merced, San Joaquin, and Kings River drainages. Of the twenty-six debris flows observed, six were examined in detail to provide specific data on this phenomenon. Triggering events for debris flows in the southern Sierra Nevada include intense rainfall, rain-on-snow storms, and seasonal melting of heavy snowpacks. Movement typically occurs at depths between 0.3 and 5 m below ground surface. This is representative of depths for the three interfaces associated with initiation of movement: (1) at the base of the root zone, (2) at the contact of well-weathered and less-weathered soil, and (3) at the contact between soil and unweathered bedrock. Measurement of debris flow velocity based on indirect methods found values ranging from 2.6 m/s to 7.2 m/s (9 km/h to 26 km/h). Recurrence intervals based on radiocarbon dates are between 425 and 500 years BP.

Roads and other land use investments have suffered damage from debris flow activity during the 1982 to 1993 period. Stump Springs road in the San Joaquin river drainage required US$1.4 million to repair damage primarily from 1982 debris flows. The threat to life arising from debris flows is illustrated by the consequences of vegetation losses from wildfire near El Portal, California. The projected population growth with associated increased infrastructure raises concern for greater impacts from future debris flow occurrence in the southern Sierra Nevada.

1. Introduction

A major storm in the southern Sierra Nevada began at 5:00 p.m. on December 9, 1937. Precipitation fell mainly as rain and ended at 7:00 p.m. on December 11 (Munson, 1937). The rainfall included two high intensity periods of 2 inches (51 mm) for one-half hour and 1.5 inches (38 mm) for one hour in the Kings River drainage (Gleason, 1938). In the Big Creek watershed, a tributary to the Kings River, the California Forest and Range Experiment Station of the USDA Forest Service maintained weir and gauge instrumentation on eight small subwatersheds ranging in size between 4 and 15 hectares. Total rainfall in the Big Creek watershed from the storm was 12.3 inches (312 mm) (Munson, 1937). Gleason (1938) noted that the rainfall occurred when only eight inches (203 mm) of snow were present on the summits. Structures housing streamflow instrumentation near the mouths of the subwatersheds were

SSDI 0169-555X(94)00024-L

Fig. 1. Channel above gauging station in Watershed No. 3 at Big Creek. Note debris flow scar in right background (dark sinuous form on ridge) and another at the head of the drainage which contributed to flow in main channel. Note that the darker material deposited along the channel in the foreground shows the limits of the debris flow as it passed. It shows some superelevation on the curves in the channel.

destroyed or severely damaged by flooding which carried ''considerable debris''.

While the damage to the dams and instrumentation on the subwatersheds in Big Creek was attributed to flooding, it is clearly the result of debris flows. Munson (1937) noted that ''land slumps'' occurred in the upper parts of the subwatershed. Photographs document shallow slope movements which lead into the channels (Fig. 1). In the channel above the gauging station in watershed 3, the passage of the debris down the channel appears to have followed its own course rather than remaining strictly confined to the channel banks. Other photographs show that the channels scoured to bedrock (Fig. 2). The bulk of the debris was described as having ''originated in the channel bottoms and sides.'' It was noted that ''the small dams were pounded terrifically by large boulders, some weighing as much as 5000 pounds, that literally battered out the centers of dams 1, 2, and 2A.'' (Fig. 2). At subwatershed 6, the passage of the debris left a 4-foot high debris levee which redirected flow and reduced damage to the dam (Fig. 3). The presence of debris levees, the large size of transported boulders, the channel scouring, the variance of the path or track of the flow relative to the channel banks, and the shallow landslide sources in the upper watershed are all evidence of debris flows rather than flood waters.

In the succeeding 45 years, no study of debris flows in the southern Sierra Nevada was undertaken. The author began gathering observations of debris flows in 1982 and continued through 1993. These observations do not represent a systematic examination of this phenomenon. Rather, it is the accumulation of data from events which caused some damage to public works or property. It seems likely that the 1937 debris flows in Big Creek would have gone unreported except for the damage caused at a research facility. Similarly, the more recent debris flows described in this paper affected roads, hydroelectric facilities, or threatened the public. Debris flow hazard is increasing from greater population and utilization of the Sierra Nevada. This hazard vulnerability will increase in the future with more urbanization in the Sierran foothills and on private land holdings within the Sierra Nevada.

Debris flows were examined in the Kings River, San Joaquin, Merced, and Toulumne River drainages (Fig. 4). Twenty-six individual features were examined by the author. Table 1 notes six debris flows, the drainage basin where they occurred, and the year they occurred. These six debris flows were selected for intensive study because they: (1) were initiated on natural slopes, (2) had significant data collected about them, and (3) were generally free from the influence of road or similar ground-disturbing activities.

Three of these six debris flows are described in detail as representative of the triggering events, failure mechanisms, and downslope effects observed during the last ten years. The data are offered as an initial assessment of the geomorphology of debris flows in the southern Sierra Nevada. It is hoped that this information will contribute to the understanding of debris flows and provide a starting point for future detailed studies to improve hazard zonation for the Sierra Nevada.

Fig. 2. Two views of the weir and stilling basin on watershed No. 1 at Big Creek. (a) View taken October 19, 1937 shows the facility. (b) View taken December 14, 1937 shows the destruction of the facility and scouring of the channel to bedrock resulting from debris flow passage. Note the small building on the middle left of each view for comparison.

Fig. 3. A four-foot high debris levee left after passage of the debris flow provided some protection to the gauge houses in watershed No. 6 in Big Creek.

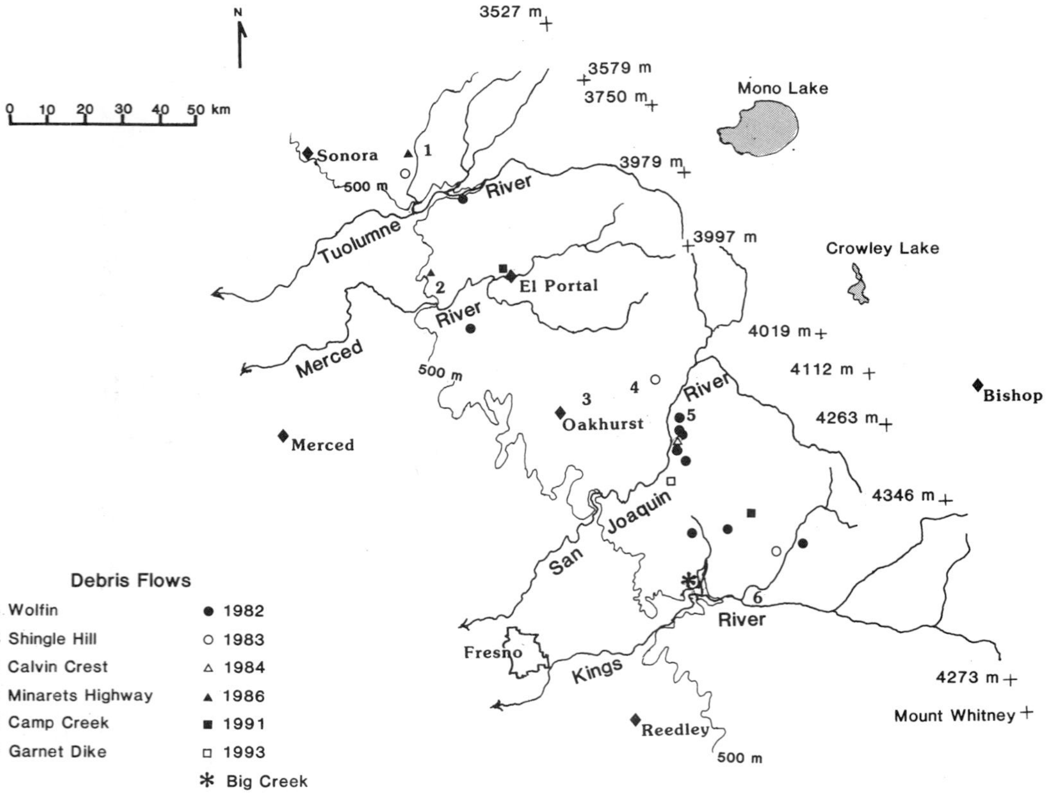

Fig. 4. A location map showing the Sierra Nevada and the main channels for the Toulumne, Merced, San Joaquin, and Kings Rivers. The locations of the six debris flows are shown by their number from Table 1. Geometric symbols show the location and year of occurrence of the remaining twenty debris flows.

Table 1
Location and year of occurrence for debris flows examined in detail

Drainage basin	Debris flow name	Year of occurrence
Toulumne River[a]	Wolfin	1986
Merced River	Shingle Hill	1986
San Joaquin River	Calvin Crest	1983
	Minarets Highway	1986
	Camp Creek	1982
Kings River	Garnet Dike	1982

[a]Drainage basins and debris flows generally reported in their order from north to south.

2. Camp Creek — a 1982 debris flow

A major storm system passed over the southern Sierra Nevada on April 10–11, 1982. It produced a rain-on-snow event responsible for triggering numerous landslides including a debris flow in Camp Creek, a tributary to the San Joaquin River (DeGraff et al., 1984). Rainfall, nearly continuous for the two-day period, included several intense periods. Additionally, the snowpack present in the Sierra Nevada was partially melted to add water to the storm runoff. Clearly, saturated conditions necessary to initiate debris flows existed.

The Camp Creek debris flow was examined within a few weeks after its occurrence. The source of this debris flow was a slightly convex, northwest-facing divide in the upper part of the watershed. Camp Creek watershed had undergone a major wildfire in 1939. Reforestation efforts in 1963 had established conifer cover on the lower part of the watershed. The debris flow originated at the upper edge of the reforested area. A few large Jeffrey Pine trees, scattered brush and herbaceous vegetation cover the slope above the initiation point.

The debris flow originated at an elevation of 1690 m (5545 ft). A scar was found which measured 15 m in length and 5 m in width. The headscarp was about 0.5 m in height. The lateral margins varied from 0.5 m near the headscarp to 0.2 m near the lower end.

Along the axis of movement, scar morphology indicated that a wedge-shaped block had detached on a planar surface. Evidence of the sliding movement was the abrupt, sharp appearance of the headscarp and margins (Fig. 5). Also, many small roots were sheared off on the surface of the failure plane which roughly coincided with the base of the root zone in the coarse-grained soil. Laboratory analysis of soil from the headscarp found that it was composed approximately of 75% fine gravel and sand and 25% silt and clay. The underlying bedrock is granodiorite and most fragments present at the site were derived from the local bedrock.

The Camp Creek slide was almost immediately mobilized into a debris flow as it moved down the steep (50%) slope. This is evident from the conditions of the scar track which included small debris levees adjacent to the track and overtopping of fluid onto the ground surface with erosion restricted to the central part of the track.

About 53 m below the scar, the debris flow entered an ephemeral channel which crossed the slope at an obtuse angle to the debris flow track. The bulk of the debris flow executed a sharp turn to enter and follow the channel which leads to the main channel of Camp Creek. A large mass of soil, woody debris, and rocks was deposited on the downslope side of the channel at this point. Between this point and Camp Creek, the channel gradient ranges from 30 to 50%. The debris flow scoured the channel into a U-shape down to bedrock along much of its extent (Fig. 6). Small, discontinuous levee deposits of matrix-supported rocks and woody debris and fine-grained material were found along the channel (Fig. 7).

At 146 m below the scar, this ephemeral channel joins the main channel of Camp Creek. The debris flow deposited prominent levees, which showed the superelevation achieved by the flow mass as it moved through the sharp turn to enter the main channel. The debris flow impacted the Stump Springs road at 166 m from the scar. The erosive force of the flow removed over 50% of the road fill as it passed. It continued for another 2275 m to enter Mammoth Pool reservoir on the San Joaquin River. Debris from the passage of Camp Creek debris flow was found along the margins of the structure diverting the channel into the reservoir. It consisted of sandy debris with "tree limbs and stumps" (J.S. Burns, CA Div. Dam Safety, written commun., 1983).

Garnet Dike debris flow in the Kings River drainage also occurred in 1982. The detailed information collected from this feature is included in later sections of this paper.

Fig. 5. View upslope of the scar created by initiation of the Camp Creek debris flow in April 1992. Note sharp margins where initial block displaced on planar surface. Shallow depth (0.3 m) of failure plane can be compared to person standing in background.

Fig. 6. View upgradient along the tributary channel to Camp Creek showing the U-shape and exposed bedrock resulting from scouring as the debris flow passed.

Fig. 7. View of debris levee along the tributary channel to Camp Creek. Note fine-grained material is predominate around larger material (matrix-support) and woody debris is present.

3. Calvin Crest — a 1983 debris flow

The winter of 1982–83 produced an unusually deep snowpack throughout much of the Sierra Nevada. The trans-Sierra highway over Tioga Pass in Yosemite National Park and many Sierra wilderness trails normally open by early summer remained blocked by snow. Snow survey measurements found the snowpack persisting later into the year leading to an unprecedented July snow survey in the Kings River drainage. Snow courses were found to have snow depths of 132 to 155 cm. Similar results were found in the San Joaquin River drainage. The widespread nature of this above-normal precipitation in California is evident by 1983 measurements representing 190% of normal statewide precipitation. The normal is based on the 50 year average between 1931 and 1981 (DWR, 1987).

In July 1983, a debris flow occurred on national forest land adjacent to the Calvin Crest Conference Center near Oakhurst, California. The debris flow originated at an elevation of 762 m (2500 ft). It was located on a 30%, southwest-facing slope which is the broad head of a small drainage basin. The slope has an open stand of mixed oak and Jeffrey pine with an understory of herbaceous vegetation. Where the 30% slope flattens to 10%, there is evidence of seasonal groundwater seepage at a number of points in the vicinity of the debris flow. The area where the debris flow occurred is underlain by granitic bedrock (tonalite of Bass Lake). A few thousand feet upslope is the contact with remnant meta-sedimentary bedrock capping the hill top. Near this contact, a number of seeps and wet meadows are present.

On July 4, 1993, a camper from Calvin Crest was near the location where the debris flow later occurred and observed water discharging from a depression several feet in diameter, near the base of the 30% slope. The debris flow is assumed to have occurred on July 5, 1993. This assumption is based on reports by campers at Calvin Crest, of feeling ground vibrations and of windows rattling in the cabins. Many campers were under the impression that an earthquake had occurred somewhere in the area. Noticeable shaking of the ground has been reported, although infrequently, by observers of debris flows (Costa, 1984).

The debris flow was discovered by hikers from Calvin Crest on July 6. It was reported to the local Forest Service office where reports of muddy water in the stream flowing from the area near Calvin Crest had also been reported. The author observed visible movement of soil blocks near the head of the debris flow on July 7. Much slower deformation resulted in additional

Fig. 8. View downslope from near the head of the Calvin Crest debris flow. The deposit can be seen in the background.

Fig. 9. Closer view of the deposit as seen on the second day after initial movement. This material had the consistency of wet cement and formed a lobate margin within the confining trees on the flat. Note the impact on the tree along the channel and material deposited against its upslope side.

movement of small blocks from the headscarp through July 14.

By July 7, the debris flow was about 200 m long and 22 m wide. The debris flow track was slightly narrower midway between the head of the flow and the deposit (Fig. 8). At the head, the scarp was 2.0 m high. Estimated depth along the flow path ranged from 1 to 2 m. The debris flow mass was deposited among small trees

Fig. 10. View across the track of the Calvin Crest debris flow showing the levees formed by overturned material along the narrow part of the track. Slickensides are present on the smooth surface within the track.

covering a flat area downslope. The deposit had the consistency of very wet cement (Fig. 9). Water discharged at a rate of 0.01 m^3/s from the end of the deposit. Groundwater flowed from the upper scarp and other points along the debris flow track. This made the bottom of the track too soft and muddy to be examined more closely. It remained saturated until the following winter. In succeeding years, grass grew over the debris flow scar and flow path. Despite revegetation, it remains fairly wet to damp throughout most of the year.

The rapid movement of the debris flow was evident from a number of facts. The force of the flow as it entered the trees and formed a deposit was sufficient to uproot several trees and tilt others. The passage of the flow through the narrower part of the track produced levees by pushing the adjacent slope up much the same way a plow turns the soil in a furrow (Fig. 10). Slickensides were visible on the surface of the overturned soil. Nearer the deposit, levees of matrix-supported woody debris and fine-grained material were present. Muddy splash marks were found on trees next to the flow track at heights 0.5 to 1.5 m above the inferred surface of the debris flow.

The movement seemed to be a product of groundwater conditions resulting from above-average recharge rather than a storm event. No precipitation was received during the previous twenty-four days at the South Entrance of Yosemite National Park, the nearest station to the debris flow (NOAA, 1983a, b). Precipitation totaled only 15.5 mm for the thirty-five days prior to the Calvin Crest debris flow. The premovement observation of groundwater flow from a depression at the base of the 30% slope suggests high pore water pressures were present in the slope materials. The sudden appearance of the debris flow the day following observation of the seepage and associated ground vibration, indicates that a part of the slope above this discharge moved catastrophically to create the debris flow track and deposit. The scarp created by this movement removed support of the slope above and permitted several additional retrogressive movements to form the debris flow as seen on July 7.

The depth of movement was examined later with seismic profiles at the head of the feature. A low velocity layer 1.2 m thick overlies a higher velocity layer which extends at least 9 m deeper. The different layers were not visually evident in the adjacent headscarp. A rather uniform sandy material with the appearance and texture of the underlying granitic rock was present. It was found to be sufficiently weathered so that one could readily dig into the material by hand. However, it became more difficult to excavate lower on the exposure. This change was close to the boundary between the first and second seismic layers. This indicates that the depth of movement was controlled by the depth of weathering in the granitic rock and its associated control on groundwater flow.

4. Shingle Hill — 1986 debris flows

A major frontal storm system crossed the central Sierra Nevada in mid-February 1986. The southern edge of this storm triggered three debris flows on the north-facing slopes of Shingle Hill near Greeley Hill, California, within the Merced River drainage. The storm precipitation fell in the form of rain. No snow accumulation was present on Shingle Hill which ranges from about 640 m (2099 ft) to over 945 m (3100 ft). A rural county road along the base of the hill was open the evening of February 17. On the morning of February 18, deposits were blocking the road at two locations. A third deposit was found in an ephemeral channel a few tens of feet upgradient from the road. The debris flows occurred the night of February 17 or early the morning of February 18, a time which coincided with a high intensity rainfall period.

Shingle Hill is underlain by metasedimentary rock which is highly fractured and consists mainly of phyllite and metaconglomerate (Calaveras Formation?). This bedrock is mantled by a regolith containing abundant coarse fragments and a high percentage of fines derived from local bedrock.

The debris flows are identified from north to south as A, B and C. Unlike the debris flows described earlier, only debris flow A originated on an undisturbed slope. Debris flows B and C originated on steep-slopes where all trees had been removed by cable harvest methods the previous year. Except for this loss of vegetation, the slope was otherwise undisturbed. The two largest, debris flows A and B, created the deposits on the road. Both debris flows were initiated in swales at the heads of first-order ephemeral drainages. The debris flows revealed bedrock hollows under these swales similar to those described by Dietrich et al. (1986). Debris flow B had two distinct source areas immediately above a fork in the channel. Movement took place at the boundary of the soil and underlying hard bedrock in the source areas for debris flow A and debris flow B (Fig. 11).

Debris flow C occurred on a planar slope adjacent to an ephemeral channel. Movement took place within the soil near the base of the root zone. Table 2 shows the slope at the source area, the estimated volume, and average depth to failure plane for the three debris flows.

Debris flows A and B appeared to have moved rapidly down channels which were scoured to bedrock at

Fig. 11. View upslope of the source area for debris flow A at Shingle Hill. Note the in-place bedrock exposed in the floor of the scar.

Table 2
Source area characterisitics for 1986 debris flows at Shingle Hill

Debris Flow	Slope steepness (%)	Avg. depth to failure plane (m)	Est. vol. (m^3)
Debris Flow A	55	1.1	126
Debris Flow B	60[a]	0.6	147
	50	0.9	68
Debris Flow C	65	0.5	16

[a]Debris Flow B consists of two distinct source areas in adjacent swales. The larger source area is the northwestern one.

some places along the debris flow tracks. The channels had a U-shaped cross-section with erosion being greatest in the central part of the track. Debris levees and isolated deposits of matrix-supported debris trapped on the upslope sides of trees were present along the margins. Where the channels curved, there was evidence of superelevation by the debris flow passage as seen by the levees and limits of muddy deposits (Fig. 12). Some large trees and brush in or close to the channel were uprooted by debris flows A and B. Their passage did not disturb the leaves and needles under muddy deposits at the edge of the debris flow track.

Debris flow C also appeared to have moved rapidly. However, it entered a relatively straight channel section with a gentle gradient. The evidence of rapid movement was a small runup deposit on the slope opposite from the side where it entered the channel and some levee deposits.

Deposits from debris flows A and B were removed to clear the road before they could be examined. The deposit for debris flow C consisted of angular rock fragments with tree limbs and other woody debris in a mass of fine-grained material. This deposit appeared to be 11 to 15 m^3 in size. This volume is close to the estimated volume of the source area for debris flow C.

Movement of debris flows A and B was the result of pore water pressures in the soil overlying less permeable bedrock. It is possible the fractures in the bedrock contributed some subsurface flow to the soil. This has been documented in other swales where debris flows have occurred (Wilson and Dietrich, 1987). No evidence of surface water discharge to the source areas was found in the field. Macropores in the soil and possible soil pipes were visible in the exposed soil at the heads of debris flows A and B. It may be inferred that movement was facilitated by groundwater movement along these conduits. Pierson (1983) documented a similar mechanism for debris flows in New Zealand.

Data were collected for two other debris flows which occurred in 1986; the Wolfin debris flow (Tuolumne

Fig. 12. View downslope along the track followed by debris flow A at Shingle Hill. The limits of the flow are evident from the channel cross-section and muddy area and show superelevation around the curve in the pre-existing channel.

River drainage) and Minarets Highway debris flow (San Joaquin River drainage). This information will be discussed later in this paper.

5. Debris flow geomorphology

5.2. Physical characteristics

Physical evidence along flow tracks provides clear indications of debris flow rather than water flood origin for those features observed in the southern Sierra Nevada during the past ten years. Costa and Jarrett (1981) noted the importance of distinguishing between debris flows and water floods for understanding geomorphic activity in mountainous areas.

Debris flows in the Tuolumne, Merced, San Joaquin and Kings River drainages tend to be very fluid. It is noted earlier that the deposit of the Calvin Crest debris flow had the consistency of wet concrete. This fluidity contributes to the distance travelled by debris flows. Where the channel in which the debris flow moves is generally steep, the debris flow may travel a long distance. The Camp Creek debris flow which travelled 2441 m from its source area to Mammoth Pool Reservoir represents the maximum distance observed. The sixty meters travelled by the Calvin Crest debris flow represents the minimum distance observed. At Calvin Crest and many other locations, the distance travelled was limited by low gradient reaches of the channel or significant widening of the channel.

Debris flow levees and terminal lobes were associated with the observed debris flows. The levees were often discontinuous along the channel margins. With the more fluid debris flows, it was limited to a several centimeter thickness of fine grained material bordering the eroded channel. The Calvin Crest debris flow splashed fine-grained debris onto trees standing next to its track. At other locations, this fine-grained material was intermixed with large woody debris, cobbles, and occasional boulders. The 1982 Garnet Dike debris flow occurred in the Kings River drainage several miles from Big Creek. Fig. 13 shows the bouldery levee created in the channel just upslope from its deposit.

The material in debris flow deposits is a mixture of sandy-silty material surrounding larger fragments of rock or wood. The larger material is supported in the matrix of this finer material. There is no sorting or stratification visible in these deposits.

Vegetation is usually eroded from the central parts of the debris flow track. The impact on large trees is evident at the margins of the debris flow and occasion-

Fig. 13. View of the bouldery levee created by passage of the Garnet Dike debris flow in 1982. The person standing to the left is within the debris flow track. The other individual is standing on undisturbed ground at the edge of the levee.

ally in the central areas. Fig. 14 shows a tree engulfed at the base with rocks and debris. This is consistent with the observation that debris flows damage vegetation in the direct flow path, but do little or no damage to vegetation, except burial, along the margins of the flow path (Costa, 1988). Along the low gradient section of the channel followed by the Minarets Highway debris flow, several centimeter thick mud deposits that had dried could be lifted to reveal conifer needles and leaves undisturbed by passage of the debris flow. As Costa and Jarrett (1981) note, undisturbed forest litter on the margins of a flow path is diagnostic of a debris flow rather than a water-flood event.

5.2. Velocities

Indirect measurements were made to determine the velocity of debris flows in the southern Sierra Nevada. The velocity for several debris flows was estimated from their superelevation in bends or tendency to reach higher elevations on the outside of the channel bend than on the inside (Costa, 1984). It required measurement of the channel slope, channel width, the radius of curvature of the channel bend, and elevation difference between the flow surface on the inside and outside of the bend. The velocity for other debris flows was estimated from runup elevation or velocity-head (Webb et al., 1988). This involved measuring the difference between runup elevation where a debris flow impinged on a tree oriented perpendicular to the flow direction and the unobstructed flow-surface elevation.

Table 3
Velocities of southern Sierra Nevada debris flows

Debris flow name	Velocity		Measurement method
	(m/s)	(km/h)	
Wolfin	6.9	25	superelevation
Shingle Hill A	6.1	22	superelevation
Shingle Hill B	4.7	17	velocity head
Calvin Crest	5.7	20	velocity head
Minarets Highway	2.5[a]	9	superelevation
	4.4	16	superelevation
Camp Creek	7.2	26	superelevation
Garnet Dike	5.0	18	velocity head

[a]Lesser velocity represents measurement nearest initiation point for the debris flow. The other measurement is the maximum velocity achieved downchannel prior to entering a lower gradient reach.

Table 3 lists the drainages, debris flow velocities, and measurement method. The average maximum velocity for these observed debris flows is 5.7 m/s or 20 km/h. The estimated velocities for southern Sierra Nevada debris flows are consistent with the velocities

Fig. 14. Debris at the base of a tree from passage of the Garnet Dike debris flow. The flow passed around the base of the small oak with most of the flow passing down channel on the right side. Note that despite the size of the material, little damage or abrasion of the tree trunk took place.

compiled from literature sources by Costa (1984).

The 1986 Minarets Highway debris flow in the San Joaquin River drainage left deposits along curves in the channel in several locations. The slope along this channel ranged from 47 to 21%. At 240 m below the source area, a velocity of 2.5 m/s was obtained. When the debris flow reached 360 m downslope from the source area, it was moving at 3.6 m/s. At the base of this slope about 490 m downslope, it reached its maximum velocity of 4.4 m/s as it entered a lower gradient reach of the channel.

5.3. Deposits

Landslide mapping conducted in the Merced, San Joaquin, and Kings River drainages indicates debris flows to be the most common type of landslide. However, it is typically the debris flow scar where the flow initiated and the channelized path which are recognized. The debris flow deposits are relatively rare features. This rarity may be partly attributable to the difficulty in mapping small, vegetation-covered deposits. The Calvin Crest debris flow represents a feature where the deposit occurred on the hillslope. Examination of this deposit in 1993 found a grassy flat with a sparse tree cover similar in appearance to other flats nearby. Its identification as a debris flow deposit is dependant on association with the more evident scar and track leading to it. Examination of aerial photos at a scale of 1:15,840 taken in 1987 would cause this deposit to be overlooked without first-hand knowledge of its location.

Another reason for rarity of debris flow deposits may be that few deposits remain to be found. The absence of deposits seems attributable to several causes. First, few stream terraces or alluvial fills exist where debris flow deposits may be preserved for long periods of time. A 1982 debris flow in the Kings River drainage a few miles from big Creek did leave a deposit where a small terrace was present (Fig. 15). This suggests that more deposits might be found if more extensive terrace and alluvial fill deposits existed.

Second, many debris flows travel down channels and enter larger streams. Their greater discharge entrains the debris flow material and transports it downstream as fluvial materials. The 1986 Minarets Highway debris flow was observed within a few days of occurrence. The track for this flow was easily followed to its junction with a major stream. At that time the stream was a raging torrent above bankful discharge. No trace of the

Fig. 15. The deposit from the Garnet Dike debris flow on a stream terrace bordering the South Fork of the Kings River. A person standing on the deposit provides scale. Note the steep front of the main deposit with the finer-material washed out along the margins. The channel followed by the debris flow is visible in the background.

debris flow material was evident. Examination of the junction later during low flow yielded no evidence of the debris flow deposit. The Camp Creek debris flow had a similar fate except it was deposited as fluvial material in the reservoir behind Mammoth Pool dam.

Third, smaller debris flow deposits left in channels are eroded and re-deposited as fluvial deposits further downstream. Most of the 1986 Shingle Hill C deposit remained in the ephemeral channel following its occurrence. Much of the deposit was still present in 1987 and 1988. By 1992, only the larger woody debris and a few large cobbles marked its location. The smaller particle-size material had been eroded and removed by various storm flows.

6. Debris flow origins

6.1. Triggering events

During the last ten years, debris flows in the southern Sierra Nevada were triggered by intense rainfall, rain-on-snow events, and snow melt. Intense rainfall events are typically frontal systems which affect large areas rather than being isolated storm cells. The 1937 storm event is a good example. Storage gauges at the Big Creek research facility recorded total rainfall from December 9 through December 12 of slightly over 12 inches (305 mm) (Gleason, 1938). Table 4 shows the elevation and precipitation recorded for this period at a number of stations. The values are comparable to those measured at Big Creek. The greatest precipitation was received on December 10.

Table 4
Precipitation record for the December 9–11, 1937 storm

Station	Elevation (m)	Total precip. (mm)	Drainage basin
Crane Valley[a]	1067	374	San Joaquin
Auberry	625	234	San Joaquin
Huntington Lake	2134	258	San Joaquin
Ash Mountain	488	276	Kings
Big Creek[b]	594	305	Kings
Balch Camp	396	261	Kings
Gen. Grant[c]	2065	383	Kings
Giant Forest[c]	1938	413	Kaweah
Springville	1234	372	Kaweah

[a]Stations are generally listed in their order from north to south; precipitation and elevation taken from Gleason (1938).
[b]Storage gage in the Big Creek research facility.
[c]Stations within present-day Sequoia–Kings Canyon National Park

Campbell (1975) demonstrated the importance of antecedent moisture and precipitation intensity in triggering debris flows. Review of storm data shows that most stations had received half or more of the total precipitation from the 1937 storm by the evening of December 10 (Gleason, 1938). This would indicate that antecedent moisture requirements were likely met for this area by that time. This conclusion is consistent with the field observation made by Forest Service technician W.J. King at the height of the storm that "literally sheets of water poured from grass-covered hillsides in Big Creek" (Gleason, 1938).

High intensity rainfall after the antecedent moisture conditions were satisfied is assumed to have happened during the 1937 storm event. Gleason (1938) noted a period between 2:00 a.m. and 2:30 a.m. on December 11 when rain fell at a rate of 4.12 inch/h (105 mm/h) at Big Creek. Later in the day, another intense period lasting an hour began at 3:00 p.m. with a rate of 1.7 inch/h (43 mm/h). While the exact time of the debris flows in Big Creek is unrecorded, it seems likely these intense rainfall periods could have served as triggers for their occurrence.

The February 1986 storm event is a more current example of widespread rainfall. This frontal storm was centered through the northern and central Sierra Nevada. The storm primarily affected the Tuolumne and Merced, the two northernmost drainages in the area under study.

The role of the 1986 storm in initiating the debris flows at Shingle Hill can be evaluated by the precipitation records for the Groveland Ranger District located a several kilometers away at about the same elevation (NOAA, 1986). The station shows that the area received no precipitation in the five days immediately before the storm began on February 12. In fact, only 46 mm of precipitation was received in the ten days prior to the storm.

Fig. 16 shows a cumulative precipitation curve for the period February 11–21, 1986. Nearly 50% of the total storm precipitation was received by the start of February 17. This seems to indicate satisfaction of antecedent moisture requirements at Shingle Hill. The debris flows occurred the night of February 17 or early morning of February 18. This coincides with the period

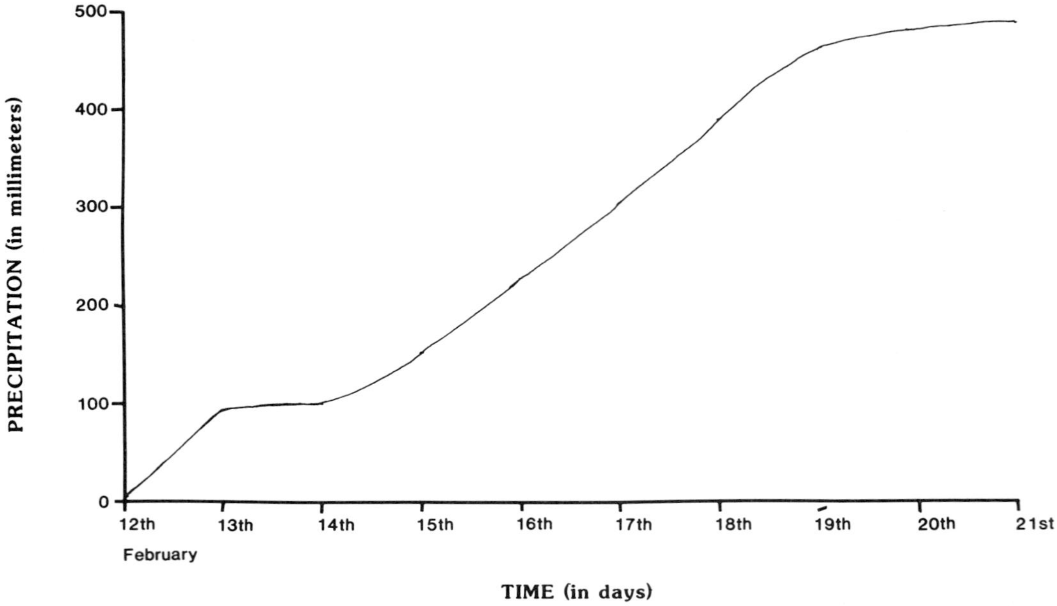

Fig. 16. Cumulative total daily precipitation for the February 11–21, 1986 storm event from data collected at the Groveland Ranger Station several miles from Shingle Hill.

of highest recorded rainfall during the storm event. Total precipitation was 76 mm for both February 16 and February 17. The greatest amount received was 86 mm on February 18. The instrumentation used does not permit identification of intense rainfall periods during the storm. The coincidence of the debris flows on the days with the greatest amount of rainfall for a 24-hour period would suggest that the highest rainfall intensities may have also occurred at this time.

Rain-on-snow events were recognized as significant factors in recent years. Studies in the Cascade Mountain of Oregon and Washington established the increased runoff and flooding potential associated with rainfall on the warm snowpacks of the Cordilleran (Harr, 1980). The relationship between the occurrence of specific landslides and rain-on-snow is often difficult to establish. Their presence is commonly removed from roads or other points where observers might establish the time of occurrence. The Camp Creek debris flow and others occurring along Stump Springs road in 1982 are the only debris flows clearly associated with a rain-on-snow event. Based on nearby stations closely approximating conditions near the Camp Creek debris flow, precipitation at the site ranged between 155 mm/h to 224 mm/h with maximum intensities of 14 to 18 mm/h.

Fig. 17 shows the cumulative precipitation recorded at the Teakettle Experimental Forest in the Kings River drainage and at the Crane Valley hydroelectric project in the San Joaquin River drainage from early April 10, 1982 to late April 11, 1982. The similarity between total precipitation and periods of intense rainfall for these stations and those closer to the Camp Creek debris flow location demonstrates the widespread nature of this storm event. Because temperatures remained above 0°C through most of the storm, the warm snowpack present in the Sierra Nevada underwent a loss between 0.4 and 0.5 m. These losses are computed to be equivalent to an additional 130 mm of runoff.

The 1983 debris flow at Calvin Crest and several others during that period seem to be attributable to above-normal precipitation rather than being the consequence of specific storm events like the April 10–11, 1982 rain-on-snow event or the February 12–21, 1986 storm event. Measurements at the Central Sierra Snow Laboratory (CSSL) operated by the USDA Forest Service illustrate the nature of the 1983 snowpack. While the 1983 water year was only the second highest in total precipitation recorded at CSSL, it set a record for maximum snowpack water equivalent (J.A. Bergman, CSSL, written commun., 1987). Water years 1982 and 1983 set a new record as the two wettest succeeding

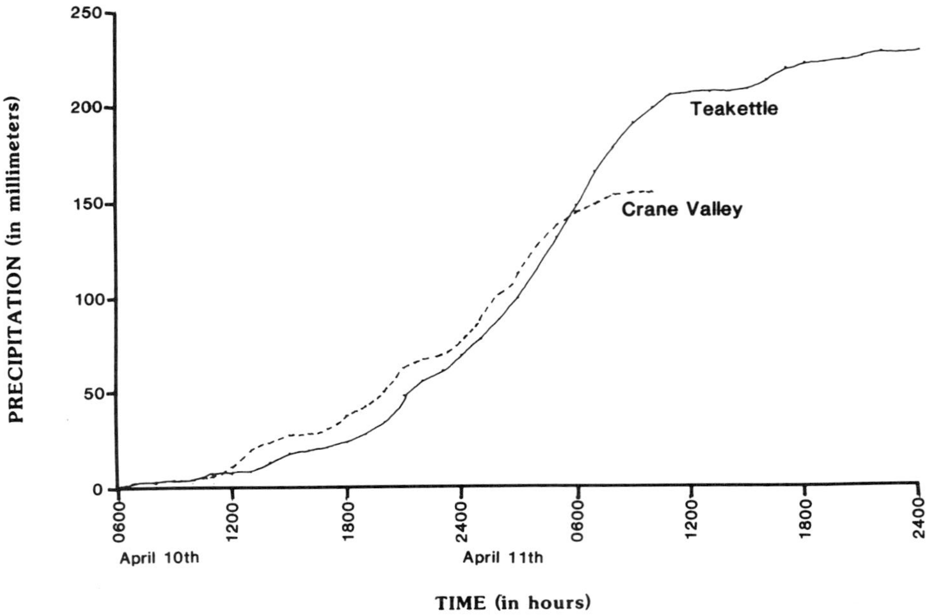

Fig. 17. Cumulative precipitation recorded for April 10–11, 1992 at the Teakettle Experimental Forest (about 40 km southeast of Camp Creek) and Crane Valley hydroelectic facility (about 20 km west of Camp Creek).

years at CSSL and other stations in the Sierra Nevada snow zone.

Jahns (1969) noted the relationship of landslide activity in California to long-term precipitation trends. He also pointed out the tendency for greater landslide activity during the early phase of a wet cycle. Fleming and Schuster (1985) attributed the widespread landslide activity in Utah which occurred in 1983 and 1984 to this same climatic circumstance. Beginning in 1980, precipitation was in excess of the long-term average in the southern Sierra Nevada. In 1983, three large landslides began moving in the San Joaquin drainage near Stump Springs road. Two of these features were existing, inactive landslides which were reactivated. The third was a large landslide initiated at the head of a first-order stream (DeGraff et al., 1987). Investigation found movement on all three were controlled by groundwater conditions. Re-activation of movement on two large, inactive landslides, development of a large debris slide, and initiation of Calvin Crest debris flow coincided with a record amount of precipitation in 1983 at the beginning of a wet cycle affecting this area. While not proving that debris flows in the Sierra Nevada are influenced by a relationship similar to the one expressed by Jahns (1969) for southern California, it suggests a possiblility which deserves further investigation.

6.2. Failure mechanisms

In an examination of hillslope elements important to debris flow initiation, Reneau and Dietrich (1987) noted that hillslopes can be divided into hollows, noses and side slopes. There was no clear correlation between source area for debris flow initiation and hillslope elements for debris flows in the southern Sierra Nevada. This lack of correlation existed even for debris flows occurring at the same location and triggered by the same event. At Shingle Hill, debris flows A and B both occurred in hollows near the heads of first order channels. Debris C occurred on a planar, side slope. An example of a debris flow occurring on a nose, an interfluve where the slope is planar to convex, is the Camp Creek debris flow.

Costa (1984) notes that transformation of a moving block of soil to a debris flow is thought by some investigator to result from contractive behavior leading to liquifaction and by others to occur though dilatancy. Ellen and Fleming (1987) described the basic differences between these behaviors and examined how transformation from slide to flow could be understood by a steady-state soil mechanics model. This model relates soil void ratio and effective confining stress to explain whether contractive or dilative behavior could be expected. No examination of soils involved with debris flow occurrence in the southern Sierra Nevada was undertaken as part of the observations reported in this paper. However, debris flows observed in the southern Sierra Nevada tend to travel long distances, leave little material along their path, and generally leave deposits of smooth consistency. These observed behaviors are consistent with contractive failure leading to transformation of the initial soil mass to a debris flow slurry (Ellen and Fleming, 1987).

Assuming contractive behavior for failure, movement would initially involve displacement of a rigid slab under high pore pressures and reduced shear strength followed by remolding of the soil particles into a debris flow. The debris flows observed in the southern Sierra Nevada provide evidence of both initial movement as rigid slabs and high pore pressures at the time of mobilization.

Camp Creek debris flow provided a clear indication that initial movement involved sliding of a rigid mass which almost immediately became a viscous slurry. The morphology of the source areas for debris flows A, B. and C at Shingle Hill is consistent with this type of mechanism. Sheared roots were present on the plane where the initial movement occurred. Sharp, irregular boundaries are found on the margins of the source areas in contrast to the smooth border along the flow track downgradient.

Initial movement of a rigid mass was evident for the Wolfin debris flow in the Tuolumne river drainage. This 1986 debris flow began as a 30 m long by 22 m wide slab on the 60% slope adjacent to an existing intermittent channel. The main scarp was 3 m high with marginal scarps averaging 1.5 m in height. Sheared roots were visible in the scarps outlining the source area. The mass entered perpendicular to the direction of the channel and immediately began moving down channel in a clear indication that remolding into a flowing mass took place. Debris flow levees and other evidence of debris flow passage were found only a few tens of meters farther down channel where velocity measurements were made based on superelevation through existing channel bends.

Evidence of the initial failure occurring under high pore pressures is supported by field observation. One observation supporting this idea of failure is the coincidence of debris flows with intense or prolonged rainfall, rain-on-snow events, or following melt of above-average snowpack conditions. This indirect evidence of high pore pressures is corroborated by direct observation of groundwater conditions. Significant, even copious, amounts of groundwater are often seen seeping from debris flow scars. Groundwater discharge is concentrated in the upper part of the debris flow within the headscarp, the area where the rigid slab originated, or just below it. Calvin Crest debris flow represents a slope movement where these seepage conditions were observed shortly after failure. Discharge was from the headscarp and along the upper half of the exposed scar. Such a discharge was observed at Camp Creek and at several debris flows along Stump Springs Road in 1982 and persisted for weeks (DeGraff et al., 1984). Seepage on adjacent undisturbed slopes is typically absent.

Initial movement of the rigid slab requires detachment along a plane. This plane represents a boundary where conditions permit detachment. The debris flows examined over the last ten years illustrate three different boundaries where initial movement can occur. The three boundaries are (1) at the base of the root zone, (2) at the contact of well-weathered and less-weathered soil, and (3) at the contact between soil and unweathered bedrock. These boundaries reflect differences in hydraulic conductivity which promote the development of higher pore pressures in the overlying material.

Initial movement on the Camp Creek debris flow and Shingle Hill debris flow C was very shallow and coincided with the base of the plant root zone. The root zone represents the main concentration of plant roots and is usually confined to the upper one meter or less of the soil. This concentration was visible in the margins of the source area. Some large roots extended below the plane where movement occurred. A significant discontinuity in hydraulic conductivity will exist between the soil in the root zone and the underlying denser, relatively root-free soil. This promotes development of a temporary perched zone of saturation in the root-zone soil mass. The tensile strength provided by the roots in this soil mass enables it to remain rigid until initial detachment and downslope movement occurs. This type of movement is termed a soil slip by Campbell (1975).

A similar discontinuity between soil layers with differing hydraulic conductivity occurs between weathered and less-weathered soil or saprolite. Experience in the southern Sierra Nevada suggests this is usually 1 to 3 m below the ground surface. The Calvin Crest debris flow involved this boundary condition along the plane where movement occurred. While little visible evidence was present to detect this boundary, differences in both seismic velocity and ease of excavation illustrates the presence of two materials with differing density. This density difference would result in differing hydraulic conductivity and favor development of a perched zone of saturation during periods of high infiltration or downslope seepage.

Obviously, a difference in hydraulic conductivity exists between soil and underlying bedrock. Shingle Hill debris flows A and B represent complete evacuation of the soil mass overlying hard, unweathered bedrock, a depth between 1 to 3 m below ground surface. In addition to the difference in hydraulic conductivity, the soil-filled bedrock hollows promoted convergence of groundwater flow within the overlying permeable soil (Dietrich et al., 1993).

6.3. Recurrence

The recurrence of debris flows at specific locations is significant for establishing the degree of hazard present. The historic record for the Sierra Nevada is too short and unreliable to provide an accurate assessment. Radiocarbon dating can establish recurrence over a sufficiently long period to be meaningful and represents site specific information.

One of the few published radiocarbon dates for debris flows in the Sierra Nevada is provided by a study done by Reneau et al. (1986). Samples from two hollows in the foothills of the Sierra Nevada within the Kern River drainage yielded ages of 17,100 ± 1000 yr BP and 11,285 ± 565 yr BP. Both sites were from basal colluvium estimated to be about 1 m thick in hollows within granitic bedrock.

The Shingle Hill debris flows yielded quite different radiocarbon dates compared to those of Reneau et al. (1986). Debris flow B, the feature with two source areas, fully evacuated the soil from two shallow swales in the underlying metasedimentary bedrock. The irregular, fractured bedrock was observed to have charcoal fragments lodged in some of the exposed fractures. One sample (GR-516-86-1) was obtained about 30 cm above the soil–bedrock contact. It was a piece of charred wood in a soil mass remaining after movement. The soil was held in place by sheared roots which remained anchored in the underlying bedrock. The soil was in the part of the swale common to the two source areas. The ^{13}C corrected age for this charcoal was 425 ± 70 yr BP. The second sample consisted of small bits of charcoal and charred wood collected from the irregularities exposed in the larger swale evacuated by debris flow B. The ^{13}C corrected age for this charcoal was 505 ± 210 years BP. The greater uncertainty for this date seems attributable to variability resulting from the many pieces making up the sample. From the closeness of the dates obtained from the two samples, it seems reasonable to conclude that the last time the swales were completely evacuated was between 355 and 715 yr BP.

At Shingle Hill, ephemeral channels have (1) a generally straight alignment, (2) older trees (greater than 70 to 100 years in age) absent from in or immediately adjacent to the channel, and (3) a U-shaped cross-section. This includes both the channels where the 1986 debris flows occurred and four unaffected channels adjacent to them. Broad, fan-shaped deposits of coarse, unstratified material are present at the mouths of several channels. This morphology is interpreted as evidence for more frequent debris flow occurrence involving partial evacuation of the contributing swales which serve as their sources. If this is correct, the recurrence time of 400 to 500 years obtained from the radiocarbon dating represents a maximum recurrence interval for debris flow occurrence.

The 11,265 and 17,100 yr dates obtained by Reneau et al. (1986) and the 425 and 505 yr dates obtained at Shingle Hill suggest that debris flow recurrence may vary by as much as two orders of magnitude. The Shingle Hill recurrence interval is comparable to those reported by Costa (1984). However, there is clearly a need for more radiocarbon dating to establish whether different recurrence intervals truly exist in the southern Sierra Nevada.

7. Impact on human activities

Until now, the impact of debris flow on human activities in the southern Sierra Nevada has been limited.

Debris flows have contributed additional sediment to streams feeding into hydroelectric reservoirs such as Mammoth Pool. However, there is no evidence of significant impacts to long-term storage capacity or sediment damage to generating equipment. A significant effect is the cost of repairing damaged roads. Repair to Stump Springs road for the damage from the 1982 Camp Creek debris flow cost US$65,000. National Forests in the southern Sierra Nevada suffered nearly US$1,553,000 worth of damage from landslides in 1982. Another US$557,000 was incurred in 1983. Debris flows represent the principal landslide type causing this impact.

Loss of life and property damage due to debris flows is minor. The principal example is the March 1991 storm event at El Portal, California. This small community was originally established in the Merced River canyon to serve the needs of workers on the local railroad and those working at timber harvest and mining. More recently, it provides residences for employees of Yosemite National Park and its concessionaires. In August 1990, a major wildfire burned parts of Yosemite National Park and the adjacent Stanislaus National Forest. The burned area included the small, steep watersheds which empty into El Portal. A routine assessment of possible landslide hazard resulting from loss of vegetation was carried out (DeGraff and Lewis, 1989) Previous landslide mapping indicated a large number of past debris flow scars present on the slopes of the burned watersheds. Several closed debris basins were constructed on two drainages where the severity of vegetation loss from the fire, indications of past debris flow activity, and presence of houses at the mouths of these ephemeral channels represented a high risk for debris flow damage.

The 1991 water year was the fifth year of drought conditions in the Sierra Nevada. A major storm occurred on February 27 to March 4, 1991 which triggered debris flows from the drainages above El Portal. No precipitation was received in the El Portal area for 21 days prior to this storm. Table 5 shows the daily precipitation total for two stations near El Portal. The sewage station is about one mile downstream and Yosemite Valley is several miles upstream. After several days of significant rainfall, the greatest amount of daily precipitation was received on March 3. Interviews with residents of El Portal disclosed that a period of intense rainfall occurred shortly after 11:00 p.m. on

Table 5
Daily precipitation from the February 27 to March 4, 1991 storm at two stations near El Portal, California

Date	Precipitation (mm)	
	El Portal Sewage Plant	Yosemite Valley
Feb. 27	23[a]	16[b]
Feb. 28	38	41
Mar. 1	33	51
Mar. 2	37	15
Mar. 3	45	79
Mar. 4	30	62

[a]L. Johnson, Yosemite NP, pers. commun., March 1991.[b]NOAA,

Sunday, March 3. The time was firmly established by a number of residents who watch the late television news. At that time water and debris was seen spilling from the closed basin at Chapel Lane. This basin had been observed to contain little debris earlier that afternoon and water was flowing through a small culvert where a road crosses the channel upstream from the basin. Following the intense rainfall, muddy water was seen flowing down Chapel Lane and the closed basin was filled. A total of 100 cubic yards of debris was trapped in the basin. Excavation of the basin revealed matrix-supported cobbles and rock fragments along with woody debris. Only a 15-cm thick layer capping the deposit was stratified. The culvert on the road upstream from the basin was plugged with cobbles, gravel, and finer sediment. The debris flow path along the channel and across the road was visible from the cobbles, woody debris, and finer-materials outlining its margins.

Smaller debris flows occurred on two other drainages leading into El Portal at the same time the basin at Chapel Lane trapped its debris flow. At these drainages the site conditions had not permitted construction of debris basins. At one location, water, mud, and occasional cobbles flowed against a house and passed between the house and an outbuilding. Fifteen minutes earlier, only water was seen flowing through this location. The debris flow was described by the residents as swift enough to "carry away a small child". A few tens of meters to the east, residents at another house felt vibrations that seemed greater than the impact of the intense rainfall. Turning on their outside lights, they saw water and debris issuing from a channel and flowing across their backyard. A deposit about 0.3 m thick was formed against the back of their house.

The debris flows at El Portal illustrate the potential for damage to property and threat to life which exists in the southern Sierra Nevada. The Chapel Lane debris basin built as part of burned area emergency rehabilitation and the small volume of the debris flows from the other two drainages kept costs to a minimum.

8. Conclusion

The low population density in the Sierra Nevada has limited the effects of debris flow occurrence. However, the foothill communities of the Sierra Nevada continue to grow at a rapid pace. An increased number of roads, pipelines, and electrical transmission lines are likely to be constructed crossing the Sierra Nevada. This increase in people and property represents greater vulnerability to losses from debris flows in the future. Avoiding these losses can only result from land use and design practices which limit the risk from debris flows.

Risk reduction requires an understanding of both what is vulnerable to loss and the degree of hazard present. This paper represents an initial examination of what is known about debris flow hazard in the southern Sierra Nevada. Unless more detailed evaluation of debris flow hazard is undertaken, the risk from debris flow losses will increase without any means for effective reduction.

References

Campbell, R.H., 1975. Soil slips, debris flows, and rainstorms in the Santa Monica Mountains and vicinity, California. U.S. Geol. Surv. Prof. Pap., 851: 51 pp.

Costa, J.E., 1984. Physical geomorpology of debris flows. In: J.E. Costa and P.J. Fleisher (Editors), Developments and Applications of Geomorphology. Springer, Berlin, pp. 268–317.

Costa, J.E., 1988. Rheologic, geomorphic, and sedimentologic differentiation of water floods, hyperconcentrated flows, and debris flows. In: V.R. Baker, R.C. Kochel and P.C. Patton (Editors), Flood Geomorphology. Wiley, New York, pp. 113–121.

Costa, J.E. and Jarrett, R.D., 1981. Debris flows in small mountain stream channels of Colorado and their hydrologic implications. Bull. Assoc. Eng. Geol., 18: 309–322.

DeGraff, J.V. and Lewis, D.S., 1989. Using past landslide activity to guide post-wildfire mitigation. In: R.J. Watters (Editor), Engineering Geology and Geotechnical Engineering. Balkema, Rotterdam, pp. 65–75.

DeGraff, J.V., Dellavalle, P. and Gemperle, R., 1987. Movement and stability analysis of the Camp Creek landslide, Sierra National Forest, east-central California. In: J.V. DeGraff (Editor), Landslide Activity in the Sierra Nevada During 1982 and 1983. Earth Resources Monogr. 12. USDA Forest Service, Pacific Southwest Region, San Francisco, pp. 63–74.

DeGraff, J.V., McKean, J., Watanabe, P.E. and McCaffery, W.F., 1984. Landslide activity and groundwater conditions: insights from a road in the central Sierra Nevada, California. Transp. Res. Record, 965: 32–37.

Dietrich, W.E., Wilson, C.J. and Reneau, S.L., 1986. Hollows, Colluvium, and Landslides in Soil-mantled Landscapes. In: A.D. Abrahams (Editor), Hillslope Processes. Allen and Unwin, New York, pp. 361–388.

Dietrich, W.E., Wilson, C.J., Montgomery, D.R. and McKean, J., 1993. Analysis of erosion thresholds, channel networks, and landscape morphology using a digital terrain model. J. Geol., 101: 259–278.

DWR, 1987. California Water Supply Outlook. Department of Water Resources, Division of Flood Management, Sacramento, 5 pp.

Ellen, S.D. and Fleming, R.W., 1987. Mobilization of debris flows from soil slips, San Francisco Bay Region, California. In: J.E. Costa and G.F. Wieczorek (Editors), Debris Flows/Avalanches: Processes, Recognition, and Mitigation. Reviews in Engineering Geology, VII. Geol. Soc. America, Boulder, pp. 31–40.

Fleming, R.W. and Schuster, R.L., 1985. Implications of the current wet cycle to landslide in Utah. In: D.S. Bowles (Editor), Delineation of Landslide, Flash Flood, and Debris Flow Hazards in Utah. Utah Water Research Laboratory, Utah State Univ.

Gleason, C.H., 1938. Field examination, lower Kings watershed after storm of december 9–11, 1937. California Forest and Range Experiment Station, USDA Forest Service, 20 pp.

Harr, D., 1980. Some characteristics and consequences of snowmelt during rainfall in western Oregon. J. Hydrol., 53: 277–304.

Jahns, R.H., 1969. California's ground-moving weather. Engineering and Science Magazine, California Institute of Technology, Pasadena, 8 pp.

Munson, S.M., 1937. Report on flood damage to Big Creek watershed installations, Kings River Branch, Storm fo December 9–11, 1937. California Forest and Range Experiment Station, USDA Forest Service, 16 pp.

NOAA, 1983a. Climatological Data, California, June 1983. National Oceanic and Atmospheric Administration, 87: 16–19.

NOAA, 1983b. Climatological Data, California, July 1983. National Oceanic and Atmospheric Administration, 87: 16–19.

NOAA, 1986. Climatological Data, California, February 1986. National Oceanic and Atmospheric Administration, 90: 16–17.

NOAA, 1991a. Climatological Data, California, February 1991. National Oceanic and Atmospheric Administration 95: 13.

NOAA, 1991b. Climatological Data, California, March 1991. National Oceanic and Atmospheric Administration 95: 13.

Pierson, T.C., 1983. soil pipes and slope stability. Q. J. Eng. Geol., 16: 1–11.

Reneau, S.L. and Dietrich, W.E., 1987. The importance of hollows in debris flow studies; examples from Marin County, California. In: J.E. Costa and G.F. Wieczorek (Editors), Debris Flows/Avalanches: Processes, Recognition, and Mitigation. Reviews in

Engineering Geology, VII. Geol. Soc. of America, Boulder, pp. 165–180.

Reneau, S.L., Dietrich, W.E., Dorn, R.I., Berger, C.R. and Rubin, M., 1986. Geomorphic and paleoclimatic implications of latest Pleistocene radiocarbon dates from colluvium-mantled hollows, California. Geology, 14: 655–658.

Webb, R.H., Pringle, P.T., Reneau, S.L. and Rink, G.R., 1988. Monument Creek debris flow, 1984: Implications for formation of rapids on the Colorado River in Grand Canyon National Park. Geology, 16: 50–54.

Wilson, C.J. and Dietrich, W.E., 1987. The contribution of bedrock groundwater flow to storm runoff and high pore pressure development in hollows. In: R.L. Beschta, T. Blinn, G.E. Grant, G.G. Ice and F.J. Swanson (Editors), Erosion and Sedimentation in Pacific Rim. IAHS–AISH Publ. 165, pp. 49–59.

ELSEVIER

Geomorphology 10 (1994) 253–264

GEOMORPHOLOGY

Lichen dating of coseismic landslide hazards in alpine mountains

William B. Bull, John King, Fanchen Kong, Thomas Moutoux, William M. Phillips

Geosciences Department, University of Arizona, Tucson, AZ, 85721, USA

(Received March 8, 1994; revised March 29, 1994; accepted April 1, 1994)

Abstract

Studies of rockfalls and block slides provide insight about seismic shaking hazards in alpine mountains subject to earthquakes. Large samples of the longest axis of the largest lichen on each block can be used to identify regional landslide events; lichen-size distributions for many sites cluster consistently at the same sizes. The coseismic rockfall lichenometry model can be used to (1) date and locate prehistorical earthquakes, (2) document regional frequency of earthquakes, and (3) describe regional patterns of seismic shaking. Determination of colonization time, great-growth phase, and especially uniform phase rates of lichen growth are essential for dating regional landslide events.

Rocks that tumble downhill during historical earthquakes allow accurate calibration of the growth rate for *Rhizocarpon* subgenus *Rhizocarpon*; these plentiful fresh substrates are dated to their day of formation. An initial calibration of uniform growth rate, based on historical and tree-ring dated landslides was fine tuned by assigning earthquake dates to mean values of lichen-size peaks for regional rockfall events that increase in abundance towards epicenters of historical earthquakes. Calibration for each new site is unnecessary in the Southern Alps of New Zealand and in the Sierra Nevada of California because several species of yellow rhizocarpons within each climatic region have constant growth rates that are independent of altitude or substrate lithology.

1. Introduction

Unstable slopes in rugged alpine mountains are sources of detritus for a variety of mass-movement processes; many are generated by earthquakes (Keefer, 1984). Debris flows and snow avalanches commonly result from unusual storms, as well as from occasional earthquakes. Rockfalls may be most common during earthquakes. Clouds of rockfall-generated dust were plainly visible in the low hills near Los Angeles during magnitude 4.5 aftershocks of the 1994 Northridge earthquake. Lofty, tectonically active mountain ranges such as the Southern Alps of New Zealand and the Sierra Nevada of California register the effects of local and distant seismic shaking with landslides. A coseismic (earthquake generated) event is distinctive; rockfalls occur on seismically sensitive landforms throughout a region and rockfall abundance increases towards the earthquake epicenter.

This is a study of rocks that tumble down hills from crumbly outcrops and debris slopes, sometimes at distances of 500 km from earthquake epicenters. Gradual accumulation of blocks on talus, footslopes, and terraces creates deposits whose lichen sizes record times when fresh rock surfaces were exposed for colonization by lichens. Incremental accumulation of rockfall blocks contrasts with other landslide processes, such as a cataclysmic rock avalanche where massive collapse of a mountain creates an enormous block field in an instant of geologic time. Most historical rockfall accumulations in the Southern Alps have a strong coseismic signature; pulses of rockfall activity are recorded by

SSDI 0169-555X(94)00026-N

clustering of lichen sizes. Repositories of rockfall blocks in the northern South Island have received surprisingly few blocks since the most recent large earthquake in 1968. In contrast, some block fields in the central Sierra Nevada continue to be modified by snow avalanches, which may add fresh blocks or disturb the deposit each decade.

We have two goals in presenting this progress report. One is to demonstrate that key assumptions for a new method in lichenometry based on 33,000 lichen-size measurements in the Southern Alps also apply to the much different climatic and lithologic setting of the Sierra Nevada. This article compares growth rates for *Rhizocarpon* subgenus *Rhizocarpon* (yellow rhizocarpons) in both mountain ranges. The focus is primarily on post-1850 coseismic rockfalls whose fresh substrates are valuable for lichen-growth calibration purposes and for demonstrating the effects of seismic shaking caused by large historical earthquakes on the plate-boundary fault systems of New Zealand and California. We also document geomorphic evidence for earthquakes previously reported in the Sierra Nevada (Holden, 1898; Rinehart, 1982, Ellsworth, 1990), report new evidence for landslides caused by San Andreas fault earthquakes, and examine frequency of prehistorical regional landslide events.

Processes of erosion and deposition are dated in many parts of the world by measuring sizes of *Rhizocarpon* subgenus *Rhizocarpon* that are 20 to more than 3000 yr old (Beschel, 1961; Benedict, 1967; Curry, 1969; Denton and Karlen, 1973; Miller and Andrews, 1972; Birkeland, 1981; Locke et al., 1979; Innes, 1983, 1984, 1985a, b). Traditional lichenometry methods used by the above authors commonly represent the age of a deposit with the five largest lichens on the entire deposit. We use the largest lichen on each rockfall block, so our sample sizes range from 20 to more than 1000. Several species of yellow rhizocarpons are common in the Sierra Nevada, especially on the western slopes of the range (Tucker and Jordan, 1979; Hale and Cole, 1988). Although our work in progress uses several lichen genera — *Acarospora chlorophana*, *Lecanora sierrae*, *Lecidea atrobrunnea* and *Parmelia glabra* — we concentrate on the yellow rhizocarpons here because they are the lichens discussed by the above authors. They grow slowly for long times. Coseismic landslides in Tajikistan have been dated with lichenometry by Smirnova and Nikonov (1990). Porter and Orombelli (1981) used lichenometry to evaluate landslide hazards in the Italian Alps.

Uses of lichenometry would be greatly enhanced, and its limitations better understood, by more accurate calibration of growth rates. The coseismic rockfall model used here emphasizes use of historical substrates dated to the day and prehistorical landslide substrates that have potential of being tree-ring dated to the season of a year (Alestalo, 1971; Shroder, 1978; Hupp, 1983).

Two mountain ranges with different climates and rock types provide diverse data for our coseismic rockfall studies. Granitic rocks in the strongly seasonal subhumid to humid and strongly seasonal frigid Sierra Nevada of California provide 1 to 10 m joint blocks with minimal matrix of small fragments. Fractured greywacke in the weakly seasonal extremely humid and moderately seasonal frigid Southern Alps of New Zealand provides abundant rock fragments smaller than 1 m as well as larger blocks. Classification of climate is after Bull (1991, table 2.1). The combination of steep slopes underlain by brittle or jointed rock types that are subject to moderately frequent large earthquakes lend themselves well to our type of analysis. We start with a summary of key conclusions from work in the Southern Alps.

2. Southern Alps

The coseismic rockfall lichenometry model has been developed mainly in New Zealand since 1989 with the discovery that block fields accumulating on hillside benches and stream terraces record pulses of rockfalls during historical earthquakes. Field measurement and data-analysis methods are discussed in detail by Bull and Brandon (1994). Their paper concentrates on probability density estimates of lichen sizes measured with a digital caliper, factors affecting lichen growth, and calibration and validation of the coseismic rockfall model. Replicate pairs of lichen-size measurements made 1 to 104 days apart differed by 0 to 3 mm, but 53% of second measurements were within 0.00 to 0.25 mm of the first.

Potential for differences of lichen growth rates caused by variations in altitude and substrate lithology for different species of *Rhizocarpon* subgenus *Rhizocarpon* was tested by combining polymodal frequency distributions of lichen sizes. The sites occur in a 20,000

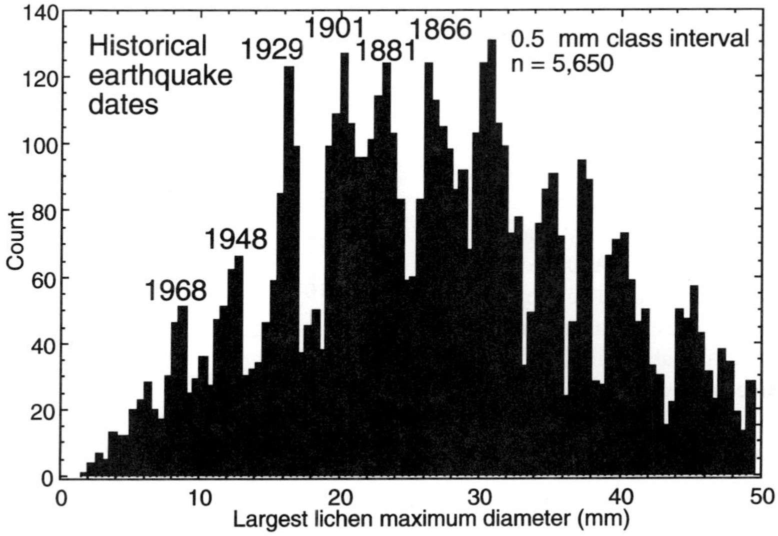

Fig. 1. Frequency distribution of combined data sets for 20 sites with substrate lithologies of quartzitic greywacke sandstone, syenite, and argillaceous greywacke sandstone. Altitude ranges from 380 to 1620 m, and mean annual precipitation from less than 500 mm to more than 5000 mm in 20,000 km^2 of the northern part of the South Island of New Zealand. $n=5650$. This data set was used to calibrate uniform phase lichen growth rate of *Rhizocarpon* subgenus *Rhizocarpon*.

km^2 region with a 1240 m altitude range; some are near the coast and others are near the crest of the Southern Alps. The combined data set has 5650 measurements. Variable lichen growth rates between sites would cause merging of frequency distribution peaks. Lichen-size peaks that record historical rockfall events are the same at the 20 widely separated sites in the northern South Island. Clear separation of high, but narrow peaks (1 mm = 6 yr) that rise far above the background level (Fig. 1) show that effects of altitude are minimal despite the 1240 m altitude range. In contrast to more advanced plants, many species of yellow rhizocarpons at different altitudes, and on different substrates, appear to grow at the same rate above and below tree line.

One does not need a new calibration at each site because lichen-growth rates are virtually constant within a large climatic region. Similar colonization times and growth rates for different types of yellow rhizocarpons greatly simplify calibration procedures.

A composite probability density curve was computed as sums of gaussians for individual measurements. The curve for the 5 to 32 mm size range of the regional data set of Fig. 1 (3866 measurements) was decomposed into subpopulations that describe individual rockfall events (See Figs. 2, 10 for examples). Mean lichen sizes were regressed against dates of historical earthquakes to obtain the regional calibration equation:

$$y = -0.16x + 328 \quad (1)$$

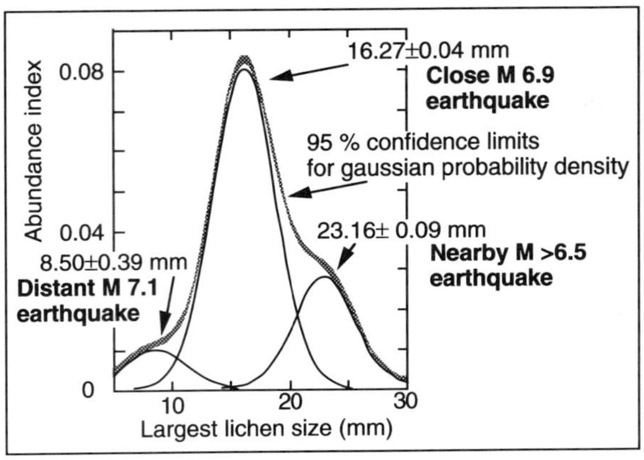

Fig. 2. Lichen sizes on rockfall blocks at Rough Creek, Arthur's Pass National Park, New Zealand. Probability density is portrayed as a composite curve with 95% confidence lines that sums the gaussians represented by individual measurements. Decomposition of the curve reveals three subpopulations. Each resulted from a historical earthquake (Table 1). Relative abundance of rockfalls is largest for the subpopulation with a 16 mm mode, which was caused by the nearby Arthur's Pass earthquake of March 20, 1929.

where y is the modal size of largest lichen maximum diameters in mm, and x is the calendric year; $r^2 = 0.99$. Eq. (1) describes uniform growth for the northern South Island of New Zealand after passage through the colonization and great-growth (rapid, exponentially declining) phases, so the number of calendric years for a lichen of a given size is the sum of the three phases. Uniform growth is 16 mm per century and lichens whose substrates were exposed in 1 A.D. would have a modal size of 328 mm. A standard reference year, such as 1950 for radiocarbon, is not necessary. Large sample sizes result in remarkably similar equations describing uniform phase lichen-growth rates from frequency distribution histograms, continuous curves generated by overlapping window widths, and composite probability density curves computed as sums of gaussians.

The Rough Creek data set illustrates the procedure for estimating mean lichen sizes of component coseismic rockfall events, and a seismic shaking index based on relative rockfall abundance. The composite probability density curve (Fig. 2) has tight 95% confidence limits; two shoulders suggest a polymodal density distribution. Decomposition of the curve reveals three subpopulations. The dominant peak has a mean lichen size of 16.27 ± 0.04 mm (95% confidence level), which is the same as the regional assigned lichen size for the 1929 earthquake (Table 1). Two smaller peaks record coseismic rockfall events that appear to be associated with the 1881 Hurunui and 1968 Inangahua earthquakes.

Areas under the three gaussian curves are useful indices of relative seismic shaking. Relative seismic shaking indices, such as those described for the Rough Creek site (Table 1) may be used to describe regional seismic shaking patterns associated with historical, or prehistorical earthquakes. Relative rockfall abundance decreases systematically with increasing distance to the earthquake epicenters presumed to be associated with the three regional rockfall events (Fig. 3). Decomposition of the 20 site data set (Fig. 1) yielded mean lichen sizes for the three regional rockfall events that are the same as those at the Rough Creek site (Table 1), which was not included in the regional data set. Each new lichenometry site tests the coseismic rockfall model assignment of dates for historical earthquakes from lichen-size peaks that record regional rockfall events. These New Zealand derived aspects of the model are directly applicable to the Sierra Nevada of California.

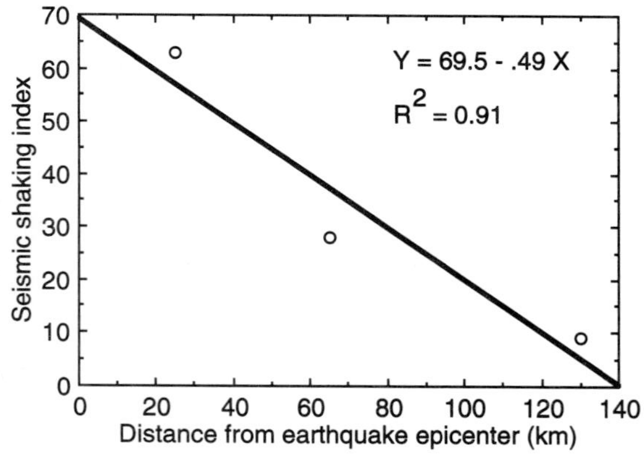

Fig. 3. Decrease of relative seismic shaking index with distance to earthquake epicenters, Rough Creek lichenometry site, New Zealand.

Table 1
Lichen sizes associated with historical coseismic rockfalls at the Rough Creek site, New Zealand

Associated historical earthquake	Approximate distance to earthquake epicenter (km)	Mean lichen size for Rough Creek (with 2σ error in mm), and mean size for regional data set [•]	Area under gaussian subpopulation curve, in % of total area
1881 Hurunui	65	23.16 ± 0.09 [23.17]	28
1929 Arthur's Pass	25	16.27 ± 0.04 [16.29]	63
1968 Inangahua	130	8.50 ± 0.39 [8.87]	9

3. Sierra Nevada

3.1. Site selection

Potential lichenometry sites in the Sierra Nevada need to be inspected carefully in order to assess their value for resolving specific questions. Landforms below cliffy ridges in the central part of the range typically consist of debris cones derived from avalanche chutes and planar talus slopes derived from triangular facets between the chutes. The debris cones generally should be avoided, unless one wishes to emphasize dating of both nonseismic and seismic snow avalanches and debris flows as well as rockfall events. Measurements from the talus slopes generally contain a stronger paleoseismic signal, with varying degrees of snow avalanche noise. Loose, bouldery glacial moraines lacking interstitial fine materials can be ideal for recording seismic shaking. They are created as highly unstable masses of loose detritus shoved to the front or sides of advancing glaciers or by melting of ice. Moraines generally lack the height needed to be a source of avalanches, and deposits consisting of well sorted boulders are not likely sources of debris flows or water-laid deposits during exceptional rainstorms. Many of our sites are along the crest of the range on small moraines of the past 1000 years. Unfortunately, moraines proved to be less reliable than in New Zealand. Good quality yellow rhizocarpons are not common at windy, harsh-climate sites above 3000 m, and areas of lichen snowkill occur where snowfields persist for several years.

Some data sets are not contaminated by debris flows or snow avalanches. Locally, 5 to 20 m diameter rockfall blocks roll out past the toe of a talus sheet. Slabs from the overhanging portions of these blocks become detached along joint planes at later dates. New substrates for lichens to colonize are created on these break-away slabs, and on the exposed joint faces from which they were detached. Seismic shaking and frost wedging are about the only nonrandom processes to affect this class of seismically sensitive lichenometry site. Initial tests indicate that data from break-away slabs may be one of several useful tools for identifying which peaks in the regional lichen-size distribution record seismic events.

The drier (semi-arid to subhumid) Kern River Canyon in the southern Sierra Nevada (Fig. 4) has several characteristics conducive for lichenometry

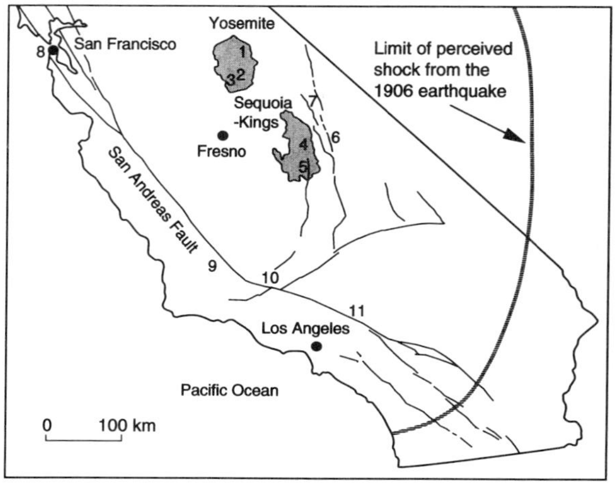

Fig. 4. Map of central and southern California showing San Andreas fault system, Yosemite and Sequoia–Kings Canyon National Parks in the Sierra Nevada, locations of lichenometry study sites, and approximate locations of epicenters of large historical earthquakes. Numbers are for: 1, "The Slide"; 2, Liberty Cap; 3, Middle Brother; 4, Knapp's Rock; 5, Kern River Canyon; 6, 1872 earthquake, 7, 1980 earthquake; 8, 1906 earthquake; 9, 1857 earthquake, 10, 1952 earthquake, and 11, 1812 earthquake.

studies of high quality. Microclimates in a deep glacial valley favor growth of several genera of lichens, including *Rhizocarpon* subgenus *Rhizocarpon*. Joints are common as a result of 6 to 13 km of right-lateral displacement of plutonic and roof pendant rocks by a strike-slip fault that now is inactive (Moore and Du Bray, 1978); they provide excellent planar surfaces that promote growth of circular lichens. Initial studies reveal fewer snow avalanches that complicate analyses of polymodal plots for sites farther north in the range. Evidence for snow avalanches, such as damage to plants, was virtually absent following the record snowpack winter of 1992–93. Extermination of lichens by consecutive years of burial under snow appears minimal. Lichen-size measurements that record earthquakes are more likely because the Canyon is closer to the San Andreas fault, and the epicenter of the 1872 Owens Valley earthquake is only 40 km away.

3.2. Lichen sizes of regional rockfall events

Mean or modal lichen sizes can be used to date mass movements of single or composite deposits. Single-event deposits have unimodal, normal frequency distributions of largest lichen maximum diameters (Fig. 5). The wide range of largest lichen sizes on adjacent

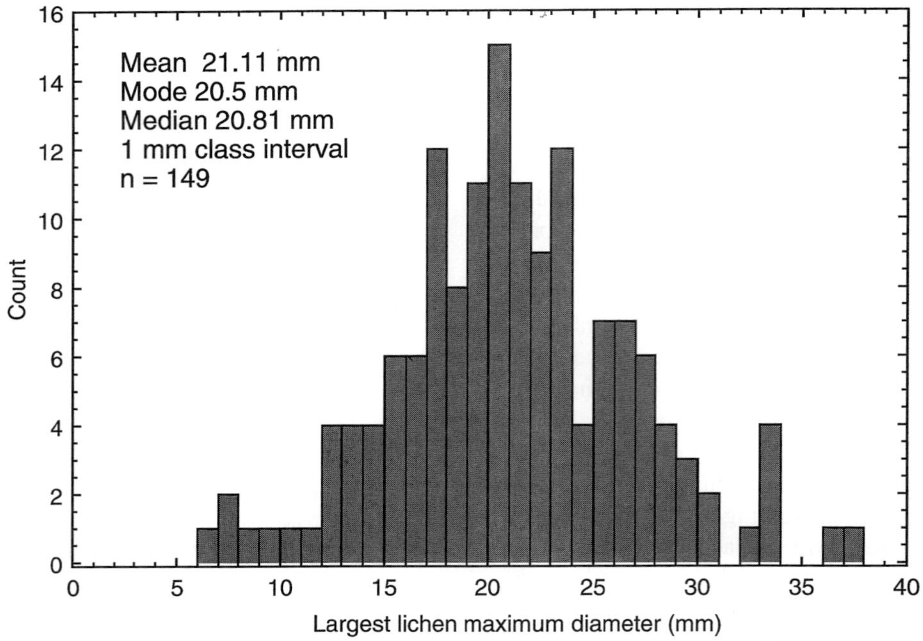

Fig. 5. Unimodal, normal probability density distribution of largest lichen maximum diameters on the blocks of a prehistorical rock slide from Liberty Cap, Yosemite National Park, gives the impression of being a single-event deposit. The presence of the 20 mm modal lichen size at many sites in the central and southern Sierra Nevada is indicative of a regional mass movement event (see Figs. 6, 7, 9, and 10).

blocks does not allow an observer to recognize a single-event deposit until the probability density distribution is plotted. Multiple-event deposits have highly polymodal frequency distributions that are strongly skewed towards the larger sizes (Fig. 7).

The Middle Brother site illustrates the difficulty of obtaining replicate lichen-size measurements for a site. Bunnel (1911) was with the first non-native American party to enter Yosemite Valley and describes where the northside trail had been buried by recently fallen rocks near Middle Brother. The excellent catalog of Yosemite National Park landslides by Wieczorek et al. (1992) describes the "pre-1851" rockfall as very large.

We tried to date the "pre-1851" landslide event by using lichenometry. Lichen-size measurements were made in two adjacent areas of a block field that had the appearance of being a single-event deposit derived from the highly jointed unstable cliffs towering above. The data appear to have little in common (Fig. 6), because this is a multiple-event deposit where younger rockfalls have splashed onto and partly concealed older landslide detritus. The 1872 coseismic rockfall event is absent from quadrat A but is present as a 11.5 mm mode in the quadrat B plot. The combined data set was modeled as a composite probability density curve. Three of the subpopulation ages, at the 95% confidence level, are 1902 A.D. $\pm$7.9 yr, 1860 A.D. $\pm$4.4 yr, and 1813 A.D. $\pm$2.3 yr. These landslide events may have been generated by distant San Andreas fault earthquakes of 1906 (250 km), 1857 (330 km), and 1812 (420 km). Keefer's (1984) work shows that these distances are reasonable. The 1813 (1812?) event may have created the deposit that appeared fresh to the 1851 explorers, but in 1851 the composite landslide deposit may have had a substantial component remaining from an earlier landslide that now is recorded by the 20 mm lichen size peak. This earlier event is regional in extent and has a lichenometry age of about 1783 A.D.

Middle Brother has a well deserved reputation for being unstable. The composite landslide deposit at its base records numerous coseismic rockfalls, as well as nonseismic events. The next magnitude Mw 8 earthquake on the San Andreas fault, even if 400 km away, may trigger a fresh collapse. A New Zealand magnitude

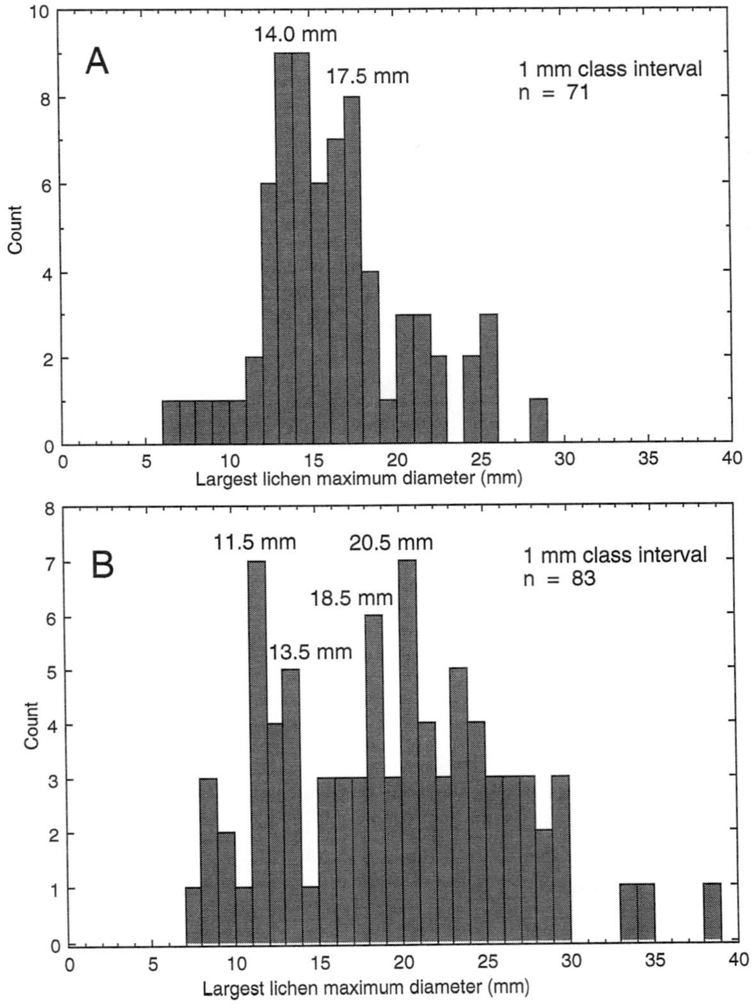

Fig. 6. Lichen-size distributions on blocks below Middle Brother, a crumbly cliff rising above Yosemite Valley. Lichen sizes were measured by two observers in adjacent areas that gave the false appearance of being a single-event deposit. (A) Measurements of observer 1 plot as a frequency distribution with a prominent double peak. (B) Measurements of observer 2 plot as a highly polymodal frequency distribution.

Mw 8+ earthquake in 1855 caused abundant rockfalls at distances of 500 km in the Southern Alps.

Combined data sets from several nearby sites are even more polymodal. The overall rise and fall of the probability density distribution of Fig. 7 portrays a lichen life-expectancy curve that reflects growth rates and disturbances to the lichen–bryophyte community growing on the blocks at the base of a talus slope. Older rockfall blocks in a deposit have cumulatively more opportunities to be broken or overturned by incoming blocks. It also becomes less likely that the largest lichen on old blocks can be measured as single thallus that does not merge with its neighbors. Most of the lichen-size peaks in the Kern River data set occur at Yosemite (Figs. 5 and 6) and Kings River Canyon (Fig. 10) sites. Relative ages of many regional rockfall events are readily apparent even before calibration of lichen growth rates.

3.3. Calibration of yellow rhizocarpon growth rates

Accurate calibration of growth rates for *Rhizocarpon* subgenus *Rhizocarpon* in the Sierra Nevada is possible because of (1) precise measurements of lichen size

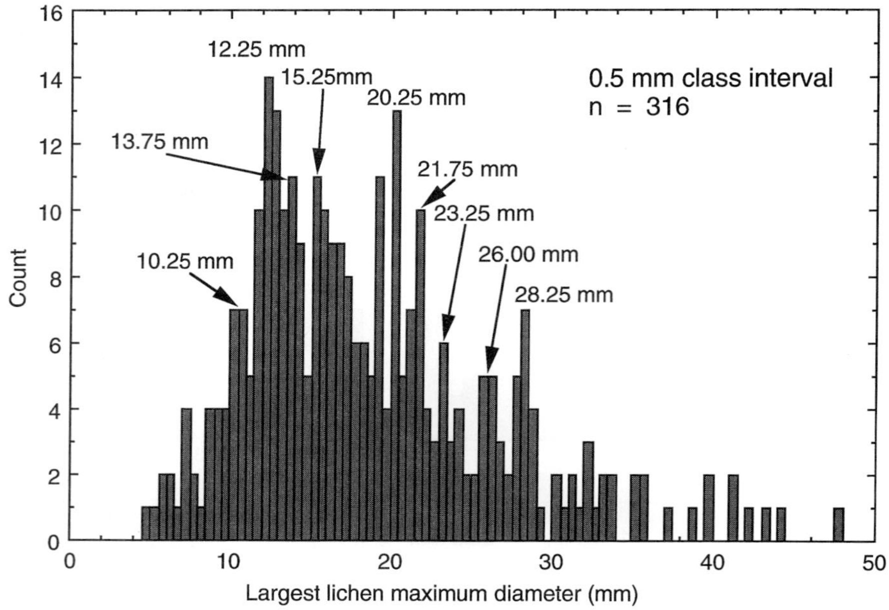

Fig. 7. Polymodal frequency distribution of largest lichen maximum diameters on blocks on several talus slopes along 5 km of the west side of the Kern River Canyon, Sequoia–Kings Canyon National Park. Nine regional rockfall event modes of the past 300 years are labeled.

with digital calipers — a 2σ measurement error of about 3.0 ± 0.5 mm, (2) tight clustering of lichen sizes (high precision) at sites whose rockfalls record historical earthquakes (Figs. 7, 9), (3) historical coseismic rockfall events dated to the day, and (4) tree-ring dated landslide events dated to the year or season. The following calibration is provisional and will be modified mainly by future tree-ring dating of landslides, such as the regional rockfall event identified by the 20 mm lichen-size peak.

Two historical sites and two dendrolichenometry sites constrain the duration of colonization time plus great-growth phase and estimate the uniform phase growth rate for *Rhizocarpon* subgenus *Rhizocarpon* (Fig. 8A). The sum of colonization time and great-growth phase is only 50 years, or less. Lichen-size measurements at 10 to 50 year old sites indicate that colonization time is 15 ± 5 yr and uniform growth begins after yellow rhizocarpons have attained a size of 6 ± 2 mm. Uniform growth is about 9.4 mm per century and lichens whose substrates were exposed in 1 A.D. would have a modal size of about 188 mm.

The initial calibration was then used to estimate sizes of yellow rhizocarpons for the times of five large historical earthquakes. This easily allows assignment of earthquake dates to frequency–distribution peaks that record regional rockfall events. A second regional calibration is created (Fig. 8B). This application of the coseismic rockfall model allows utilization of representative data from diverse sites at different altitudes. The regression control points are mean lichen sizes for subpopulations of decomposed composite probability density curves for nine sites in the central and southern Sierra Nevada. Alternatively, a combined nine-site data set can be analyzed. Uniform growth is about 10.4 mm per century and lichens whose substrates were exposed in 1 A.D. would have a modal size of about 207 mm.

The two calibration approaches are combined for our present best estimate of the uniform phase growth rate (Fig. 8C):

$$y = -0.096x + 192 \tag{2}$$

Yellow rhizocarpons in the Sierra Nevada grow about 9.6 ± 0.8 mm per century and lichens whose substrates were exposed in 1 A.D. would have a modal size of about 192 ± 15 mm.

Our estimates of growth rates for Sierra Nevada *Rhizocarpon* subgenus *Rhizocarpon* are twice as fast as those of previous workers; they did not have the advantages of digital calipers, tree-ring dated landslides, and

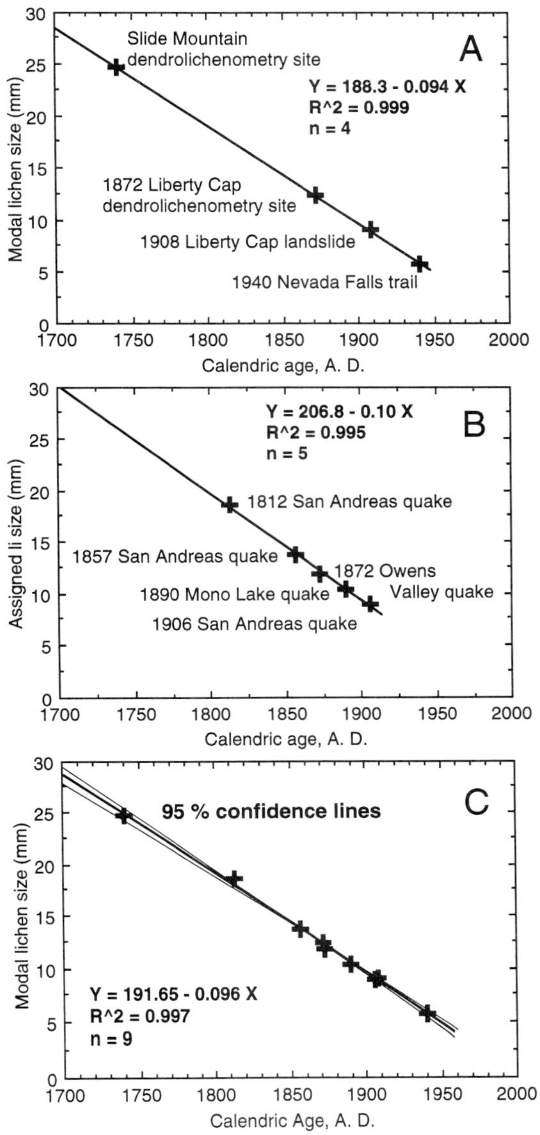

Fig. 8. Calibration of the uniform growth phase for *Rhizocarpon* subgenus *Rhizocarpon* in the central and southern Sierra Nevada. (A) Regression data from a historical 1940 trail exposure and 1908 block slide, and tree-ring dated 1872 block slide and 1740 rock avalanche (''The Slide'' in Yosemite National Park, Phillips et al., 1994). Site altitudes range from 1600 to 2800 m. (B) Regression using times of historical earthquakes that have been assigned to mean lichen sizes for regional rockfall events at nine sites ranging in altitude from 1200 to 3500 m. (C) Combined regression methods of calibrations (A) and (B) define provisional uniform phase growth rate for use in dating unknown lichen-size populations on Sierra Nevada substrates.

the large sample sizes used in the coseismic rockfall model. Curry (1969) estimated that uniform growth was between 3.8 and 4.5 mm per century and that lichenometry could be used to date surfaces as old as 3000 yr. Our present largest modal size for the Sierra Nevada of 76 mm suggests an age of only 800 years. It appears that lichenometry rarely will be useful for dating Sierra Nevada events older than 1000 yr, but it has excellent potential as a surface-exposure dating tool for the past 500 yr.

Regionally uniform lichen growth rates are not unique to New Zealand. Constant growth rates, and presumably similar colonization time and great-growth phases, permit use of lichen-size data for calibration purposes from sites at different altitudes (Fig. 8). Decomposition of the composite probability density curves provided estimates of means of lichen-size subpopulations for nine rockfall sites in the Sierra Nevada. A frequency distribution of these mean values reveals that many regional rockfall events are prominent at more than half of the sites even though a narrow 0.5 mm class-interval is used. Each bar that describes a major regional rockfall event in Fig. 9 is clearly separated from its neighbors and represents a time span of only 5 years. The coseismic rockfall model appears to have a dating accuracy of ± 5 yr, or better. Yellow rhizocarpons on rockfall blocks in the Sierra Nevada of California clearly have regionally constant uniform growth in the 520 to 3500 m altitude range of our study sites, but growth is slower than in New Zealand. As in New Zealand, yellow rhizocarpons appear to grow in a narrow range of conditions, and adjust to different altitudes and climatic settings by growing in slightly different microenvironments on the rockfall blocks.

3.4. Regional landslide events

Tight clustering of lichen-size peaks at many sites throughout the Sierra Nevada (Fig. 9) requires a regional cause. Climatic as well as seismic events may add fresh blocks to talus, but seismic causes are most likely to disturb glacial moraine and break-away slab lichenometry sites. Our limited sample size of 125 suggests that regional landslide events are frequent, but variable. Few significant events appear to have occurred between about 1790 and 1840 but the centuries before and after this relatively quiet period averaged an event every 20 years. Rockfalls caused by three

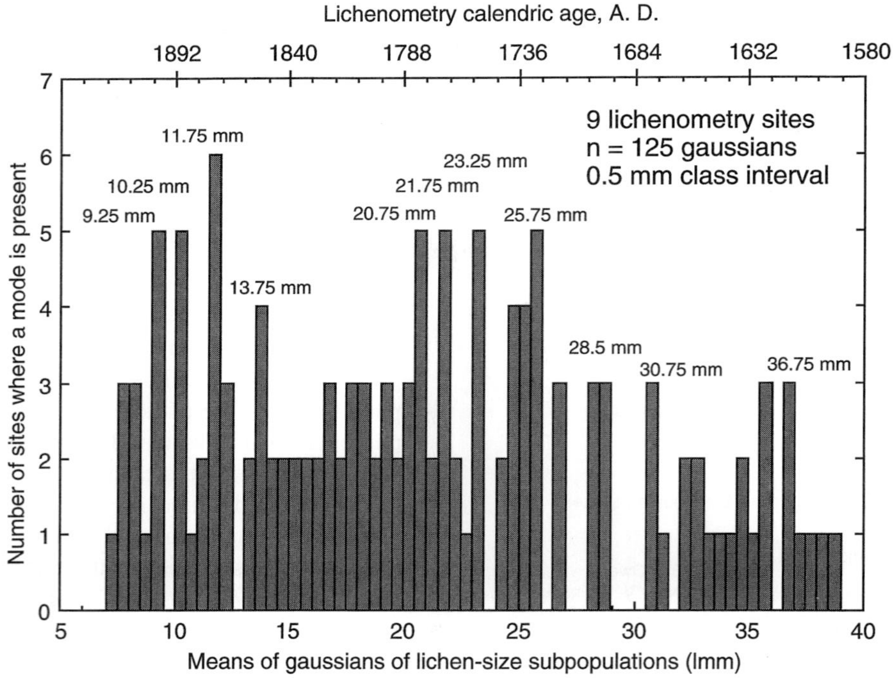

Fig. 9. Abundance of modal sizes at nine lichenometry sites where lichen-size data was modeled as composite probability density curves that were decomposed into their constituent subpopulations. Seven of the subpopulations occur at 5 or 6 of the possible 9 sites. Regional rockfall events caused by historical earthquakes include 9.25 mm (1906 A.D.), 10.25 mm (1890 A.D.), 11.75 mm (1872 A.D.), and 13.75 mm (1857 A.D.). A class interval of 0.5 mm represents about 5 years.

magnitude Mw 6 1980 earthquakes have yet to appear in the lichenometry record. The 1952 magnitude Mw 7.4 Kern earthquake may be represented by the 6 mm peak of Fig. 7.

The effects of seismic shaking decrease away from the epicenter of the magnitude Mw 7.6 1872 Owens Valley earthquake. The subpopulations of lichen sizes at the Knapp's Rock site (Fig. 10) have peak lichen sizes of 11.99, 16.55, 20.28, and 23.26 mm. Fig. 9 indicates that these record regional rockfall events. The 6.9 mm peak appears to be a random, local rockfall event. The prehistorical earthquake presumed to be associated with the 23 mm peak may have been distant or weak. The large seismic shaking index for the 12 mm peak reflects the strong 1872 earthquake, whose epicenter was only 50 km away. Seismic shaking indices decrease away from the epicenter of the 1872 earthquake (Fig. 11), but is still recorded by rockfalls at distances of 200 km and may be locally apparent at distances of more than 250 km. Not all outcrops and deposits collapse during an earthquake, even near the

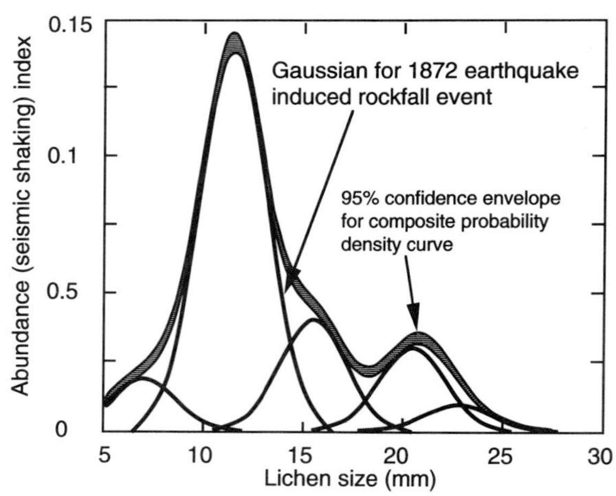

Fig. 10. Decomposition of the composite probability density curve for lichen-size measurements at the Knapp's Rock rockfall site, Sequoia–Kings Canyon National Park records several regional rockfall events.

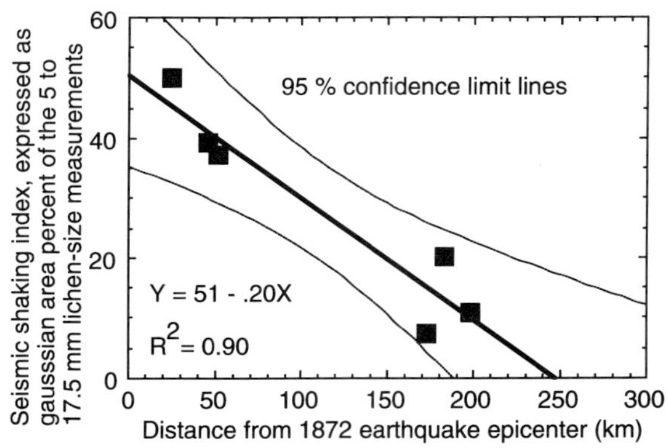

Fig. 11. Decrease in seismic shaking index away from the 1872 earthquake (Lubetkin and Clark, 1988) suggests that this large earthquake may have caused rockfalls that could be recognized with lichen-size measurements at distances of roughly 250 km from the Owens valley epicenter.

epicenter. So, seismic shaking index values of zero for three sites were omitted from the Fig. 11 regression.

Measurements of lichen sizes for other genera appear to verify the results of the yellow rhizocarpons analyses presented here. Each genera has a different colonization time, great-growth phase, and uniform-phase growth rate. *Acarospora chlorophana* grows almost as slowly as *Rhizocarpon* subgenus *Rhizocarpon*. Its presence in the rain shadow along the eastern side of the Sierra Nevada, and even along the Lone Pine fault zone (1872 surface rupture) provides a source of data for future analyses in areas that are too dry for yellow rhizocarpons.

4. Conclusions

Lichen dating of landslides in alpine mountains provides key information about geomorphic processes and frequency of hazards associated with mass movements. Lichen growth rates, based on calibration sites that are dated to the day or year, can provide substrate exposure ages for seismic or nonseismic processes of the past 500 years more accurately than radiocarbon age determinations.

Single event landslide deposits are not common in either the fractured greywacke of the Southern Alps or the plutonic rocks of the Sierra Nevada; even small block slides may be substantially modified by subsequent rockfalls. Clustering of lichen sizes in polymodal probability density distributions illustrates the composite nature of alpine landslide accumulations, including rockfalls, block slides, and rock avalanches. Distal portions of exceptionally large landslides are more likely to be pristine. Continuing instability of these and other alpine mountain ranges is partly a response to deglaciation (O'Connor and Costa, 1993).

Consistent clustering of lichen sizes in a combined data set from many sites suggests a regional cause for rockfalls and landslides such as seismic shaking. Eye witness accounts such as that by John Muir (1912) clearly describe the widespread landslide havoc wrought by the Owens Valley earthquake of 1872. Future magnitude 7 earthquakes on the range-bounding faults along the east side of the range can be expected to cause extensive landslide hazards. Sierra Nevada residents and visitors should also be aware of potential landslide hazard from more distant San Andreas fault earthquakes. Lichenometry of rockfall deposits describes the impact of earthquakes even at sites that are 200 to 400 km from earthquake epicenters. The Sierra Nevada, including heavily visited Yosemite and Sequoia–Kings Canyon National Parks, should experience many rockfalls and landslides from the anticipated next magnitude 8 earthquake emanating from the San Andreas fault.

The contrast in lichen growth rates between the Southern Alps and Sierra Nevada study areas suggests influence by a regional factor — most likely climate. Wetter regions seem to favor faster growth of yellow rhizocarpons (Porter, 1981). Several key topics need elucidation. What is the extent of the two regions described in this article that are characterized by constant growth, and are the boundaries between constant-growth regions abrupt or fuzzy? Do the fluctuations in climate during the past millennium have minimal influence on growth of *Rhizocarpon* subgenus *Rhizocarpon*, or do they add uncertainty to lichenometry age estimates? What is the influence, if any, of the ongoing episode of global warming?

The coseismic rockfall model seems sufficiently promising to warrant application in other areas where histories of earthquakes and landslides of the past millennium are sought. Appropriate study areas include the Transverse Ranges of southern California, which straddle the San Andreas fault; Cascadia and Alaska that have potential for great subduction zone earthquakes; and the active plate boundaries along the west side of South America.

Acknowledgements

We thank New Zealand farmers, national park officials in California, and many university and governmental scientists who generously provided advice, equipment, and permissions. The work was funded by the Geophysics Program of the National Science Foundation and the National Geographic Society. Jerome DeGraff provided helpful review comments.

References

Alestalo, J., 1971. Dendrochronolgical interpretation of geomorphic processes. Fennia, 105: 1–140.

Birkeland, P.W., 1981. Soil data and the shape of the lichen growth-rate curve for the Mount Cook area. N. Z. J. Geol. Geophys., 23: 443–445.

Benedict, J.B., 1967. Recent glacial history of an alpine area in the Colorado Front Range, U.S.A. 1. Establishing a lichen growth curve. J. Glaciol., 6: 817–832.

Beschel, R.E., 1961. Dating rock surfaces by lichen growth and its application to glaciology and physiography (lichenometry). In: G.O. Raasch (Editor), Geology of the Arctic, 2. Univ. of Toronto Press, Toronto, pp. 1044–1062.

Bull, W.B., 1991. Geomorphic Responses to Climatic Change. Oxford Univ. Press, New York, 326 pp.

Bull, W.B. and Brandon, M.T., 1994. Lichenometry of regional rock-fall events, Southern Alps, New Zealand. Geol. Soc. Am., submitted.

Bunnel, L.H., 1911. Discovery of Yosemite and the Indian War of 1851 which led to that event. 4th ed. G.W. Gerlicher, Los Angeles, 315 pp. (reprinted 1990, Yosemite National Park, Yosemite Association, 315 pp.)

Curry, R.R., 1969. Holocene climatic and glacial history of the central Sierra Nevada, California. In: S.A. Schumm and W.C. Bradley (Editors), Geol. Soc. Am. Spec. Pap., 123: 1–47.

Denton, G.H. and Karlen, W., 1973. Lichenometry; its application to Holocene moraine studies in southern Alaska and Swedish Lappland. Arct. Alp. Res., 5: 347–372.

Ellsworth, W.L., 1990. Earthquake history, 1769–1989. In: R.E. Wallace (Editor), The San Andreas Fault System, California. U.S. Geol. Surv. Prof. Pap., 1515: 153–187.

Hale, M.E. and Cole, M., 1988. Lichens of California. Univ. California Press, Berkeley, 254 pp.

Holden, E.S., 1898. A catalog of earthquakes on the Pacific coast, 1769–1897. Smithsonian Inst., Washington, DC, 253 pp.

Hupp, C.R., 1983. Geo-botanical evidence of late Quaternary mass wasting in block field areas of Virginia. Earth Surf. Process. Landforms, 8: 439–450.

Innes, J.L., 1983. The development of lichenometric dating curves for Highland Scotland. Trans. R. Soc. Edinb. Earth Sci., 74: 23–32.

Innes, J.L., 1984. The optimal sample size in lichenometric studies. Arct. Alp. Res., 16: 233–244.

Innes, J.L., 1985a. Lichenometry. Prog. Phys. Geogr., 9: 187–254.

Innes, J.L., 1985b. A standard Rhizocarpon nomenclature for lichenometry. Boreas, 14: 83–85.

Keefer, D.K., 1984. Landslides caused by earthquakes. Geol. Soc. Am. Bull., 95: 406–421.

Locke, W.W., III, Andrews, J.T. and Webber, P.J., 1979. A manual for lichenometry. Br. Geomorphol. Res. Group Tech. Bull., 26: 47 pp.

Lubetkin, L.K.C. and Clark, M.M., 1988. Late Quaternary activity along the Lone Pine fault, eastern California. Geol. Soc. Am. Bull., 100: 755–766.

Miller, G.H. and Andrews, J.T., 1972. Quaternary history of northern Cumberland peninsula, Baffin Island, North West Territories, Canada. Part VI: Preliminary lichen growth curve. Geol. Soc. Am. Bull., 83: 1133–1138.

Moore, J.G. and Du Bray, E., 1978. Mapped offset on the right-lateral Kern Canyon fault, southern Sierra Nevada, California. Geology, 6: 205–208.

Muir, J., 1912. The Yosemite. Century Company, 284 pp. (reprinted 1962, Garden City, New York, Anchor Books, Doubleday and Company, 225 pp.).

O'Connor, J.E. and Costa, J.E., 1993. Geologic and hydrologic hazards in glacierized basins in North America resulting from 19th and 20th century global warming. Nat. Hazards, 8: 121–140.

Phillips, W.M., Bull, W.B. and Montoux, T., 1994. Tree-ring dating of "The Slide", Yosemite National Park, USA. Arct. Alp. Res., submitted.

Porter, S.C., 1981. Lichenometric studies in the Cascade Range of Washington; establishment of *Rhizocarpon geographicum* growth curves at Mount Rainier. Arct. Alp. Res., 13: 11–23.

Porter, S.C. and Orombelli, G., 1981. Alpine rockfall hazards. Am. Sci., 69: 67–75.

Rinehart, C.D., 1982. Earthquakes — Sierra Nevada Mountains, California and Nevada. Genny Smith Books, Los Altos, 62 pp.

Shroder, J.F., 1978. Dendrogeomorphological analysis of mass movements on Table Cliffs Plateau, Utah. Quat. Res., 9.: 168–185.

Smirnova, T.Y. and Nikonov, A.A., 1990. A revised lichenometric method and its application dating great past earthquakes. Arct. Alp. Res., 22: 375–388

Tucker, S.C. and Jordan, W.P., 1979. A catalog of California lichens. The Wasmann J. Biol., 36: 1–105.

Wieczorek, G.F., Snyder, J.B., Alger, C.S. and Isaacson, K.A., 1992. Rock falls in Yosemite Valley, California. U.S. Geol. Surv. Open-File Rep., 92-387: 136 pp.

ELSEVIER

Geomorphology 10 (1994) 265–284

GEOMORPHOLOGY

The importance of earthquake-induced landslides to long-term slope erosion and slope-failure hazards in seismically active regions

David K. Keefer

U.S. Geological Survey, Menlo Park, CA 94025, USA

Received February 8, 1994; accepted April 4, 1994

Abstract

This paper describes a general method for determining the amount of earthquake-induced landsliding that occurs in a seismically active region over time; this determination can be used as a quantitative measure of the long-term hazard from seismically triggered landslides as well as a measure of the importance of this process to regional slope-erosion rates and landscape evolution. The method uses data from historical earthquakes to relate total volume of landslide material dislodged by an earthquake to the magnitude, $\boldsymbol{M}$, and seismic moment, M_o, of the earthquake. From worldwide data, a linear-regression relation between landslide volume, V, and M_o is determined as: $V = M_o / 10^{18.9(\pm 0.13)}$, where V is measured in m^3 and M_o is in dyn-cm.

To determine the amount of earthquake-generated landsliding over time, this relation is combined with data on seismic-moment release for a particular region, which may be derived from either earthquake-history or fault-slip data. The form of the M_o–V relation allows the rate of production of earthquake-induced landslides over time to be determined from total rate of seismic-moment release without regard to the distribution of individual events, thus simplifying and generalizing the determination.

Application of the method to twelve seismically active regions, with areas ranging from 13,275 to 2,308,000 km^2, shows that erosion rates from earthquake-induced landslides vary significantly from region to region. Of the regions studied, the highest rates were determined for the island of Hawaii, New Zealand, western New Guinea, and the San Francisco Bay region of California. Significantly lower rates were determined for Iran, Tibet, the Sierra Nevada–Great Basin region of California, and central Japan (for the time period from 715 AD to the present). Intermediate rates were determined for Peru, southern California, onshore California, Turkey, and central Japan (for the time period from 1586 AD to the present).

To determine the relative, long-term importance of seismically triggered landslides, these erosion rates are compared to erosion rates calculated for other slope processes and to rates calculated from fluvial sediment discharge. Comparisons with other slope processes indicate that earthquake-induced landslides are the predominant agents of slope erosion on the island of Hawaii, in the San Francisco Bay region, and in western New Guinea. For Hawaii, the San Francisco Bay region, and Sierra Nevada–Great Basin region of California, the erosion rates calculated for earthquake-induced landslides also exceed the regional erosion rates calculated from fluvial sediment discharge.

1. Introduction

The ability of earthquakes to generate landslides has been known for thousands of years. Damaging earthquake-induced landslides have been documented from at least as early as 1789 BC in China (Hansen and Franks, 1991) and 373 or 372 BC in Greece (Seed, 1968). Studies of individual seismic events and anal-

SSDI 0169-555X(94)00025-M

yses of the historical earthquake record (Keefer, 1984; Keefer and Wilson, 1989) have shown that large earthquakes can generate tens of thousands of landslides over thousands of square kilometers, dislodging as much as several billion cubic meters of material from slopes. In many recent earthquakes, landslides have been a leading cause of damage and casualties (Kobayashi, 1981; Keefer, 1984).

In contrast to relatively extensive documentation of landslides in particular earthquakes, little is currently known about the long-term hazard of such landslides or their importance in slope erosion, either in absolute terms or relative to other, nonseismic, agents of erosion. The purposes of this paper are to present a model for determining the contribution of earthquake-induced landslides to long-term slope erosion and landslide hazard and to apply this model to several regions of high seismicity. The model is based on a relation between earthquake size and total landslide volume, determined from worldwide data, combined with data on earthquake occurrence, determined specifically for each region.

1.1. Previous work

Early speculations on the long-term importance of earthquake-induced landslides include those of John Muir (1912) and François Matthes (1930), both of whom were impressed by massive talus deposits in Yosemite Valley, California. Muir (p. 57), recording observations of a trip that began in 1868, concluded that the talus was largely the result of earthquake-induced rock falls: "The avalanche taluses, ... of the Valley ... instead of being slowly weathered from the cliffs like ordinary taluses, ... were all formed suddenly and simultaneously by a great earthquake that occurred at least three centuries ago." His conclusion was reinforced when he experienced the Owens Valley earthquake of March 26, 1872 (magnitude 7.75), while in Yosemite Valley (p. 59): "At half-past two o'clock ..., I was awakened by a tremendous earthquake ... I was now convinced before a single boulder had fallen that earthquakes were the talus-makers and positive proof soon came. ... The Eagle Rock on the south wall ... gave way and I saw it falling in thousands of the great boulders I had so long been studying." Away from Yosemite Valley, Muir (1919, quoted in Matthes, 1930) concluded that more than 90 percent of all the talus in the Sierra Nevada of California was generated by earthquakes.

In contrast, Matthes (1930, p. 106) concluded that, while earthquakes were important in talus formation, most talus was the result of more gradual and ongoing processes: "It is indeed probable that a considerable part of the rock waste in the Yosemite Valley is of earthquake origin ... But ... rock waste is being shed from the cliffs in quantities by no means negligible at the present time, as the result of the normal weathering ... and of the recurrent action of such destructive agencies as snow avalanches and torrential rains... To credit earthquakes with five-tenths of the total amount of rock waste in the valley, therefore, would probably be liberal." Recent studies in Yosemite Valley by Wieczorek et al. (1992) have demonstrated that at least 12 percent, by volume, of the historical landsliding — and 53 percent of that definitely attributable to a specific cause — was produced by earthquakes.

Estimates of long-term rates of erosion from earthquake-induced landslides have also been made for areas in New Zealand and New Guinea. In New Zealand, Adams (1980) studied long-term erosion rates from an area of 154,000 km^2 on the South Island. From analyses of fluvial sediment discharge and of landslides produced by three large earthquakes, he concluded that 40 to 67 percent of all the material eroded from that region originated in earthquake-induced landslides. Additional analyses, by Pearce and Watson (1986), indicated that, while earthquake-induced landslides were the main regional sediment-supply mechanism, much of the landslide material probably remained in storage for long periods of time, before being transported out of the area by rivers.

In New Guinea, Simonett (1967) studied erosion in the Torricelli Mountains, shaken by a magnitude 7.9 earthquake in 1935. He converted the volume of landslides generated by that earthquake to amounts of denudation, which ranged from 74 to 400 mm for areas underlain by various types of bedrock. He then combined these results with an estimate of earthquake recurrence and determined that the average, long-term erosion rate due to earthquake-triggered landslides in that region was between 1000 and 1430 m^3/km^2-yr (equivalent to 1000 to 1430 mm/1000 yr of ground-surface lowering, or denudation), which he compared to an erosion rate of 230 m^3/km^2-yr, calculated for adjacent mountain areas of low seismicity.

Citing a revised estimate of earthquake recurrence, Pain and Bowler (1973) increased the estimated erosion rate from earthquake-generated landslides in the Torricelli Mountains to 2200 m^3/km^2-yr. They also studied a 1970 earthquake (magnitude 7.1) in the Adelbert Range of New Guinea and estimated a long-term erosion rate from earthquake-induced landslides of 580 m^3/km^2-yr, compared to between 200 and 400 m^3/km^2-yr from other processes. In addition, they showed that landslide material from the 1970 earthquake was rapidly transported out of the area by rivers, with about half of the material moving to the sea within the first six months after the earthquake.

Keefer and Wilson (1989) presented data on total volumes of landslide material generated by ten earthquakes in North and South America, India, New Zealand, and New Guinea and calculated a linear-regression equation, relating total volume of landslide material, V, to earthquake moment magnitude, $\boldsymbol{M}$, as:

$$\log_{10} V = 1.44\boldsymbol{M} - 2.34 \quad (1)$$

where V is in m^3.

They also suggested that V could be more conveniently related to another measure of earthquake size, the seismic moment, M_o, which is defined as the product of fault slip, rupture area, and rigidity of materials along the fault (Kanamori, 1977):

$$M_o = \mu d' s \quad (2)$$

where μ is the rigidity, d' is the average fault slip, and s is the rupture area. As derived by Hanks and Kanamori (1979), M_o is related to $\boldsymbol{M}$ by the formula:

$$\boldsymbol{M} = 2/3 \log_{10} M_o - 10.7 \quad (3a)$$

where M_o is in dyn-cm. With terms rearranged, this equation can be rewritten as:

$$\log_{10} M_o = 1.5\boldsymbol{M} + 16.05 \quad (3b)$$

or, in exponential form:

$$M_o = 10^{1.5\boldsymbol{M} + 16.05} \quad (3c)$$

M_o is now generally regarded as being among the most accurate measures of earthquake size, particularly for large earthquakes ($\boldsymbol{M} > 7.5$). For $\boldsymbol{M} \leqslant 7.5$, the moment magnitude, $\boldsymbol{M}$, is approximately equivalent to the Richter surface-wave (M_s) or local (M_L) magnitude (Kanamori, 1977; Hanks and Kanamori, 1979).

Based on the similarity between the coefficients of $\boldsymbol{M}$ in Eq. (1) (1.44) and in Eq. (3b) (1.5), Keefer and Wilson (1989) suggested that landslide volume, V, could be related to seismic moment, M_o, by a least-squares linear regression with a slope of unity as:

$$\log_{10} V = \log_{10} M_o - 19.1\ (\pm 0.93) \quad (4)$$

where V is in m^3 and M_o is in dyn-cm.

1.2. Scope of paper

In the present paper, I use data now available for fifteen historical earthquakes to derive a similar, but revised, relation between seismic moment and landslide volume. I next combine this relation with data on seismic-moment release from twelve regions to determine total volumes of earthquake-induced landslide material produced over time. These values are then normalized over the area of each region to calculate areally averaged rates of slope erosion. Finally, these rates of slope erosion are compared to rates calculated for other processes to estimate the relative contribution of earthquake-generated landslides in each region. By using appropriate data on earthquake occurrence, the method thus illustrated may be used to estimate the rates of seismically-generated landsliding and resulting slope erosion for any region that contains widespread slopes susceptible to landslides.

2. Occurrence and volume of landslides in historical earthquakes

2.1. Occurrence of landslides in earthquakes

Earthquakes generate a wide variety of landslides, which Keefer (1984) classified into three main categories and fourteen individual types (Table 1). The most abundant of these landslides are rock falls, rock slides, and disrupted soil slides — shallow, highly to very highly disrupted, masses of rock and (or) soil that typically originate on steep slopes, travel at high speed, and form deposits at or near the bases of slopes (Fig. 1; Table 1). Other types of earthquake-generated landslides include slower-moving and deeper-seated slumps and block slides; large, complex rock avalanches; soil avalanches; soil falls; slow earth flows; and fluid, fast-moving soil flows (Table 1).

Table 1
Characteristics and abundance of earthquake-induced landslides (modified from Keefer, 1984)

Name	Type of movement	Internal disruption[a]	Water content[b]				Speed[c]	Depth[d]	Abundance in earthquakes
			D	U	PS	S			
Disrupted Slides and Falls									
Rock falls	Bounding, rolling, free fall	High or very high	X	X	X	X	Extremely rapid	Shallow	Very abundant
Rock slides	Translational sliding	High	X	X	X	X	Rapid to extremely rapid	Shallow	Very abundant
Rock avalanches	Complex, involving fall, sliding, and (or) flow	Very high	X	X	X	X	Extremely rapid	Deep	Uncommon
Soil falls	Bounding, rolling, free fall	High or very high	X	X	X	X	Extremely rapid	Shallow	Moderately common
Disrupted soil slides	Translational sliding	High	X	X	X	X	Moderate to rapid	Shallow	Very abundant
Soil avalanches	Translational sliding with subsidiary flow	Very high	X	X	X	X	Very rapid to extremely rapid	Shallow	Abundant
Coherent slides									
Rock slumps	Rotational sliding	Slight or moderate	?	X	X	X	Slow to rapid	Deep	Moderately common
Rock block slides	Translational sliding	Slight or moderate	?	X	X	X	Slow to rapid	Deep	Uncommon
Soil slumps	Rotational sliding	Slight or moderate	?	X	X	X	Slow to rapid	Deep	Abundant
Soil block slides	Translational sliding	Slight or moderate	?	?	X	X	Slow to very rapid	Deep	Abundant
Slow earth flows	Translational sliding with minor internal flow	Slight			X	X	Very slow to moderate, with very rapid surges	Generally shallow; occasionally deep	Uncommon
Lateral spreads and flows									
Rapid soil flows	Flow	Very high	?	?	?	X	Very rapid to extremely rapid	Shallow	Moderately common
Soil lateral spreads[e]	Translation on zone of liquefied or sensitive material	Generally moderate; occasionally slight or high			X	X	Very rapid	Variable	Abundant
Subaqueous landslides[e]	Complex, generally lateral spreading and (or) flow; occasionally sliding	Generally high or very high; occasionally moderate or slight			X	X	Generally rapid to extremely rapid; occasionally slow to moderate	Variable	Uncommon

[a]Internal disruption: ''slight'' signifies landslide consists of one or a few coherent blocks; ''moderate'' signifies several coherent blocks; ''high'' signifies numerous small blocks and individual soil grains and rock fragments; ''very high'' signifies nearly complete disaggregation into individual soil grains or small rock fragments.
[b]Water content: D = dry; U = moist but unsaturated; PS = partly saturated; S = saturated.
[c]Speed: very slow = 0.6–1.5m/yr; slow = 1.5 m/yr–1.5m/mo; moderate = 1.5m/mo–1.5m/day; rapid = 1.5 m/day–0.3m/min; very rapid = 0.3 m/min–3 m/sec; extremely rapid = > 3m/sec (Terminology from Varnes, 1973).
[d]Depth: ''shallow'' signifies thickness generally < 3 m; ''deep'' signifies depth generally > 3 m.
[e]Not included in determinations of total volume.

Fig. 1. Oblique aerial photograph showing slopes in the Santa Susana Mountains of southern California affected by the January 17, 1994 Northridge, California earthquake (*M* 6.6). Light-toned areas are sites of recurrent landslide activity (primarily rock falls and disrupted soil slides). Most of these areas contain deposits of landslides generated by the earthquake (lightest-toned material). This earthquake generated tens of thousands of landslides throughout an area of about 10,000 km^2.

Table 2
Earthquake localities, dates, magnitudes, seismic moments, and volumes of landslide material

Locality	Date	Magnitude	Seismic moment (dyn-cm)[a]	Vol. of landslide material (m^3)	Refs.[b]
Arthur's Pass, New Zealand	1929 Mar. 9	6.9	2.51E+26	5.90E+07	[1]
Buller, New Zealand	1929 Jun. 17	7.6	2.82E+27	1.30E+09	[1]
Torricelli Mtns., New Guinea	1935 Sep. 20	7.9	7.94E+27	2.15E+08	[2]
Assam, India	1950 Aug. 15	8.6	**1.00E + 29**	4.70E+10	[3]
Daly City, Calif.	1957 Mar. 27	5.3	1.00E+24	6.70E+04	[4]
Inangahua, New Zealand	1968 May 23	7.1	5.01E+26	5.20E+07	[1]
Peru	1970 May 31	7.9	**1.00E + 28**	1.41E+08	[5]
Papua New Guinea	1970 Oct. 31	7.1	5.01E+26	2.80E+07	[6]
Guatemala	1976 Feb. 4	7.6	**2.60E + 27**	1.16E+08	[7]
Darien, Panama	1976 Jul. 11	7.0	3.55E+26	1.30E+08	[8]
Mammoth Lakes, Calif.	1980 May 25	6.2	**2.40E + 25**	1.20E+07	[9]
Coalinga, Calif.	1983 May 2	6.5	**5.31E + 25**	1.94E+06	[10]
El Salvador	1986 Oct. 10	5.4	1.41E+24	3.78E+05	[11]
Ecuador	1987 Mar. 5	7.2	**6.40E + 26**	9.53E+07	[12]
Loma Prieta, Calif.	1989 Oct. 17	7.0	**3.55E + 26**	7.45E+07	[13]

[a]Seismic moments in bold are from original sources listed in references or from Kanamori (1977) or Hanks and Johnston (1992). Other seismic moments are calculated from magnitudes using Eq. (3c).
[b]References: [1] Adams (1980); [2] Simonett (1967); [3] Mathur (1953); [4] Bonilla (1960); [5] Plafker et al. (1971); [6] Pain and Bowler (1973); [7] Harp et al. (1981); [8] Garwood et al. (1979); [9] Harp et al. (1984); [10] Harp and Keefer (1990); [11] Rymer (1987); [12] Nieto et al. (1991); [13] Keefer, in press.

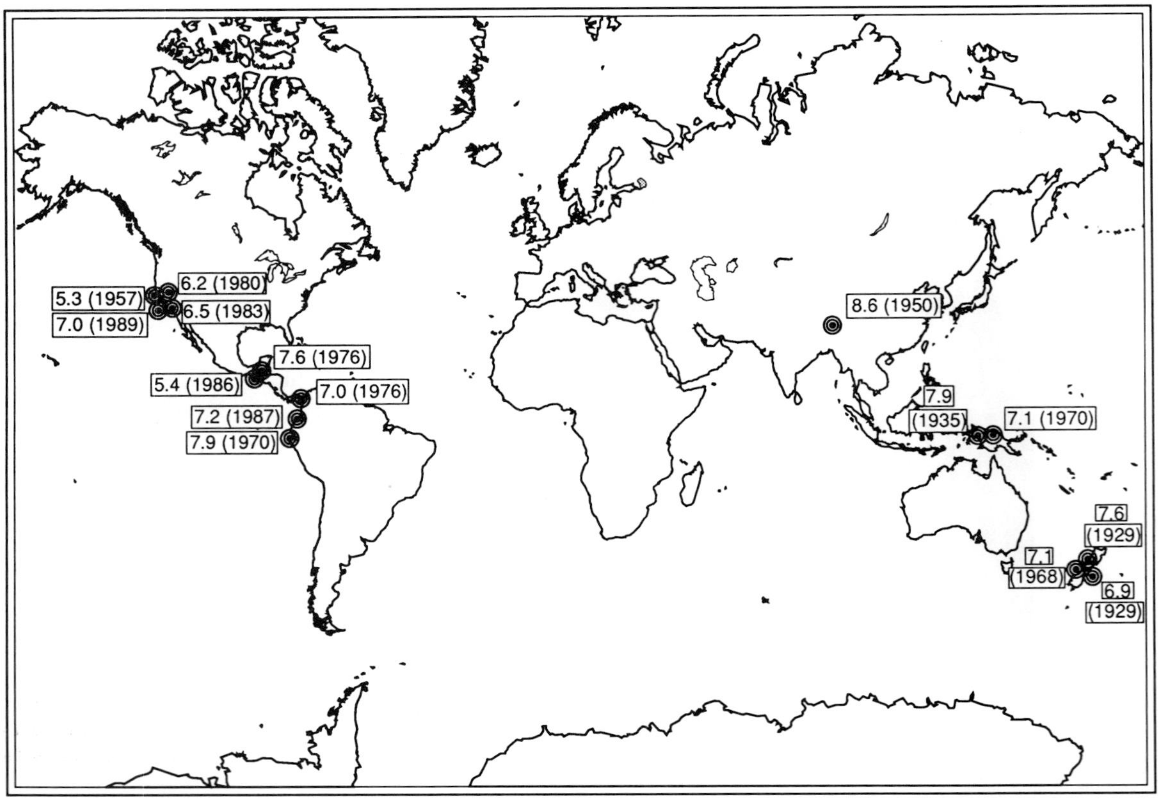

Fig. 2. Locations, magnitudes, and years of occurrence of fifteen earthquakes for which total volumes of landslide material have been determined. Earthquake parameters and landslide data are listed in Table 2.

Earthquakes also produce two types of landslides caused by soil liquefaction: lateral spreads and subaqueous landslides. Liquefaction-induced landslides are not included in the volume determinations and method discussed in this paper because they occur in environments significantly different from those of the other landslides (most commonly, gently sloping alluvial floodplains, coastal areas, and offshore) and because volume estimates are not generally available for them.

Previous work (Keefer, 1984) has shown that the numbers of landslides and geographic extent of landslide occurrence are strongly dependent on $\boldsymbol{M}$, and that the smallest earthquakes likely to cause landslides are about $\boldsymbol{M}=4$. Even in areas with abundant susceptible slopes, earthquakes with $\boldsymbol{M}<5$ typically produce only a few landslides, whereas events with $\boldsymbol{M}>7.5$ typically produce thousands or tens of thousands of landslides.

Using data from 37 earthquakes, Keefer and Wilson (1989) characterized the average area of landslide occurrence in an earthquake by the relation:

$$\log_{10}A_e = \boldsymbol{M} - 3.46 \quad (5 < \boldsymbol{M} < 9.2) \tag{5}$$

where A_e is the total area affected by landslides in km^2. These areas encompass all landslide sites from a given earthquake and include peripheral areas where landslides are few in number and separated by relatively large distances. Typically, more than 95 percent of the landslides produced by an earthquake are concentrated in less than half of that total area (D.K. Keefer, unpubl. data). Thus, for example, an ''average'' earthquake of $\boldsymbol{M}=8.0$ would produce landslides throughout an area of about 35,000 km^2, with most of those landslides occurring within an area of less than 17,500 km^2.

2.2. *Relating total landslide volume to earthquake magnitude and seismic moment*

Total volumes of landslide material generated by fifteen historical earthquakes have been determined by several investigators from detailed field investigations

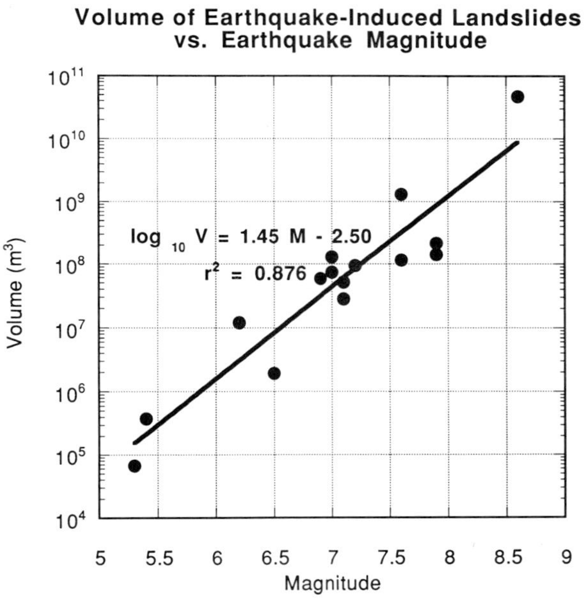

Fig. 3. Relation between earthquake magnitude and total landslide volume. Dots are data for fifteen earthquakes in Table 2. Solid line is least-squares linear regression.

and interpretation of aerial photographs (Table 2). These determinations sample a wide variety of regions, climates, geologic conditions, and tectonic settings (Fig. 2). All fifteen earthquakes were shallow (focal depth, $h < 60$ km) and occurred in areas of high historical seismicity, where slopes susceptible to landslides are also widespread. All of the earthquakes except the 1970 Peru and 1976 Panama events were onshore; the two latter events had epicenters short distances offshore, thus limiting the potential areas of subaerial landsliding to perhaps half those for equivalent onshore events.

In spite of the wide variety of geologic, climatic, and tectonic conditions represented by these data, the correlation between $\boldsymbol{M}$ and total landslide volume, V, is high (Fig. 3). A least-squares linear-regression yields the relation:

$$\log_{10} V = 1.45\boldsymbol{M} - 2.50 \quad (6)$$

$(5.3 \leqslant \boldsymbol{M} \leqslant 8.6;\ r^2 = 0.876;\ n = 15)$

where V is in m^3. The regression is statistically significant at a confidence level greater than 99.9 percent. Eq. (6) also differs little from Eq. (1), determined by Keefer and Wilson (1989).

Using either independent determinations of seismic moment, M_o, or the conversion between M_o and $\boldsymbol{M}$ (Eq. 3), V was also related to M_o (Fig. 4). A least-squares regression through these data yields the relation (dashed line in Fig. 4):

$$\log_{10} V = 0.95 \log M_o - 17.7 \quad (7)$$

$(1.4 \times 10^{24} \leqslant M_o \leqslant 10^{29};\ r^2 = 0.873;\ n = 15)$

where V is in m^3 and M_o is in dyn-cm. This regression is also statistically significant at a confidence level greater than 99.9 percent.

The slope coefficient in Eq. 7 (0.95) is so close to unity that it can be deleted, which simplifies the determination of landslide volume over time. A least-squares regression constrained to have a slope of unity yields the relation (solid line in Fig. 4):

$$\log_{10} V = \log M_o - 18.9\ (\pm 0.13) \quad (8a)$$

or:

$$V = M_o / 10^{18.9(\pm 0.13)} \quad (8b)$$

$(1.4 \times 10^{24} \leqslant M_o \leqslant 10^{29};\ r^2 = 0.870;\ n = 15)$

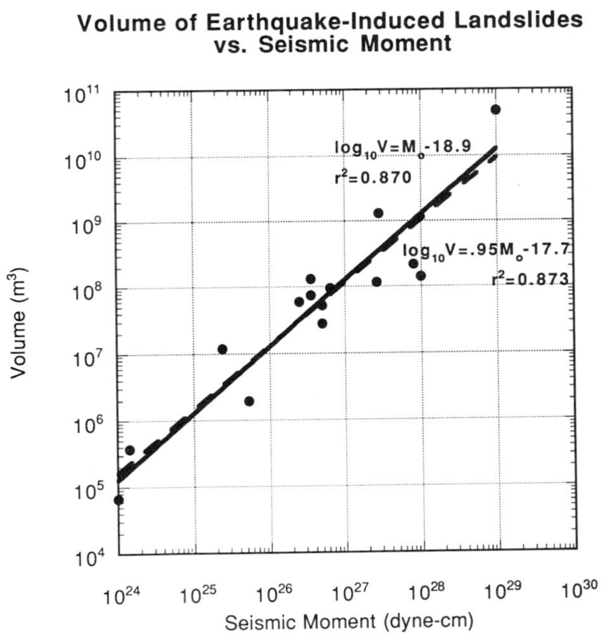

Fig. 4. Relation between seismic moment of earthquakes and total landslide volume. Dots are data for fifteen earthquakes in Table 2. Dashed line is unconstrained, best fit, least-squares linear regression. Solid line is least-squares linear regression constrained to have a slope of unity.

where V is in m^3 and M_o in dyn-cm.

The standard error of this regression (defined as the standard deviation divided by the square root of n) is ± 0.13; this value is relatively low, which indicates that Eq. (8) can be reasonably applied to a wide variety of environments.

3. Determining rates of seismic-moment release, earthquake-induced landslide occurrence, and seismically generated slope erosion over time

3.1. Seismic-moment release and volume of earthquake-induced landslides produced over time

Rates of seismic-moment release over time, M_o', may be determined from a catalog of historical seismicity for a region (generally converting $\boldsymbol{M}$ to M_o using Eq. 3c), or from fault areas and slip rates, using Eq. (2). Both types of determinations are used here to determine M_o' for various of the twelve regions discussed below.

Determination of M_o' from historical seismicity is illustrated here for Peru and adjacent areas, for which a catalog of historical seismicity from 1900 through 1984 has been compiled by Espinosa et al. (1985); the catalog was supplemented by M_o determinations by Kanamori (1977) for five of the largest earthquakes. For earthquakes of $\boldsymbol{M} \geqslant 6$, the catalog appears relatively complete back in time to 1900, while for smaller events, the catalog appears complete for the period 1963–1984. Thus, in calculating annual frequency of events, a period of 84 years was used for earthquakes $\boldsymbol{M} \geqslant 6$, and a period of 22 years was used for smaller earthquakes (Table 3). Where magnitudes in the catalog were given as body-wave magnitudes (m_b), they were converted to surface-wave magnitudes, M_s, which are approximately equal to $\boldsymbol{M}$ for earthquakes with $\boldsymbol{M} \leqslant 7.5$ (Kanamori, 1977; Hanks and Kanamori, 1979), using the formula of Richter (1958):

$$M_s = 1.59 m_b - 3.97$$

Only shallow ($h < 60$ km) and intermediate-depth (60 km $\leqslant h < 300$ km) events were included in the calculations because previous analyses of worldwide data (Keefer, 1984; Ambraseys, 1988) did not show any landslides to have been triggered by deep-focus earthquakes.

The data from Espinosa et al. (1985) and Kanamori (1977) were used to construct a magnitude–frequency relation for earthquake recurrence in Peru and adjacent regions (Fig. 5), of the standard form:

$$\log_{10} F = a - b\boldsymbol{M}$$

where F is the number of earthquakes per unit time with magnitude $\geqslant \boldsymbol{M}$, and a and b are empirically determined coefficients. Only events of $\boldsymbol{M} \geqslant 4$ were used because landslides are rarely generated by smaller earthquakes (Keefer, 1984). A linear-regression line through the Peruvian data (Fig. 5) yields the relation:

$$\log_{10} F = 4.272 - 0.676\boldsymbol{M}$$

$$(\boldsymbol{M} \geqslant 4.0;\ r^2 = 0.977;\ n = 39)$$

By converting $\boldsymbol{M}$ to M_o, using Eq. (3c) and performing the stepwise subtraction $F_{\boldsymbol{M}} - F_{\boldsymbol{M}+0.1}$ for each $\boldsymbol{M}$, the annual seismic-moment release associated with each $\boldsymbol{M}$ may be determined (column 9 in Table 3). By summing these values from $\boldsymbol{M} = 4$ through the maximum $\boldsymbol{M}$ in the record (8.3), the total annual seismic-moment release for the region may be determined as $M_o' = 1.35 \times 10^{27}$ dyn-cm/yr (column 9 in Table 3). From this value of M_o', the mean rate of production of earthquake-induced landslide material, V', is determined from Eq. (8b) as (column 10 in Table 3):

$$V'_{(\mathrm{Peru})} = 1.35 \times 10^{27} / 10^{18.9}$$

or:

$$V'_{(\mathrm{Peru})} = 1.70 \times 10^{8}\ \mathrm{m}^3/\mathrm{yr}$$

One general characteristic of the V' determination for all regions, illustrated by these data from Peru, is that most of the total volume of earthquake-induced landslides over time is generated by a few large events; in this region, only 0.1 percent of the landslide volume over time was associated with events of $\boldsymbol{M} < 5.0$, and only 1 percent was associated with events of $\boldsymbol{M} < 6.0$, whereas 92 percent was attributable to events of $\boldsymbol{M} \geqslant 7.0$ (Fig. 6). This relation has two implications: first, the determination of V' is relatively sensitive to the F and $\boldsymbol{M}$ of the largest events in a region, and, second, for practical purposes, events of $\boldsymbol{M} < 6$ (or, in some regions, even $\boldsymbol{M} < 7$) may be disregarded with only a negligible effect on the calculated V'.

Magnitude–frequency relations of the form $\log_{10} F = a - b\boldsymbol{M}$, were determined in a similar manner,

Table 3
Annual occurrence of shallow and intermediate-depth earthquakes, seismic-moment release, and volume of earthquake-induced landslides for Peru and adjacent regions

Magnitude	Seismic moment (dyn-cm)	No. of earthquakes						Annual seismic moment regression[d] (dyn-cm/yr)[e]	Annual vol. of landslides regression[d] (m^3/yr)[f]
		1901–1984[a]	1963–1984[a]	per year[b]	per year ≥*M* (counted)[c]	per year ≥*M* curve-fit[d]	per year of *M* curve-fit[d]		
4	1.26E+22		166	7.55	35.51	36.92	5.32	6.70E+22	8.44E+03
4.1	1.78E+22		141	6.41	27.96	31.60	4.56	8.10E+22	1.02E+04
4.2	2.51E+22			0.00	21.55	27.04	3.90	9.80E+22	1.23E+04
4.3	3.55E+22		115	5.23	21.55	23.14	3.34	1.18E+23	1.49E+04
4.4	5.01E+22		86	3.91	16.33	19.81	2.86	1.43E+23	1.80E+04
4.5	7.08E+22			0.00	12.42	16.95	2.44	1.73E+23	2.18E+04
4.6	1.00E+23		53	2.41	12.42	14.51	2.09	2.09E+23	2.63E+04
4.7	1.41E+23			0.00	10.01	12.41	1.79	2.53E+23	3.18E+04
4.8	2.00E+23		42	1.91	10.01	10.62	1.53	3.06E+23	3.85E+04
4.9	2.82E+23		26	1.18	8.10	9.09	1.31	3.70E+23	4.65E+04
5	3.98E+23		23	1.05	6.92	7.78	1.12	4.47E+23	5.62E+04
5.1	5.62E+23		20	0.91	5.87	6.66	0.96	5.40E+23	6.80E+04
5.2	7.94E+23		8	0.36	4.96	5.70	0.82	6.53E+23	8.22E+04
5.3	1.12E+24		11	0.50	4.60	4.88	0.70	7.89E+23	9.93E+04
5.4	1.58E+24		13	0.59	4.10	4.17	0.60	9.54E+23	1.20E+05
5.5	2.24E+24		2	0.09	3.51	3.57	0.52	1.15E+24	1.45E+05
5.6	3.16E+24		19	0.86	3.42	3.06	0.44	1.39E+24	1.75E+05
5.7	4.47E+24		5	0.23	2.55	2.62	0.38	1.69E+24	2.12E+05
5.8	6.31E+24		3	0.14	2.33	2.24	0.32	2.04E+24	2.56E+05
5.9	8.91E+24		6	0.27	2.19	1.92	0.28	2.46E+24	3.10E+05
6	1.26E+25	30		0.36	1.92	1.64	0.24	2.98E+24	3.75E+05
6.1	1.78E+25	4		0.05	1.56	1.40	0.20	3.60E+24	4.53E+05
6.2	2.51E+25	3		0.04	1.51	1.20	0.17	4.35E+24	5.48E+05
6.3	3.55E+25	21		0.25	1.48	1.03	0.15	5.26E+24	6.62E+05
6.4	5.01E+25	5		0.06	1.23	0.88	0.13	6.36E+24	8.00E+05
6.5	7.08E+25	23		0.27	1.17	0.75	0.11	7.68E+24	9.67E+05
6.6	1.00E+26	4		0.05	0.89	0.64	0.09	9.29E+24	1.17E+06
6.7	1.41E+26	1		0.01	0.85	0.55	0.08	1.12E+25	1.41E+06
6.8	2.00E+26	20		0.24	0.83	0.47	0.07	1.36E+25	1.71E+06
6.9	2.82E+26	9		0.11	0.60	0.40	0.06	1.64E+25	2.07E+06
7	3.98E+26	12		0.14	0.49	0.35	0.05	1.98E+25	2.50E+06
7.1	5.62E+26	4		0.05	0.35	0.30	0.04	2.40E+25	3.02E+06
7.2	7.94E+26	2		0.02	0.30	0.25	0.04	2.90E−25	3.65E+06
7.3	1.12E+27	7		0.08	0.27	0.22	0.03	3.50E+25	4.41E+06
7.4	1.58E+27	3		0.04	0.19	0.19	0.03	4.24E+25	5.33E+06
7.5	2.24E+27	2		0.02	0.15	0.16	0.02	5.12E+25	6.45E+06
7.6	3.16E+27	2		0.02	0.13	0.14	0.02	6.19E+25	7.79E+06
7.7	4.47E+27			0.00	0.11	0.12	0.02	7.48E+25	9.42E+06
7.8	6.31E+27	2		0.02	0.11	0.10	0.01	9.05E+25	1.14E+07
7.9	8.91E+27	3		0.04	0.08	0.09	0.01	1.09E+26	1.38E+07
8	1.26E+28			0.00	0.05	0.07	0.01	1.32E+26	1.66E+07
8.1	1.78E+28	1		0.01	0.05	0.06	0.01	1.60E+26	2.01E+07
8.2	2.51E+28	2		0.02	0.04	0.05	0.01	1.93E+26	2.43E+07
8.3	3.55E+28	1		0.01	0.01	0.05	0.01	2.34E+26	2.94E+07
8.4					0.00	0.04			
Total				35.51				1.35E+27	1.70E+08

[a]Data from Espinosa et al. (1985), supplemented by data in Kanamori (1977) for five of the largest earthquakes (1901–1984 only).
[b]Calculated from data in column 4 for earthquakes of *M* 4.0–5.9; from data in column 3 for *M* 6.0–8.3.
[c]Calculated from data in column 5.
[d]Calculated from regression line in Fig. 5
[e]Annual seismic moment is value in column 2 multiplied by value in column 8.
[f]Calculated from data in column 9 using Eq. (8b), without standard error term.

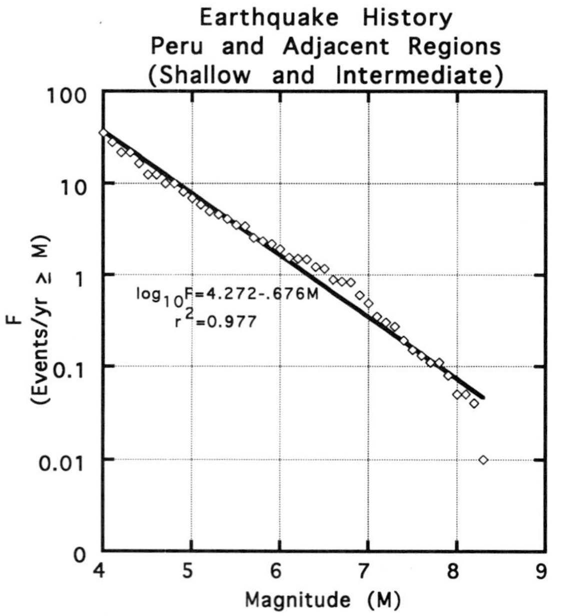

Fig. 5. Earthquake magnitude–frequency data for shallow and intermediate-depth events in Peru and adjacent regions, determined from data in Espinosa et al. (1985), supplemented by data on large earthquakes in Kanamori (1977). Diamonds are data by magnitude units of 0.1 (See Table 3). Solid line is least-squares linear regression.

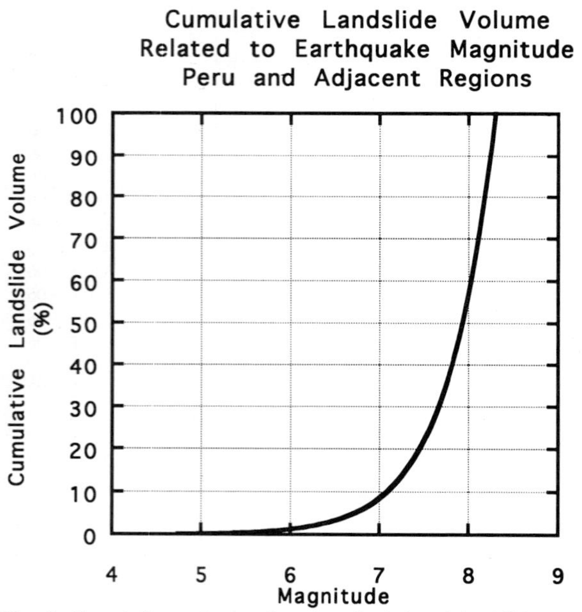

Fig. 6. Cumulative volume of earthquake-induced landslides as related to magnitude, M, in Peru and adjacent regions, from regression line in Fig. 5 and Eq. (8b). Note that, for this region, 99 percent of the calculated volume of material was produced by earthquakes of $M \geqslant 6.0$, and 92 percent was produced by earthquakes of $M \geqslant 7.0$.

from earthquake catalogs, for the San Francisco Bay and Sierra Nevada–Great Basin regions of California and for the island of Hawaii. A magnitude–frequency relation was likewise derived for Turkey using data of Ambraseys (1988), but in this case a nonlinear curve was fitted to the data. Values of M'_0 determined by others from historical seismicity were used for onshore California (Wesnousky, 1986); central Japan (Kanaori et al., 1992); and Iran, western New Guinea and Tibet (Ekström and England, 1989). M'_0 for New Zealand was determined from the magnitude–frequency relation of Adams (1980). Results of these determinations are shown in Table 4.

For Iran and Japan, where records of historical seismicity are particularly long, values of M'_0 were determined for two time periods each; in these regions, M'_0 for the shorter, more recent periods (400 years for central Japan, and 70 years for Iran) were higher than values computed from the entire 1300-year-long records by factors of 3.4 and 1.6, respectively (Table 4).

For six regions, M'_0 was determined from fault areas and long-term slip rates (Table 4). Such determinations may be used for calculations of V', without regard to the distribution of M or M_0 among individual events, because Eq. (8a) has a slope of unity and because small earthquakes, which typically do not cause landslides, also produce only a negligible amount of the total M'_0 in seismically active regions (Kanamori, 1977). M'_0 from fault-area and slip-rate data for the San Francisco Bay region, Sierra Nevada–Great Basin region, southern California, and onshore California were taken from determinations by Wesnousky (1986). Similar determinations for central Japan were taken from Kanaori et al. (1992). The slip-rate value of M'_0 shown for New Zealand in Table 4 is a composite value, calculated using the maximum M from fault-area and slip data (8.4) combined with historical-seismicity data for smaller events (Adams, 1980). For the five regions for which M'_0 was determined using both historical seismicity and fault-slip methods, the ratio of the two determinations varies from 0.57 to 3.4 (Table 4).

For each of the twelve regions considered, rates of production of earthquake-induced landslide material, V', were calculated from values of M'_0 using Eq. (8b) (Table 5). Mean V' was calculated by using the mean M'_0 (column 3 in Table 5) in Eq. (8b), without the standard error term. The minimum V' for each region

Table 4
Annual seismic-moment release from historical seismicity and fault slip

Region	M'_o from historical seismicity (dyn-cm/yr)	M'_o from fault slip (dyn-cm/yr)	Ratio[a]	References
San Francisco Bay Region, Calif.	5.03E+25	1.43E+25–2.96E+25	2.29	Wesnousky, 1986; Dieterich et al., 1990; U.S. Geological Survey, 1990
Sierra Nevada–Great Basin, Calif.	2.23E+25	6.94E+24	3.22	Real and Toppozada, 1978; Wesnousky, 1986; Stover and Coffman, 1993
Southern California		3.37E+25		Wesnousky, 1986
California (onshore)	1.60E+26	2.30E+26	0.70	Wesnousky, 1986
Island of Hawaii	5.83E+25			Stover and Coffman, 1993
New Zealand	5.03E+26	8.83E+26	0.57	Adams, 1980
Western New Guinea	1.20E+27			Ekström and England, 1989
Turkey	2.47E+26			Ambraseys, 1988
Iran (since 700 AD)	8.00E+25			Ekström and England, 1989
Iran (1901–1981)	1.30E+26			Ekström and England, 1989
Tibet	4.40E+25			Ekström and England, 1989
Central Japan (since 715 AD)	9.00E+24			Kanaori et al., 1992
Central Japan (since 1586 AD)	3.03E+25	7.00E+24–1.08E+25	3.40	Kanaori et al., 1992
Peru and adjacent regions	1.35E+27			Espinosa et al., 1985; Kanamori, 1977

[a]Ratio M'_o (seismicity) / M'_o (fault slip).

was calculated by using the minimum M'_o (column 2 in Table 5) in Eq. (8b) with the positive value of the standard error (+0.13) included; the maximum V' was calculated by using the maximum M'_o (column 4 in Table 5) in Eq. (8b) with the negative value of the standard error (−0.13) included.

3.2. Rates of slope erosion calculated from volumes of earthquake-induced landslides

To calculate rates of slope erosion from earthquake-induced landslides, values of V' for each region (columns 5–7 in Table 5) were divided by the area of each region (column 8 in Table 5), to determine areally averaged slope-erosion rates (columns 9–11 in Table 5):

$$D' = V'/A \tag{9}$$

where D' is the slope-erosion rate from earthquake-induced landslides, and A is the area of the region. These results can be expressed in units of volume divided by area and time (m^3/km^2-yr) or, equivalently in units of length divided by time (mm/1000 yr, or mm/ka), representing rates of equivalent ground-surface lowering, or denudation. (Note that 1 m^3/km^2-yr = 1 mm/ka.)

Areas in Table 5, which range from 13,275 km^2 for the island of Hawaii to 2,308,000 km^2 for Peru and adjacent regions, are plan areas, uncorrected for slope. If slope areas were used, and if, for example, it were assumed that all ground within a region had a slope of 35°, then the area would be increased by a factor of 1/cos 35°, or 1.22. Erosion rates normalized for slope area would thus be smaller than those normalized for plan area, but in most or all regions, this difference is relatively small for practical purposes.

Mean slope-erosion rates, D', for the twelve areas studied range widely, from about 5 m^3/km^2-yr for Tibet to 553 m^3/km^2-yr for the island of Hawaii (Table 5; Fig. 7). In addition to Hawaii, mean D' for New Zealand, western New Guinea, and the San Francisco Bay region of California are relatively high ($D' > 200$ m^3/km^2-yr), while D' values for Iran, Tibet, the Sierra Nevada–Great Basin region of California, and central Japan (calculated from 715 AD through the present) are lower by more than an order of magnitude ($D' < 20$ m^3/km^2-yr). D' values for onshore California, Peru, Turkey, central Japan (since 1586 AD), and southern

Table 5
Annual seismic-moment release, volumes of earthquake-induced landslides, and erosion rates from earthquake-induced landslides

Region	Min.[a] M'_o (dyn-cm/yr)	Mean[b] M'_o	Max.[a] M'_o	Min. V' $(m^3/yr)^c$	Mean V' $(m^3/yr)^c$	Max. V' $(m^3/yr)^c$	Area (km^2)	Min. erosion rate[d] $(m^3/km^2\text{-yr})$ (mm/1000 yr)	Mean erosion rate[d] $(m^3/km^2\text{-yr})$ (mm/1000 yr)	Max. erosion rate[d] $(m^3/km^2\text{-yr})$ (mm/1000 yr)
San Francisco Bay Region, Calif.	1.43E+25	3.61E+25	5.03E+25	1.34E+06	4.55E+06	8.55E+06	19,770	67.67	230.16	432.23
Sierra Nevada–Great Basin, Calif.	6.94E+24	1.46E+25	2.23E+25	6.48E+05	1.84E+06	3.79E+06	101,183	6.40	18.20	37.47
Southern California	3.37E+25	3.37E+25	3.37E+25	3.15E+06	4.25E+06	5.73E+06	69,626	45.22	61.00	82.29
California (all onshore)	1.60E+26	1.95E+26	2.30E+26	1.49E+07	2.45E+07	3.91E+07	411,013	36.33	59.73	95.03
Island of Hawaii	5.83E+25	5.83E+25	5.83E+25	5.44E+06	7.34E+06	9.90E+06	13,275	409.73	552.70	745.58
New Zealand	5.03E+26	6.93E+26	8.83E+26	4.69E+07	8.72E+07	1.50E+08	268,675	174.58	324.72	557.94
Western New Guinea	1.20E+27	1.20E+27	1.20E+27	1.12E+08	1.51E+08	2.04E+08	537,522	208.35	281.05	379.13
Turkey	2.47E+26	2.47E+26	2.47E+26	2.31E+07	3.12E+07	4.20E+07	1,054,558	21.90	29.55	39.86
Iran (since 700 AD)	6.00E+25	8.00E+25	1.00E+26	5.60E+06	1.01E+07	1.70E+07	1,648,000	3.40	6.11	10.30
Iran (1901–1981)	1.30E+26	1.30E+26	1.30E+26	1.21E+07	1.64E+07	2.21E+07	1,648,000	7.36	9.93	13.40
Tibet	4.40E+25	4.40E+25	4.40E+25	4.11E+06	5.54E+06	7.47E+06	1,042,357	3.94	5.31	7.17
Central Japan (since 715 AD)	9.00E+24	9.00E+24	9.00E+24	8.40E+05	1.13E+06	1.53E+06	66,320	12.66	17.08	23.05
Central Japan (since 1586 AD)	7.00E+24	1.96E+25	3.03E+25	6.53E+05	2.47E+06	5.15E+06	66,320	9.85	37.21	77.59
Peru and adjacent regions	1.35E+27	1.35E+27	1.35E+27	1.26E+08	1.70E+08	2.29E+08	2,308,000	54.63	73.69	99.41

[a]From Table 4.
[b]Calculated from values in Table 4; where range of values is given for fault-slip data middle of range is averaged with value from seismic history to calculate mean.
[c]Minimum V' = Minimum $M'_o/10^{19.03}$; Mean V' = Mean $M'_o/10^{18.9}$; Maximum V' = Maximum $M'_o/10^{18.77}$.
[d]Erosion rate = V'/Area.

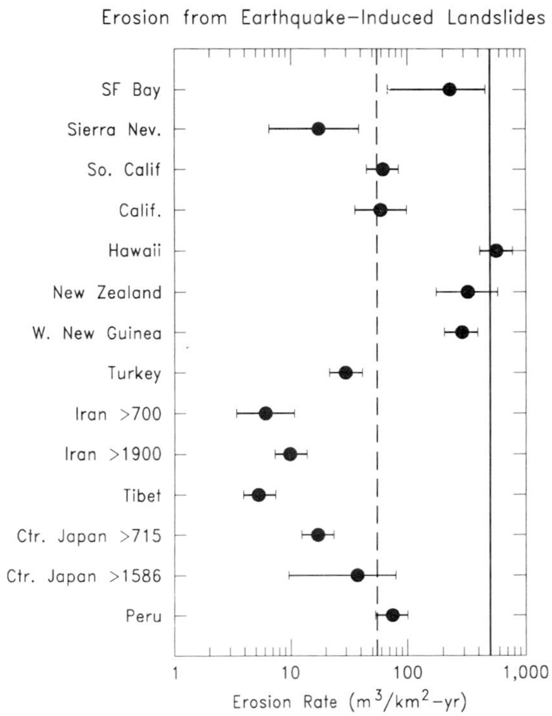

Fig. 7. Rates of erosion from earthquake-induced landslides in twelve regions. Dots are mean values; error bars show ranges determined statistically, as described in text and shown in Table 5. Erosion rates for Iran and central Japan determined from seismic data for two time periods in each region. Vertical dashed line is areally weighted, mean erosion rate for earthquake-induced landslides in the twelve regions (56 m^3/km^2-yr). Vertical solid line is worldwide mean erosion rate for all processes on steep slopes (500 m^3/km^2-yr) determined by Young (1969).

California are intermediate (20 m^3/km^2-yr $< D' <$ 200 m^3/km^2-yr).

These determinations are subject to several limitations, approximations, and potential sources of error, including the following:

The values of V' from which the slope-erosion rates are calculated are based on a regression equation involving worldwide data (Table 2, Fig. 4, and Eq. 8). The low standard error of this regression (± 0.13) indicates that it is a good approximation for use in most or all seismically active regions. However, if landslide-volume data were available for enough earthquakes in a particular region or particular geologic environment, V' and D' calculated from the more specific data would probably be more precise than those calculated from the worldwide data.

Values of M_0' used to calculate V' and D' can be determined from either historical seismicity or long-term fault-slip data, and M_0' values determined by the two methods for a given region commonly differ (by as much as a factor of 3.4 for the regions in Table 4). Furthermore, rates of seismicity vary with time, as shown in the data for Iran and Japan (Tables 4 and 5). The calculated values of D', of course, are no more accurate or precise than the values of M_0' from which they are calculated.

The method and the calculated values of V' and D' are only accurate for regions in which slopes susceptible to landslides are widespread. Such conditions are common in areas affected by most large, onshore, plate-boundary earthquakes, but may be less common in intraplate areas.

D' values are derived from the corresponding values of V' by dividing by the area of the region as a whole. The smaller regions in Table 5 (for example, the island of Hawaii, or the San Francisco Bay region) were chosen as regions of particularly high seismicity, but the larger regions (Tibet, Iran, Turkey, and Peru) also include areas of relatively low seismicity; thus areally averaged slope-erosion rates for these larger regions are lower than would be calculated for the subregions within them with the highest seismicity.

The earthquake data used to calculate the worldwide regression, Eq. 8, are limited to events of $5.3 \leqslant \boldsymbol{M} \leqslant 8.6$, whereas the largest earthquake in the historic record is $\boldsymbol{M} = 9.5$ (Kanamori, 1977). Extrapolating Eq. (8) to smaller events ($4.0 \leqslant \boldsymbol{M} < 5.3$) causes negligible errors because V' values associated with such small events are negligible. However, potential errors associated with extrapolating to events with $\boldsymbol{M} > 8.6$ are undetermined.

Despite these limitations, the method yields results that are consistent with the available data on earthquakes and on comparable erosion rates calculated for other processes. For example, the range of 5 to 553 m^3/km^2-yr, calculated for slope-erosion rates from earthquake-induced landslides (Table 5; Fig. 7), is slightly lower than the range of 10 to 1200 m^3/km^2-yr, calculated for total erosion rates on steep slopes worldwide by Young (1969). The areally weighted, mean erosion rate for earthquake-induced landslides in the twelve regions studied is 56 m^3/km^2-yr (Fig. 7), which is about one-ninth the worldwide mean of 500 m^3/km^2-yr for all processes on steep slopes determined by

Young (1969). Regional comparisons between erosion rates from earthquake-induced landslides and rates calculated for other processes are discussed in the next section, below.

4. Comparison of erosion rates from earthquake-induced landslides to erosion rates from other processes: the importance of earthquakes in slope evolution and slope-failure hazard

Data on erosion rates for areas in and adjacent to the twelve seismically active regions were compiled from a variety of sources (Table 6). In most cases, comparative erosion-rate data were available for the seismic regions themselves, but, in a few cases, data were from adjacent areas of lower seismicity.

4.1. Comparison of erosion rates from earthquake-induced landslides to erosion rates from other slope processes

Erosion rates most directly comparable to rates from earthquake-induced landslides are those calculated for other landslide processes. Such data are available from eight of the regions considered, either from direct measurement of slope movements or from dating of measured accumulations of colluvial material (Table 6; Fig. 8). For processes other than earthquake-induced landslides, ranges of erosion rates shown in Fig. 8 represent total ranges in individual measurements within a region. Where sufficient data were available, means were also calculated. These means were not weighted for area except in Southern California, where data are from debris basins, built to contain debris flows from small watersheds.

Fig. 8 also shows the rates of erosion from earthquake-induced landslides, D', from Fig. 7, with ranges representing the standard errors, determined statistically as discussed above. The comparisons in Fig. 8 indicate significant variation among regions in the relative contribution of earthquake-induced landslides to overall slope erosion, with the seismic process being predominant in some regions and a relatively small contributor in others.

In the San Francisco Bay region of California, the mean calculated erosion rate from earthquake-induced landslides, 231 m^3/km^2-yr, is more than four times higher than the mean for other slope processes (Fig. 8a). The ranges for seismic and other slope processes overlap (Fig. 8a), but the significantly higher mean for earthquake-induced landslides suggests that they predominate in this region.

For the Sierra Nevada–Great Basin region, data on slope erosion are available for Yosemite Valley. The minimum erosion rate there for earthquake-induced landslides, as determined from historical observations by Wieczorek et al. (1992), is slightly higher than the range in D' calculated herein for the region as a whole (Fig. 8a). Because Yosemite Valley is walled by spectacularly high and steep slopes, the rate there is likely to be higher than for the region as a whole, and, therefore, the calculated regional D' and the rate determined from observations in Yosemite Valley are consistent with each other. The mean regional erosion rate from earthquake-induced landslides, 18.2 m^3/km^2-yr, is about 5 percent of the erosion rate for all landslides in Yosemite Valley (348 m^3/km^2-yr; Table 6).

In southern California, the mean erosion rate from earthquake-induced landslides is moderately high (61 m^3/km^2- yr), but erosion rates, calculated from debris-basin data in the Transverse Ranges (mean of 1244 m^3/km^2-yr), are exceptionally high, and thus the contribution from earthquakes appears to be relatively low, about 5 percent of the rate from other slope processes (Fig. 8a). However, parts of the Transverse Ranges have historical erosion rates among the highest recorded anywhere (Scott and Williams, 1978), and these comparative rates almost certainly represent maximum rates for the region. Therefore, throughout the region, the relative contribution of earthquake-induced landslides may be somewhat higher than indicated by these data.

Throughout California, the mean erosion rate from earthquake-induced landslides is about 11 percent of the mean calculated for other slope processes (Fig. 8a). Together with the data from the three subregions discussed above, this indicates that earthquake-induced landslides are moderately significant to locally predominant contributors to slope erosion throughout California.

The highest mean erosion rate for earthquake-induced landslides in any region studied was 553 m^3/km^2-yr, calculated for the island of Hawaii (Table 5; Figs. 7, 8b). Comparative erosion rates are not available for that island itself, and so data are taken from the nearby island of Oahu, which has significantly less seismic activity (Table 6). The earthquake-related rate for Hawaii is 3.5 times higher than the maximum rate of long-term slope erosion on Oahu (Fig. 8b), and earthquake-induced landsliding is thus almost certainly a predominant process on the island of Hawaii, as is consistent with the dynamics of this active volcanic environment.

In western New Guinea, the high mean erosion rate from earthquake-induced landslides, 281 m^3/km^2-yr, is slightly higher than the rate for landslides not related to earthquakes, 227 m^3/km^2-yr (Table 6, Fig. 8b), indicating that the seismic process predominates in slope erosion there. Earthquake-related erosion rates for two mountain ranges of particularly high seismicity, calculated by Simonett (1967) and Pain and Bowler (1973), are higher than the earthquake-related rate for western New Guinea as a whole by factors of 2 and 8 (Fig. 8b), as is consistent with the inclusion of areas of lower seismicity in the latter calculation. In New Zealand, the evidence shows that earthquake-induced landslides are also important in slope erosion, as the range in erosion rates calculated for them is nearly the same as the range in rates determined for other slope processes (Fig. 8b).

In central Japan, the range of erosion rates calculated for earthquake-induced landslides is lower than the range of rates determined for other slope processes (Table 6; Fig. 8b), but the reported range in rates of other slope processes is an order of magnitude (100 to 1000 m^3/km^2-yr). Thus, the erosion rate from earthquake-induced landslides may be as little as 1 percent or as much as 78 percent of the rate for other slope processes.

4.2. Comparison of erosion rates from earthquake-induced landslides to erosion rates calculated by the fluvial discharge method

The most widely available data on erosion, or "denudation," rates are those calculated from the discharge of fluvially transported, bedload, suspended, and dissolved sediment. These rates are not directly comparable to the erosion rates calculated for earthquake-induced landslides because landslide movement is typically only the first in a series of processes by which material may be removed from slopes and eventually transported out of a region by fluvial action; much landslide material does not immediately enter the fluvial system but rather is reworked by subsequent landslides or other processes, perhaps several times, before entering a river. Previous studies have shown that material from landslides and related slope-movement processes not directly deposited in river channels can remain in storage for periods of time ranging from a few months or less to thousands of years (Pain and Bowler, 1973; Pearce and Watson, 1986; Meade et al., 1990). Thus, comparisons of erosion rates calculated by the fluvial discharge method to those calculated for earthquake-induced landslides in a given region only provide a general relative index for evaluating the effects of the latter process in that region.

In calculating the fluvial erosion rates from a region (Table 6, Fig. 9), the average annual sediment discharge from a river system was divided by the area of that system's drainage basin. Where adequate data from several drainage basins were available for a region, a mean, weighted by the areas of the individual basins, was calculated. Ranges of fluvial erosion rates shown in Fig. 9 represent the ranges of values calculated for individual drainage basins within each region (Table 6). These fluvial data integrate contributions from all weathering and erosion processes in a drainage basin that contribute material to the fluvial system. However, most of these data are from relatively short periods of observation, typically a few years to decades, and thus may not include significant contributions from severe but infrequent events, such as major storms or large earthquakes. These data also may not reflect long-term changes in discharge due to changes in climate, vegetation, or land use.

As with slope-erosion data (Fig. 8), comparisons of erosion rates calculated for earthquake-induced landslides to those calculated for fluvial discharge show significant variations among the regions. In three regions — San Francisco Bay, the island of Hawaii, and the Sierra Nevada–Great Basin — the mean erosion rate calculated for earthquake-induced landslides is higher than the rate calculated for fluvial discharge. In five additional regions — onshore California, Turkey, western New Guinea, Peru, and New Zealand — the mean erosion rate calculated for earthquake-induced landslides is a substantial fraction (between about 20 and 65 percent) of the erosion rate calculated from fluvial discharge. In four other regions studied — southern California, Iran, central Japan, and Tibet —

Table 6
Erosion rates for seismic regions calculated for other processes or by other methods

Comparative area	Process	Erosion rate (m^3/km^2-yr) (mm/1000 yr)	Reference
San Francisco Bay Region, California			
Oakland Hills	Slope erosion–erosion plots	100–150	Booker et al., 1993
Skyline Blvd., Berkeley Hills	Long-term erosion in colluvial hollow	79	Reneau, 1988
San Pedro Ridge, Marin County	Long-term erosion in colluvial hollows	7.0–57	Reneau, 1988
Marin County (other than SPR)	Long-term erosion in colluvial hollows	30–111	Reneau, 1988
Lone Tree Creek Basin	All slope erosion	410	Lehre, 1981
Lone Tree Creek Basin	Landslides	70	Lehre, 1981
Lone Tree Creek Basin	Fluvial – total sediment (long-term)	89	Lehre, 1981
Lone Tree Creek Basin	Fluvial – total sediment (short-term)	275	Lehre, 1981
San Lorenzo River Basin	Fluvial – total sediment	222	Nolan and Marron, 1988
Alameda Creek Basin	Fluvial – suspended sediment	46	Ritter, 1967
Napa River Basin	Fluvial – suspended sediment	104	Ritter, 1967
Sierra Nevada–Great Basin, California			
Yosemite Valley	Landslides – seismically induced	> 42	Wieczorek et al., 1992
Yosemite Valley	Landslides – total	348	Wieczorek et al., 1992
Lake Tahoe Basin	Fluvial – suspended sediment	14	Nolan and Hill, 1991
San Joaquin River Basin	Fluvial – suspended sediment	3	Ritter, 1967
Sacramento River Basin	Fluvial – suspended sediment	13	Ritter, 1967
Sierra Nevada	Fluvial – suspended sediment	< 30	Marchand, 1971
Sierra Nevada foothills	Fluvial – total sediment	43	Marchand, 1971
Southern California			
Transverse Ranges	Slope-debris basin sedimentation	519–4618	Scott and Williams, 1978
San Gabriel Mtns.	Fluvial – reservoir sedimentation	600–4090	Bull, 1991
Other California			
Eel River Basin	Fluvial – suspended sediment	772	Ritter, 1967
Eel River Basin	Fluvial – suspended sediment	150–750	Marchand, 1971
Mad River	Fluvial – suspended sediment	490	Ritter, 1967
Trinity River	Fluvial – suspended sediment	150	Ritter, 1967
Redwood Creek	Fluvial – total sediment	503	Nolan and Marron, 1988
Van Duzen River Basin	Fluvial – total sediment	969	Kelsey, 1980
Hawaii			
Honolulu, Oahu	Debris flows – nonseismically induced	30–160	S.D. Ellen, pers. commun., 1993
Lanai	Fluvial – valley aggradation	61–105	Moberly, 1963
New Zealand			
New Zealand, South Island	Landslides – seismically induced	490–1470	Adams, 1980
New Zealand – all	Landslides – general	250–1000	Saunders and Young, 1983
New Zealand, South Island	Fluvial – total sediment	1300–10,000	Adams, 1980
Western New Guinea			
Adelbert Range	Landslides – seismically induced	575	Pain and Bowler, 1973
Torricelli Mtns.	Landslides – seismically induced	1000–1429	Simonett, 1967
Western Bewani Mtns.	Landslides – nonseismically induced	227	Simonett, 1967
Adelbert Range	Fluvial – total sediment	800–1000	Pain and Bowler, 1973

Comparative area	Process	Erosion rate (m^3/km^2-yr) (mm/1000 yr)	Reference
Turkey			
Black Sea Basin	Fluvial – total (?) sediment	63	Saunders and Young, 1983
Iran			
Caucasus Mtns.	Fluvial – total (?) sediment	200	Saunders and Young, 1983
Tibet			
Himalayas	Fluvial – total (?) sediment	500–5000	Saunders and Young, 1983
Central Japan			
Tanazawa Mountains	Vegetated slopes	100	Tanaka, 1976
Tanazawa Mountains	All slope processes	1000	Tanaka, 1976
Japan – general	Landslides	1000	Saunders and Young, 1983
Central Japan	Fluvial – reservoir sedimentation	384–6312	Yoshikawa, 1974
Peru and adjacent regions			
Pilcomayo–Bermejo Basin	Fluvial – total sediment	50–1259	Guyot et al., 1990a
Bolivian Andes	Fluvial – total sediment	11-274	Guyot et al., 1990b

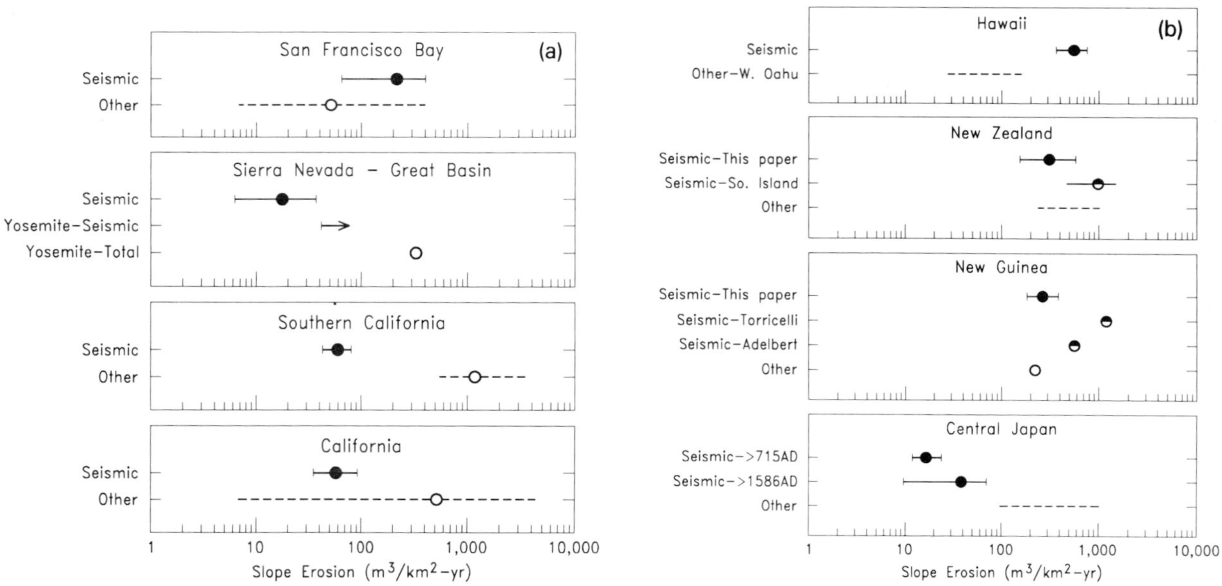

Fig. 8. Rates of erosion from earthquake-induced landslides compared to erosion rates determined for other slope processes in eight regions. Dots with error bars show means and statistically determined ranges of erosion rates from earthquake-induced landslides, calculated using the method described in this paper (See Table 5 and Fig. 7). Half-open circles show erosion rates for earthquake-induced landslides determined by others; solid line for New Zealand indicates range of determinations of Adams (1980), and arrow for Yosemite Valley indicates minimum value from historical observations reported by Wieczorek et al. (1992). Circles are means, and dashed lines are ranges in erosion rates calculated for other slope processes (See Table 6). (a) Results for onshore California and subregions within California (San Francisco Bay, Sierra Nevada–Great Basin, and southern California). (b) Results for Hawaii, New Zealand, New Guinea, and central Japan.

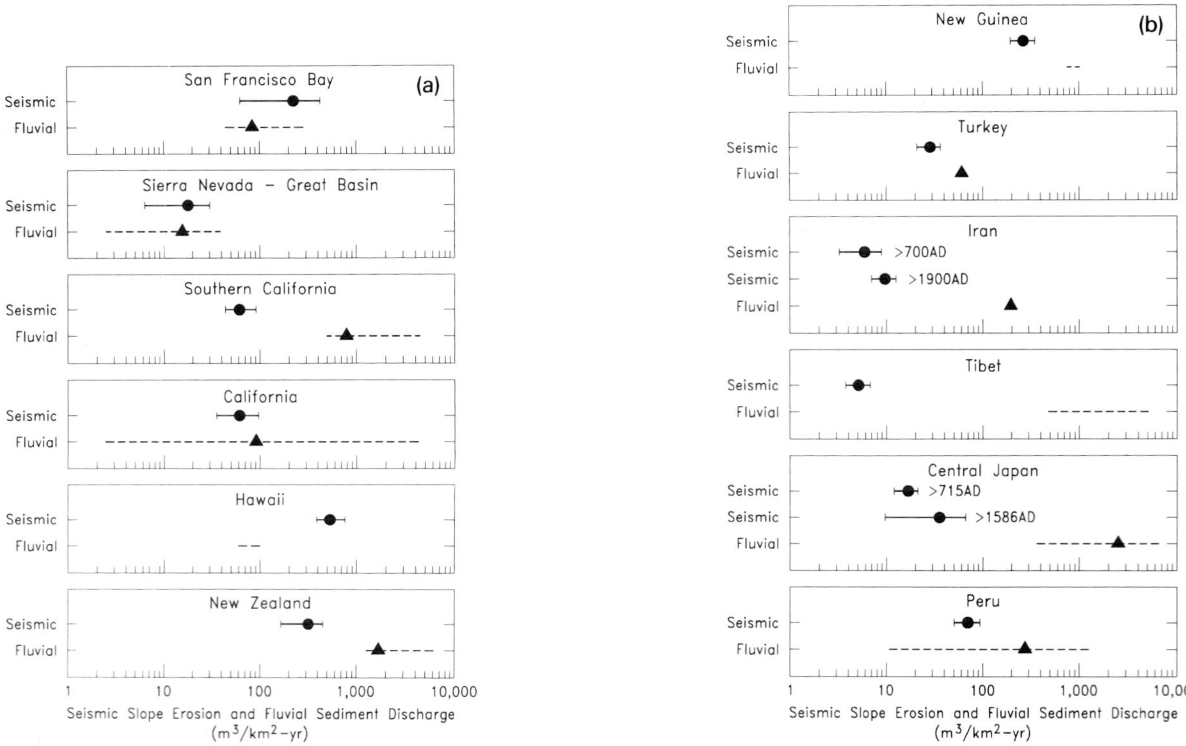

Fig. 9. Rates of erosion from earthquake-induced landslides compared to erosion rates determined from fluvial sediment discharge in twelve regions. Dots with error bars show means and statistically determined ranges of values for erosion rates from earthquake-induced landslides, calculated using the method described in this paper (See Table 5 and Fig. 7). Filled triangles are mean erosion rates (weighted for drainage-basin area) calculated from fluvial sediment discharge; dashed lines are ranges in values for individual drainage basins within each region. (a) Results for onshore California, subregions within California (San Francisco Bay, Sierra Nevada–Great Basin, and southern California), Hawaii, and New Zealand. (b) Results for New Guinea, Turkey, Iran, Tibet, central Japan, and Peru.

by contrast, the mean erosion rates from earthquake-induced landslides are much lower, less than 10 percent of the erosion rates calculated from fluvial discharge.

5. Conclusions

Data on total volumes of landslides triggered by each of fifteen historical earthquakes show strong correlations with earthquake magnitude, ***M***, and seismic moment, M_o, despite the wide variety of environments in which these earthquakes occurred. The total volume of landslide material produced by an earthquake, V, may be closely approximated by the relation:

$$V = M_o / 10^{18.9(\pm 0.13)}$$

$$(1.4 \times 10^{24} \leqslant M_o \leqslant 10^{29};\ r^2 = 0.870;\ n = 15)$$

where V is in m^3, and M_o in dyn-cm.

Combination of this relation with regional data on seismic-moment release over time provides a quantitative estimate of the long-term contribution of earthquake-induced landslides to slope erosion and slope-failure hazards.

Three measures of the importance of earthquake-induced landslides to regional erosion and hazard are proposed: the rate of earthquake-induced landslide generation itself, the comparison of this rate with the erosion rate from other slope processes, and the comparison of this rate with the erosion, or denudation, rate calculated from fluvial sediment discharge.

Application of this method to twelve seismically active regions shows significant regional variations in all these measures. Mean erosion rates for earthquake-induced landslides vary from about 5 to 553 m^3/km^2-yr, (equivalent to 5 to 553 mm/ka of ground-surface lowering), with the highest rates (> 200 m^3/km^2-yr) being calculated for the island of Hawaii, New Zealand, western New Guinea, and the San Francisco Bay region of California. Compared to erosion rates from other slope processes, earthquake-induced landsliding appears to predominate in all of these areas except possibly New Zealand. In Hawaii and the San Francisco Bay region, as well as in the Sierra Nevada–Great Basin region of California, the erosion rate from earthquake-induced landsliding also exceeds the erosion rate calculated from fluvial discharge of sediment.

These estimates may be used to evaluate quantitatively the risk from earthquake-induced slope failures in a given region. For example, all three measures suggest that on the island of Hawaii and in the San Francisco Bay region, earthquake-induced landslides are a predominant process; thus, decisions concerning mitigation of slope-failure hazards there could be directed specifically at seismically triggered failures. Another possible use of the method is in estimating the sediment input from earthquake-induced landslides in determining sediment budgets or other measures of overall erosion in particular watersheds.

The method described here may be applied to any seismically active region where slopes susceptible to landslides are widespread, provided that a regional earthquake history or rate of seismic-moment release may be determined. The method incorporates several approximations and is subject to several limitations in accuracy and precision. Despite these limitations, however, the method yields results consistent with available measurements in several regions and provides initial quantitative estimates of both the absolute amount and relative importance of earthquake-induced landslides to erosion in seismically active regions. As additional data become available, the method may be refined to incorporate, for example, differing M_0 versus V relations for specific regions or specific geologic conditions.

Acknowledgements

Kenneth R. Lajoie, Richard F. Madole, and Marie Morisawa reviewed the manuscript of this article and provided many helpful suggestions. Kenneth M. Nolan provided data and insights on erosion rates calculated by the fluvial discharge method. Their assistance is gratefully acknowledged.

References

Adams, J., 1980. Contemporary uplift and erosion of the Southern Alps, New Zealand. Geol. Soc. Am. Bull., 91(1): 2–4, 91(2): 1–114.

Ambraseys, N.N., 1988. Engineering seismology. Earthquake Eng. Struct. Dyn., 17: 1–105.

Bonilla, M.G., 1960. Landslides in the San Francisco South Quadrangle, California. U.S. Geol. Surv. Open-File Rep., 44 pp.

Booker, F.A., Dietrich, W.E. and Collins, L.M., 1993. Runoff and erosion after the Oakland firestorm. Calif. Geol., 46: 159–173.

Bull, W.B., 1991. Geomorphic Response to Climate Change. Oxford Univ. Press, New York, 326 pp.

Dieterich, J.H., Allen, C.R., Cluff, L.S., Cornell, C.A., Ellsworth, W.L., Johnson, L.R., Lindh, A.G., Nishenko, S.P., Scholz, C.H., Schwartz, D.P., Thatcher, W. and Williams, P.L., 1990. Probabilities of large earthquakes in the San Francisco Bay Region, California. U.S. Geol. Surv. Circ., 1053: 51 pp.

Ekström, G. and England, P., 1989. Seismic strain rates in regions of distributed continental deformation. J. Geophys. Res., 94B: 10,231–10,257.

Espinosa, A.F., Casaverde, L.A., Michael, J.A., Alva-Hurtado, J., and Vargas-Neumann, J., 1985. Earthquake Catalog of Peru. U.S. Geol. Surv. Open-File Rep., 85-286: 618 pp.

Garwood, N.C., Janos, D.P. and Brokaw, N., 1979. Earthquake-caused landslides: a major disturbance to tropical forests. Science, 205: 997–999.

Guyot, J.L., Calle, H., Cortes, J. and Pereira, M., 1990a. Transport de matières dissoutes et particulaires des Andes vers le Rio de La Plata par les tributaires boliviens (rios Pilcomayo et Bermejo) du Rio Paraguay. Hydrol. Sci. – J. Sci. Hydrol., 35: 653–665.

Guyot, J.L., Roche, M.A., Noriega, L., Calle, H. and Quintanilla, J., 1990b. Salinities and sediment transport in the Bolivian highlands. J. Hydrol., 113: 147–162.

Hanks, T.C. and Johnston, A.C., 1992. Common features of the excitation and propagation of strong ground motion for North American earthquakes. Bull. Seismol. Soc. Am., 82: 1–23.

Hanks, T.C. and Kanamori, H., 1979. A moment magnitude scale. J. Geophys. Res., 84B: 2348–2350.

Hansen, A. and Franks, C.A.M., 1991. Characterisation and mapping of earthquake triggered landslides for seismic zonation. In: Proc. 4th Int. Conf. on Seismic Zonation, Stanford, California, Vol. 1, pp. 149–195.

Harp, E.L. and Keefer, D.K., 1990. Landslides triggered by the earthquake. In: M.J. Rymer and W.L. Ellsworth (Editors), The Coalinga, California, Earthquake of May 2, 1983. U.S. Geol. Surv. Prof. Pap., 1487: 335–348.

Harp, E.L., Wilson, R.C. and Wieczorek, G.F., 1981. Landslides from the February 4, 1976, Guatemala Earthquake. U.S. Geol. Surv. Prof. Pap., 1204-A: 35 pp.

Harp, E.L., Tanaka, K., Sarmiento, J. and Keefer, D.K., 1984. Landslides from the May 25–27, 1980, Mammoth Lakes, California, earthquake sequence. U.S. Geol. Surv. Misc. Invest. Ser. Map I-1612, scale 1:62,500.

Kanamori, H., 1977. The energy release in great earthquakes. J. Geophys. Res., 82: 2981–2987.

Kanaori, Y., Kawakami, S. and Yairi, K., 1992. Space–time distribution patterns of destructive earthquakes in the inner belt of central Japan (part 2): moment-release rates and earthquake prediction. Eng. Geol., 32: 113–122.

Keefer, D.K., 1984. Landslides caused by earthquakes. Geol. Soc. Am. Bull., 95: 406–421.

Keefer, D.K. (Editor), in press. The Loma Prieta, California, Earthquake of October 17, 1989 — Landslides. U.S. Geol. Surv. Prof. Pap., 1551-C.

Keefer, D.K. and Wilson, R.C., 1989. Predicting earthquake-induced landslides, with emphasis on arid and semi-arid environments. In: P.M. Sadler and D.M. Morton (Editors), Landslides in a Semi-Arid Environment with Emphasis on the Inland Valleys of Southern California, Vol. 2. Inland Geological Society of Southern California Publications, Riverside, CA, part 1, pp. 118–149.

Kelsey, H.M., 1980. A sediment budget and an analysis of geomorphic process in the Van Duzen River basin, north coastal California, 1941–1975: summary. Geol. Soc. Am. Bull., 91(1): 190–195.

Kobayashi, Y., 1981. Causes of fatalities in recent earthquakes in Japan. J. Disaster Sci., 3: 15–22.

Lehre, A.K., 1981. Sediment budget of a small California Coast Range drainage basin near San Francisco. In: T.R.H. Davies and A.J. Pearce (Editors), Erosion and Sediment Transport in Pacific Rim Steeplands. Proc. Symp. Christchurch, New Zealand. Int. Assoc. Hydrol. Sci., IAHS–AISH Publ. 132, pp. 123–139.

Marchand, D.E., 1971. Rates and modes of denudation, White Mountains, eastern California. Am. J. Sci., 270: 109–135.

Mathur, L.P., 1953. Assam earthquake of 15th Aug., 1950 — a short note on factual observations. In: M.B. Ramachandra Rao (Editor), A Compilation of Papers on the Assam Earthquake of August 15, 1950. National Geophysical Research Institute, Hyperabad, India, The Central Board of Geophys. Publ., 1: 56–60.

Matthes, F.E., 1930. Geologic history of the Yosemite Valley. U.S. Geol. Surv. Prof. Pap., 160: 137 pp.

Meade, R.H., Yuzyk, T.R. and Day, T.J., 1990. Movement and storage of sediment in rivers of the United States and Canada. In: M.G. Wolman and H.C. Riggs (Editors), Surface Water Hydrology. The Geology of North America, O-1. Geol. Soc. of America, pp. 255–280.

Moberly, R., Jr., 1963, Rate of denudation in Hawaii. J. Geol., 71: 371–375.

Muir, J., 1912. The Yosemite. Doubleday, New York, 1962 Reprint ed., 225 pp.

Nieto, A.S., Schuster, R.L. and Plaza-Nieto, G., 1991. Mass wasting and flooding. In: R.L. Schuster (Editor), The March 5, 1987, Ecuador Earthquakes, Mass Wasting and Socioeconomic Effects. National Research Council, National Academy Press, Washington D.C., Natural Disaster Studies, 5: 51–82.

Nolan, K.M. and Hill, B.R., 1991. Suspended-sediment budgets for four drainage basins tributary to Lake Tahoe, California and Nevada, 1984–1987. U.S. Geol. Surv. Water-Resour. Invest. Rep., 91-4054: 40 pp.

Nolan, K.M. and Marron, D.C., 1988. Stream-channel response to the storm in the Santa Cruz Mountains. In: S.D. Ellen and G.F. Wieczorek (Editors), Landslides, Floods, and Marine Effects of the Storm of January 3–5, 1982, in the San Francisco Bay Region, California. U.S. Geol. Surv. Prof. Pap., 1434: 245–264.

Pain, C.F. and Bowler, J.M., 1973. Denudation following the November 1970 earthquake at Madang, Papua New Guinea. Z. Geomorphol. Suppl., 18: 92–104.

Pearce, A.J. and Watson, A.J., 1986, Effects of earthquake-induced landslides on sediment budget and transport over a 50-yr period. Geology, 14: 52–55.

Plafker, G., Ericksen, G.E. and Fernández Concha, J., 1971. Geological aspects of the May 31, 1970, Perú earthquake. Bull. Seismol. Soc. Am., 61: 543–587.

Real, C.R. and Toppozada, T.R., 1978. Earthquake catalog of California January 1, 1900–December 31, 1974. Calif. Div. Mines Geol. Spec. Publ., 52: 15 pp.

Reneau, S.L., 1988. Depositional and erosional history of hollows: application to landslide location and frequency, long-term erosion rates, and the effects of climatic change. Ph.D. Diss., Univ. of California, Berkeley, 328 pp.

Richter, C.F., 1958. Elementary Seismology. Freeman, San Francisco, 768 pp.

Ritter, D.F., 1967. Rates of denudation. J. Geol. Educ., 15: 154–159.

Rymer, M.J., 1987. The San Salvador earthquake of October 10, 1986 — geologic aspects. Earthquake Spectra, 3: 435–463.

Saunders, I. and Young, A., 1983. Rates of surface processes on slopes, slope retreat and denudation. Earth Surf. Process. Landforms, 8: 473–501.

Scott, K.M. and Williams, R.P., 1978. Erosion and sediment yields in the Transverse Ranges, southern California. U.S. Geol. Surv. Prof. Pap., 1030: 38 pp.

Seed, H.B., 1968. Landslides during earthquakes due to soil liquefaction. Am. Soc. Civ. Eng. J. Soil Mech. Found. Div., 94: 1053–1122.

Simonett, D.S., 1967. Landslide distribution and earthquakes in the Bewani and Torricelli Mountains, New Guinea — a statistical analysis. In: J.N. Jennings and J.A. Mabbutt (Editors), Landform Studies from Australia and New Guinea. Cambridge Univ. Press, Cambridge, pp. 64–84.

Stover, C.W. and Coffman, J.L., 1993. Seismicity of the United States, 1568–1989 (Revised). U.S. Geol. Surv. Prof. Pap., 1527: 418 pp.

Tanaka, M., 1976. Rate of erosion in the Tanzawa Mountains, central Japan. Geogr. Ann., 58A: 155–163.

U.S. Geological Survey, 1990. The next big earthquake in the Bay Area may come sooner than you think. U.S. Geol. Surv., Menlo Park, CA, 23 pp.

Varnes, D.J., 1978. Slope movement types and processes. In: R.L. Schuster and R.J. Krizek (Editors), Landslides — Analysis and Control. Nat. Acad. Sci. Transport. Res. Board Spec. Rep., 176: 11–33.

Wesnousky, S.G., 1986. Earthquakes, Quaternary faults, and seismic hazards of California. J. Geophys. Res., 91B: 12,587–12,631.

Wieczorek, G.F., Snyder, J.B., Alger, C.S. and Isaacson, K.A., 1992. Rock falls in Yosemite Valley, California. U.S. Geol. Surv. Open-File Rep., 92-387A: 185 pp.

Yoshikawa, T., 1974. Denudation and tectonic movement in contemporary Japan. Bull. Dept. Geogr., Univ. Tokyo, 6: 1–14.

Young, A., 1969. Present rate of land erosion. Nature, 224: 851–852.

ELSEVIER

Geomorphology 10 (1994) 285–315

GEOMORPHOLOGY

Large mid-Holocene and late Pleistocene earthquakes on the Oquirrh fault zone, Utah

Susan S. Olig*, William R. Lund, Bill D. Black

Utah Geological Survey, 2363 South Foothill Drive, Salt Lake City, UT 84109-1491, USA

Received March 20, 1994; revised April 5, 1994; accepted April 6, 1994

Abstract

The Oquirrh fault zone is a range-front normal fault that bounds the east side of Tooele Valley and it has long been recognized as a potential source for large earthquakes that pose a significant hazard to population centers along the Wasatch Front in central Utah. Scarps of the Oquirrh fault zone offset the Provo shoreline of Lake Bonneville and previous studies of scarp morphology suggested that the most recent surface-faulting earthquake occurred between 9000 and 13,500 years ago. Based on a potential rupture length of 12 to 21 km from previous mapping, moment magnitude (M_w) estimates for this event range from 6.3 to 6.6. In contrast, our results from detailed mapping and trench excavations at two sites indicate that the most-recent event actually occurred between 4300 and 6900 yr B.P. (4800 and 7900 cal B.P.) and net vertical displacements were 2.2 to 2.7 m, much larger than expected considering estimated rupture lengths for this event. Empirical relations between magnitude and displacement yield M_w 7.0 to 7.2. A few, short discontinuous fault scarps as far south as Stockton, Utah have been identified in a recent mapping investigation and our results suggest that they may be part of the Oquirrh fault zone, increasing the total fault length to 32 km. These results emphasize the importance of integrating stratigraphic and geomorphic information in fault investigations for earthquake hazard evaluations.

At both the Big Canyon and Pole Canyon sites, trenches exposed faulted Lake Bonneville sediments and thick wedges of fault-scarp derived colluvium associated with the most-recent event. Bulk sediment samples from a faulted debris-flow deposit at the Big Canyon site yield radiocarbon ages of 7650 ± 90 yr B.P. and 6840 ± 100 yr B.P. (all lab errors are $\pm 1\sigma$). A bulk sediment sample from unfaulted fluvial deposits that bury the fault scarp yield a radiocarbon age estimate of 4340 ± 60 yr B.P. Stratigraphic evidence for a pre-Bonneville lake cycle penultimate earthquake was exposed at the Pole Canyon site, and although displacement is not well constrained, the penultimate event colluvial wedge is comparable in size to the most-recent event wedges. Charcoal from a marsh deposit, which overlies the penultimate event colluvium and was deposited during the Bonneville lake cycle transgression, yields an AMS radiocarbon age of $20{,}370 \pm 120$ yr B.P. Multiple charcoal fragments from fluvial deposits faulted during the penultimate event yield an AMS radiocarbon age of $26{,}200 \pm 200$ yr B.P. Indirect stratigraphic evidence for an antepenultimate event was also exposed at Pole Canyon. Charcoal from fluvial sediments overlying the eroded free-face for this event yields an AMS age of $33{,}950 \pm 1160$ yr B.P., providing a minimum limiting age on the antepenultimate event.

Ages for the past two events on the Oquirrh fault zone yield a recurrence interval of 13,300 to 22,100 radiocarbon years and estimated slip rates of 0.1 to 0.2 mm/yr. Temporal clustering of earthquakes on the nearby Wasatch fault zone in the late Holocene does not appear to have influenced activity on the Oquirrh fault zone. However, consistent with findings on the Wasatch fault zone and with some other Quaternary faults within the Bonneville basin, we found evidence for higher rates of activity during interpluvial periods than during the Bonneville lake cycle. If a causal relation between rates of strain release along faults

*Presently at Woodward-Clyde Federal Services, 500 12th St., Oakland, CA 94607, USA.

SSDI 0169-555X(94)00027-O

and changes in loads imposed by the lake does exist, it may have implications for fault dips and mechanics. However, our data are only complete for one deep-lake cycle (the past 32,000 radiocarbon years), and whether this pattern persisted during the previous Cutler Dam and Little Valley deep-lake cycles is unknown.

1. Introduction

The Oquirrh fault zone lies near the eastern boundary of the Basin and Range province in central Utah (Fig. 1). It has long been recognized as a potential source for large earthquakes (Gilbert, 1890), which would not only adversely affect the city of Tooele, Tooele Army Depot, and the USPCI hazardous waste incinerators, but also the more populous Wasatch Front to the east. For example, although downtown Salt Lake City is roughly 30 km east of the fault trace, ground motion estimates for a surface-faulting earthquake on the Oquirrh fault zone exceed 0.2 g using Campbell's (1987) empirical relation for Utah. These motions could induce liquefaction and would probably seriously damage older buildings and bridges.

Even though the Oquirrh fault zone has been mapped (Everitt and Kaliser, 1980; Barnhard and Dodge, 1988; Solomon, 1993), little is known about the prehistoric behavior of the fault, which has hampered comprehensive seismic-hazard evaluations in central Utah (e.g., Arabasz et al., 1989; Youngs et al., 1987). This study investigated the paleoseismology of the Oquirrh fault zone primarily to provide better information for characterizing the fault as a potential earthquake source in hazard evaluations.

We were also interested in comparing faulting activity on the Oquirrh fault zone with the chronology of late Pleistocene Lake Bonneville and with faulting patterns on nearby faults, such as the Wasatch fault zone, a large range-bounding normal fault roughly 35 km to the east, and the West Valley fault zone, a short fault zone within Salt Lake Valley, located between the Oquirrh fault zone and the Wasatch fault zone (Fig. 1). Many researchers have suggested that there may be a causal relation between changes in loads imposed by Lake Bonneville and patterns of late Quaternary faulting in the Bonneville basin (Gilbert, 1890; Swan et al., 1983; Keaton et al., 1987; Machette et al., 1992; McCalpin et al., 1992). However, faulting patterns and their apparent relation to lake chronology vary considerably within the basin and differing explanations have been proposed to explain the patterns.

For example, extensive paleoseismic investigations of the central segments of the Wasatch fault zone show evidence for temporal clustering of earthquakes during the late Holocene and a possible causal relation between temporal changes in slip rates and deep cycles of Lake Bonneville, with high rates occurring during interpluvial periods and possibly related to unloading of the crust (Machette et al., 1991, 1992). Keaton et al. (1987) found evidence for greater rates of activity along the West Valley fault zone during interpluvial periods, also suggesting a causal relation between loads imposed by Lake Bonneville and decreased faulting activity. In comparison, McCalpin et al. (1992) and McCalpin and Forman (1991) found evidence for higher rates of activity during the Bonneville lake cycle along the Hansel Valley and East Cache fault zones in the northern Bonneville basin. More paleoseismic data on faults within the Bonneville basin are needed to further define possible relations between faulting and lake chronology.

Our study included interpretation of aerial photographs, detailed geologic mapping of surficial deposits at trench sites, profiling fault scarps, excavating and mapping trenches at two sites, and radiocarbon dating key stratigraphic units. Our results emphasize the importance of integrating data from mapping, trenching, and scarp profiling in fault investigations because: the most-recent surface-faulting event was younger than previously estimated based on scarp-morphology data alone, and net vertical displacements for faulting events were much larger than would be expected from the mapped fault trace length.

2. Geologic setting

The Oquirrh fault zone extends for about 21 km from Lake Point to south of Tooele along the east side of Tooele Valley at the base of the Oquirrh Mountains (Fig. 2). This west-dipping normal fault varies in strike from north-northwest to northeast. A variety of different names have been used for the Oquirrh fault zone including: the Oquirrh marginal fault (Everitt and Kal-

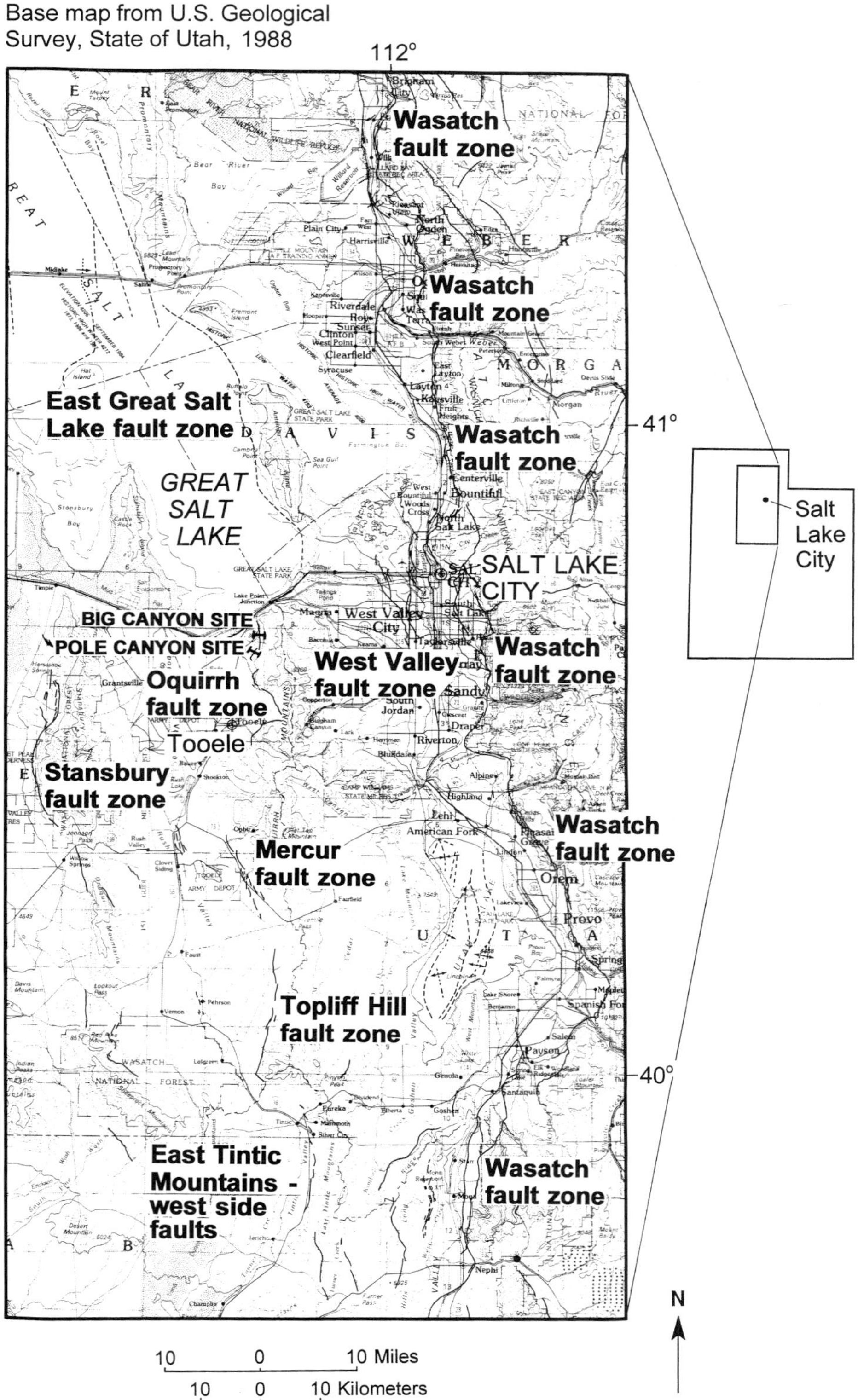

Fig. 1. Regional Quaternary fault map of central Utah showing location of the Oquirrh fault zone (modified from Hecker, 1993).

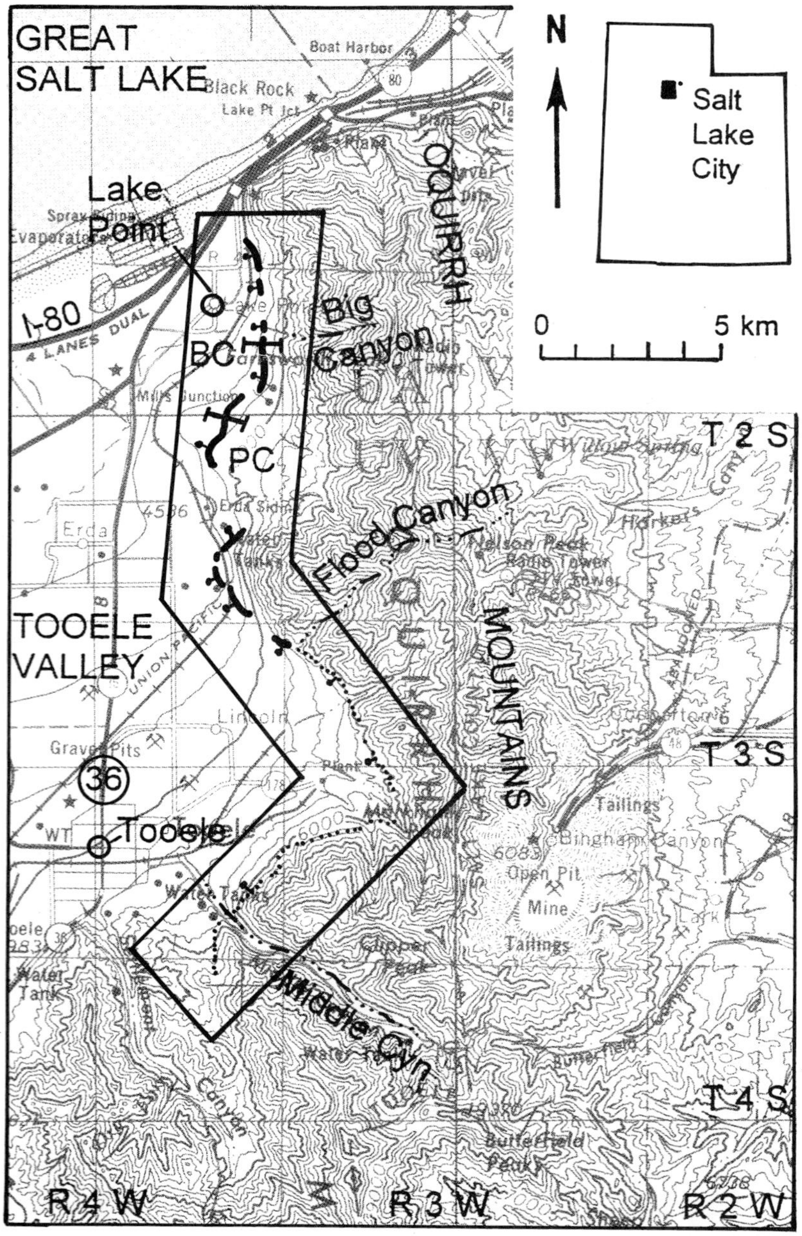

Fig. 2. Map of the Oquirrh fault showing trench site locations: *PC* – Pole Canyon; *BC* – Big Canyon (modified from Barnhard and Dodge, 1988). Heavy lines represent fault scarps on alluvium with bars and balls on downthrown side. Dotted lines represent fault traces at the bedrock–alluvium contact.

iser, 1980), the northern Oquirrh fault zone (Barnhard and Dodge, 1988; Hecker, 1993); and the Oquirrh fault zone (Barnhard, 1988; Olig et al., 1993). The latter name is retained here for simplicity and because the northern Oquirrh Mountains is not a formal place name.

Major west-dipping, range-bounding Quaternary faults along strike with the Oquirrh fault zone include the East Great Salt Lake fault zone to the north (Viviveros, 1986; Pechmann et al., 1987) and the Mercur and Topliff Hills fault zones, and faults along the East Tintic

Mountains to the south (Bucknam and Anderson, 1979a; Everitt and Kaliser, 1980; Barnhard and Dodge, 1988) (Fig. 1). Because these faults have a similar sense of displacement (down to the west) and form a north–south trending, albeit somewhat discontinuous, zone of faults through central Utah, they could be considered as individual segments in a large fault zone that extends for 205 km from Rouzel Bay in Great Salt Lake south to Furner Pass. Indeed, in their hazard evaluation for the Wasatch Front, Youngs et al. (1987) grouped these faults together as one zone that they called the Oquirrh Mountains fault zone. However, neither Anderson and Miller (1979) nor Hecker (1993) chose to group all of these faults together as one zone in their compilations and we have followed this nomenclature here.

This part of central Utah was characterized by roughly east–west extension during the Quaternary, as indicated by dominantly normal slip on numerous faults that generally strike north (Fig. 1; Zoback, 1983; Hecker, 1993). For example, the Oquirrh fault zone is roughly 35 km west of the Salt Lake City segment of the Wasatch fault zone. The Wasatch fault zone is the most active Quaternary fault in Utah (Hecker, 1993) and it marks the eastern boundary of the Basin and Range province (Machette et al., 1991). The area is also within the central part of the Intermountain seismic belt, a north-trending zone of shallow, diffuse intraplate seismicity that extends from Montana, through central Utah, to northern Arizona (Smith and Sbar, 1974; Smith and Arabasz, 1991). The non-surface-rupturing earthquakes in the Intermountain seismic belt (usually less than magnitude 6.5) are generally difficult to associate with particular Quaternary faults (Arabasz et al., 1992). However, seismicity and well-bore breakouts along the Wasatch Front indicate extension is still occurring in a west-northwest to westerly direction (Zoback, 1983, 1989; Eddington et al., 1987; Bjornadottir and Pechmann, 1989), consistent with extension along Quaternary faults.

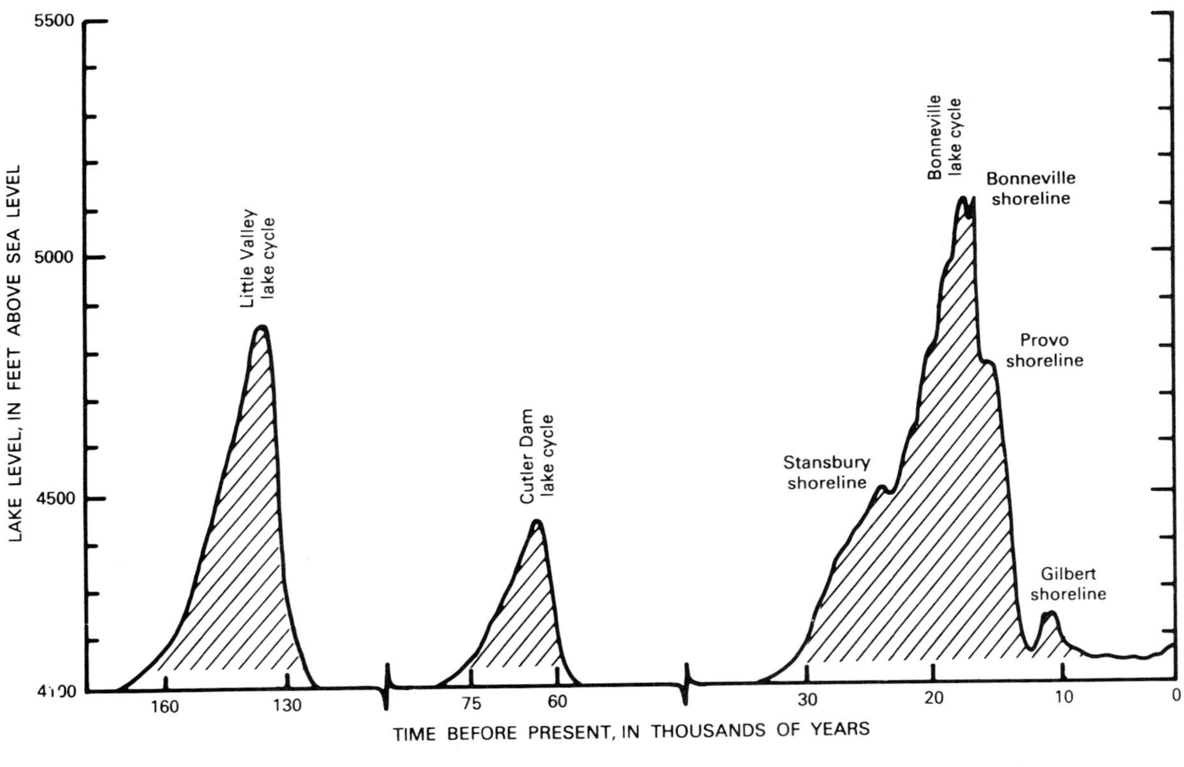

Fig. 3. Time–altitude diagram of deep-lake cycles in the Bonneville basin during the past 170,000 years (from Machette et al., 1992 and originally modified from Currey and Oviatt, 1985).

Late Quaternary deposition along the western base of the Oquirrh Mountains was dominated by alluvial-fan sediments being shed off the mountains and deposition of lacustrine sediments (Solomon et al., 1992). Tooele Valley lies within the Bonneville basin, which was repeatedly inundated by paleolakes during the late Quaternary. Fig. 3 depicts the altitude through time of the three most recent deep-lake cycles: the Bonneville, Cutler Dam (or Hansel Valley), and the Little Valley. The chronology of these lake cycles is fairly well understood (e.g., Currey, 1982; Scott et al., 1983; Currey and Oviatt, 1985; McCoy, 1987; Oviatt et al., 1987; 1992; McCalpin et al., 1992), particularly for Lake Bonneville. Thus, key stratigraphic and geomorphic features associated with these lake cycles can be used to help constrain the timing of faulting events during the last 100,000 years (e.g., Olig et al., 1986; Machette, 1988; Machette et al., 1992; McCalpin et al., 1992). It should be noted that the detailed chronology for the Bonneville Lake cycle is based on numerous radiocarbon ages (Oviatt et al., 1992) and so this chronology is in radiocarbon years and not calendar-corrected for the variable production of radiocarbon through time (Stuiver and Quay, 1979).

3. Previous work

Previous studies of the Oquirrh fault zone focused on the surficial expression of the fault. Gilbert (1890, p. 352) first identified west-facing fault scarps on Lake Bonneville deposits along the west side of the northern Oquirrh Mountains. Bucknam (1977) first mapped the fault in his regional study. Everitt and Kaliser (1980) mapped the fault in more detail and they recognized two sections: a northern section that included fault scarps in alluvium, and a southern section that included a fault contact between bedrock and alluvium along the range front. In their regional study of fault scarp morphologies, Barnhard and Dodge (1988) also identified northern and southern sections of the fault on the same basis.

Our investigation focused on the northern section of the fault, which consists of a discontinuous zone of fault scarps on unconsolidated deposits that extend from east of Lake Point south to Flood Canyon (Fig. 2). These scarps are generally below the Provo shoreline, but Barnhard and Dodge, (1988) identified three localities where fault scarps climb upslope and offset the abrasion platform of the Provo shoreline, constraining the most recent faulting to be younger than about 14,000 radiocarbon-years ago (Oviatt et al., 1992).

The southern section is characterized by a fault contact between bedrock and alluvium, extending for 8 km from Flood Canyon south to Pine Canyon, where the fault makes a near right-angle bend and continues southwest for 10 km to a canyon south of Middle Canyon (Fig. 2). It is more difficult to constrain the timing of most-recent faulting on this southern section of the fault because it lies entirely above the Bonneville shoreline and only very young and very old deposits are preserved along the range front. Solomon (1993) mapped some short fault scarps preserved on older Quaternary gravels, but he found no scarps offsetting younger alluvial fans at canyon mouths along the southern section of the fault.

Barnhard and Dodge (1988) measured topographic profiles to characterize scarp morphology and help constrain fault behavior. Scarp heights ranged between 2.9 and 10.8 m and surface offsets between 1.3 and 7.3 m. Many of the scarps were compound, representing displacement during more than one surface-faulting earthquake. Barnhard and Dodge (1988) compared a plot of scarp heights versus the maximum slope angle for the Oquirrh fault zone with similar plots for other scarps of known ages in the Bonneville basin. Based on this and crosscutting relations with the Provo shoreline, they suggested that the latest faulting on the Oquirrh fault zone occurred between 9,000 and 13,500 years ago, and probably was closer to 13,500 years ago. However, results from our trenching, discussed in the following sections, indicate that the latest faulting was actually much younger.

4. Sequence of deposition and faulting

4.1. Big Canyon site

Surficial geology

The Big Canyon trench site is roughly 2 km southeast of Lake Point and 0.3 km west of the mouth of Big Canyon (Fig. 2). Here the fault comprises a west-facing main scarp and a smaller east-facing antithetic scarp, forming a graben 50 to 150 m wide (Figs. 4 and 5). The altitude near the base of the main fault scarp is

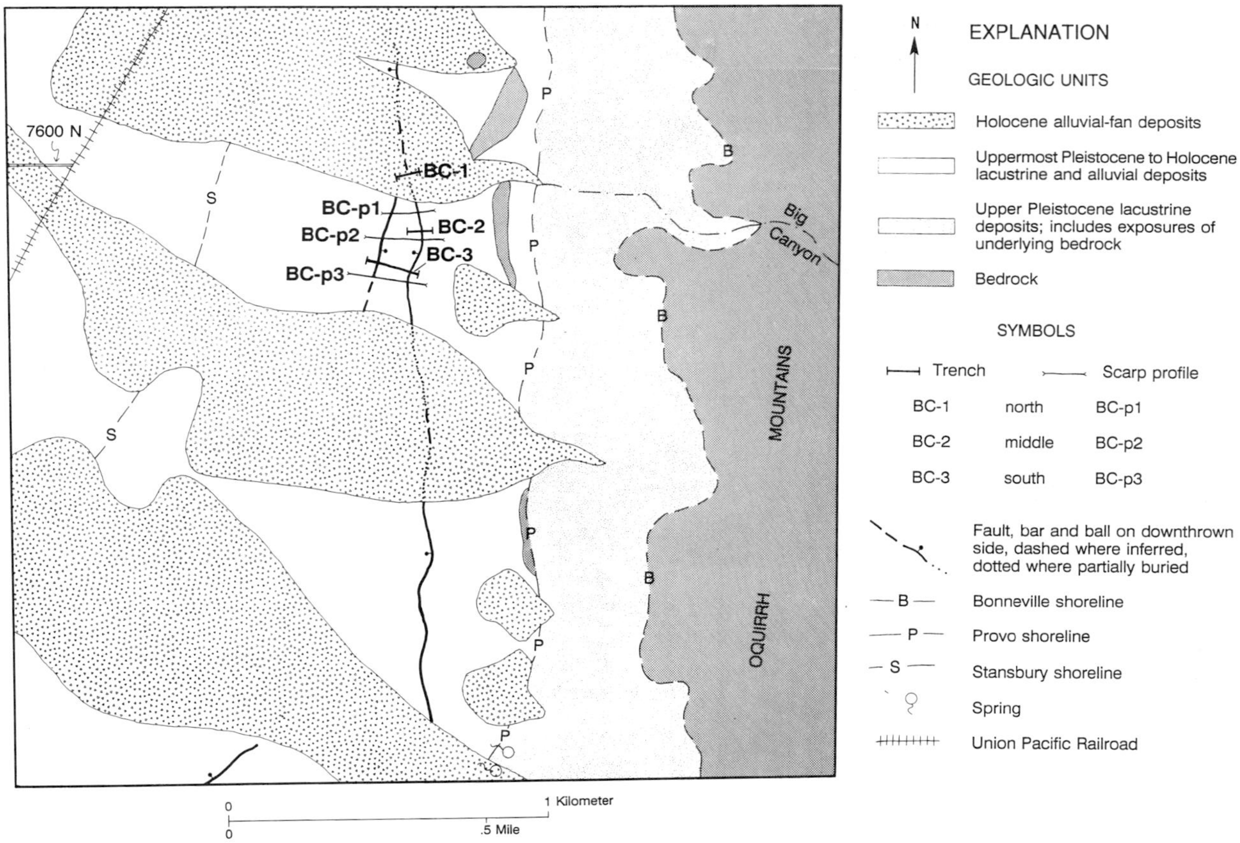

Fig. 4. Generalized geologic map of the Big Canyon site.

about 1427 m; roughly 60 m below the Provo shoreline (Figs. 4 and 5). Using Oviatt et al.'s (1992) simple linear method to adjust for isostatic rebound yields an adjusted altitude of about 1403 m. The time-altitude diagram for the Bonneville lake cycle (Fig. 3) indicates that the lake transgressed across the site about 20,000 radiocarbon-years ago, subsequently formed the Bonneville shoreline, then dropped from the Bonneville highstand to the Provo level during the Bonneville Flood about 14,500 radiocarbon-years ago, and finally regressed from the site about 13,000 radiocarbon-years ago (Scott et al., 1983; Currey and Oviatt, 1985; Oviatt et al., 1992). Since that time, deposition near the Big Canyon site has been dominated by the construction of alluvial fans at the mouths of drainages and the reworking of lacustrine and alluvial deposits elsewhere on the piedmont, with some areas such as the graben also trapping significant amounts of eolian sediment. At the site, both the main and antithetic fault scarps offset lacustrine and reworked lacustrine deposits, but are buried to the north by younger alluvial-fan deposits from Big Canyon. Thus, the timing of latest faulting on these scarps is younger than 13,000 radiocarbon years, but older than the youngest fan alluvium.

Profiles of fault scarps on reworked lacustrine deposits at the Big Canyon site yield scarp heights ranging from 12 to 18 m, maximum slope angles of 24° to 32°, and surface offsets of 4.0 to 6.8 m (Fig. 6). Complex profiles with a break-in-slope between the crest and base, along with the large surface offsets indicate that the main scarp is probably compound and represents cumulative displacement from more than one faulting event. Although scarp measurements might suggest 4.0–6.8 m of net vertical displacement occurred in the last 13,000 radiocarbon years, highly eroded scarp crests and the concentration of boulders along the scarp (Fig. 5) suggest that a pre-existing scarp was partially modified by transgression of the Bonneville lake cycle,

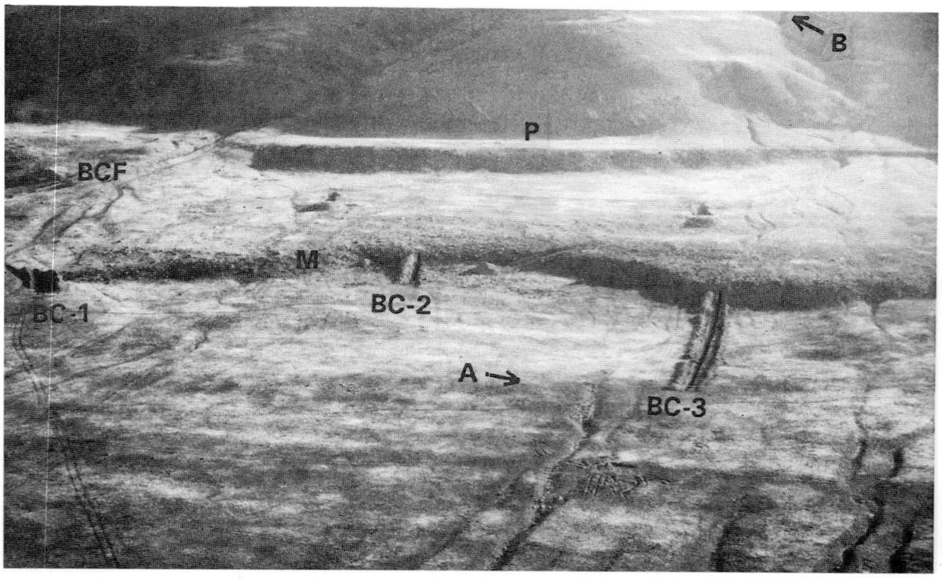

Fig. 5. Oblique aerial view of the Big Canyon site (looking east) showing locations of trenches (BC-1, BC-2, and BC-3) relative to key geomorphic features: *M* – main fault scarp; *A* – antithetic fault scarp; *BCF* – alluvial fan at the mouth of Big Canyon; *P* – Provo shoreline; *B* – Bonneville shoreline.

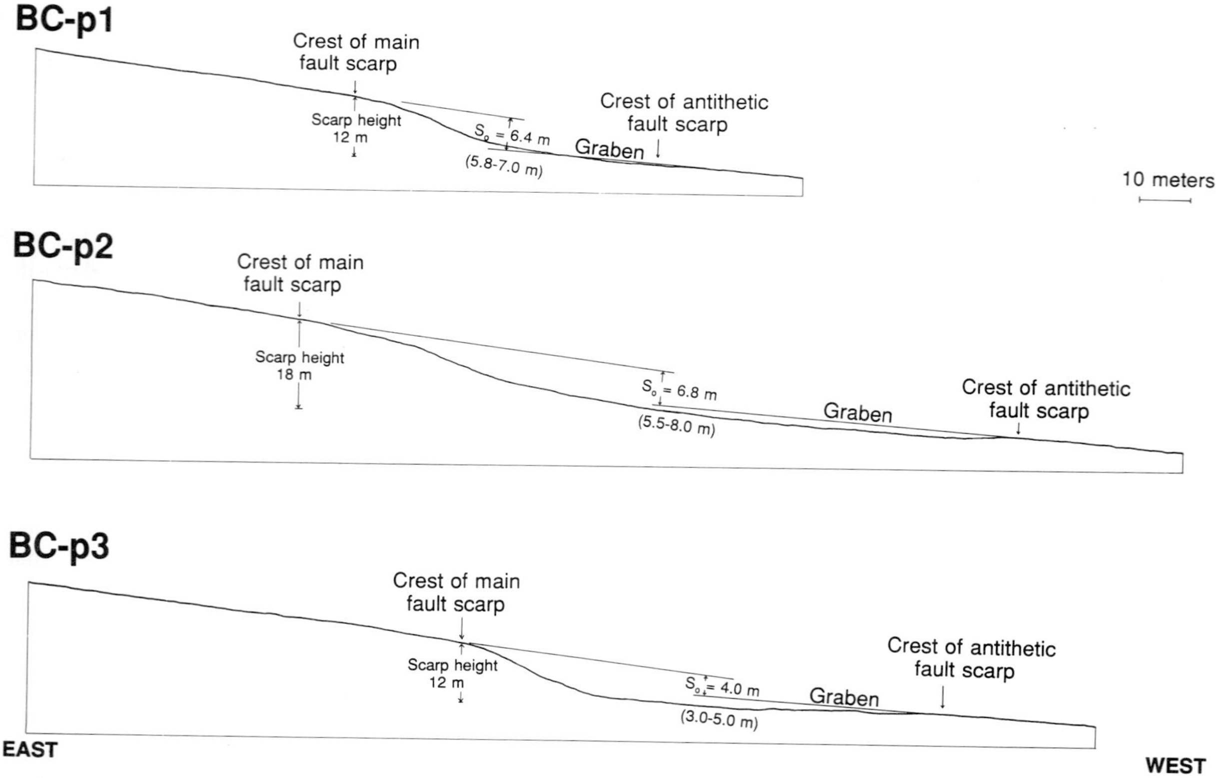

Fig. 6. Topographic scarp profiles measured at the Big Canyon site. Preferred estimates of surface offset (S_o) are based on a surface reconstruction using a break in slope halfway between the crests of the main and antithetic fault scarps. Range of values (in parentheses) is the S_o at 1/3 and 2/3 the distance between the crests of the main and antithetic fault scarps.

with the boulders probably representing a beach lag deposit. Thus, not all of the measured surface offset would have actually occurred during the last 13,000 radiocarbon years.

Trench exposures

Three trenches were excavated at the Big Canyon site: trench BC-3 cut across the entire graben, trench BC-2 only crossed the main fault, and trench BC-1 was in unfaulted Big Canyon fan alluvium that buried the main fault scarp (Figs. 4 and 5). Trenches BC-2 and BC-3 exposed very similar stratigraphy with evidence for one surface-faulting event since the Bonneville lake cycle (Figs. 7, 8, 9, and 10). In both trenches, a well-defined main fault zone had down-dropped a debris-flow deposit overlying lacustrine sediments in the hanging wall, against older alluvial-fan sediments in the footwall (Figs. 8 and 10). An unfaulted wedge-shaped deposit of scarp-derived colluvium was overlying the debris-flow deposit at the base of the scarp in the hanging wall. A thin layer of unfaulted colluvium and alluvium draped the entire scarp.

In trench BC-2, the main fault consisted of a single splay, striking N3–5°W, dipping 85°W, and was characterized by well-developed shear fabric (preferential alignment of elongate clasts) (Fig. 10). In trench BC-3, the main fault was about 2 m wide and consisted of a 20-cm wide major zone, also characterized by well-developed shear fabric, and four smaller subsidiary faults within the hanging wall (Fig. 8). The major zone was striking N17–22°E and dipping 79–84°W. Drag folding of hangingwall sediments was observed in both trenches, but was particularly pronounced in trench BC-3. The antithetic fault zone in trench BC-3 consisted of three, near-vertical splays oriented from N5°E to N15°E, and forming a zone about 1 m wide.

We interpret lacustrine sediments exposed near the fault in the hanging wall (units 8–10, Figs. 8 and 10) to be a nearly complete package of Bonneville lake cycle sediments based on their geomorphic and stratigraphic position, lithologic characteristics, the presence of ostracods in unit 10, and the presence of gastropods in unit 9. In both trenches, unit 10 contains a thin, nearly continuous sand horizon (Figs. 8 and 10). Horizons such as this, with medium to coarse particles of sand or ostracods, are common at this stratigraphic position throughout the Bonneville basin and probably were deposited during the Bonneville Flood (D.R. Currey, pers. commun., 1992).

Unconformably overlying unit 10 were well-rounded, tufa-cemented gravels (unit 11) that could almost be traced throughout the entire length of trench

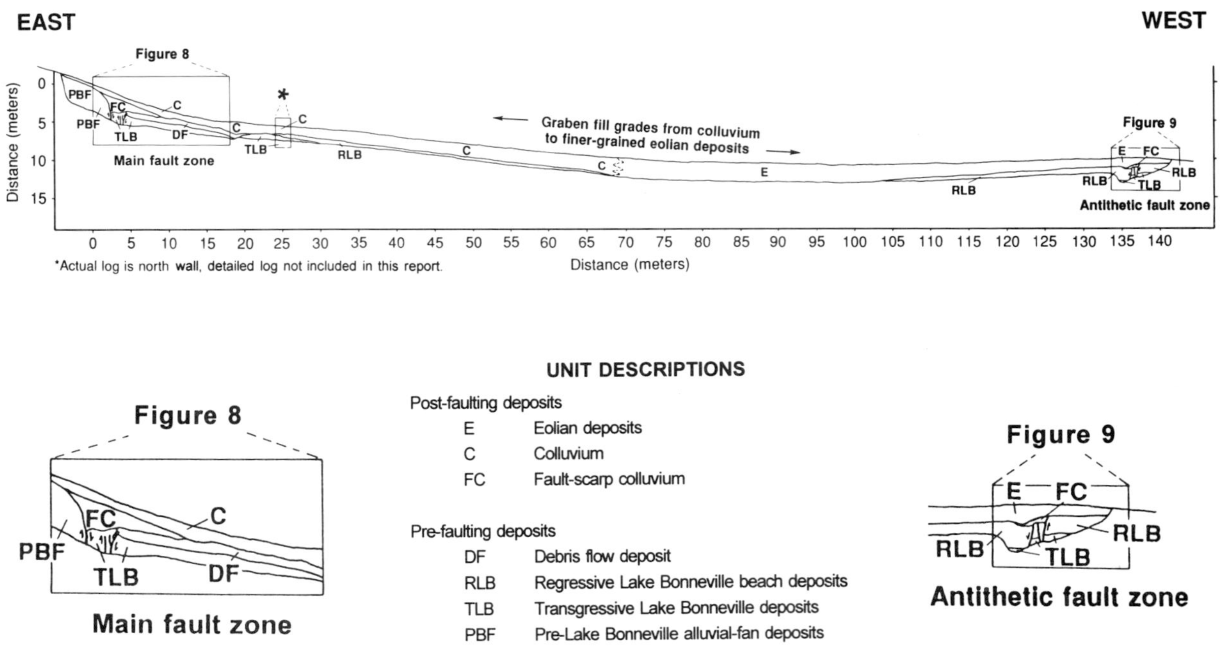

Fig. 7. Diagram of trench BC-3 at the Big Canyon site.

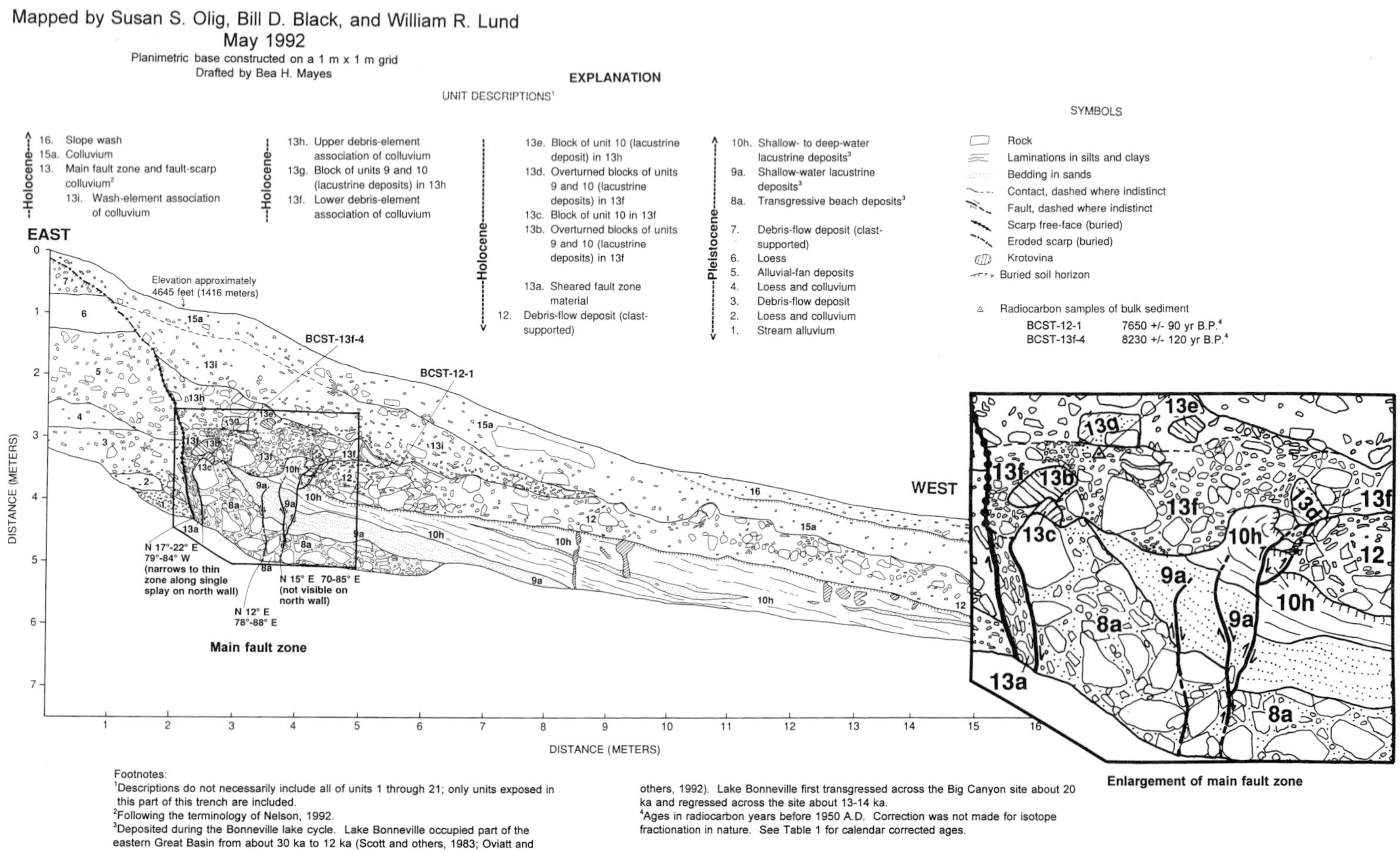

Fig. 8. Log of the main fault zone in trench BC-3 at the Big Canyon site.

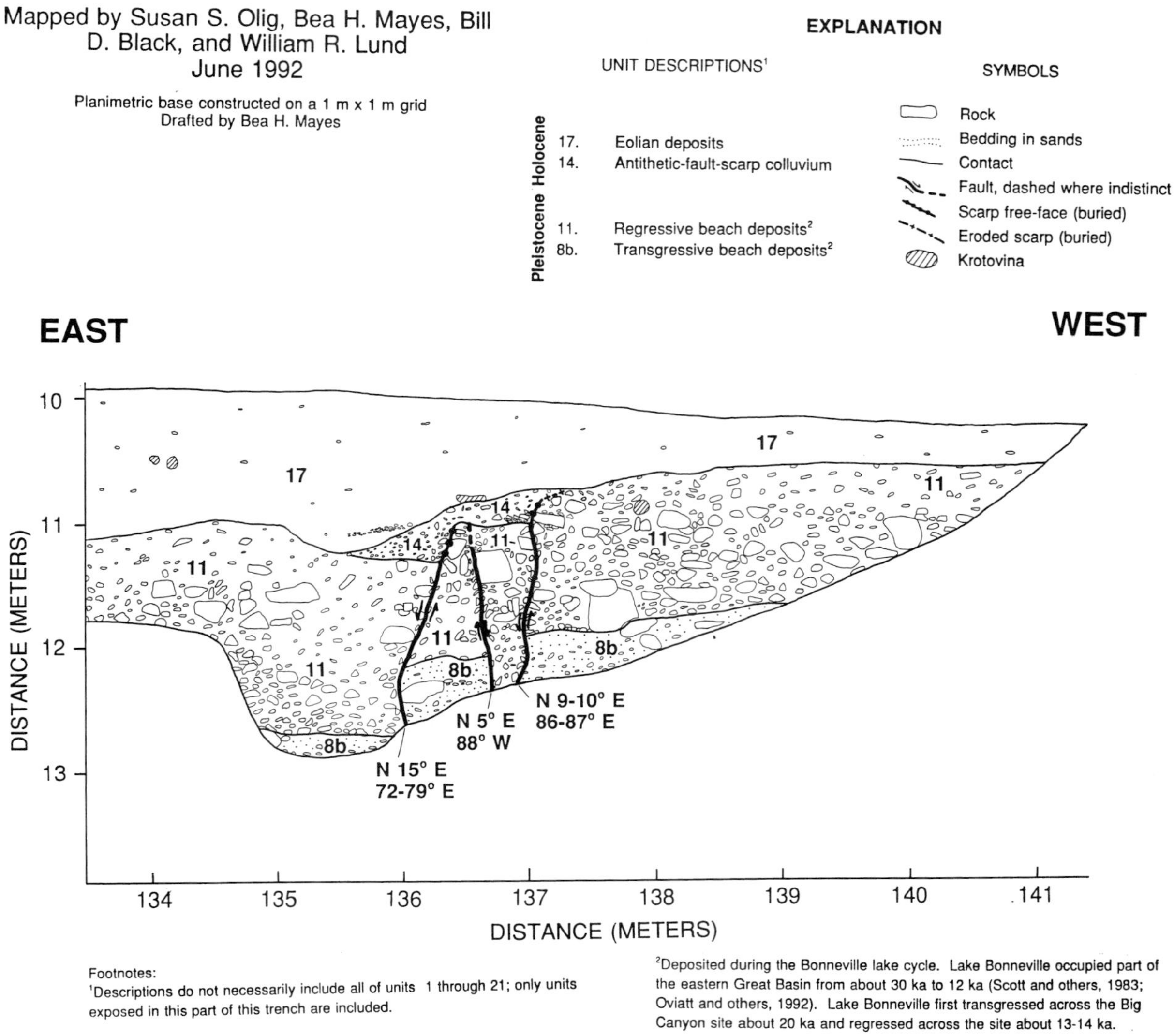

Fig. 9. Log of the antithetic fault zone in trench BC-3 at the Big Canyon site.

BC-3 (Fig. 7). Unit 11 is interpreted to be regressive beach deposits of the Bonneville lake cycle. These are the youngest deposits exposed in the west end of trench BC-3 that are offset by the antithetic fault zone (Fig. 10); they were probably deposited about 13,000 radiocarbon-years ago (Fig. 3). A buried, weakly developed Bt horizon was observed on parts of units 10 and 11, indicating that they were subaerially exposed for at least some time before unit 10 was buried by the overlying debris-flow deposit (unit 12, Figs. 8 and 10). The soil profile is variably truncated, with the A horizon and part to all of the B horizon eroded away.

Also within unit 10, trench BC-2 exposed evidence for soft-sediment deformation including convoluted bedding and injection features. Sand and pea-size gravel, injected at high angles to bedding in unit 10, formed an extremely irregular and convoluted contact (Fig. 10, between stations 2 and 6). Clasts larger than 50 cm were also incorporated into unit 10. This deformation clearly post-dates deposition of the sand horizon in unit 10 and probably predates subaerial exposure of unit 10, indicating the deformation occurred between 13,000 and 14,500 radiocarbon-years ago.

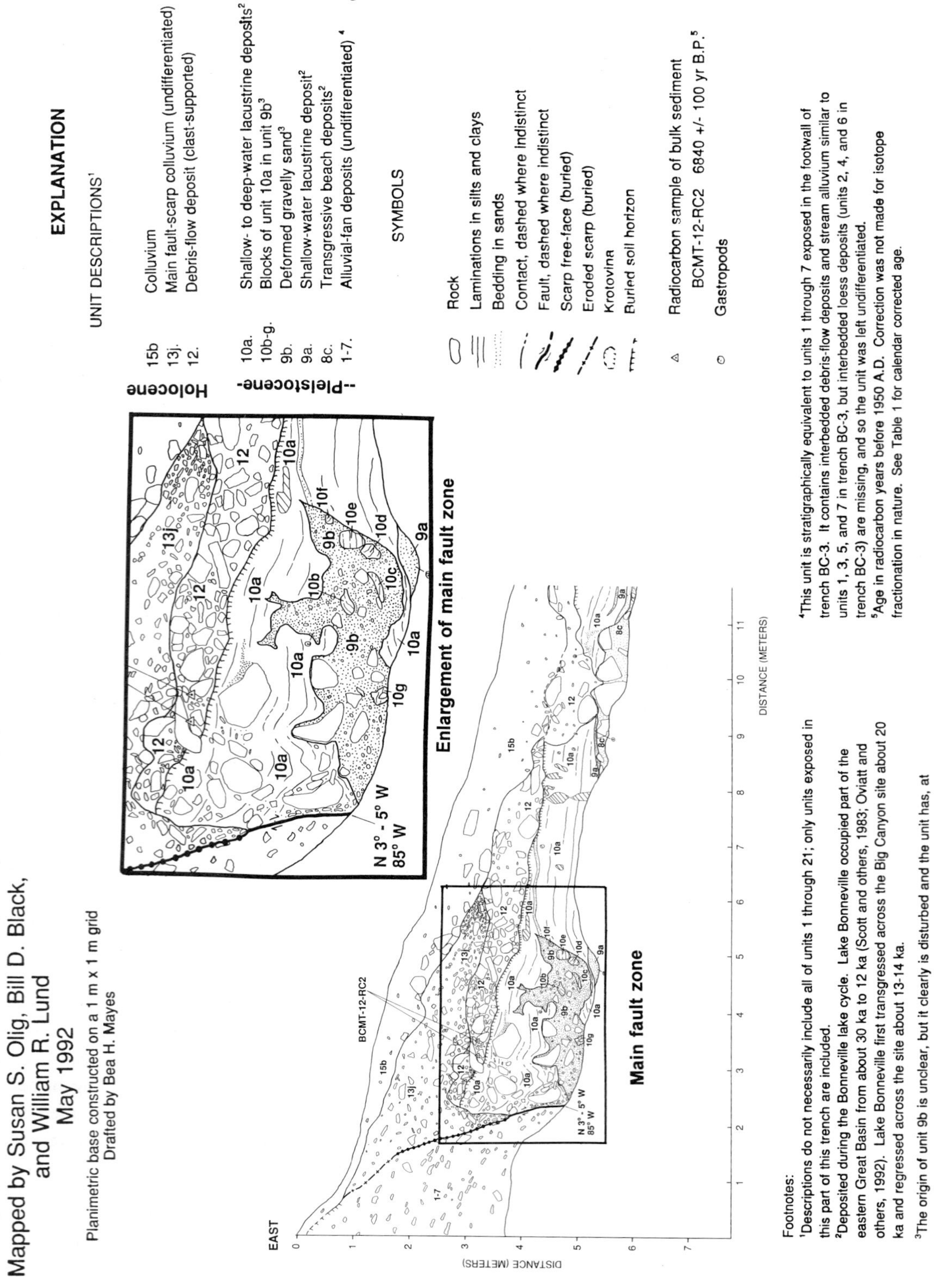

Fig. 10. Log of the main fault zone in trench BC-2 at the Big Canyon site.

The characteristics of the deformation and its localized distribution (i.e., it was not observed in trench BC-2) suggests it could have been caused by a debris-flow entering Lake Bonneville, probably during the Provo highstand. Alternatively, it is possible that the deformation could be related to liquefaction induced by an earthquake that apparently did not cause surface rupture on the Oquirrh fault zone (based on observations in trench BC-3 and at the Pole Canyon trench site). In fact, this event may not have occurred on the Oquirrh fault zone. Although liquefaction of material as coarse as gravel is rare, it has been observed, such as in the 1983 M_s 7.3 Borah Peak, Idaho, earthquake (L. Youd, pers. commun., 1992).

Unit 12 was the youngest unit offset by the main fault in trenches BC-2 and BC-3 (Figs. 8 and 10). This debris-flow deposit and underlying Bonneville lake cycle sediments were drag-folded up against the main fault in both trenches. Although the matrix of the debris-flow deposit contained disseminated charcoal flecks and other organic material, we did not observe an A horizon developed on this unit. We believe that the variable mottled texture and distribution of organic material throughout this unit is associated with a reworked soil incorporated during mobilization of the flow. If this is true, the organic material in the matrix of this deposit could possess a significant inherited age and be much older than the debris-flow event. Bulk sediment samples from unit 12 yield radiocarbon age estimates of 6840 ± 100 yr B.P. (trench BC-2) and 7650 ± 90 yr B.P. (trench BC-3) (Table 1). We believe that the difference of roughly 800 radiocarbon years between these estimates does not necessarily indicate that the deposits were laid down during different debris-flow events, rather that the difference reflects the age span of detrital organic material from the reworked soil incorporated into the debris flow. Detailed studies of modern soil profiles developed on alluvial-fan sediments along the Wasatch Front suggest that mean residence times can be as long as 800 years for A horizons and 2000 years for C horizons (Stafford and Forman, 1993). Thus, an age span of at least 800 years appears reasonable for the carbon incorporated into unit 12 from throughout a pre-existing soil profile.

Because both radiocarbon ages for unit 12 are probably older than the actual debris-flow event, the youngest ages provides the best constraint on the maximum age for most recent faulting on the Oquirrh fault zone. The calibrated age, corrected for variations of ^{14}C production in the atmosphere (Stuiver and Quay, 1979), is 7600^{-100}_{+300} cal B.P. (rounded to the nearest century — see Table 1 footnotes for details on the calibration method used and a caveat on σ).

After the most-recent event, a thick wedge (2.6 to 2.8 m) of colluvium accumulated at the base of the main fault scarp and partially buried the free-face (Figs. 8 and 10). Smaller colluvial wedges (<0.5 m thick) were also deposited at the base of the antithetic fault scarp (Fig. 9). Stratigraphic and geomorphic relations strongly suggest that the most-recent event on the antithetic fault zone is contemporaneous with the most-recent event on the main fault zone at the Big Canyon site. Fault-scarp derived colluvium exposed in trench BC-3 included distinct upper and lower debris-element associations (units 13b through 13h) and a wash-element association (unit 13i), exhibiting many typical characteristics of deposits derived from normal fault scarps (Nelson, 1992). Unit 13j in trench BC-2 was relatively more homogeneous and without distinctive slope and debris facies. It did have a 1.1-m long "heal", interpreted to be a filled fissure at the base of the free-face.

A bulk sediment sample from an organic-rich section of unit 13f in trench BC-3 (Fig. 8) yielded a radiocarbon age of 8230 ± 120 yr B.P. (Table 1). However, unit 13f contains components of many older reworked deposits (for example units 13b through 13e). Because this age estimate is older than the ages for the underlying unit (12), we believe that this sample contained significant amounts of older detrital carbon reworked into the colluvium, and therefore does not provide the best minimum limiting age for most recent faulting.

Trench BC-1 exposed a 4-m-thick package of unfaulted, interbedded debris-flow deposits and alluvium (Fig. 11) that clearly bury the main fault scarp (Fig. 5). A bulk sediment sample from unit 19 yielded a radiocarbon age of 4340 ± 60 yr B.P. and a calibrated age of 4900 ± 100 cal B.P. (Table 1). Similar to the other bulk sediment samples, this age provides a maximum limiting estimate for the actual age of the deposit due to the age span of the detrital carbon in the sample. However, there is at least a half a meter of older unfaulted deposits underlying unit 19; therefore, we believe that the age estimate from unit 19 provides a reasonable minimum age limit for most-recent faulting

Table 1
Radiocarbon ages for samples collected from trenches across the Oquirrh fault zone

Location (site and trench)[a]	Field and laboratory numbers	Sample material[b] and unit sampled	Radiocarbon age[c] and laboratory error (yr B.P.)	Calibrated age[d] and range of 1σ (cal B.P.)	Remarks
Big Canyon					
Trench BC-1	BCNN-19-1 BETA 60645	Bulk sediment, matrix of unfaulted stream alluvium, Unit 19	4340 ± 60	4900 (4800–5000)	Minimum limiting age for the most recent event
Trench BC-2	BCMT-12-RC2 BETA 55559	Bulk sediment, matrix of faulted debris-flow deposit, Unit 12	6840 ± 100	7600 (7500–7900)	Best maximum limiting age estimate for the most recent event
Trench BC-3	BCST-12-1 BETA 55560	Bulk sediment, matrix of faulted debris-flow deposit, Unit 12	7650 ± 90	8400 (8200–8600)	Maximum limiting age for the most recent event
	BCST-13f-4 BETA 55561	Bulk sediment, matrix of fault-scarp colluvium, debris facies, Unit 13f	8230 ± 120	9200 (8900–9400)	Probably contains older detrital carbon
Pole Canyon					
Trench PC	OFPC-RC2 BETA 63545	Charcoal, Lake Bonneville transgressive marsh deposit, Unit D1a	20370 ± 120	Too old	Minimum limiting age for the penultimate event
Trench PC	OFPC-RC3 BETA 63546	Multiple small flecks of charcoal, fine-grained stream alluvium, Unit B2	26200 ± 220	Too old	Best maximum limiting age for the penultimate event. Average age for charcoal in stream deposit.
Trench PC	OFPC-RC5 BETA 65456	Single fleck of charcoal, fine-grained stream alluvium, Unit B2	33950 ± 1160	Too old	Best minimum limiting age for the antepenultimate event. Shows range in age of charcoal in Unit B2.

[a]Samples from the Big Canyon site were analyzed using the conventional gas-proportional method and all samples from the Pole Canyon site were analyzed using an accelerator mass spectrometer.
[b]Bulk sediment samples are the fine (<125 μm) organic fraction concentrated using methods after Kihl (1975) by the Sedimentology Laboratory at INSTAAR, University of Colorado.
[c]Radiocarbon years before 1950 A.D. with a half-life of 5568 years and 95% of activity of the National Bureau of Standards oxalic acid used as modern standard. Lab errors represent 1σ. Ages for charcoal samples were corrected for fractionation to a base of $\delta^{13}C = -25\%$. Ages for bulk sediment samples not corrected for isotope fractionation.
[d]All ages for bulk sediment samples were rounded to the nearest century. Calibration procedure after Stuiver and Reimer (1993) using a lab error multiplier of 2, the 20-year atmospheric record (dataset 1), a carbon age span of 300 years (the maximum permissible), and the intercept method. Actual age spans for bulk sediment samples are probably closer to 1000 years, so σ may be larger than the stated values.

even though the actual age of unit 19 might be slightly younger than 4,900 years.

Net vertical tectonic displacement (NVTD) is a measure of vertical slip across a fault zone produced by a surface-faulting earthquake (Schwartz, 1988). It consists of apparent slip minus the effects of stratal backtilting and antithetic faulting. Determining net slip per event provides an estimate of individual earthquake size and allows a comparison between successive surface-faulting earthquakes on a fault zone. A scarp profile can be used to estimate the net vertical tectonic displacement across a fault zone if the same undeformed surface can be identified on both the up- and downthrown sides of the fault (Bucknam and Anderson, 1979b).

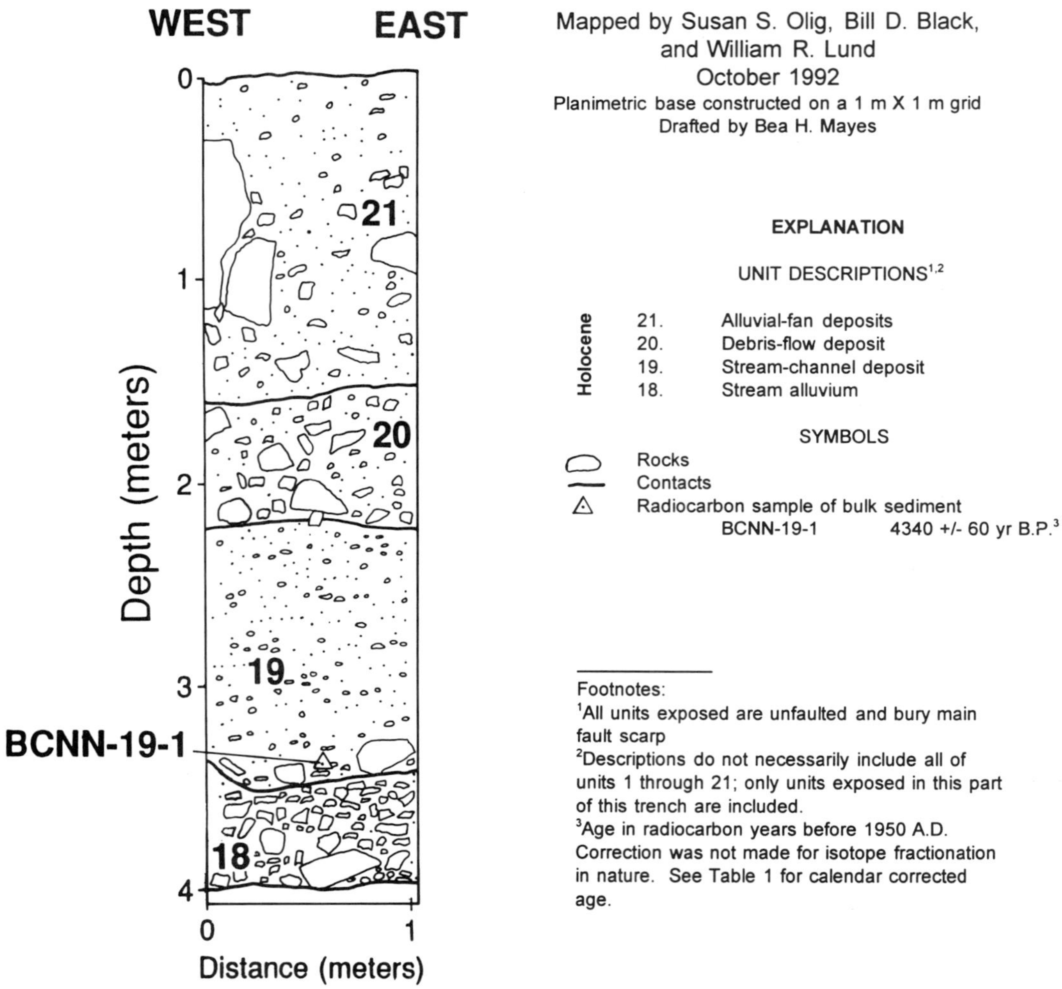

Fig. 11. Log of part of trench BC-1 showing stratigraphic relations of unfaulted Big Canyon alluvial-fan deposits.

Unfortunately, NVTD for the most-recent event at the Big Canyon site cannot be directly measured from trench exposures or scarp profiles. However, by making some assumptions about scarp height, NVTD can be estimated from trench exposure and scarp profile data. Models of normal-fault scarp formation and degradation suggest that scarp heights are typically twice the wedge thickness as long as slopes of surfaces offset are not too steep and recurrence intervals are not too short, resulting in characteristically thinner wedges (Ostenaa, 1984). Thicknesses of colluvial wedges range from 2.6 m (trench BC-3) to 2.8 m (trench BC-2) at the Big Canyon site. Assuming the associated scarp heights for this event were twice the wedge thickness yields a range of 5.2 to 5.6 m. Antithetic faulting and backtilting in the hanging wall along with offset of a sloping ground surface have accentuated scarp height, whereas drag folding has decreased scarp height in relation to the actual NVTD at the Big Canyon site.

We calculate that for an 80° dipping fault and 6° dipping slope, scarp height is about 0.1 m greater than surface offset. We measured 0.8 m of total vertical separation for the base of unit 11 across the antithetic fault zone in trench BC-3. The base of unit 11 can also be used to estimate backtilting. This contact dips 3° west in pit exposures about 100 m east of the main fault scarp. In trench BC-3 this contact dips 0° within the graben east of the antithetic fault, suggesting about 3°

of backtilting to the east. Using an inflection point between stations 75 and 80 (Fig. 7) to remove 3° of backtilting for the base of unit 11 yields 3.0 to 3.2 m. Finally, we measured 0.9 to 1.0 m of dragfolding on the tops of units 8 and 10.

Thus, starting with estimated scarp heights of 5.2 to 5.6 m, subtracting 0.1 m to account for the difference between surface offset and scarp height on a sloping surface, subtracting 0.8 m of antithetic faulting, subtracting 3.0 to 3.2 m of backtilting, and adding 0.9 to 1.0 m of drag folding yields 2.0 to 2.7 m of estimated NVTD for the most-recent event at the Big Canyon site. Similarly, our best estimate for scarp height is 5.2 m (because the wedge in trench BC-3 did not have a heal), subtracting 0.1 m to account for the difference between surface offset and scarp height on a sloping surface, subtracting 0.8 m of antithetic faulting and 3.1 m of backtilting and adding 1.0 m of drag folding yields

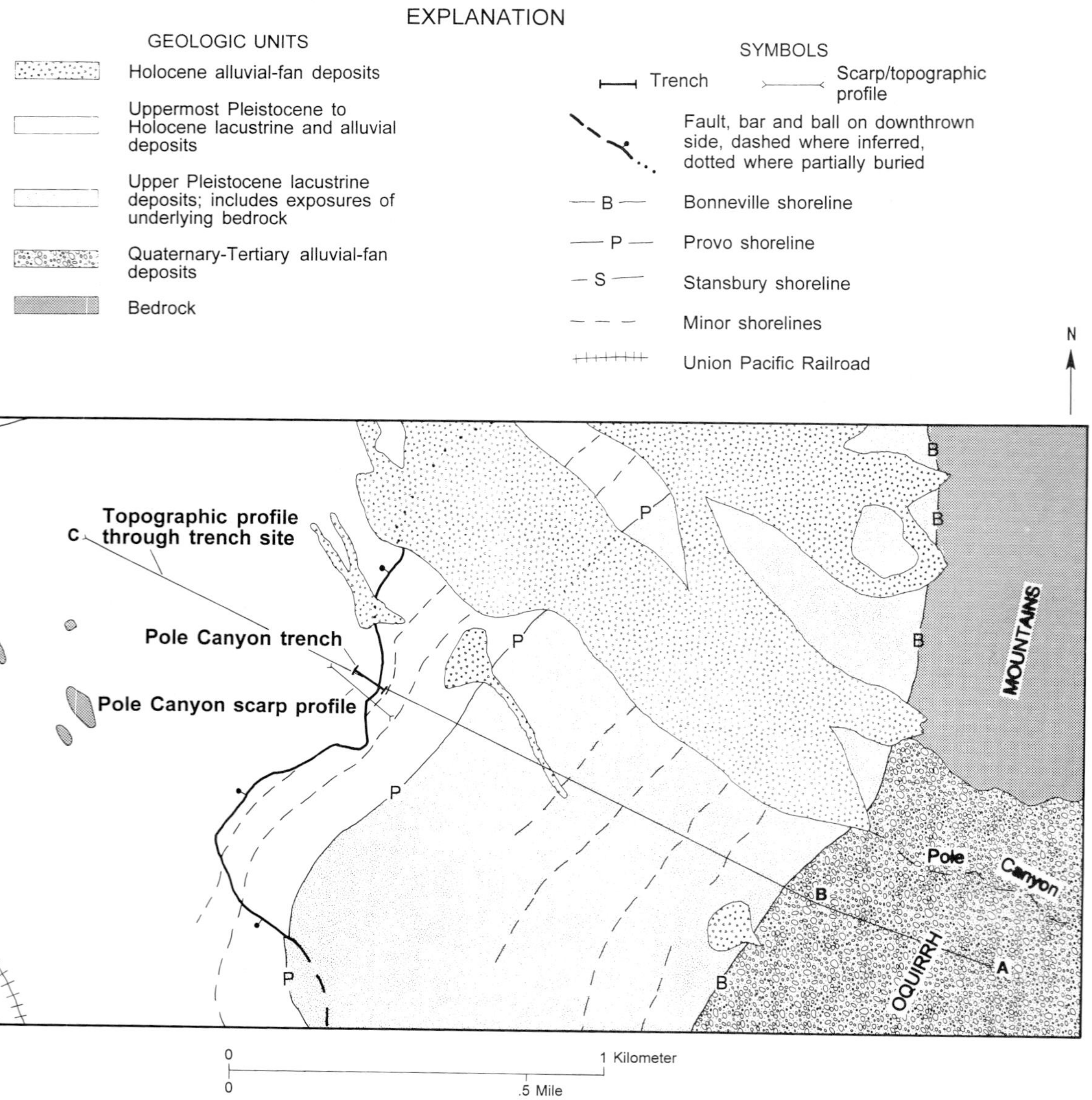

Fig. 12. Generalized geologic map of the Pole Canyon site.

2.2 m for our best estimate of NVTD at the Big Canyon site.

In summary, evidence from surficial mapping, scarp profiling, and three trenches at the Big Canyon site indicates that the most recent surface-faulting earthquake on the Oquirrh fault zone occurred between 4300 and 6900 yr B.P. (4800 and 7900 cal B.P.) and resulted in 2.0 to 2.7 m (best estimate 2.2 m) of NVTD.

4.2. Pole Canyon site

Surficial geology

The Pole Canyon trench site lies on a broad, wave-cut surface at the base of the Oquirrh Mountains 2.7 kilometers southwest of the Big Canyon trench site, and 1.7 kilometers northwest of the mouth of Pole Canyon (Fig. 2). Surficial geologic deposits in the area consist of coarse-grained Lake Bonneville sediments, post-Bonneville alluvial fans, and areas of undifferentiated lacustrine and alluvial deposits (Solomon, 1993; Fig. 12).

The Oquirrh fault zone is represented at the Pole Canyon site by a single, sinuous, west-facing scarp nearly 15 m high (Everitt and Kaliser, 1980). Neither east-facing antithetic scarps nor a graben are evident at the ground surface. The elevation at the base of the fault scarp is about 1448 m, placing it roughly 40 m below the Provo shoreline of Lake Bonneville. In addition, numerous other shorelines of Lake Bonneville parallel topographic contours at elevations both above and below the trench site. The fault scarp is easily distinguishable from shorelines because it commonly does not parallel topographic contours and it offsets several of the shorelines, including the Provo shoreline just south of the trench site. Currey and Burr's (1988) linear model of threshold-controlled shorelines of Lake Bonneville indicates that the lake regressed from the Pole Canyon site shortly after 14,200 radiocarbon-years ago, providing a maximum limiting age on the most recent faulting.

The fault scarp at the Pole Canyon trench site is geomorphologically, tectonically, and stratigraphically complex, making tectonic interpretations of the topographic profile difficult (Fig. 13). A small (1 .5 m high) regressive shoreline is etched into the fault scarp a few meters east of the trench (Figs. 12 and 13) and stratigraphic relations in the trench (discussed in the next section) show that a minimum of 1.5 m of material was eroded from the scarp as Lake Bonneville transgressed across the site. In fact, a topographic profile drawn from the mountain front through the trench site shows that as much as 20+ m of pre-Bonneville alluvial-fan material was eroded from the vicinity of the trench site by the lake. Both the Bonneville and Provo shorelines east of the trench site are retrograde features marked by prominent erosional escarpments (Currey, 1982). The eroded material was probably transported by longshore currents along the mountain front and deposited in the Stockton Bar, a large Lake Bonneville depositional feature south of Tooele (Gilbert, 1890; Burr and Currey, 1988).

After eroding the pre-lake land surface, Lake Bonneville deposited a veneer of mostly coarse-grained sediments on the wave-cut surface as it transgressed to and then regressed from its highstand (Bonneville shoreline). Since the Bonneville lake cycle, the degraded and sediment-draped scarp was then further modified by the most-recent surface-faulting earthquake and by subsequent erosion and mantling of the scarp by reworked lacustrine deposits and slopewash.

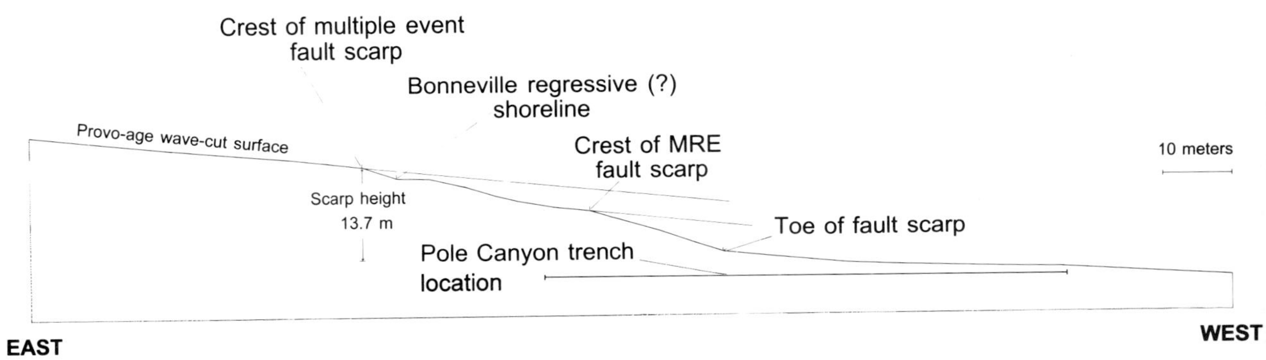

Fig. 13. Topographic profile measured at the Pole Canyon site.

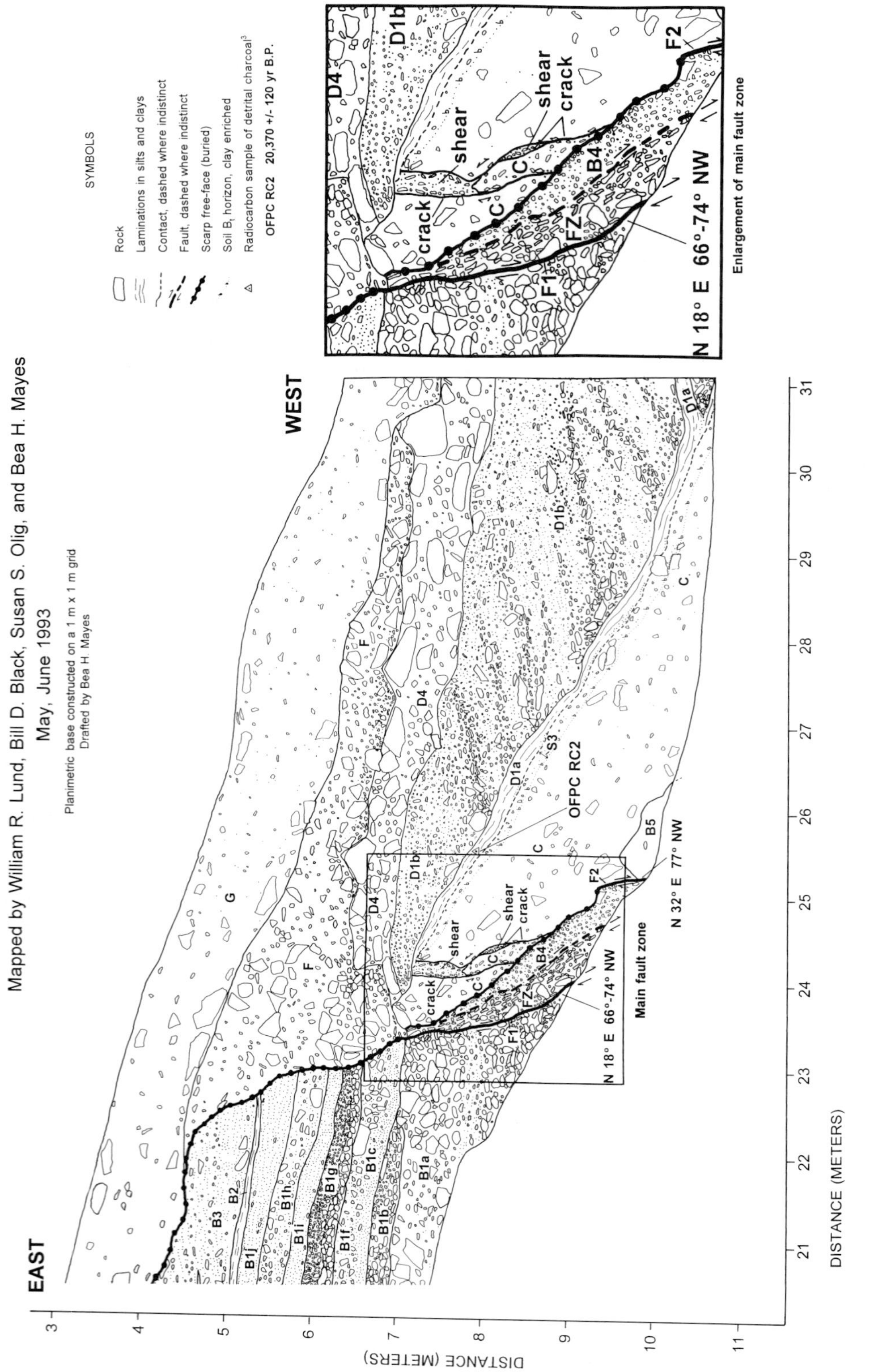

Fig. 14. Part of the log of the trench at the Pole Canyon site showing the main fault zone and colluvial-wedge deposits associated with the most-recent (package F) and penultimate (package C) surface faulting events on the Oquirrh fault zone.

EXPLANATION

UNIT DESCRIPTIONS[1]

<-Holocene->

STRATIGRAPHIC PACKAGE G - Post-Most Recent Event Colluvium

G Gravelly, sandy clay/Gravelly, clayey sand

STRATIGRAPHIC PACKAGE F - Most Recent Event Colluvial Wedge

F Clayey, sandy gravel

STRATIGRAPHIC PACKAGE FZ - Sheared Fault-Zone Material

FZ Sandy gravel with cobbles

<-----Pleistocene----->

STRATIGRAPHIC PACKAGE D - Lacustrine Deposits of the Bonneville Lake Cycle[2]

D4 Regressive beach deposit
- Bouldery, cobbly, sandy gravel

D1 Nearshore transgressive deposits - downthrown side of fault
- D1b Sandy gravel with cobbles (prograding bar)
- D1a Silty clay with sand (marsh deposit)

STRATIGRAPHIC PACKAGE C - Penultimate Event Colluvial Wedge

S3 Paleosol, formed on unit C

C Gravelly, clayey sand and clayey, sandy gravel

STRATIGRAPHIC PACKAGE B - Post-Antepenultimate/Pre-Penultimate Event Fluvial Deposits

B5 Sediments beneath penultimate event colluvial wedge
- Sandy, clayey silt

B4 Sediments downfaulted with penultimate event colluvial wedge during most recent event
- Gravelly sand

B3 Coarse-grained sediments interbedded with unit B2
- Sand with gravel

B2 Fine-grained sediments filling a buried stream channel
- Silty clay with sand stringers

B1 Coarse-grained fluvial deposits truncated, interbedded with or buried by unit B2
- B1j Gravelly sand
- B1i Sand with gravelly sand stringers
- B1h Gravelly sand with cobbles
- B1g Gravel
- B1f Cobbly gravel with boulders
- B1c Gravelly sand with clay and silt stringers
- B1b Gravel
- B1a Cobbly, sandy gravel

Footnotes:

[1]Descriptions do not necessarily include all of units A through G; only units exposed in this part of this trench are included.

[2]Deposited during the Bonneville lake cycle. Lake Bonneville occupied part of the eastern Great Basin from about 30 ka to 12 ka (Scott and others, 1983; Oviatt and others, 1992). Lake Bonneville first transgressed across the Pole Canyon site about 20 ka and regressed across the site about 13-14 ka.

[3]Age in radiocarbon years before 1950 A.D. includes correction for isotope fractionation, and was obtained using an atomic mass spectrometer. See Table 1 for radiocarbon age estimates for other samples collected from the Pole Canyon trench, but not shown in this portion of the log.

Trench exposure

A single 76-m-long, 11-m-deep trench was excavated at the Pole Canyon site, exposing direct stratigraphic and structural evidence for two surface faulting events (Fig. 14), and indirect stratigraphic evidence for a third surface faulting event. We extended the trench from the top of the steep (20°) lower portion of the scarp westward onto the downthrown side of the fault zone (Fig. 13) to detect possible evidence for antithetic faults, a buried graben, or stratal backtilting, but none was found.

Deposits in the Pole Canyon trench were grouped into seven stratigraphic packages, A through G, each representing either a unique depositional environment during late Pleistocene or Holocene time, or providing stratigraphic evidence for a surface-faulting earthquake. Some stratigraphic packages consist of a single geologic deposit (eg. package G), whereas others were subdivided into units (eg. B1, B2, B3) that exhibit similar genetic, textural, or stratigraphic characteristics. Some units were further divided into subunits based chiefly on textural characteristics (eg. B1a, B1b, B1c).

Stratigraphic package A includes the pre-antepenultimate earthquake deposits in the trench and post-event deposits related to the antepenultimate event. These deposits lie at the base of the east end of the trench (several meters east of station 21 in Fig. 14) and were the oldest sediments exposed. Unit A1 consists of horizontally bedded alluvial and eolian deposits truncated by a buried paleoslope. The paleoslope is draped by colluvial and eolian deposits of unit A2. Fluvial deposits of unit A3 both incise unit A2 in places and mantle the paleoslope in other places. Two truncated buried soils (S1 and S2) are present on stratigraphic package A at different locations in the trench. The buried soils are probably equivalent in age, but can not be correlated within the trench exposure. Both paleosols exhibit weakly developed, clay-enriched Bt horizons, indicating that the unconformity separating stratigraphic packages A from the overlying deposits represents a former stable land surface that existed for a considerable time (possibly thousands of years).

The relations exhibited by the geologic units in stratigraphic package A are interpreted to represent a degraded scarp free face (paleoslope cut on unit A1) draped by post-fault colluvium (unit A2) and incised and further buried by later fluvial deposits (unit A3). The paleosols then developed on the stable colluvial surface. These relations provide indirect evidence for an antepenultimate (third oldest) surface-rupturing earthquake on the Oquirrh fault zone.

Stratigraphic package B unconformably overlies stratigraphic package A. It consists of a pre-Bonneville sequence of interbedded, coarse- and fine-grained fluvial deposits that postdate the antepenultimate event and predate the penultimate (second oldest) surface-faulting earthquake on the fault zone. Unit B1 is comprised of several, often interbedded, well-sorted, coarse-grained, gravel beds (subunits B1a through B1j; Fig. 14) deposited by a swiftly moving stream. All of these gravel beds are either stratigraphically below or are truncated by unit B2, a fine-grained deposit of laminated silt and clay occupying a paleostream channel incised into unit B1. Two samples of detrital charcoal from unit B2 yield ages of 26,200 ± 220 yr B.P. (OFPC-RC3) and 33,950 ± 1160 yr B.P. (OFPC-RC5; Table 1). Unit B3 is a coarse-grained fluvial unit interbedded with unit B2. Unit B4 is a remnant of stratigraphic package B downfaulted to the west with the penultimate-event colluvial wedge by the most recent surface-faulting earthquake. Unit B5 is a fine-grained silty and clayey deposit believed to be a continuation of unit B2 on the downthrown side of the fault zone. The interpretation of stratigraphic package B as a late Pleistocene fluvial sequence is based on its lithologic characteristics, stratigraphic position below Lake Bonneville deposits, and the radiocarbon ages.

Stratigraphic package C is the colluvial wedge that formed following the penultimate surface-faulting earthquake at the Pole Canyon site. The wedge consists of material eroded from the upper part of the penultimate-event scarp and deposited at the base of the scarp. Although it contains considerable sand and gravel, stratigraphic package C has a high percentage of silt and clay matrix that gives the wedge a generally fine-grained appearance. The source of the silt and clay is believed to be unit B2, which was exposed in the scarp free face following the penultimate event. The colluvial wedge rests unconformably on a fine-grained deposit (unit B5), that is textually similar to unit B2 and likely is a continuation of that unit on the downthrown side of the fault. However, because a direct correlation cannot be made across the fault zone between the two units, they have been given separate unit designations.

A paleosol (S3) is present at the top of stratigraphic package C. It consists of a weak, clay-enriched Bt horizon overlying a Bk horizon. The Bt horizon exhibits thin clay skins and prismatic soil structure; the Bk horizon shows stage II $CaCO_3$ accumulation. The presence of the paleosol shows that the penultimate-event colluvial wedge was exposed at the ground surface for a considerable time before being buried by stratigraphic package D.

Stratigraphic package D consists of lacustrine deposits of the Bonneville lake cycle. Unit D1 is comprised of nearshore transgressive sediments deposited on the downthrown side of the fault zone. Subunit D1a, a pale-yellow silty clay layer a few tens of centimeters thick, rests directly on stratigraphic package C (the penultimate-event colluvial wedge). The clay layer is thought to be the "green clay," a lake-marginal marsh deposit commonly found at the base of the Bonneville Alloformation (D.R. Currey, pers. commun., 1993). Charcoal from subunit D1a (OFPC-RC2) yielded a radiocarbon age estimate of 20,370 ± 120 yr B.P., which constrains the timing of Lake Bonneville's transgression across the site.

Subunits D1b and D1c consist of coarse-grained sand, gravel, and cobbles deposited in a high-energy, nearshore environment. Subunit D1b exhibits crude bedding that dips steeply (50–60°) toward the fault zone near the bottom of the deposit, but becomes progressively less steep near the top of the deposit (Fig. 14). The change in dip is a primary depositional feature. It suggests that subunit D1b was a prograding bar that formed and eventually filled a shallow pond or lagoon as Lake Bonneville transgressed toward and eventually across the fault scarp. A few meters west of the scarp, subunit D1a splits into two parts, most of it continues beneath unit D1b. However, a thin stringer extends westward and upward into unit D1b parallel to bedding at an angle of about 35°. The split represents a pause in Lake Bonneville's transgression across the site, during which fine-grained sediment was deposited in the pond that existed between the fault scarp and the temporarily stable prograding bar. Once Lake Bonneville resumed its transgression, the pond quickly filled with coarse-grained sediment burying the fine-grained stringer.

As it transgressed across the penultimate-event scarp, Lake Bonneville encountered and began to erode the fine-grained sediment of unit B2 on the upthrown side of the fault. Stratigraphic relations in the trench show that a minimum of 1.5 m of unit B2 was removed by the lake before coarse-grained unit D2 was deposited disconformably on the wave-cut surface. Unit D3 is of a layer of lacustrine, brown silty clay deposited conformably on the nearshore sands and gravels of unit D2. The unit may represent deep-water deposition at the trench site as Lake Bonneville transgressed to the Bonneville shoreline. When the lake reached its highstand, the Pole Canyon site was under more than 135 m of water (Fig. 3). Alternatively, unit D3 may represent deposition in the calm water of a lagoon or other protected environment prior to the lake reaching its highest elevation. This alternative is suggested because unit D3 is only locally present in the trench and pinches out to the west against unit D2.

The waning phase of Lake Bonneville is represented by unit D4, a very coarse-grained, regressive beach deposit. Numerous large boulders and local areas of extremely strong (concrete-like) tufa cementation make unit D4 a distinctive and easily recognized unit in the Pole Canyon trench. It is the only geologic deposit in the trench that can be conclusively identified on both sides of the fault zone. Unit D4 was deposited across whatever remained of the penultimate-event scarp shortly after 14,200 years ago as Lake Bonneville regressed from the Provo shoreline under closed-basin conditions. No evidence of soil development, representing post-Bonneville exposure at the ground surface was found on unit D4.

Stratigraphic package E is a poorly sorted, post-Bonneville, subaerial deposit consisting of brown, clayey, sandy gravel. The deposit's genesis is uncertain, but it is probably the remnant of a debris flow that issued from the Oquirrh Mountains and flowed across part of the site. Stratigraphic package E is only present in the trench in a small area on the upthrown side of the fault zone. It rests directly on the Lake Bonneville regressive beach deposit (unit D4) and is truncated on the west by the eroded free face of the most-recent-event scarp. Because of its geomorphic position, we believe that the absence of stratigraphic package E on the downthrown side of the fault zone is likely due to nondeposition in that area rather than to erosion. Those relations show that stratigraphic package E post-dates Lake Bonneville and could either pre-date or post-date the most-recent surface-faulting earthquake on the Oquirrh fault zone.

Stratigraphic package F is the colluvial wedge that formed following the most-recent surface-faulting earthquake at the Pole Canyon site. The wedge consists of material eroded from the upper part of the most-recent-event scarp and deposited at the base of the scarp. This colluvial wedge is much coarser grained then the penultimate-event colluvial wedge because the geologic deposits exposed in the most-recent-event scarp were mostly coarse-grained units of stratigraphic packages B and D. The most-recent-event wedge rests unconformably on the regressive beach deposits of unit D4. There is no evidence of soil formation on the wedge.

Stratigraphic package G consists of post-most-recent-event colluvium mantling the fault scarp. The colluvium represents a return to slopewash deposition across the scarp. The colluvium did not begin to accumulate until the combination of scarp erosion and deposition at the scarp base reduced the angle of the most-recent-event scarp to a more stable configuration. Other than the most-recent-event colluvial wedge, stratigraphic package G is the only geologic deposit in the trench that postdates all surface faulting at the site.

The Oquirrh fault zone is exposed in the Pole Canyon trench as a roughly north–south trending (N18–32°E), high-angle (66–77°) normal-slip fault displacing unconsolidated deposits of stratigraphic packages B, C, and D down to the west (Fig. 14). The fault zone and related tectonic colluvial-wedge deposits provide direct evidence for a post-Lake Bonneville surface-faulting earthquake and a pre-Lake Bonneville event on the Oquirrh fault zone. Faults F1 and F2 are related to the most-recent and penultimate surface-faulting earthquakes, respectively (Fig. 14). During the most-recent event, a small remnant of stratigraphic package B (unit B4), was displaced down to the west with the penultimate-event colluvial wedge (stratigraphic package C) along fault F1 (Fig. 14). Limitations on the depth of the trench constrained what could be deduced regarding displacement along fault F2 during the penultimate event. Unit B5 beneath the penultimate-event colluvial wedge is textually similar to unit B2 on the upthrown side of the fault zone. However, the exposure of unit B5 was small and, even with additional hand excavation in the fault zone, a direct correlation could not be made between the two units.

Although structural evidence for a third, older surface-faulting earthquake was not exposed in the trench, there is strong indirect stratigraphic evidence for an antepenultimate surface-faulting earthquake on another fault splay that would be east of faults F1 and F2 exposed in the trench. Horizontally bedded geologic deposits comprising unit A1 are truncated by a paleoslope dipping 15–30° westward toward the Oquirrh fault zone. The paleoslope is mantled by units A2 and A3, and two paleosols (S1 and S2) are present on the mantling deposits. A short distance further west and stratigraphically higher in the trench, a similar relationship exists between stratigraphic packages B and G. There, colluvium of stratigraphic package G mantles the degraded most-recent-event scarp free face eroded across horizontally bedded deposits of stratigraphic package B. The similarity between the stratigraphic relations in the two areas of the trench is striking. Based on that similarity, stratigraphic package A is interpreted as an eroded and colluvium-mantled scarp free face formed following a surface-faulting earthquake. Buried soils S1 and S2 formed during a long period of exposure at the ground surface before stratigraphic package A was buried by the fluvial deposits of stratigraphic package B. Stratigraphic and structural relations in the trench show that stratigraphic package B was later offset by the penultimate surface-faulting earthquake. Therefore, the earthquake responsible for the geological relations seen in stratigraphic package A predates the penultimate event and represents a third surface-faulting earthquake on the fault zone.

The Pole Canyon trench provided neither stratigraphic relations nor dateable organic material that would allow the timing of the most-recent event on the Oquirrh fault zone to be more closely constrained than was possible at the Big Canyon trench site (see the previous section on the Big Canyon site). The most-recent-event colluvial wedge in the Pole Canyon trench rests directly on Lake Bonneville regressive beach deposits (unit D4), demonstrating that the event is post-Bonneville in age. Close examination of the contact between the two units shows no evidence of soil development on the beach deposit, thus eliminating the possibility of obtaining a radiocarbon age estimate on bulk soil organics. The lack of soil development and the absence of subaerial deposits between the beach deposit and the colluvial wedge implies that the most-recent event happened soon after Lake Bonneville regressed from the site. However, evidence from the Big Canyon site shows that the most-recent surface-faulting earthquake actually occurred several thousand years after the Lake Bonneville regression. The lack of illuviation in unit D4 at the Pole Canyon site that was variably observed in unit 10 at the Big Canyon could be related to the intense tufa-cementation preventing translocation of silts and clays, retarding pedogenic processes at the Pole Canyon site.

The three radiocarbon age estimates from the Pole Canyon trench help constrain the timing of the penultimate event on the Oquirrh fault zone (Table 1). All of these ages were too old to calendar calibrate. Age estimates for unit B2 predate the penultimate event and provide an upper limit for the time of faulting. Sample OFPC-RC3 consisted of several small flecks of detrital charcoal collected from an approximately 2 square m area of unit B2. The flecks combined to create a single sample which gave an "average" radiocarbon age estimate of 26,200 ± 120 yr B.P. (Table 1). Sample OFPC-RC4 consisted of a single, larger piece of charcoal from the same area of unit B2 and it yielded a radiocarbon age of 33,950 ± 1160 yr B.P. The two age estimates show that the detrital charcoal in unit B2 extends across an age range of several thousand years. This is not unexpected since unit B2 is a channel-fill deposit of a paleostream that likely drained from the Oquirrh Mountains and the charcoal in unit B2 undoubtedly was derived from a variety of sources of different ages in the stream's drainage basin. Therefore, the deposit itself is somewhat younger than the charcoal it contains and 26,200 ± 220 yr B.P. is the best maximum limiting age for the penultimate event.

Sample OFPC-RC2 from unit D1a post-dates the penultimate event and provides a minimum limiting age of 20,370 ± 120 yr B.P. (Table 1) for the penultimate event. This is consistent with independent determination of the age of unit D1a and the transgression of Lake Bonneville based on the altitude of the site (Fig. 3). Therefore, the penultimate event on the Oquirrh fault zone occurred sometime within the approximate 6,200 radiocarbon year window between 26,440 and 20,250 yr B.P. (one-sigma laboratory error limits added to OFPC-RC3 and subtracted from OFPC-RC2 to give the maximum time window).

Stratigraphic and pedologic relations that might otherwise further constrain the penultimate event within the radiocarbon time window are contradictory and of no help in determining the time of faulting. Paleosol S3 developed on the penultimate-event colluvial wedge (stratigraphic package C) and exhibits a weak clay-enriched Bt horizon. The presence of a paleosol with this degree of pedogenic development implies that the penultimate-event wedge was exposed at the ground surface for considerable time (possibly thousands of years) before being buried by Lake Bonneville deposits. That relation argues for placing the penultimate event toward the distal part of the radiocarbon time window. Conversely, the penultimate-event colluvial wedge lacks the colluvial mantle that typically develops as a fault scarp erodes and assumes a more stable configuration (note stratigraphic package G's relation to the most-recent-event colluvial wedge). Instead, the penultimate-event wedge is directly overlain by Lake Bonneville deposits, implying that the lake transgressed across the wedge before a stable colluvial cover had time to accumulate. That relation argues for placing penultimate event toward the proximal part of the radiocarbon time window and closer to the Lake Bonneville transgression. It is possible that the transgressing lake washed away any colluvium overlying the penultimate-event wedge; however, that scenario fails to explain the presence of paleosol S3 and the apparent long period of soil formation that followed the penultimate event. Because the two lines of evidence are contradictory, it is not possible to select a "best estimate" for the timing of the penultimate event other than to place it in the middle of the radiocarbon time window at 23,350 yr B.P. Assigning this timing to the penultimate event is arbitrary, but it minimizes the error between the best estimate and the actual time of faulting in either a plus or minus direction.

Little evidence exists to constrain the timing of the antepenultimate earthquake. Stratigraphic relations in the trench show that the event predates unit B2, which yielded a radiocarbon age estimate on charcoal of 33,950 ± 1160 yr B.P. Unit B2 is younger than the detrital charcoal it contains, but is clearly older than Lake Bonneville. Unit B2 unconformably overlies stratigraphic package A, which includes the geologic evidence for the antepenultimate event. The unconformity is marked by two paleosols (S1 and S2) that exhibit weakly developed, clay-enriched Bt horizons. The paleosols suggest that a long soil-forming interval may have followed the antepenultimate event before stratigraphic package A was buried by stratigraphic package B. Clearly, the antepenultimate event is older than unit B2; how much older is uncertain, but the unconformity and associated paleosols between stratigraphic packages A and B show that it may be thousands or even tens of thousands of years.

Because the fault scarp at the Pole Canyon site was significantly modified by Lake Bonneville, NVTD cannot be determined from the scarp profile alone. However, NVTD can be estimated for the most-recent and penultimate events using stratigraphic relations in the trench and the size of the colluvial wedges associated with those events.

NVTD for the most-recent event can be estimated using the tufa-cemented, regressive beach deposit (unit D4), which provides a distinctive marker horizon on both sides of the fault zone. We measure 3.3 m of vertical stratigraphic separation down to the west of the base of unit D4. However, some of that offset is from drape of unit D4 across the pre-Lake Bonneville fault scarp. Reconstruction of stratigraphic package D, indicates a 0.6 m difference in elevation (down to the west) across the fault at the base of unit D4 prior to the most-recent event. That difference represents the pre-event drape of unit D4 across the scarp and must be subtracted from the total separation (3.3 m) to arrive at a best estimate of net slip for the most-recent event of 2.7 m. The thickness of the colluvial-wedge deposit, 2.2 m, provides a minimum bound on the estimated NVTD for the most-recent event, whereas the vertical stratigraphic separation of unit D4 provides a maximum estimate of 3.3 m. The absence of antithetic faulting and stratal backtilting at the Pole Canyon site made adjustments for those features unnecessary when calculating the net vertical tectonic displacement across the fault zone.

The Pole Canyon trench did not provide stratigraphic relations necessary to directly measure NVTD produced by the penultimate event. However, a rough estimate of net slip can be made by comparing the thicknesses of the penultimate- and most-recent-event colluvial wedges and the calculated net slip for the most-recent event. The most-recent-event colluvial wedge is 2.2 m thick, which is 82 percent of the best estimate for net vertical displacement (2.7 m) associated with the most-recent event and 67 percent of the maximum estimate (3.3 m). By comparison, the penultimate-event colluvial wedge is 1.9 m thick. If it is assumed that the thickness of the penultimate-event colluvial wedge also represents roughly 82 percent of the net vertical displacement produced by that event, then net slip for the penultimate surface-faulting earthquake would be about 2.3 m. The thickness of the colluvial-wedge deposit, 1.9 m, provides a minimum estimate of NVTD for the penultimate event, whereas assuming the penultimate wedge thickness represents 67 percent of the NVTD yields 2.9 m for a maximum estimate of the NVTD for the penultimate event.

5. Paleoseismic characteristics

5.1. Earthquake timing, recurrence and slip rates

Based on evidence from the Big Canyon and Pole Canyon sites, Fig. 15 summarizes timing information for the most recent, penultimate, and antepenultimate surface-faulting events on the Oquirrh fault zone. We only provide a range of ages for these events because we believe the possible large age spans of detrital carbon for most of the radiocarbon ages, particularly the bulk sediment samples, precludes specifying a best estimate with any confidence. The age ranges are (1) 4300–6900 yr B.P. for the most-recent event; (2) 20,300–26,400 yr B.P. for penultimate event; and (3) >32,800 yr B.P. for the antepenultimate event. These age ranges yield a minimum recurrence interval of 13,300 radiocarbon years and a maximum of 22,100 radiocarbon years, between the penultimate and most-recent events. There are a few notable observations from these results: (1) the most-recent event was actually 2000 to 9000 years younger than previous age estimates based on scarp morphology; (2) no surface faulting occurred on the Oquirrh fault zone during the Bonneville lake cycle; and (3) the elapsed time since the most-recent event is less than the recurrence interval, however, a comparison with only one interval is somewhat inconclusive.

Using our best estimates for NVTD from the Pole Canyon and Big Canyon sites yields slip rates ranging from 0.1 mm/yr (2.2 m/22,100 yr) to 0.2 mm/yr (2.7 m/13,300 yr) for the interval between the most recent and penultimate events. Using our best estimates for NVTD from the Pole Canyon site yields slip rates ranging from 0.19 mm/yr (5.0 m/26,400 yr) to 0.25 mm/yr (5.0 m/20,300 yr) for the time since the past two surface-faulting events. For comparison, average Holocene slip rates for the central segments of the Wasatch fault zone are 1 to 2 mm/yr (Machette et al., 1991),

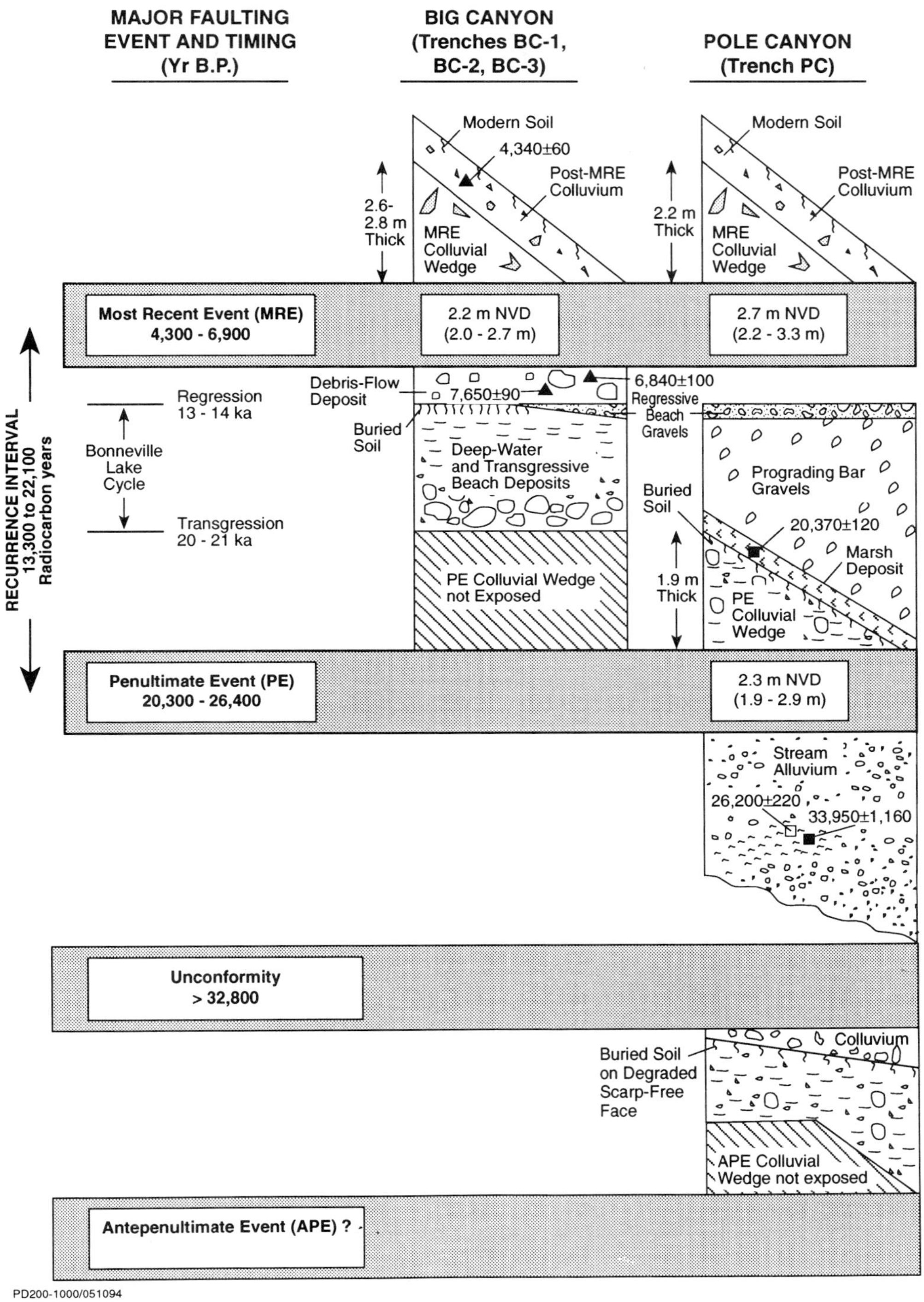

Fig. 15. Schematic representation of Holocene and latest Pleistocene faulting chronology for the Oquirrh fault zone based on evidence at the Big Canyon and Pole Canyon trench sites. Timing and recurrence is based on stratigraphic and geomorphic relations, and radiocarbon ages. Symbols representing radiocarbon samples include triangles for bulk sediment samples, solid boxes for single fragment charcoal samples, and an open box for a sample of several charcoal fragments. All ages in yr B.P. (radiocarbon years before 1950 A.D.).

whereas slip rates for other Quaternary faults in Utah usually range between 0.01 and 0.5 mm/yr (Hecker, 1993).

5.2. Paleomagnitude estimates and implications from comparisons of fault length and displacement

Since Tocher (1958) first developed an empirical relation between earthquake size and surface-rupture length, many relations have been developed to estimate magnitude from fault parameters (see DePolo and Slemmons, 1990 for discussion). Table 2 compares magnitude estimates based on fault length, displacement per event, and a combination of length and displacement for the Oquirrh fault zone. Estimates vary considerably depending on the relation used (see Table 2 footnotes for details on relations and input parame-

Table 2
Magnitude estimates for surface-faulting earthquakes on the Oquirrh fault zone

Based on	Moment magnitude (M_w)		Surface-wave magnitude (M_s)	
	Refs: [1][a]	[2][b]	[3][c]	[4][d]
Length	6.3 ($L_s=12$ km)		6.5–6.6 ($L_s=12$ km; $L_t=15$ km)	
	6.6 ($L_s=21$ km)		6.8–7.0 ($L_s=21$ km; $L_t=28$ km)	
	6.8 ($L_s=32$ km)		7.0–7.2 ($L_s=32$ km, $L_t=40$ km)	
Displacement	7.2 ($d=2.4$ m)			
	7.0 ($D=2.7$ m)		7.3 ($D=2.7$ m)	
	7.2 ($D=4.8$ m)		7.5 ($D=4.8$ m)	
Length and displacement		6.8 ($L_s=21$ km, $d=2.4$ m)		6.9 ($L_s=21$ km, $D=2.7$ m)
		6.9 ($L_s=32$ km, $d=2.4$ m)		7.2 ($L_t=40$ km, $D=4.8$ m)

References: [1] Wells and Coppersmith (1994); [2] Hanks and Kanamori (1979); [3] Bonilla et al. (1984); Mason and Smith (1993).
[a]Relations for all types of slip: $M_w=5.08+1.16\log L_s$ ($s=0.28$); $M_w=6.23+0.28\log d$ ($s=0.39$); $M_w=6.69+0.74\log D$ ($s=0.40$).
[b]$M_w=0.66\log M_o-10.7$, where M_o is seismic moment and $M_o=\mu dA$; μ is the shear modulu, d is average displacement, and A is the rupture plane area; we assume $\mu=3.3\times10^{11}$ dyn/cm^2, a fault dip of 55°, and the seismogenic depth is 15 km.
[c]Ordinary least-squares relations for western North America: $M_s=5.17+1.237\log L$ ($s=0.324$); $M_s=6.98+0.742\log D$ ($s=0.367$).
[d]$M_s=6.1+0.47\log(D\times L_s)$.
[e]All lengths measured from Barnhard and Dodge (1988) and Solomon (1993), L_s is straight line distance and L_t is along trace distance. The minimum L_s (12 km) is for the northern section, scarps on alluvium from Lake Point to Flood Canyon. The intermediate L_s (21 km) also includes the southern section, the fault contact between bedrock and alluvium from Flood Canyon to south of Middle Canyon. The maximum L_s (32 km) and L_t (40 km) was measured from Black Rock to Stocton to include a short scarp mapped by Solomon (1993) on Lake Bonneville sediments near Stockton not previously recognized as part of the Oquirrh fault zone.
[f]d is average displacement determined from an average of three observations of NVTD at the Big Canyon and Pole Canyon sites (2.2, 2.3, and 2.7 m); D is maximum displacement taken as twice the average displacement (4.8 m) as per observations of Wells and Coppersmith (1994) or as the maximum observed NVTD (2.7 m).

ters), but estimates based on displacement are consistently larger than estimates based on length. As might be expected, our magnitude estimates based on both length and displacement fall between the length-based and displacement-based estimates (Table 2).

The discrepancy between the length and displacement data is even more evident when one uses either Bonilla et al.'s (1984) or Wells and Coppersmith's (1994) relations to estimate surface-rupture length from displacement. Length estimates vary from 48 km (for $D=2.7$ m) to 71 km ($d=2.4$ m), four to almost six times the measured straight line distance of the northern section of the Oquirrh fault zone, and two to three times the measured straight line distance for the entire fault.

There are three obvious reasons for the apparent discrepancy between fault length and displacement on the Oquirrh fault zone: (1) we have overestimated displacement, (2) we have underestimated length, or (3) the Oquirrh fault zone does not fit available empirical models, possibly because of differences in the shear modulus, stress drop, or rupture surface. In making our comparisons, we have been careful to use fault measurements appropriate for the specific relation (for example using maximum [D] versus average [d] displacement) and we have allowed for the possibility that the 2.7 m of NVTD observed at Pole Canyon was actually the maximum displacement for the fault. Therefore, we do not believe that we have significantly overestimated displacement.

It is possible that the fault rupture extended farther north beyond Lake Point for a short distance under Great Salt Lake or farther south beyond Middle Canyon and associated fault scarps have been eroded and/or buried. In fact, in his detailed mapping, Solomon (1993) identified short discontinuous fault scarps on older alluvial-fan deposits near Silcox Canyon, roughly 5 km southwest of the southernmost extent of the bedrock–alluvium fault trace and a very short fault scarp on Lake Bonneville sediments near the town of Stockton. Tooker and Roberts (1992) identified many additional faults in alluvium in the Stockton area, but Solomon (1993) interpreted these to be shorelines associated with Lake Bonneville or stream terraces.

Although we did not study the section of the fault south of Flood Canyon and timing of faulting is not well-constrained for this section, our displacement data for the northern section indicates a much longer rupture segment than 12 km is likely for the Oquirrh fault zone. Thus, we infer that it is likely that the northern and southern sections together form a single rupture segment, and it is possible that prehistoric ruptures of the Oquirrh fault zone extended as far south as Stockton and northward beneath Great Salt Lake; however, further investigation of timing of events on the southern section is needed to test this preliminary hypothesis.

6. Summary and discussion

Direct stratigraphic and geomorphic evidence along the northern section of the Oquirrh fault zone indicates: (1) the most recent surface faulting event occurred between 4300 and 6900 yr B.P. (4800 and 7900 cal B.P.), (2) the penultimate event occurred between 20,300 and 26,400 yr B.P., (3) estimates of NVTD range between 1.9 and 3.3 m with best estimates of 2.2 and 2.7 m for the most-recent event and 2.3 m for the penultimate event, (4) the recurrence interval between the last two events is between 13,300 and 22,100 radiocarbon years, and (5) calculated slip rates are 0.1 to 0.2 mm/yr for this interval. Indirect stratigraphic evidence for the antepenultimate event suggests it occurred before $33{,}950 \pm 1160$ yr B.P.

Although our results only directly pertain to the northern section of the Oquirrh fault zone, they do have some implications for rupture segmentation. Estimates of fault length based on the observed displacements are 4 to 6 times the length of the northern section and 2 to 3 times the length of the northern and southern sections combined. This suggests the northern and southern sections are part of the same rupture segment and a few isolated, short, discontinuous fault scarps as far south as Stockton (Solomon, 1993) could also be part of the Oquirrh fault zone; however, further work is needed along the southern section of the fault to test this preliminary hypothesis. Comparing our results with available information on the Mercur fault zone to the south suggests that the Mercur and Oquirrh fault zones have behaved as independent rupture segments, at least since the Bonneville lake cycle. Barnhard and Dodge (1988) reinterpreted Everitt and Kaliser's (1980) trench log for an excavation across a scarp of the Mercur fault zone and they found that the most recent faulting occurred prior to the Bonneville lake cycle.

Available paleoseismic information is inconclusive regarding a possible rupture segment boundary between the Oquirrh fault zone and the East Great Salt Lake fault zone to the north. Timing of the most recent activity along the East Great Salt Lake fault zone is not well-constrained, but geomorphic and indirect stratigraphic evidence indicates that faulting occurred post-Bonneville lake cycle and perhaps even less than 6800 years ago (D.R. Currey, written communication, 1994). Thus, this faulting could have been contemporaneous with the most-recent event on the Oquirrh fault zone. However, Pechmann et al. (1987) estimated Quaternary slip rates of 0.4 to 0.7 mm/yr for the East Great Salt Lake fault zone based on vertical subsidence rates determined from seismic and drill-hole data, suggesting a greater rate of activity for the East Great Salt lake fault than the Oquirrh fault zone.

Estimates of M_w based on displacements range from 7.0 to 7.2 and are consistently higher than M_w estimates based on length (6.3 to 6.8) for the Oquirrh fault zone. In comparison, magnitude estimates based on length and displacement are slightly larger than length-based estimates and are within the range of one standard deviation (typically 0.3 to 0.4 for the relations used). The variability of displacement along a rupture (e.g., Crone et al., 1987) typically makes estimates of paleomagnitudes based on displacements difficult and often undesirable because of the limited number of observations (DePolo and Slemmons, 1990). It is worthwhile to note that even though we only have a very limited number of observations, measured NVTDs are similar, suggesting they are closer to an average than an extreme. However, even if our maximum observed NVTD is taken as a maximum displacement for the entire fault, which is unlikely, the estimated M_w is still 7.0, about one standard deviation larger than length-based estimates. The Oquirrh fault zone does not appear unique in this regard. Mason and Smith (1993) found that paleomagnitudes based on displacement were typically 20% larger than paleomagnitudes based on length for Quaternary normal faults in the Intermountain seismic belt. Mason (1992) suggested that underestimation of fault length could be a common cause for the discrepancy. Regardless, until further investigation can clarify this discrepancy, it would be prudent to not use magnitude estimates based on length alone in characterizing the Oquirrh fault for earthquake-hazard evaluations.

Temporal clustering of earthquakes on the central segments of the Wasatch fault zone during the late Holocene (between 400 and 1500 years ago; Machette et al., 1991, 1992) does not appear to have influenced activity on the Oquirrh fault zone. Out of the past four events on the Salt Lake City segment, only the antepenultimate event (5400 ± 300 cal B.P.; Lund, 1993) may have been contemporaneous with the most-recent event on the Oquirrh fault zone (4800–7900 cal B.P.).

Finally, the timing of the past two events on the Oquirrh fault zone shows a higher rate of activity during interpluvial periods than during the Bonneville lake cycle. However, the record for events is only complete for the past 32,000 radiocarbon years and we do not know if this pattern persisted through the previous Cutler Dam and Little Valley lake cycles. Trench exposures along the Granger fault of the West Valley fault zone indicated a similar pattern of activity with evidence for post-Bonneville lake cycle faulting (in the past 12,000–13,000 years) and evidence for no faulting during the Bonneville lake cycle (Keaton et al., 1987). More regionally, Machette et al. (1992) found higher post-Bonneville slip rates along the central segments of the Wasatch fault zone (0.4–1.8 mm/yr) compared to lower rates recorded in pre-Bonneville lake cycle alluvium (0.1–0.3 mm/yr), however, the pre-Bonneville data are limited. They speculated that Lake Bonneville may have decreased seismic activity and slip along the Wasatch fault zone possibly due to increased confining pressures associated with loads from the lake and thick sediment packages deposited by the lake, and that accelerated rates of faulting activity during the regressive phase of lake cycles and interpluvial periods might be associated with a decrease in the confining pressures and resulting isostatic uplift in the Bonneville basin. If this is true, it may have some implications in regards to fault dips and mechanics.

In his analysis using a simple elastic model of loads imposed by ice sheets on the crust, Johnston (1989) pointed out that in an extensional regime the effect of the load on fault stability varies and depends on the dip of the fault. The load decreases fault stability for dips greater than 45° and increases stability for dips less than 45°. Thus, if this simple elastic model is applicable, decreased fault activity during the Bonneville lake cycle might imply overall shallow fault dips ($<45°$). Alternatively, isostatic rebound from loads imposed by Lake Bonneville has long been recognized (Gilbert,

1890) and is well documented (Crittenden, 1963; Currey, 1982), suggesting that a visco-elastic model, where the load depresses a brittle crust into an underlying ductile medium, might be more appropriate. In his analysis for this model, Johnston (1989) showed that if isostasy is assumed, the load not only increases vertical stresses, but also increases horizontal stresses. Depending on the boundary conditions, the change in vertical stresses will be greater than or equal to the change in horizontal stress. Therefore, the effect of the load on fault stability would still be dependent on fault dip and if a visco-elastic model is applicable, decreased faulting during the Bonneville lake cycle would still imply fault dips of less than 45°.

Many researchers have discussed both the mechanical paradox for failure on low-angle normal faults (assuming Mohr–Coulomb failure criteria) and the lack of observations of such failures in the historical seismicity record (eg. Smith and Bruhn, 1984; West, 1992). We do not mean to imply here that observations of lower rates of faulting activity on the Oquirrh fault zone during the Bonneville lake cycle are evidence for shallow dips on seismogenic faults, rather we point out possible implications as a stimulus for developing a better understanding of faulting patterns relative to Lake Bonneville chronology. Obviously, more paleoseismic data is needed to clarify if a causal relation exists. If it does, a better understanding of this relation might provide insight into the mechanics of normal faulting.

Acknowledgements

This study benefitted from discussions with Ted Barnhard, Suzanne Hecker, Alan Nelson, and Mike Machette (U.S. Geological Survey); Don Currey (University of Utah); Jack Oviatt (Kansas State University); Ben Everitt (Utah Division of Water Resources); and Barry Solomon (Utah Geological Survey). We would also like to thank Ted Barnhard for loaning aerial photographs and Tooele County for providing excavation assistance. Bea Mayes and Janine Jarva provided field assistance and Bea Mayes and Sue Penn (Woodward–Clyde Federal Services) greatly assisted in preparation of the manuscript.

This project was partially funded by the U.S. Geological Survey Earthquake Hazard Reduction Program under award no. 1434-92-G-2218. Assistance for manuscript preparation was also provided by the Woodward–Clyde Federal Services Professional Development Fund.

References

Anderson, L.W. and Miller, D.G., 1979. Quaternary fault map of Utah, FUGRO, Inc., Long Beach, CA, 35 pp.

Arabasz, W.J., Pechmann, J.C. and Brown, E.D., 1989. Evaluation of seismicity relevant to the proposed siting of a Superconducting Supercollider (SSC) in Tooele County, Utah. Utah Geol. Miner. Surv. Misc. Publ., 89-1: 107 pp.

Arabasz, W.J., Pechmann, J.P. and Brown, E.D., 1992. Observational seismology and the evaluation of earthquake hazards and risk in the Wasatch Front area. In: P.L. Gori and W.W. Hays (Editors). U.S. Geol. Surv. Prof. Pap., 1500-D: 36 pp.

Barnhard, T.P., 1988. Fault-scarp studies of the Oquirrh Mountains, Utah. In: M.N. Machette, (Editor), In the Footsteps of G.K. Gilbert — Lake Bonneville and Neotectonics of the Eastern Basin and Range Province. Utah Geol. Miner. Surv. Misc. Publ., 88-1: 52–54.

Barnhard, T.P. and Dodge, R.L., 1988. Map of fault scarps formed on unconsolidated sediments, Tooele 1°×2° quadrangle, northwestern Utah. U.S. Geol. Surv. Misc. Field Studies Map MF-1990, scale 1:250,000.

Bjornadottir, I.T. and Pechmann, J.C., 1989. Contemporary tectonics of the Wasatch front region, Utah, from earthquake focal mechanisms. Bull. Seismol. Soc. Am., 79: 731–755.

Bonilla, M.G., Mark, R.K. and Lienkaemper, J.J., 1984. Statistical relations among earthquake magnitude, surface rupture length, and surface fault displacement. Bull. Seismol. Soc. Am., 74: 2379–2411.

Bucknam, R.C., 1977. Map of suspected fault scarps in unconsolidated sediments, Tooele 2° sheet, Utah. U.S. Geol. Surv. Open-File Rep., 77-495, scale 1:250,000.

Bucknam, R.C. and Anderson, R.E., 1979a. Map of fault scarps on unconsolidated sediments, Delta 1°×2° quadrangle, Utah. U.S. Geol. Surv. Open-File Rep., 79-366, scale 1:250,000.

Bucknam, R.C. and Anderson, R.E., 1979b. Estimation of fault-scarp ages from a scarp-height–slope-angle relationship. Geology, 7: 11–14.

Burr, T.N. and Currey, D.R., 1988. The Stockton bar. In: M.N. Machette (Editor), In the Footsteps of G.K. Gilbert — Lake Bonneville and Neotectonics of the Eastern Basin and Range Province. Utah Geol. Miner. Surv. Misc. Publ., 88-1: 66–73.

Campbell, K.W., 1987. Predicting strong ground motion in Utah. In: L. Chap, P.L. Gori and W.W. Hays (Editors), Assessment of Regional Earthquake Hazards and Risk along the Wasatch Front, Utah, II. U.S. Geol. Surv. Open-File Rep., 87-585: L-1-90.

Crittenden, M.D., Jr., 1963. New data on the isostatic deformation of Lake Bonneville. U.S. Geol. Surv. Prof. Pap., 454-E: E1–E31.

Crone, A.J., Machette, M.N., Bonilla, M.G., Leinkaemper, J.J., Pierce, K.L., Scott, W.E. and Bucknam, R.C., 1987. Surface faulting accompanying the Borah Peak earthquake and segmentation of the Lost River fault, central Idaho. Bull. Seismol. Soc. Am., 77: 739–770.

Currey, D.R., 1982. Lake Bonneville: Selected features of relevance to neotectonic analysis. U.S. Geol. Surv. Open-File Rep., 82-1070: 29 pp.

Currey, D.R. and Burr, T.N., 1988. Linear model of threshold-controlled shorelines of Lake Bonneville. In: M.N. Machette (Editor), In the Footsteps of G.K. Gilbert — Lake Bonneville and Neotectonics of the Eastern Basin and Range Province (Geological Society of America Guidebook to Field Trip 12). Utah Geol. Miner. Surv. Misc. Publ., 88-1: 104–110.

Currey, D.R. and Oviatt, C.G., 1985. Durations, average rates, and probable causes of Lake Bonneville expansion, still-stands, and contractions during the last deep-lake cycle, 32,00 to 10,000 yrs ago. In: P.A. Kay and H.F. Diaz (Editors), Problems of and Prospects for Predicting Great Salt Lake Levels. Proc. NOAA Conf. held March 26–28, 1985, Salt Lake City. Center for Public Affairs and Administration, Univ. of Utah, pp. 9–24.

DePolo, C.M. and Slemmons, D.B., 1990. Estimation of earthquake size for seismic hazards. In: E.L. Krinitzsky and D.B. Slemmons (Editors), Neotectonics in Earthquake Evaluation, Boulder, Colorado. Geol. Soc. Am. Rev. Eng. Geol., 8: 1–27.

Eddington, P.K., Smith, R.B. and Renggli, C., 1987. Kinematics of Basin and Range intraplate extension. In: M.P. Coward, J.F. Dewey and P.L. Hancock (Editors), Continental Extensional Tectonics. Geol. Soc. London Spec. Publ., 28: 371–392.

Everitt, B.L. and Kaliser, B.N., 1980. Geology for assessment of seismic risk in the Tooele and Rush Valleys, Tooele County, Utah. Utah Geol. Miner. Surv. Spec. Stud., 51: 33 pp.

Gilbert, G.K., 1890. Lake Bonneville. U.S. Geol. Surv. Monogr., 1: 438 pp.

Hanks, T.C. and Kanamori, H., 1979. A moment magnitude scale. J. Geophys. Res., 84(B5): 2348–2350.

Hecker, S., 1993. Quaternary tectonics of Utah with emphasis on earthquake hazard characterization. Utah Geol. Surv. Bull., 127: 157 pp.

Johnston, A.C., 1989. The effect of large ice sheets on earthquake genesis. In: S. Gregersen and P.W. Basham (Editors), Earthquakes at North-Atlantic Passive Margins: Neotectonics and Postglacial Rebound. Kluwer, Boston, MA, pp. 581–599.

Keaton, J.R., Currey, D.R. and Olig, S.J., 1987. Paleoseismicity and earthquake hazards evaluation of the West Valley fault zone, Salt Lake City urban area, Utah. Utah Geol. Surv. Contract Rep., CR-93-8: 33 pp.

Kihl, R., 1975. Physical preparation of organic matter samples for ^{14}C dating (app.). In: J.T. Andrews (Editor), Radiocarbon Date List II from Cumberland Peninsula, Baffin Island, N.W.T., Canada. Arctic Alpine Res., 7: 90–91.

Lund, W.R., 1993. New information on timing of large earthquakes on the Salt Lake City segment of the Wasatch fault zone: Implications for increased hazard along the central Wasatch Front: Wasatch Front Forum. Utah Geol. Surv., 8: 12–13.

Machette, M.N. (Editor), 1988. In the Footsteps of G.K. Gilbert — Lake Bonneville and Neotectonics of the Eastern Basin and Range Province. Utah Geol. Miner. Surv. Misc. Publ., 88-1: 120 pp.

Machette, M.N., Personius, S.F., Nelson, A.R., Schwartz, D.P. and Lund, W.R., 1991. The Wasatch fault zone, Utah — Segmentation and history of Holocene earthquakes. In: P.L. Hancock, R.S. Yeats and D.J. Sanderson (Editors), Characteristics of Active Faults. J. Struct. Geol., 13: 137–150.

Machette, M.N., Personius, S.F. and Nelson, A.R., 1992. Paleoseismology of the Wasatch fault zone — a summary of recent investigations, interpretations, and conclusions. In: P.L. Gori, and W.W. Hays (Editors), Assessment of Regional Earthquake Hazards and Risk along the Wasatch Front, Utah. U.S. Geol. Surv. Prof. Pap., 1500-A: 71 pp.

Mason, D.B., 1992, Earthquake magnitude potential of active faults in the Intermountain seismic belt from surface parameter scaling. M.S. Thesis, Univ. of Utah, 110 pp.

Mason, D.B. and Smith, R.B., 1993. Paleoseismicity of the Intermountain seismic belt from late Quaternary faulting and parameter scaling of normal faulting earthquakes. Geol. Soc. Am. Abstr. Programs, 25: 115.

McCalpin, J.P. and Forman, S.L., 1991. Late Quaternary faulting and thermoluminescence dating of the East Cache fault zone, north-central Utah. Bull. Seismol. Soc. Am., 81: 139–161.

McCalpin, J.P., Robison, R.M. and Garr, J.D., 1992. Neotectonics of the Hansel Valley–Pocatello Valley corridor, northern Utah and southern Idaho. In: P.L. Gori and W.W. Hays (Editors), Assessment of Regional Earthquake Hazards and Risk along the Wasatch Front, Utah. U.S. Geol. Surv. Prof. Pap., 1500-G: 18 pp.

McCoy, W.D., 1987. Quaternary aminostratigraphy of the Bonneville basin, western United States. Geol. Soc. Am. Bull., 98: 99–112.

Nelson, A.R., 1992. Lithofacies analysis of colluvial sediments — an aid in interpreting the recent history of Quaternary normal faults in the Basin and Range province, Western United States. J. Sediment. Petrol., 63: 607–621.

Olig, S.J., Keaton, J.R. and Currey, D.R., 1986. Paleoseismicity and earthquake hazard evaluation of the West Valley fault zone, Salt Lake metropolitan area, Utah. In: Assoc. Eng. Geol. 29th Annual Meeting Abstr. Program; 59.

Olig, S.S., Lund, W.R. and Black, B.D., 1993. Evidence for mid-Holocene surface rupture on the Oquirrh fault zone, Utah. Geol. Soc. Am. Abstr. Programs, 25: 129.

Ostenaa, D., 1984. Relationships affecting estimates of surface fault displacements based on scarp-derived colluvial deposits. Geol. Soc. Am. Abstr. Programs, 16: 327.

Oviatt, C.G., McCoy, W.D. and Reider, R.G., 1987. Evidence for a shallow early or middle Wisconsin-age lake in the Bonneville basin, Utah. Quat. Res., 27: 248–262.

Oviatt, C.G., Currey, D.R. and Sack, D., 1992. Radiocarbon chronology of Lake Bonneville, eastern Great Basin, U.S.A. Paleogeogr. Palaeoclimatol. Palaeoecol., 99: 225–241.

Pechmann, J.C., Nash, W.P., Vivieros, J.J. and Smith, R.B., 1987. Slip rate and earthquake potential of the East Great Salt Lake fault, Utah. EOS Trans., 68: 1369.

Schwartz, D.P., 1988. Geology and seismic hazards: Moving into the 1990's. In: J.L. Von Thun (Editor), Earthquake Engineering and Soil Dynamics. II .Recent Advances in Ground-Motion Evaluation. Am. Soc. Civ. Eng. Proc., pp. 1–42.

Scott, W.E., McCoy, W.D., Shroba, R.R. and Rubin, M., 1983. Reinterpretation of the exposed record of the last two cycles of Lake Bonneville, Western United States. Quat. Res., 20: 261–285.

Smith, R.B. and Bruhn, R.L., 1984. Intraplate extensional tectonics of the eastern Basin-Range: Inferences on structural style from seismic refection data, regional tectonics, and therma-mechanical models of brittle–ductile deformation. J. Geophys. Res., 89: 5733–5762.

Smith, R.B. and Sbar, M.L., 1974. Contemporary tectonics and seismicity of the western United States with emphasis on the Intermountain Seismic Belt. Geol. Soc. Am. Bull., 85: 1205–1218.

Smith, R.B. and Arabasz, W.J., 1991. Seismicity of the Intermountain Seismic Belt. In: D.B. Slemmons, D. Blackwell, E.R. Engdahl and M. Zoback (Editors), Neotectonics of North America, Boulder, Colorado. Geol. Soc. Am. Decade Map, 1: 185–228.

Solomon, B.J., 1993. Quaternary geologic maps of Tooele Valley and the West Desert Hazardous Industry Area, Tooele County, Utah. Utah Geol. Surv. Open-File Rep., 296, 48 pp., 1:24,000 scale.

Solomon, B.J., Everitt, B.L., Currey, D.R., Paulick, J.S., Wu, D., Olig, S.S. and Burr, T.N., 1992. Quaternary geology and geologic hazards of Tooele and northern Rush Valleys, Utah. In: J.R. Wilson (Editor), Field Guide to Geologic Excursions in Utah and Adjacent Areas of Nevada, Idaho, and Wyoming. Utah Geol. Surv. Misc. Publ., 92-3: 179–205.

Stafford, T.W. and Forman, S.L., 1993. Radiocarbon and thermoluminescence dating of Wasatch faulting events, Gardner Canyon, Utah. Utah Geol. Surv. Contract Report, 15 pp.

Stuiver, M. and Quay, P.D., 1979. Changes in atmospheric carbon-14 attributed to a variable sun. Science, 207: 11–19.

Stuiver, M. and Reimer, R.J., 1993. Extended ^{14}C data base and revised CALIB 3.0 ^{N}C age calibration program. Radiocarbon, 35: 215–230.

Swan, F.H., III, Hanson, K.L., Schwartz, D.P. and Black, J.H., 1983. Study of earthquake recurrence intervals on the Wasatch fault, Utah. 8th Semi-Annual Tech. Rep. to U.S. Geological Survey under contract 14-08-0001-19842, 20 pp.

Tocher, D., 1958. Earthquake energy and ground breakage. Bull. Seismol. Soc. Am., 48: 147–153.

Tooker, E.W. and Roberts, R.J., 1992. Preliminary geologic map of the Stockton 7.5-minute quadrangle, Tooele County, Utah. U.S. Geol. Surv. Open-File Rep., 92-385, scale 1:24,000.

Vivieros, J.J., 1986. Cenozoic tectonics of the Great Salt Lake from seismic reflection data, M.S. Thesis, Univ. of Utah, 99 pp.

Wells, D.L. and Coppersmith, K.J., 1994. Updated empirical relationships among magnitude, rupture length, rupture area, and surface displacement. Bull. Seismol. Soc. Am., 84.

West, M.W., 1992. An integrated model for seismogenesis in the Intermountain seismic belt. Bull. Seismol. Soc. Am., 82: 1350–1372.

Youngs, R.R., Swan, F.H., Power, M.S., Schwartz, D.P. and Green, R.K., 1987. Probabilistic analysis of earthquake ground shaking hazard along the Wasatch Front, Utah. In: P.L. Gori and W.W. Hays (Editors), Assessment of Regional Earthquake Hazards and Risk Along the Wasatch Front, Utah, 2. U.S. Geol. Surv. Open-File Rep. 87-585: M-1–110.

Zoback, M.L., 1983. Structure and Cenozoic tectonism along the Wasatch fault zone, Utah. In: D.M. Miller, V.R. Todd and K.A. Howard (Editors), Tectonics and Stratigraphy of the Eastern Great Basin. Geol. Soc. Am. Mem., 157: 3–27.

Zoback, M.L., 1989. State of stress and modern deformation of the northern Basin and Range province. J. Geophys. Res., 94: 7105–7128.

ELSEVIER

Geomorphology 10 (1994) 317–329

GEOMORPHOLOGY

Global warming and permafrost in Eurasia: a catastrophic scenario

Jaromír Demek

Palacký University, Faculty of Sciences, 771 46 Olomouc, Czech Republic

Received February 10, 1994; revised April 2, 1994; accepted April 5, 1994

Abstract

Since models of global warming predict that climatic changes will be greatest in the Arctic and might already be in progress, it is prudent to attempt to understand the rapidly changing thermal régime on Eurasian cold regions. In this paper the author tries to evaluate the probable catastrophic implications of global climatic change in the cold regions of Eurasia. The impact on Man activities in permafrost regions is discussed.

1. Introduction

In Arctic and subArctic lands and high mountains of Eurasia the bedrock and surficial deposits are permanently frozen. Permafrost, defined as ground that remains frozen for more than 2 years, may occur where mean annual air temperatures (MAAT) are less than $-1°$ or $-2°C$. Although dry permafrost can exist, most of the permafrost contains ground ice. Ground ice may be present in a variety of forms and in different contents. Because permafrost is controlled by climatic conditions, it is potentially sensitive to climatic change.

Fig. 1 shows the near-ground air temperatures in the Northern Hemisphere between 1850 and 1990. The curve shows a warming trend beginning about 1880 to 1890. In the Northern Hemisphere this warming cul-

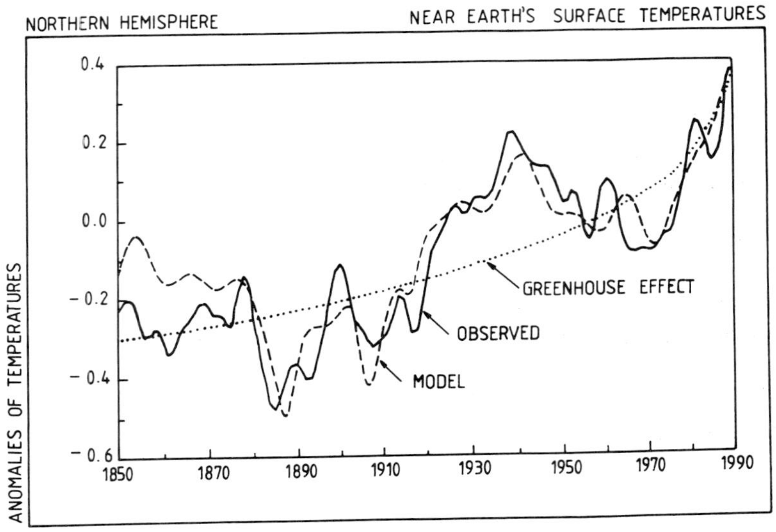

Fig. 1. Anomalies of near surface temperatures in the Northern Hemisphere between 1850 and 1990 in degrees Kelvin (modified from Schönwiese, 1993, p. 79).

SSDI 0169-555X(94)00028-P

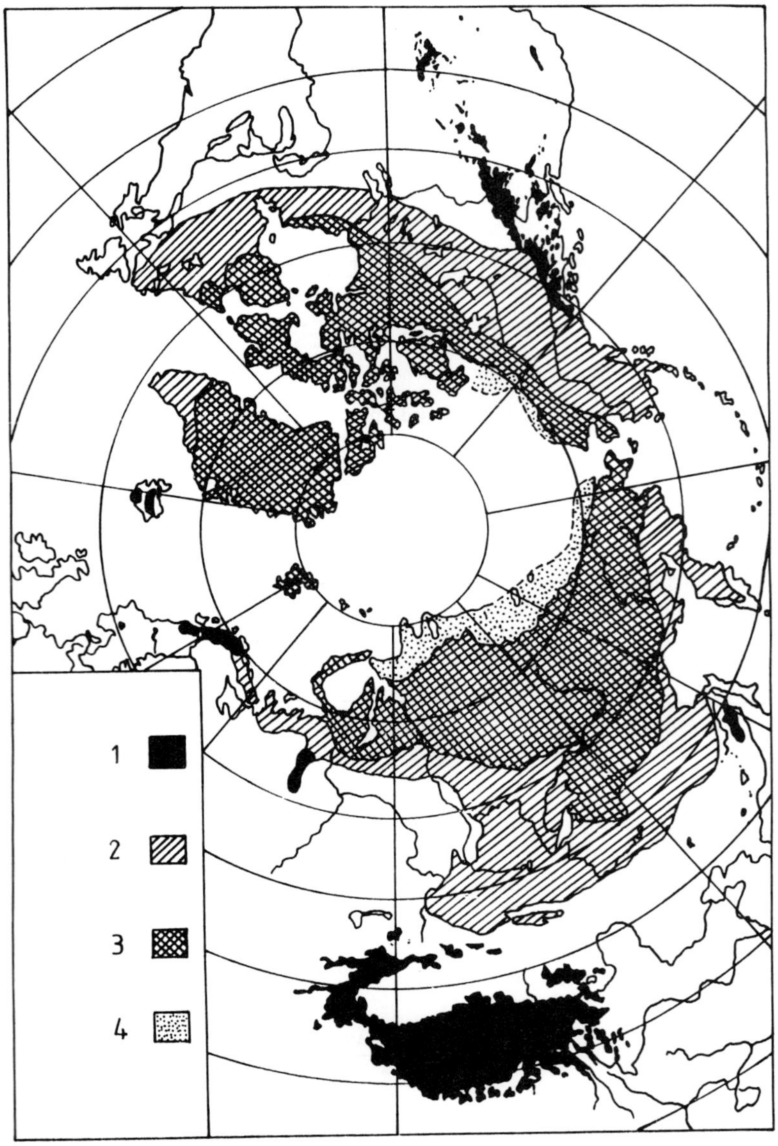

Fig. 2. Distribution of Permafrost in the Northern Hemisphere (modified from Péwe, 1963). 1: Zone of alpine permafrost, 2: zone of discontinuous permafrost, 3: zone of continuous permafrost, 4: zone of sub-sea permafrost.

minated around 1940, with subsequent cooling until the mid-1960s, followed by a recent rise to levels higher than those previously attained. Many climatologists have claimed that the increase of air temperature results from the greenhouse effect, although it is not completely clear how much is due to this effect (e.g. Flohn, 1983; Smith, 1988; Schönwiese, 1993).

Studies demonstrate, however, in almost universal agreement that warming was concentrated in the polar and sub-polar regions (Lachenbruch and Marshall, 1986) along with increased precipitation. For a doubling of the atmospheric carbon dioxide content, some models predict that annual air temperatures throughout the Arctic could increase by 3 to 6°C (Hansen et al., 1984; Smith, 1988). Warming in Arctic regions from

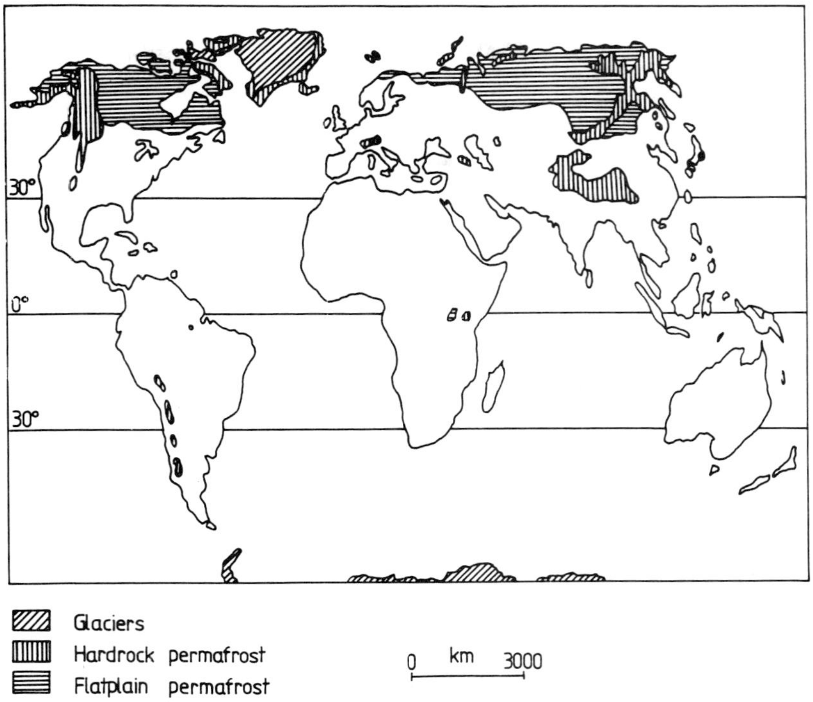

Fig. 3. Distribution of hardrock and flatplain permafrost (modified from Farley, 1991, p. 36).

1947 to 1993 was concentrated in the polar regions over northwestern North America and Siberia. The magnitude of such changes in climate of northern Eurasia would produce serious and far-reaching environmental and engineering problems in polar and sub-polar regions underlain by permafrost. The mechanical properties of permafrost are especially dependent upon temperature. Fig. 2 shows the distribution of permafrost in the Northern Hemisphere with the areas in the northern part of Eurasia underlain by permafrost. Most of this permafrost exists at temperatures warmer than −5°C. Tens of thousands of square kilometers of permafrost are even warmer than −3°C, and it can be expected that most of this would disappear under the global warming predicted.

Human activity in cold regions of Eurasia has increased greatly in recent decades, largely to develop natural resources — especially oil, gas, diamonds, gold and other minerals. Large cities, many factories, mines, oil fields, nuclear power plants, airports, roads, water reservoirs etc. have been built on permafrost. All these constructions will be endangered by catastrophic processes caused by melting of permafrost and/or by changing its mechanical properties. Catastrophic processes will be different in areas of flatplain permafrost compared to areas of hardrock or alpine permafrost (Fig. 3).

2. Degradation of permafrost

Large-scale processes of permafrost degradation can be expected in vast areas of polar lowlands underlain by syngenetic permafrost with its high content (up to 90%) of ground ice. Ground ice may be present in a variety of forms ranging from a cement in soil pores to masses of almost pure ice in veins, lenses and wedges. Fig. 4 shows typical ice layers or lenses of primary cement ice. The ice layers vary greatly in size and form, from soil to soil, such as massive ice bodies in syngenetic permafrost of the Central Yakutian Lowland (Fig. 5). Syngenetic ice wedges of secondary ice (Fig. 6) and epigenetic ice (Fig. 7) are also common. Ground ice bodies in this Lowland are situated just under the surface of the earth covered by vegetation.

Fig. 4. Typical structure of frozen deposits (right) with primary cement ground ice. Photo J. Demek.

Fig. 5. Massive ice body in syngenetic permafrost. Terrace deposits of Lena River in the Central Yakutian Lowland. Photo J. Demek.

Fig. 6. Syngenetic ice wedges of secondary ice in fluvial deposits of Aldan River. Photo J. Demek.

When degradation of permafrost occurs in such areas, the disappearance of ground ice can result in a type of terrain known as thermokarst. Thermokarst results from the disruption of the thermal equilibrium of the permafrost and an increase in the depth of the active layer. When permafrost thaws, a highly mobile liquid sludge results (Fig. 8). The melted ground, especially if excess ice exists, has a much smaller volume — often only a small fraction — compared with former permafrost. The principal effect is, therefore, the creation of collapse features. Two types of manifestation of thermokarst processes occur (Czudek and Demek, 1970; French, 1974; Embleton and King, 1975).

The first type is the downward degradation of the generally flat permafrost surface. Fig. 9 shows the evolution leading to the stage when thaw-in lakes are formed. At the beginning small surface pits develop (Ib — see Fig. 10), later funnel-shaped sinks (II), and at the end a lake basin with steep sides and a flat bottom, called an alas (IIIa — see Fig. 11). These thaw-in depressions exhibit different forms, from small circular lakes to depressions of tens of kilometers in diameter. Upon attaining a certain depth, the lake does not freeze to the bottom in winter and a talik begins to develop below the lake (IIIa). In the fourth stage (IVa) the alas lake disappears.

The second type is a lateral degradation from river banks or lake and sea shores resulting in the formation of amphiteatre-shaped hollows termed thermocirques (Fig. 12). On the sides of thermocirques massive ice bodies are exposed. Pools of water collect on the floor of these. The presence of water considerably accelerates the development of the thermokarst. The water also causes mudflows (Fig. 13).

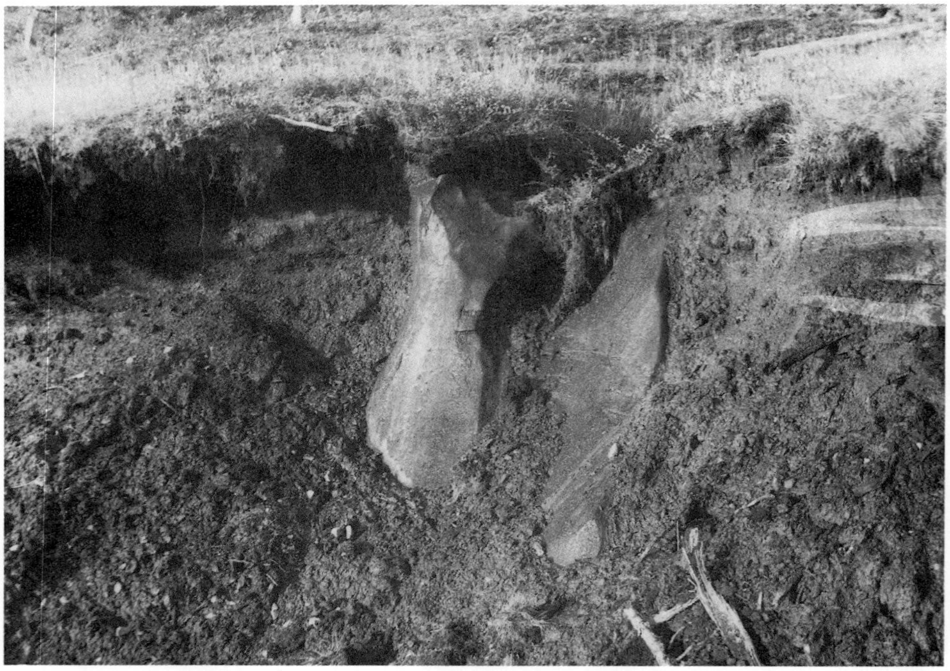

Fig. 7. Epigenetic ice wedges. Ground ice body is situated just under the tundra surface. Photo J. Demek.

Fig. 8. High mobile liquid sludge as result of the permafrost melting. Photo J. Demek.

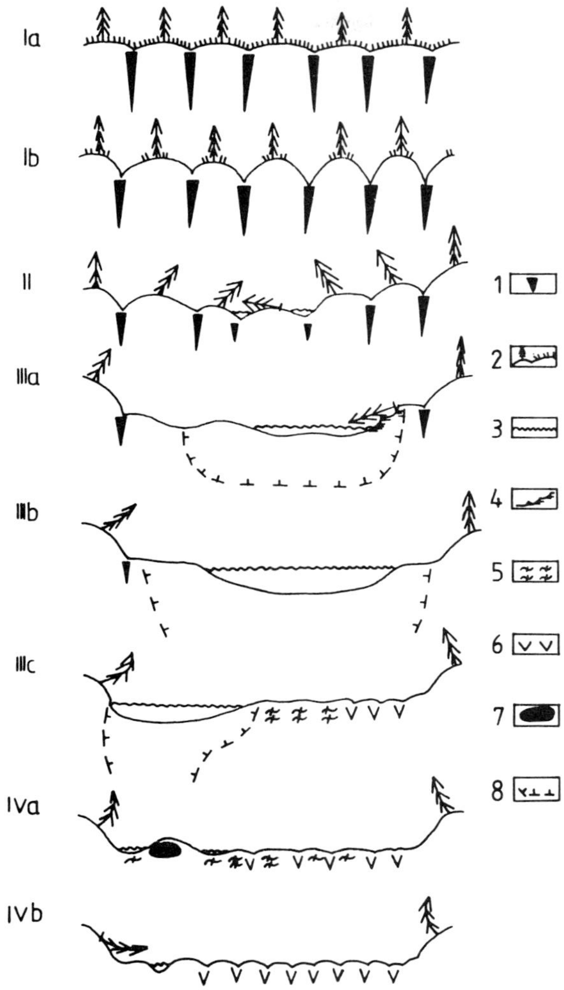

Fig. 9. Degradation of permafrost and the evolution of alases. For explanation see text (after Czudek and Demek, 1970).

All thermokarst processes may totally destroy engineering structures such as the road damaged by thermokarst processes (Fig. 14). At present this road no longer exists. The disruption of the ground from thawing and settlement is similarly catastrophic for fields and pastures. Fig. 10 shows the surface of a cattle pasture in Central Yakutian Lowland destroyed by thermokarst processes. Reclamation of such ground is nearly impossible.

Because thermokarst processes create a new lower surface form on the Eurasian Arctic Plains, the previous ground is destroyed: vegetative cover, roads and settlements. Fig. 11 shows trees sliding into a thermokarst depression.

3. Effects of permafrost degradation

Experience with large-scale thermokarst development in the Central Yakutian, Kolyma and Primorskaya Lowland during post-glacial warming can be applied to this situation. Fig. 15 shows that in Central Yakutian Lowland about 50% of the Pleistocene surface has been subjected to alas development. This apparently occurred predominantly during a warm phase of the Holocene between 9000 and 2500 B.P. Also the relief of the Primorskaya Lowland has been changed considerably by thermokarst processes. In the area of Kolyma and Primorskaya Lowland an extensive plain occurred in the Pleistocene, probably extending far into the Arctic Ocean and including the present Lyakhovskie Islands and Ayon and Vrangel Islands. In the Upper Pleistocene stratified loams with peat beds were deposited in the plain during continuous subsidence. Ice veins, 30–50 m thick and 10 m wide, occur in the loams. Because of the thermokarst, scarps 25 m high retreat at an annual rate of 15 to 18 m parallel to their front on a section several kilometers long. Lakes develop at the base, and the erosional activity of the lakes and undercutting of the scarp accelerate scarp retreat. At the present time, a small remnant of the vast Upper Pleistocene surface is preserved above the level of the lower lowland and is bordered by scarps with thick ice veins. Development of thermokarst depressions — alases and thermo-cirques — is accompanied by other catastrophic effects of thawing such as mudflows (Fig. 13), landslides and slumps.

4. Catastrophic scenario

On thawing, when excess ice is present, water will be liberated. Water accumulation in alases is an important factor accelerating thermokarst processes and resulting in the above mentioned thaw-in lakes, with shoreline features of thermo-abrasion, slumping, and subsidence (Fig. 11). In relation to this, very important changes in precipitation are predicted in the catastrophic scenario. Precipitation would be 10 to 50%

Fig. 10. Surface of cattle pasture in central Yakutia destroyed by thermokarst processes. The process will result in surface shown on Fig. 8. Photo J. Demek.

Fig. 11. Trees sliding into an alas lake. Photo J. Demek.

Fig. 12. Thermocirque in the terrace of Aldan River in Siberia. Photo J. Demek.

higher in summer and 60% higher in winter. This will supply additional water to thermokarst processes and in a feed-back loop accelerate the thermokarst processes in Arctic Eurasia. According to the author's personal observations, the present-day intensity of thermokarst processes in dry areas of eastern Siberia is limited only by the lack of free water. Thermokarst processes will be affected by increased rainfall, but more importantly, perhaps, changes in snow cover would complicate the effect of global warming. Precipitation increases of as much as 60% in autumn and early winter could be a significant factor in permafrost degradation, particularly in marginal permafrost areas with higher temperatures.

In the zone of continuous permafrost, ground temperatures that may be many degrees below 0°C, would rise with climatic warming, but probably no major changes in geographical distribution of permafrost, would occur. Meanwhile progressive deepening of the active layer would take place with melting of ground ice, and thermokarst processes would start even in these regions on the surface. In addition to permafrost degradation, any change in the temperature of the permafrost could cause a major change in the strength and deformation properties of the frozen layers, even without any thaw.

The formation of a new, lower surface of Arctic plains could result in a transgression of the cold Arctic Ocean deep into the Eurasian continent. Long sections of the Arctic Ocean coast are composed of frozen deposits with a high content of ground ice. Space images have shown that in recent decades the areal extent of sea ice is decreasing (Fig. 16). Thermo-abrasion results in cliff recession, thaw slumps and ther-

Fig. 13. Mudflow caused by the melting of permafrost. Photo J. Demek.

Fig. 14. Depression in the centre of the road caused by melting of ice wedge polygons in Sredne Sibirskoye Ploskogorye. Photo J. Demek.

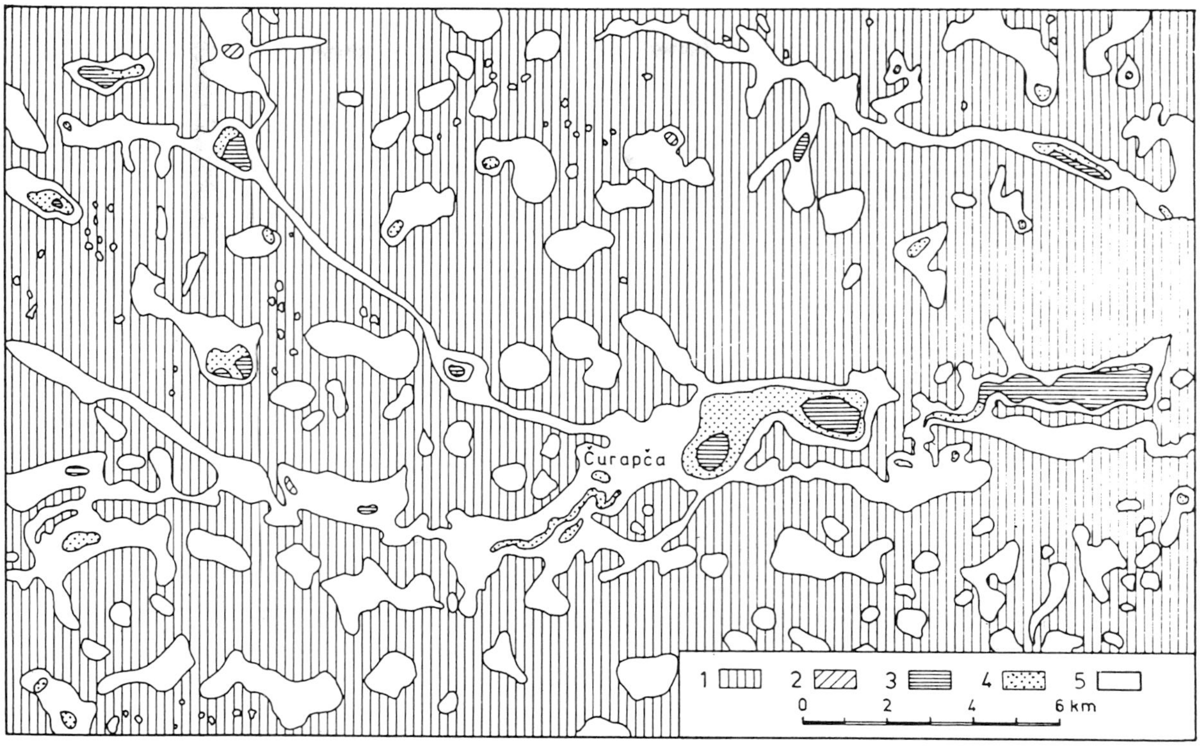

Fig. 15. Distribution of thermokarst depressions in central Yakutiya (Sacha Republic). Explanation: 1. Original lowland level with syngenetic ice wedges, 2. young lakes, 3. lakes in alases, 4. recently desiccated lakes in alases, 5. old alases.

mogullying and produces cliff recession of some hundreds of meters per year. The thermo-abrasion will be all the more efficient because the global warming will cause a rise in sea level.

A climatic warming of about 0.5° to 0.8°C is reported from many mountain areas (King et al., 1992, p. 79). In the Alps a tendency for the permafrost to thin at a rate of a few centimeters per year as a consequence of recent global warming has been shown. In the same area, temperatures of the permafrost table seem to have increased by about 0.5°C since the beginning of this century (Haeberli, 1983). Permafrost in large areas of the alpine region is predicted to melt in the future because of climatic amelioration. This is particularly the case of sporadic alpine permafrost. Areas with widespread discontinuous permafrost, however, may change into sporadic permafrost (islands of permafrost). Melting of alpine permafrost results in slope instability, causing increased landslide and rockslide activity and affecting areas that have never before been subject to this hazard. Rocks loosened by unloading or frost weathering have in the past been cemented by ground ice. In response to melting, rocks are deformed by deep-seated creep, bulging, fissures open and loosened blocks fall down. Landsliding, rocksliding or

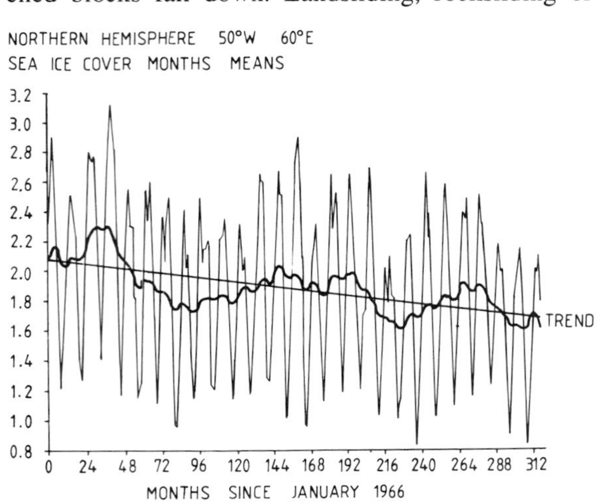

Fig. 16. Decreasing extent of sea ice cover in the Arctic Ocean (Northern Hemisphere) (after Bolle, 1993).

cambering also commence. Good examples of these processes are well known from former Pleistocene periglacial regions in Europe and Asia (Záruba and Mencl, 1969; Nemčok, 1982).

5. Thermokarst and man

The melting of the permafrost is very slow, e.g. given a rise of the mean annual temperature (MAAT) of 0.5

Fig. 17. Concrete piles used in northern engineering constructions on permafrost. Photo J. Demek.

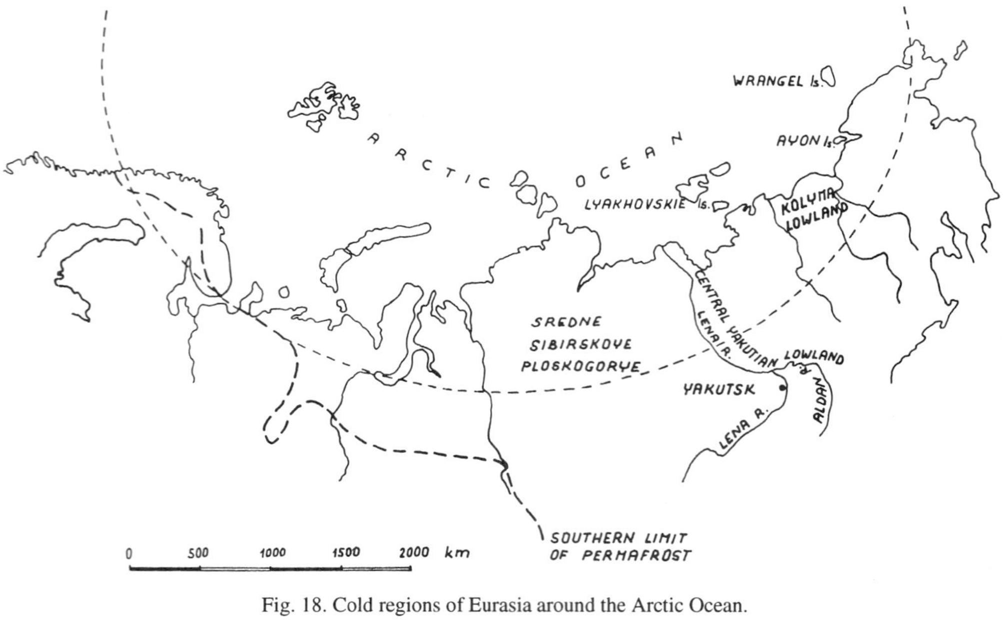

Fig. 18. Cold regions of Eurasia around the Arctic Ocean.

or 0.8°C, it will take decades for a 10 m thick permafrost to melt and centuries for a 50 m thick permafrost layer to thaw. In considering the permafrost response to climatic trends, it must be recognized, first, that permafrost responds with a long time lag and, secondly, that the response to a given climatic forcing will differ regionally in magnitude and in timing. But, even if the development of thermokarst process will be more limited in comparison with this catastrophic scenario, any change in the temperature of the permafrost could cause a major change in the strength and deformation properties of the frozen layers, even without any thaw. The mechanical properties of permafrost are highly temperature dependent. Problems will occur with the bearing capacity of piles, as the permafrost warms and adhesion forces declines. Fig. 17 shows that piles are widely used in northern engineering constructions. Warming of permafrost will undoubtedly create severe maintenance and repair problems for houses, factories, utilities and other engineering constructions founded on piles (Williams and Smith, 1989).

6. Conclusions

The global warming of the climate is most marked in the Arctic. Some temperature measurements in the permafrost indicate that the warming has probably already begun in the Arctic Lowland. It will cause the formation of numerous thermokarst forms and mass movements on slopes in areas underlain by permafrost. Coastal thermo-abrasion will give rise to rapid retreat of coastal lowlands with deposits that contain large amounts of ground ice. The consequence of global warming will be the gradual degradation of the permafrost in Eurasia (Fig. 18). We must be fully aware of the practical impact of this warming on engineering constructions, such as roads, buildings, pipelines and airfields. The usual protection against the thawing, like piles or layers of gravel on the surface, will not be available to ameliorate the consequences of changing permafrost properties.

References

Bolle, H.-J., 1993. Das System Erde. Geogr. Rundsch., 45(2): 81–87.

Czudek, T. and Demek, J., 1970. Thermokarst in Siberia and its Influence on the development of lowland relief. Quat. Res., 1(1): 103–120.

Embleton, C. and King, C.A.M., 1975. Periglacial Geomorphology. Edward Arnold, London, 203 pp.

Farley, P., 1991. That sinking feeling. Geogr. Mag., LXIII(2): 36–39.

Flohn, H., 1983. Das CO_2–Klima-Problem. Geogr. Rundsch., 35(5): 238–247.

French, H.M., 1974. The Periglacial Environment. Longman, London, 309 pp.

Haeberli, W., 1983. Permafrost–glacier relationships in the Swiss Alps, today and in the past. Proc. Third Int. Conf. on Permafrost, Vol. 1. Nat. Ac. Press, Washington, DC, pp. 415–420.

Hansen, J.E., Lacis, A., Rind, D., Russel, G., Stone, P., Fung, I., Ruedy, P. and Lerner, J., 1984. Climate sensitivity: analysis of feedback mechanism. In: J. Hansen and T. Takahashi (Editors), Climate Processes and Climate Sensitivity. Maurice Ewing Series, 5. American Geophysical Union, Washington, DC, pp. 130–163.

King, L., Gorbunov, A.P. and Evin, M., 1992. Prospecting and mapping of mountain permafrost and associated phenomena. Permafrost and Periglacial Processes, 3 (1992): 73–81.

Lachenbruch, A.H. and Marschall, B.V., 1986. Changing climate: geothermal evidence from permafrost in the Alaskan Arctic. Science, 234 (November): 689–696.

Nemčok, A., 1982. Landslides in Slovakian Carpathians. Veda, Bratislava, 319 pp. (in Slovak).

Péwe, T.L., 1983. The periglacial environment in North America during Wisconsin time. In: S.C. Porter (Editor), Late Pleistocene. Univ. of Minnesota Press, Minneapolis, pp. 157–159.

Schönwiese, Ch.-D., 1993. Globale Klimaänderungen in Vergangenheit und Zukunft. Geogr. Rundsch., 45(2): 74–80.

Smith, M.W., 1988. The significance of climatic change for the permafrost environment. Final Proc. Fifth Int. Conf. on Permafrost. Tapir Publishers, Trondheim, Norway, pp. 18–23.

Williams, P.J. and Smith, M.W., 1989. The Frozen Earth. Cambridge Univ. Press, Cambridge, 306 pp.

Záruba, Q. and Mencl, V., 1969. Landslides and Their Control. Elsevier, Amsterdam/Academia, Prague.

ELSEVIER

Geomorphology 10 (1994) 331–354

GEOMORPHOLOGY

Geological hazards, vulnerability, and risk assessment using GIS: model for Glenwood Springs, Colorado

Mario Mejía-Navarro, Ellen E. Wohl, Sherry D. Oaks

Department of Earth Resources, Colorado State University, Ft. Collins, CO 80523, USA

Received February 2, 1994; revised April 1, 1994; accepted April 5, 1994

Abstract

Glenwood Springs, Colorado, lies at the junction of the Roaring Fork and Colorado Rivers, surrounded by the steep peaks of the Colorado Rocky Mountains. Large parts of the region have had intensive sheet erosion, debris flows, and hyperconcentrated floods triggered by landslides and slumps. The latter come from unstable slopes in the many tributary channels on the mountainsides, causing concentration of debris in channels and a large accumulation of sediment in colluvial wedges and debris fans that line the river valleys. Many of the landslide and debris-flow deposits exist in a state resembling suspended animation, ready to be destabilized by intense precipitation and/or seismic activity.

During this century urban development in the Roaring Fork River valley has increased rapidly. The city of Glenwood Springs continues to expand over unstable debris fans without any construction of hazard mitigation structures. Since 1900, Glenwood Springs has had at least 21 damaging debris flows and floods; on July 24, 1977 a heavy thunderstorm spread a debris flow over more than 80 ha of the city.

This paper presents a method that uses Geographic Information Systems (GIS) to assess geological hazards, vulnerability, and risk in the Glenwood Springs area. The hazards evaluated include subsidence, rockfall, debris flows, and floods, and in this paper we focus on debris flows and subsidence. Information on topography, hydrology, precipitation, geomorphic processes, bedrock and surficial geology, structural geology, soils, vegetation, and land use, was processed for hazard assessment using a series of algorithms. ARC/INFO and GRASS GIS softwares were used to produce maps and tables in a format accessible to urban planners.

After geological hazards were defined for the study area, we estimated the vulnerability (V_e) of various elements for an event of intensity *i*. Risk is assessed as a function of hazard and vulnerability. We categorized the study area in 14 classes for planning procedures; 7 classes defined as areas suitable for human settlement, and 7 classes defined as unsuitable for building, and most effectively reserved for parks and forests.

1. Introduction

In the 125 years since gold fever brought settlers to Colorado, mass movements such as debris flows, floods, and avalanches have been persistent major natural hazards affecting life in the Colorado Rockies. Triggered by summer thunderstorms or fast snowmelt (Wieczorek et al., 1989), these events have devastated mining camps, ranches and homesites, and have repeatedly blocked transportation lines and waterways (Ives et al., 1976; Carrara, 1979). The magnitude of these hazards and the risk intensity imposed by such events grows proportionally with population density.

During this century urban development in the Roaring Fork River Valley of Colorado has increased rapidly. The city of Glenwood Springs in this valley

SSDI 0169-555X(94)00030-U

Table 1
Major floods and debris flow records for the Glenwood Springs area

Month, day	Year	Description	Reference[a]
	1903	Mud + rocks; one death; damages to railroad + others	(3)
June 23	1917	East slope of Glenwood Springs; mudflow + debris flows; large landslide along Colorado River	(1), (2), (3), (4)
June 14–22	1918	Flood on north side of Glenwood Springs by Colorado River; Colorado River discharge at 30100 cfs for several days	(1)
July 27–30	1929	Debris flows along whole valley between Glenwood Springs and Carbondale	(3)
July 28	1933	Southern border of Glenwood Springs	(3)
August 5–10	1933	Southern border of Glenwood Springs at 23rd Gulch	(3), (4)
Sept. 3	1934	Southern border of Glenwood Springs	(3), (4)
July 10	1936	Large debris and mud flows along both sides of Glenwood Springs with much damage	
Sept. 1	1936	The largest described mud flow within the city, mainly Cemetery Gulch and northern small drainages, also at 23rd Gulch	(2), (3), (4)
July 30	1937	Thunderstorm; large debris flows at Cemetery Gulch; main part of the city damaged	(2), (3), (4)
August 13	1938	Hyperconcentrated flood through Cemetery Gulch	(2), (3), (4)
July 30	1943	A major debris flow within the city; Cemetery Gulch, much damage to residences, business, and railroad	(2), (3), (4)
Summer	1947	Southeastern portion of Glenwood Springs (maybe 23rd Gulch)	(2), (3), (4)
July 1	1957	High snowmelt; Roaring Fork River floods low terraces between flood plain and Qt_1 (city location); river discharge 19,000 cfs	(1), (2)
August 31	1963	Northeast of downtown	(2)
	1963	Many unconfirmed histories of floods and flows in the southern area of Glenwood Springs during early 1960s	Community talks (old residents)
July 16	1967	Small mud and debris flows east, west and north of Glenwood	(2), (3)
June 10	1970	Springs and along Interstate Highway 70	
June	1972		
July 19	1974		
July 24	1977	Well described; severe thunderstorm and debris flows	(1), (2), (3), (4)
July 25	1977	Through 23rd and Cemetery Gulches; west side of Roaring Fork and NW of Glenwood Springs; hyperconcentrated flood	(4)
July 12	1981	23rd Gulch; big thunderstorm after unusually dry winter and spring	(2), (3), (4)

[a]References: (1) Gingery Associates Inc., 1977; (2) The Glenwood Post and The Free Weekly Newspaper; (3) ESA Geotechnical Consultants, 1982; (4) Morris, 1986.

continues to expand over unstable debris fans without any construction of hazard mitigation structures. The free-space areas have been covered by asphalt parking lots and wide new streets that serve as channels for the debris flows through vulnerable dense urban development. Since 1900, Glenwood Springs has had at least 21 damaging debris flows and floods (Table 1). On July 24, 1977, for example, a heavy thunderstorm spread a debris flow over more than 80 hectares of the city.

In cooperation with the planning office of Garfield County, Colorado, the work described here was undertaken as part of a pilot project to test the usefulness of Geographic Information Systems (GIS), when incor-

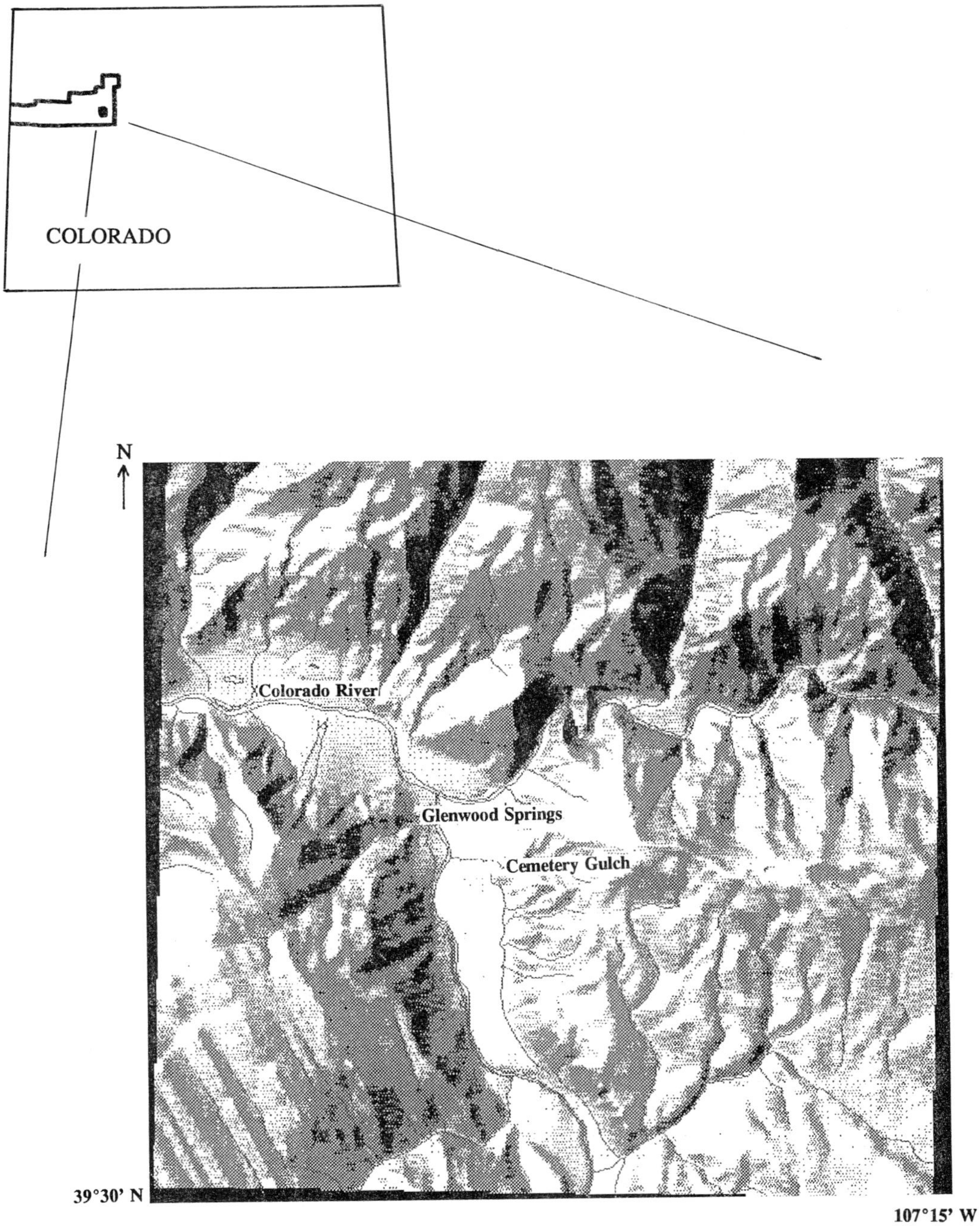

Fig. 1. Shaded relief map of the study area. Map shows almost 10 km^2 in area. Garfield County and the study area are shown on the inset map of Colorado.

porated in an Integrated Planning Decision Support (IPDS) Model (Mejía-Navarro et al., 1994), for improving urban planning and decision making in the Glenwood Springs area. The study assesses geological hazards, vulnerability, and risks. Hazard is defined as the probability that an event of an intensity equal to or greater than a category i value occurs during a period of exposure t. Geological hazards (H_i) in the study area include subsidence, landslides, debris flows, and floods. Vulnerability (V_e) is defined as the intrinsic predisposition of any element to be at risk of a mental or economic loss upon the occurrence of a hazardous event of intensity *i*. Three items were considered in the qualitative vulnerability analysis: (1) ecosystem sensitivity, (2) economic vulnerability, and (3) social structure vulnerability. The vulnerability of the area increases in proportion to the growth of population, lifeline networks, community services, and public assembly sites. Finally, risk (R_{ei}) is defined as the probability that a loss affecting an element *e* should occur as a consequence of an event of an intensity equal to or greater than *i*. Risk is a function of hazard and vulnerability, and was assessed using the equation $R_{ei}=f(H_i,V_e)$ (after Cardona, 1988). ARC/INFO and GRASS GIS softwares were used to analyze and produce maps and tables in a format accessible to urban planners, and the GIS analyses were then incorporated into an IPDS model that includes "what if scenarios" and optimization strategies. In this paper we focus on the procedures used to quantify the hazards associated with debris flows and subsidence.

We focused on 6500 hectares around the city of Glenwood Springs, an area with high urban development pressure. At 1750 m elevation, Glenwood Springs is located at the exit of the narrow canyon that the Colorado River has incised through a corner of the White River Plateau (Fig. 1). Existing maps, aerial photographs, and geotechnical reports were combined with field observations to evaluate the area. Topography, hydrology, precipitation, geomorphology, bedrock and surficial geology, structural geology, geotechnical soils analysis, vegetation, and land use were mapped onto 1:4000 to 1:24,000 scale base maps, and then processed with a series of algorithms designed to estimate hazard susceptibility. The final product of the analyses was a zoning of the area into 14 land-use suitability classes, which incorporate hazards, vulnerability, and risk parameters.

2. Previous work

Previous studies designed to evaluate the statistical relations among geological hazards (landslides, debris flows, and floods) and controlling variables (bedrock geology, topography, vegetation, and geomorphology) include those of Brabb et al. (1972), Smith (1977), Ellen (1988), Butler and Walsh (1990), Walsh et al. (1990), and Gao (1993). Pack (1985), Ellen et al. (1988), and Wieczorek et al. (1988, 1989) also correlated these factors with cumulative precipitation or precipitation intensity. Carrara (1983), Mark (1992) and Mejía-Navarro (1992) expanded the application of some of these techniques to landslide and flood susceptibility modelling by examining many more physical-attribute factors through a discriminant analysis. Mora and Vahrson (1992) analyzed hazard sensitivity with precipitation plus seismicity as triggering factors. With relatively few exceptions (Butler and Walsh, 1990; Walsh et al., 1990; Mark, 1992; Mejía-Navarro, 1992; Mora and Vahrson, 1992), most geological hazard studies have evaluated only a single controlling factor for a particular hazard (such as structural features responsible for landsliding (Mejía-Navarro, 1986), or storms responsible for debris flows), and have tended to emphasize climatic controls (Ellen et al., 1988). The work presented in this paper differs in that (a) we evaluate multiple controlling variables through the use of GIS-based weighted algorithms, and (b) we place more emphasis than has been common in previous studies on land use, geotechnical data, and both bedrock and surficial geology.

Geotechnical soil data related to void ratio, bearing capacity, moisture content, corrosive properties, plasticity index, liquid limit, erosion potential, and the Universal Soil Classification have been extracted from Soil Conservation Service reports (Harman and Murray, 1985), and multiple internal-file reports from the planning offices of Garfield County and Glenwood Springs. The latter were done by Atec Associates Inc, Woodward–Clyde Consultants, Woodward–Clevenger and Associates Inc., The Reinforced Earth Company, Gingery Associates Inc., Chen and Northern Inc., Chen and Associates Inc., and Lincoln Devore Inc. These documents were also substantiated by personal field observations.

3. Study area

3.1. Historical records

Large parts of the Roaring Fork River valley between Carbondale and Glenwood Springs have intensive

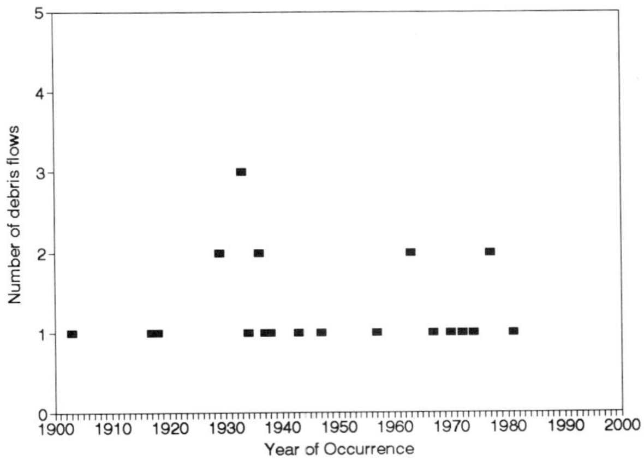

Fig. 2. Timeline of debris flows in the Glenwood Springs area.

sheet erosion, debris flows, and hyperconcentrated floods triggered by landslides and slumps. These come from the unstable slopes via the many tributary channels on the surrounding mountainsides, causing concentration of debris in channels and a high accumulation of sediments in colluvial wedges and debris fans that line the valley. Because Glenwood Springs was open ranchland until recently, there are few formal records of debris flows and floods on fans outside the limits of the original townsite. Nevertheless, it is clear from the available historical information that at least 21 significant episodes of flow have occurred on the fans in and near Glenwood Springs since the beginning of the century (Table 1, Fig. 2). This is an average of more than one mud or debris flow every five years of magnitude great enough to impact humans. About half of the documented events are either known to have affected one or more of the fans covered in this study, or can reasonably be inferred to have done so.

That the city was located in an area of such great flooding hazard may be attributed to nineteenth century optimism, ignorance of the problem on the part of the promoters, and the fact that debris flows may be sporadic events separated by several years of no activity. The presence of hot springs, the position at the junction of two of western Colorado's most important rivers, and the strategic location as the gateway to the important mining camp of Aspen far outweighed any environmental, geological, or engineering considerations in locating the townsite. Similarly, the expansion of the city onto the debris fans to the south during the last two decades has been due in part to a lull in flooding activity on those fans. That lull ended with the debris flows of 1977 and 1981.

Most historical flood and debris flow information comes either from old newspaper files or private studies done for the city of Glenwood Springs, and these usually only mention floods or debris flows on debris fans if they affected a populated area or were otherwise of unusual interest. Following a search of the available records dating back to the turn of the century, it appears that one of the first recorded debris flows occurred just south of Glenwood Springs in 1903. A newspaper item noted that a rainstorm had caused mud and rock to cover one of the railroad lines, resulting in a wreck that killed a member of the train crew. The next report dates from 1917 and reports a flow that apparently occurred in the southern part of Glenwood Springs. From the published description, it appears that this could have been either a mud or debris flow. Other records show that the Glenwood Springs–Carbondale area suffered damage during a series of storms that affected most of the central Colorado Rockies between July 27 and July 30, 1929; similar storms caused damage at various locations surrounding the city on July 28, 1933, early August 1933, and September 3, 1934. Although the newspaper descriptions are vague, it is apparent that some of these floods occurred on watersheds in what is now the southern part of Glenwood Springs.

The major debris flows and floods within the original townsite date from the late 1930s and early 1940s. A storm on July 10, 1936 blocked the highway in Glenwood Canyon at six places (Table 1). This was a prelude to a large mud flow within the city itself on September 1. Although flooding was widespread on the east side of the Roaring Fork Valley, most of the reported damage resulted from flooding on Cemetery Gulch and the many small drainages and rills east of the city (Fig. 3). Following this storm, the City began discussions with the Works Progress Administration concerning the building of a flood-control ditch along Twelfth Street, below Cemetery Gulch.

Because the west bank of the Roaring Fork River was largely undeveloped until recently, records for the area are very incomplete. The only large event of record was on July 24, 1977 (Table 1). Flows also occurred here in July 1981. In both events, debris flows crossed the new urban development and entered the river. Small debris flows occur very frequently in the source areas, perhaps during every major thunderstorm. However,

Fig. 3. Glenwood Springs from the western slope of the Roaring Fork River Valley. The Roaring Fork River is in the middle foreground, the Colorado River is at the left. Cemetery Gulch is the largest of the debris channels and fans on the eastern slope (to the right here).

these flows do not usually extend past the modern fanheads.

3.2. Climate and vegetation

Glenwood Springs has a semi-arid climate (mean annual precipitation 43 cm) characterized by occasional fast melting of heavy snowpack and/or strong, short duration spring and summer thunderstorms. Late spring to early fall cloudbursts have triggered debris flows on the upper parts of small basins and on the upper zone of debris fans throughout this century. The city is located on the widest portion of a debris fan which over the centuries has accumulated from repetitive debris flows beginning in the many channels on the surrounding mountainsides. Most of these debris fans are located at the mouth of basins with sparsely vegetated slopes, where bare soils could be expected to provide extremely low resistance to surficial erosion by rainsplash and runoff. Dense juniper forest or grass- and brush-covered slopes are less susceptible to slope failure because they are stabilized by roots.

Grass roots regulate water infiltration and provide the most strength to the soil mass above 0.3 m, whereas roots of brush and trees develop maximum strengths at deeper levels (Harman and Murray, 1985). In coarse soils, such as the "Ancient debris flows" unit of this study, the cohesion provided by roots may contribute most resistance against shallow sliding because of the small contribution of the soil frictional strength as a result of low confining pressures at shallow depths. Thus, the root strength and depth, and the vegetation cover density that controls the mass of the root system, may contribute to the differences in susceptibility of hillsides to slope failure. Vegetation is highly controlled by aspect, slope angle, and elevation (Harman and Murray, 1985). Vegetation density and height decrease from east-northeast- to west-facing slopes where vegetation is extremely sparse and is restricted to the uppermost areas (Fig. 4).

The compiled history of different catastrophic events in the study area suggests an episodic occurrence of debris flows (Fig. 2). Over the long term, vegetation stabilizes the channel bottoms. This permits storage of sediment and other debris, and prevents small flows from developing into large flows that reach the main valley bottom and the city. Eventually, a low frequency cloudburst occurs that is large enough to cause a flood

Fig. 4. The western slope of the Roaring Fork River Valley. The north-facing slope of the small drainage shown here has much more dense vegetation cover than the southeast-facing slope. Urban development is expanding toward the debris fan at the base of this drainage.

Fig. 5. Unconsolidated debris from recent small debris flows in the 23rd Gulch.

that destroys the vegetation and moves some of the stored material, resulting in a large debris flow. This large flow destabilizes the channel to the point that small and moderate floods in succeeding years likewise develop into debris flows (Fig. 5). After several years, the channel again stabilizes and the process repeats. The cycle of events from 1929 through 1947, and 1970 to 1981, combined with the absence of significant flows from 1947 to 1970, and then again the absence since 1981, appears to represent such episodic behavior.

Campbell (1975) suggested that abundant prestorm rainfall before a major storm could both decrease the amount of storm precipitation necessary to create the soil-moisture conditions conducive to debris flows, and lower the threshold of rainfall intensity for debris-flow initiation. This last condition may cause a lower quantity of sediment mobilization in a region with only high intensity, short duration storms. Because the Glenwood Springs Quadrangle does not have accurate records of rainfall distribution in the subbasins that are prone to debris flows, records are insufficient to determine the thresholds of storm rainfall that initiate debris flows. Therefore, we used interpolated data from a 1:500,000 scale isohyetal map of average annual precipitation (Colorado Climate Center, 1984), and the data from the 1977 storm. The debris flows of July 1977, a drought year, were triggered by a rainfall of nearly 2.5 cm during 30 minutes (Mears, 1977).

3.3. Geology and tectonism

The Glenwood Springs Quadrangle is characterized by a series of uplifted mountain masses deeply dissected by both fluvial and glacial processes. The area is crossed by many normal (tension) and reverse (compression) faults. The normal faults generally trend north and are at about right angles to the trend of reverse faults in the area (ATEC Associates, 1993). A group of parallel faults at the southwest corner of the area exhibits evidence of displacement in the Miocene Epoch and possibly in the Quaternary. Murray (1969) reports identifying 20 to 30 normal faults in this area which displace Miocene basalt and the overlying gravel deposits. The age of the last movement cannot be determined from the available literature.

The USGS National Earkquake Information Center (1993) reports several documented historic earthquakes believed to be centered in the Glenwood Springs vicinity, and Kirkham and Rogers (1981, 1985) suggest that these events may be associated with the faults described above. These earthquakes include an event on January 15, 1889 felt at a Modified Mercalli intensity (*MM*) of *V*, an event on January 31, 1946 felt at $MM = IV$, and an event on January 7, 1971, recorded by instrument with a calculated Richter magnitude (M) of 4.3, and estimated at $MM = V$. The faults and earthquakes suggest that there is a low-potential earthquake generation source 2 to 3 km from downtown Glenwood Springs.

Bedrock geology

Bedrock geology was analyzed only in areas where solid and consolidated rocks are exposed at the surface or overlain by a minimal thickness of unconsolidated material (less than 2 m). A sequence of quartzite, sandstone, dolomite, and limestone interbedded with very thin beds of shales is widely exposed in Glenwood Springs Canyon and the northern side of the Colorado River. Table 2 describes the bedrock beneath the lower basin (the Sawatch Quartzite through the Eagle Valley Evaporite). The Eagle Valley Evaporite is particularly important with respect to geological hazards. The unit is characterized by vertical gorges as a result of its high susceptibility to solution and erosion, and presents very serious engineering geology problems (Mallory, 1971). Solution of the evaporites results in movement of surface or ground water that can produce serious subsidence problems such as surface collapse features. The minerals of this formation contribute to chemical degradation of soil (corrosion hazard) and pollution of surface and ground water. The formation weathers deeply; moderately weathered rock occurs to depths of 10 m beneath residual soils, giving rise to deep-seated slumps and earthflows. Shallow debris slides are less common and are limited to slopes oversteepened by older scarps or by fluvial erosion. The foothills of this unit have frequent debris flows that form wide, low-gradient alluvial fans.

The Eagle Valley Evaporite tends to intertongue and grade into the red beds of the Maroon Formation which underlies most of the upper basin. The Maroon Formation weathers shallowly. Slightly weathered or fresh rock is common beneath the residual soil at depths of 0.5 to 1.5 m. Both debris slides and flows are common within the shallow regolith on steep slopes underlain by this unit, where deeper types of slope movement are

Table 2
Bedrock lithologic units of the Glenwood Springs study area, with associated geologic hazards

Name	Description
Sawatch Quartzite (Peerless Formation)	Cambrian; interbedded dolomite, sandstone, and shale
Chafee Formation	Upper Devonian; interbedded dolomite, shale, and limestone
Leadville Limestone	Mississippian; thick-bedded limestone overlying an interbedded sequence of dolomite, limestone, and sandstone
Belden Formation	Pennsylvanian; limestone and shale with some lenticular beds of sandstone and conglomerate
Eagle Valley Evaporite (Paradox Formation)	Pennsylvanian; interbedded gypsum, calcareous sandsonte, shale, halite, and hydrite; see text for discussion of hazards
Maroon Formation	Permian; arkosic sandstone with interbeds of shale, siltstone, claystone and some conglomerate and limestone; see text for discussion of hazards
Weber Sandstone	Permian; cross-bedded eolian sandstone
Chinle Formation	Triassic; interbedded and interlaminated siltstone and silt claystone
Entrada Formation	Jurassic; cross-laminated, fine to medium grained, quartzose sandstone; characterized by rockfall and colluvial wedges
Morrison Formation	Jurassic; shale with lenticular interbeds of sandstone, siltstone, and limestone; prone to sheet erosion, slumps and landslides; contains clays that tend to be bentonitic, have swelling potential, and contribute to locally unstable surface conditions because of perched water tables
Mancos Shale	Cretaceous; interbedded calcareous shale and argilliceous siltstone; prone to debris flows and landslides; high swelling potential; high permeability contrasts cause seasonal high water tables and slope instability due to water saturation; water from this unit is corrosive to concrete and metal (Fox and Associates, 1974)
Dakota Formation	Lower Cretaceous; hard quartzose sandstone and pebble conglomerate with thin interbeds of shale and siltstone

Fig. 6. The western slope of the Roaring Fork River Valley showing the abundant debris-flow channels and small fans developed on the Maroon Formation.

less common. The regular, well-defined drainage system developed over this unit provides abundant channels for debris flows, and there are large fan-shaped deposits below many of the small gullies (Fig. 6). As a result of dense fractures perpendicular to stratification (interbedded shale, siltstone, and claystone layers rich in montmorillonite and illite) and the large contrast in hydraulic conductivity between sandstone layers (> 2 m thick) and claystone layers (> 50 cm thick), locally large seepage forces can be generated in this formation, mainly at the top of more impermeable layers. The combination of seepage force, friction reduction, and weight, in conjunction with structural features, is responsible for multiple landslides, rockfall, and creep. Frost heave, active in the fractures of the sandstone layers, produces large, isolated rockfalls along the top of the hills and slides on slopes between 25° and 45° (47–100%).

The Maroon Formation is capped by the Weber Sandstone, and localized exposures of the Mesozoic units (Table 2). Late Tertiary basalt flows are widespread, capping many of the mountains and plateaus surrounding Glenwood Springs. These flows have joint and fracture systems that facilitate water percolation, weathering, and underground erosion. The basalt is susceptible to mass wasting where it overlies clays or evaporites.

Surficial geology

Ancient debris flows: pendant deposits of stratified recurrent debris flows in sequences that can be over 30 m thick occur throughout the study area (Morris and Weaver, 1978) (Fig. 7). The sediments are unconsolidated gravel with boulders up to 60 cm in diameter in a sandy–silty matrix. The matrix, in connection with sparse vegetation cover and low concentrations of sulfates and carbonates that become extremely weak when wet, keeps the deposits metastable. The deposits are characterized by very high porosity, high hydrocompactive properties and very low surficial oxidation. Cemetery Hill, which flanks Cemetery Gulch on the

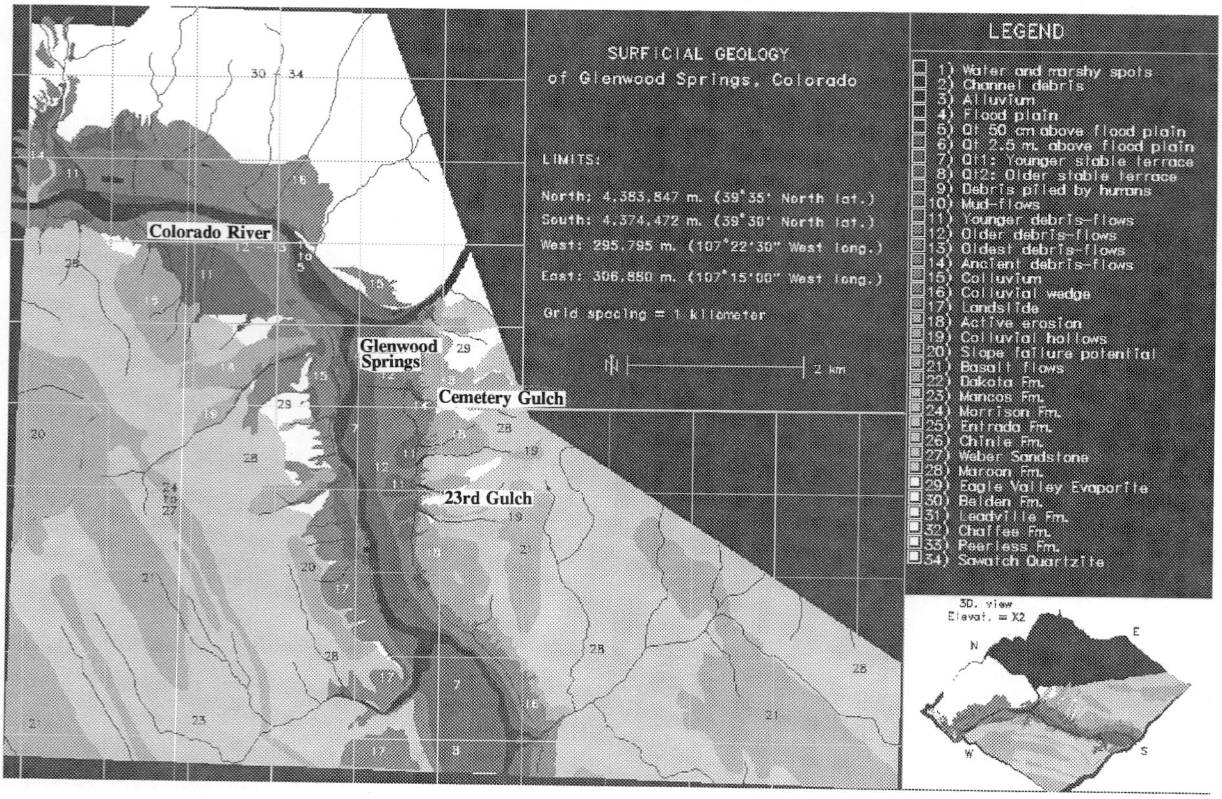

Fig. 7. Map of the surficial geology of the study area.

eastern side of the city, is a deeply eroded remnant of these debris-flow deposits, composed of angular rock fragments in a matrix of silty, slightly clayey sand. These deposits produce rockfall and are also susceptible to piping and local surface collapse.

Alluvial terraces: unconsolidated poorly-sorted alluvial deposits forming 4 terraces. The sediments commonly range in size from boulders up to 1 m in diameter of basalt, limestone, and sandstone to a silty–sandy matrix. The two high terraces (10 m and 22–25 m above the Roaring Fork River, respectively) are generally not subject to flooding, but are affected by slumping and erosion on their very unstable steep edges because of loose texture. Well-selected areas inside these geomorphic units are among the better sites for development within the study area. The two low terraces include low-lying land adjacent to major streams and have a high probability of being affected by floods. The younger alluvium underlying these surfaces can be affected by a rise of water table, flooding the basements and lower levels of any structures on them.

Pendant landslide deposits: "old" landslide deposits spread along the western slope of the Roaring Fork River (Figs. 6 and 7). These deposits are considered "pendant landslide deposits" because their location on steep slopes produces very high volumes of unconsolidated material.

Debris-flow fans: a series of coalescing debris-flow fans along the eastern slopes of the Roaring Fork River from the banks of the Colorado River south to Carbondale. These fans are located at the bottom of 12°–45° (21–100%) slopes where large debris flows occur under optimum conditions. Each lobate deposit is bouldery at the front and becomes progressively finer-grained with increasing distance behind the front. The subsurface materials encountered on debris-flow fans are somewhat unstable and subject to both vertical and lateral movement. Lateral movement consists primarily of a gradual downslope creep, while vertical movement consists of hydrocompaction or collapse of the very low density matrix. Areas underlain by these deposits present a high hazard to urban development and must be improved with channelization, permanent cleaning of the channel, and reforestation of both sides with a corridor at least 20 m wide.

Colluvial deposits: unconsolidated, poorly sorted, angular blocks in a fine-grained matrix which have highly variable porosity and permeability. Colluvial deposits have active creep processes on slopes higher than 10° (18%) and surface collapse on lower slopes, and are the sites of many cases of subsidence in the Glenwood Springs commercial sector. The deposits should be reinforced and stabilized with structural works such as piles, retaining walls and surface and underground drainage to become relatively stable and suitable for urban infrastructure.

Rockfall deposits: rockfall along the steepest slopes, usually upwards of 45° (100%) where mechanical weathering is more pronounced. Major rockfall areas occur along the steep canyon of Glenwood Springs (Fig. 7) and along the steep slopes of the Maroon Formation at the eastern border of the Roaring Fork River. The Maroon Formation, Weber Sandstone, Leadville Formation, and Sawatch Quartzite are the units most susceptible to rockfall because of their highly fractured pattern and steep slopes with almost no vegetative cover. The base of these slopes is surrounded by colluvial wedges (talus) which grow annually as a result of winter freeze–thaw processes. Rockfall can vary from the movement of small boulders ($\leqslant$50 cm in diameter) to the dislocation of large blocks of more than 2 m diameter. Development of urban or transportation infrastructure beneath rockfall areas should be avoided and if undertaken (e.g., Interstate 70 through Glenwood Springs Canyon), requires expensive and permanent maintenance.

4. Geomorphic processes

We define geomorphic processes as those that modify the landscape and its stability, producing hazards for humans and their infrastructure. Here we describe four geomorphic processes that create geological hazards in the study area.

4.1. Active sheet erosion

Areas of active sheet erosion are unvegetated slopes with little to no soil accumulation. Surfaces are characterized by very low soil moisture with low plasticity index, high porosity and high void ratio over lithologies limited by shallow impermeable clay or shale layers, allowing rainfall and runoff to easily erode the uncovered soil. These processes occur principally over the Maroon and Eagle Valley units. Most of the watersheds over these units have rapid runoff, sheet erosion, soil fertility reduction, gully and rill erosion, small slumps considered as sheet erosion because of their magnitude, siltation of streams, and floods.

4.2. Debris flows

Debris flows are the main geomorphic process affecting social and economic conditions in the Glenwood Springs area. They are generated by slumps, landslides, and intense sheet erosion which occur during heavy rainstorms. These events cause millions of dollars of damage to public and private structures during years when extremely heavy rainstorms occur (Smith, 1986). Debris flows occur on almost all the west- and southwest-facing tributaries of the Roaring Fork River. The average basin relief is 400 m, and contributing basin areas are usually very small (<60 ha). The debris flows form a line of fans along the eastern side of the main valley, beginning at elevations of 1850 m (Figs. 3 and 7). Fan gradient averages about 7° (12%) and does not exceed 15° (27%), while channel gradients average about 25° (47%) and do not exceed 60° (173%). The debris flows consist primarily of blocks weathered from the Maroon Formation in a clayey soil matrix from the basaltic soil of the upper basins and clayey silt from the Eagle Valley Evaporite. The principal clay minerals are muscovite, kaolinite, and montmorillonite, with minor chlorite (Morris, 1986).

Just above the fanhead are slopes with steep, U-shaped channels nearly 4 m deep. Small, tributary debris flows and debris avalanches enter the main channel at several points. These flows apparently occur during thunderstorms when jets of water pour off the cliffs and hit the loose colluvium. Together with rockfall debris, sediment stored in the channel, and soil derived from bank caving, these tributary events supply most of the material for the large debris flows on the lower slopes. The remains of many small debris flows lie in the channel bottoms, forming low levees and small, flat lobes of stored sediment (Fig. 5). This is especially true of the small, bouldery flows in the upper basin, where debris accumulates in underbrush and on short, low-gradient reaches. A major flood would incorporate this stored material. It also would undercut the unstable walls of gorges to cause massive bank caving. Below the highest cliffs, channels evolve into a fanhead trench that ends abruptly at the uphill edge of the Glenwood Irrigation Ditch. At this point, debris flows usually block the ditch and flow out across the fan, which is now a rapidly developing urban neighborhood.

Small debris flows are also a major mass-transport mechanism on the western slopes formed on the Eagle Valley Evaporite. Soils developed on the weathered evaporite usually form a low-density, ''popcorn''-like surface layer. When wetted, this surface layer forms miniature mudflows, up to several meters in length, that mimic the tongues, lobes, and levees of their full-size counterparts in every detail. Their appearance is similar to that formed by pouring wet ''Plaster of Paris'' on a steep slope. These miniature flows literally cover the evaporite hillsides. Moreover, these slopes are inherently unstable. The principal clay minerals in these flows are muscovite, illite, kaolinite, and montmorillonite enclosing boulders with intermediate diameters of up to 0.5 m.

4.3. Creep

In the study area, creep is manifested through cracks in walls, and wavelike surfaces with hummocky aspect created by a multiple sequence of micro-landslides (plane-scarp of 0.1 to 0.6 m width and depth). Additionally, stone lines, downslope growth of trees, tilting and breaking of structures, and accumulation of soil upslope from deep well-founded structures or fixed obstructions are also present. Measured creep movement rates range from a few mm to 10+ cm per year and can be accelerated by infiltration of water from broken water pipelines, or underground waste disposal. Creep in several portions of the study area results mainly from high infiltration rates associated with the Glenwood Ditch. This runoff-retention structure is constructed on unlined colluvial material.

4.4. Drainage

This factor is clearly controlled by lithology. Drainage networks over the Maroon Formation and associated deposits have a subparallel pattern that indicates moderate to steep slopes and elongate landforms. Over resistant rocks such as the Sawatch Quartzite, the drainage pattern is radial. The streams are principally first and second order, with torrential flooding after intense rainstorms. During rainy periods in spring and summer, the rainfall intensity normally exceeds the infiltration capacity and the overland flow traverses the surface as a broad shallow sheet or in small rills.

A common phenomenon that occurs near the lower parts of slope deposits, mainly the ancient and recent debris flows, is the subsurface stormflow that produces bank seepage in the early part of a storm. A high percentage of this throughflow uses root channels or pipes, which are linear openings having a much greater permeability than the normal soil matrix. These interflows are very dangerous, reducing bank stability, and causing subsidence in floodplains, terraces and at the foot of slopes.

A noticeable feature of high-elevation watersheds is a large and relatively flat surface morphology. The upper basin is underlain by several late Tertiary basalt flows, which cap many of the mountains and plateaus surrounding Glenwood Springs. Storm runoff collects and concentrates in the upper basin before passing over the cliffs in the lowest third of the watersheds. Below the cliffs, the channel gradient decreases and the channel opens onto a well-developed debris fan, some of which became densely urbanized after 1977.

Drainage patterns do not exist on the lower debris-flow fans; hence sheet flow is probably the primary means of runoff. The levee-bound debris channels on the upper fan segments, above the modern fanheads, are good examples of debris-flow landforms. Levees lining the channels are high, steep, and sharp-crested, and the channel walls resist erosion and resemble poor-quality concrete. Levees along the main channel, however, reflect the properties of the Maroon-derived sediment. They tend to be low, broad-crested, and uncemented. Below the fanhead, not all flow pulses follow the active paths down the fans. Instead, tongues of debris move laterally into the underbrush, where they demobilize readily in thickets of scrub oak and between houses.

5. Hazards

There is a very high hazard level associated with mass movements from both slopes of the valley, but the eastern slopes are the most important because of their dense urban development. Contributing factors include the large size of potential debris flows, urban development across entire fans, and the non-existence of canals for flow control. For example, the Cemetery Gulch Ditch is designed for water floods and is not large enough to carry a major debris flow. Furthermore, it is now overgrown with vegetation and has undersized culverts at most street crossings. Finally, the channel inlet is easily blocked by either highly concentrated debris flows or tree trunks and other organic matter. The absence of major flows in the basin since 1981 means that large quantities of stored debris are available for future events. Consequently, certain basins are likely to produce large, destructive flows whenever a sufficiently large rainstorm occurs over the watershed.

5.1. Internal slope processes

The basic descriptive data given thus far were used to quantify the stability of units to produce a zoning of settlement growth. The most important factor in these considerations is the slope, which is the basic regulating factor of geomorphic processes. The slope digital map (Fig. 8) is extremely detailed and shows thousands of very small discontinuous areas of a particular slope (approximately 1% resolution). The engineering criteria for the following description is that as slope increases, the potential for damage to the environment increases. Of particular significance is the increased potential for slope failure, erosion and sedimentation, accelerated runoff, and scarring from site grading. The nature of the underlying material will also greatly affect the potential for slope failure and erosion.

Almost all natural hillslopes in the study area are composed of stratified earth materials with different physical properties. Contrasts in the hydraulic conductivity of these layered materials can create locally large seepage forces, especially at the top of more impermeable layers, such as the steep slopes over the Maroon and Morrison Formations. There is a high potential permeability with low hydraulic head within these sandstone outcrops, where heavy blocks overlie shales with low internal friction strength and high hydraulic head at their top. Large seepage forces thus increase the hazard of slope instability. More consideration needs to be given to the contrast in permeability values between layers than to the magnitude of permeability

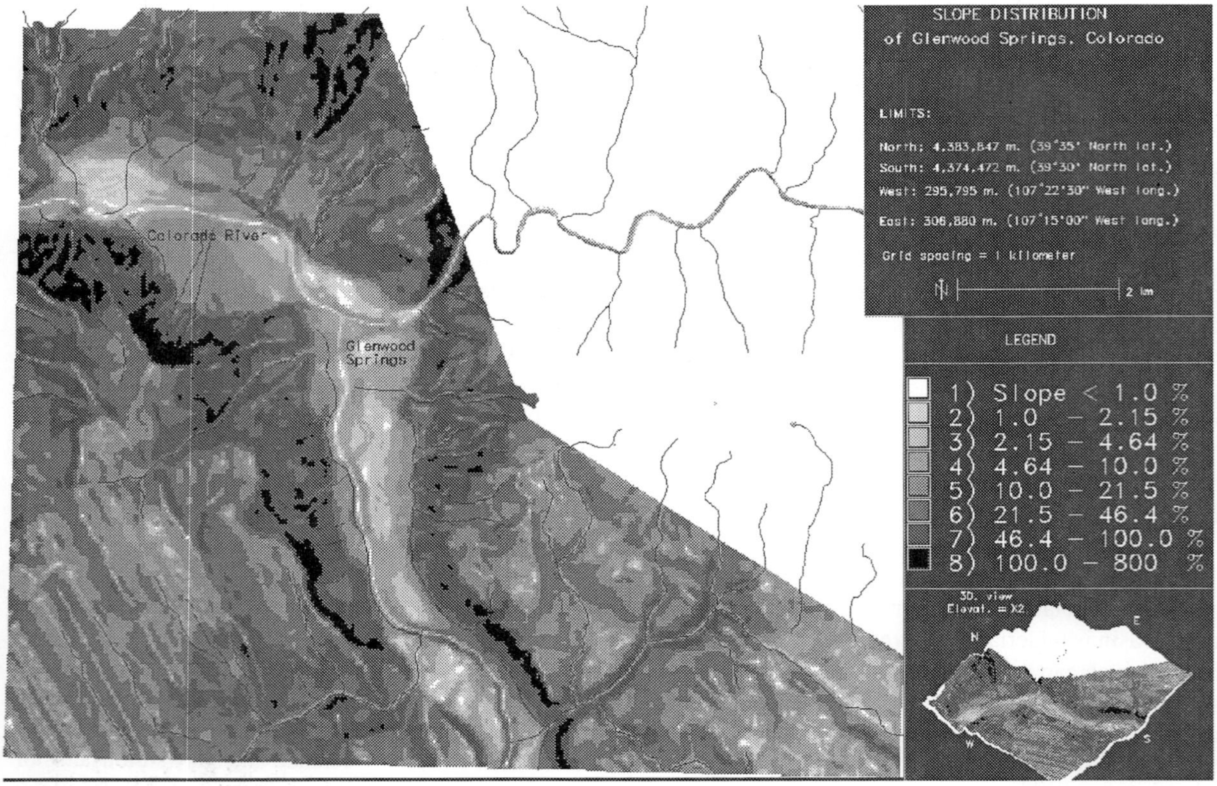

Fig. 8. Slope digital map of the study area.

in any given layer. Slope failure potential is highest where seepage forces are directed outward from the slope, immediately above the low permeability layer and at the toe of the slope in the high permeability layer (Iverson and Major, 1986; Reid and Iverson, 1992). If we assume that areas with locally high internal friction angle are more prone to slope failure, then failure might be expected to occur in convex hillslopes where seepage forces are directed outward from the steep, lower section of the slope. These convex profiles commonly occur in areas of both pendant landslides and creep movement, which are very common on both sides of the Glenwood Springs Valley.

5.2. Debris-flow hazard susceptibility

In order for debris flows to occur, certain conditions must be met: (1) sufficient available unconsolidated debris, including soil, rock, and organic material, (2) steep slopes, ranging from approximately 25° to 45° (47–100%), (3) a sufficiently high clay content in the debris, ideally with a relation of plastic index to liquid limit that locates the soil close to the "A" line of the Casagrande Classification (Lambe and Whitman, 1969), (4) hydrologic characteristics of soils that minimize infiltration relative to runoff for any precipitation range, (5) low frequency, high intensity precipitation after a long period of inactivity that allows large amounts of channel debris accumulation, and (6) the lack of infrastructures to mitigate the hazard. All of these conditions exist in the basins of the study area.

Triggers and hydro-hazard probability relationships. In order to evaluate the hazard and make recommendations about land use and mitigation, it is necessary to estimate the annual probability of debris flows similar in size to those observed during July 1977 (Fig. 9). Because the drainage basins in the study area are susceptible to debris flows, it can be assumed that a rainstorm of the intensity and duration of that which occurred on July 24, 1977 would also trigger future flows. In this particular case the probability of debris flows is the same or greater than that of the rainstorm event because a rainstorm of lesser intensity (and greater probability) may also be capable of triggering future debris flows of approximately the same magnitude. A first objective in determining the debris-flow probability is therefore that of estimating the return period of the rainstorm. The rainfall measured at the National Oceanic and Atmospheric Administration (NOAA) weather station in Glenwood Springs during the several days preceding and including the debris flows is shown in Table 3. Sufficient rainfall occurred during the three days preceding the flows to wet, although probably not saturate, the soil. It is of more significance that 21.6 mm of the July 24 rainfall came in about 30 minutes at the weather station. Rainfall intensities and amounts in the surrounding tributary basins are not well known, but judging from the comments of observers in the area, rainfall was of a magnitude similar to that recorded in town. In view of the lack of accurate data to average from more than one

Table 3
Daily precipitation in Glenwood Springs preceding and during the July 1977 debris flow

Date	Precipitation (mm)
July 18	0.51
July 19	0.00
July 20	0.00
July 21	6.10
July 22	3.80
July 23	0.00
July 24	27.40

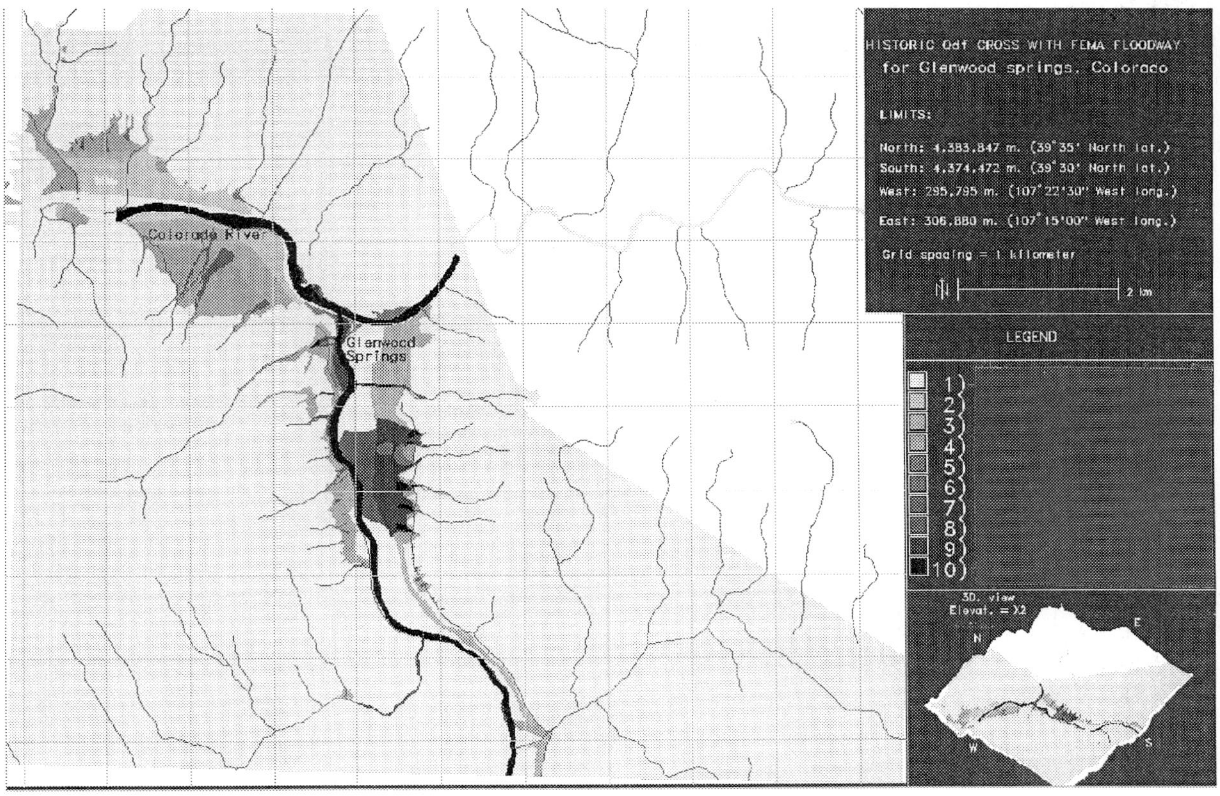

Fig. 9. Extent of the 1977 debris-flow deposits in the Glenwood Springs area, along with information on debris-flow occurrence from historical records, FEMA floodway map, and geomorphic features. Mapped units 1 through 10 indicate: (1) no records, (2) sedimentary records of debris flows, (3) frequent floods during this century, and sedimentary records of older flows, (4) frequent floods during this century, (5) few records of debris flows during this century, (6) moderate records during this century, (7) many records during this century, (8) the highest density of recorded debris flows during this century, (9) frequent floods and debris flows, and location within FEMA's 100-year flood zone, and (10) frequent debris flows, and the mapped floodway for the active channel.

Table 4
Categorization of the different physical factors considered in the algorithms for subsidence and debris flows susceptibility evaluation

Physical factors	Categories									
	1	2	3	4	5	6	7	8	9	10
slopesub (%)	< 1	100–800	1–2.15		2.15–4.64	4.64–10	46.4–100	10–21.5		21.5–46.4
slopedf (%)	< 1	1–2.15		2.15–4.64		4.64–10.00	100–800	10–21.5	21.5–46.4	46.4–100
aspect	67°–112°	22°–67°	337°–22°		292°–337°	112°–157°	157°–202°	202°–247°		247°–292°
veg.[a]	> 2500	< 2500	< 2000	< 1500	< 1000	< 900	< 800	< 700	< 650	0–500
erosK	> 0.15	> 0.2	> 0.24	> 0.28	> 0.32		> 0.37		> 0.43	> 0.53
shrswell		low				moderate				high
lusess		government build.	parks	schools	public build.	commercial	public services	farmland	residential	transport and mining
lusevul	farmland	parks		transportation	residential	commercial	government build.	public services	public build.	schools
PMP[b]				37	40	43	46	49	51	54
hgdf		A			B			C		D
hgsub			D			C		B		A
wsbuf (in m)	90–100	80–90	70–80	60–70	50–60	40–50	20–40	10–20	streams and ditches	rivers and ponds
isohyaa (in mm)		406		432		483		533		610
femahist	no reported flood or debris flows (Qdf)	only geological records	area inside 500 year flood evaluation	frequent Qdf reports	very frequent Qdf, area inside 500 year flood evaluation	inside 100 year flood evaluation, + geological records	inside 100 year flood evaluation, + + rare Qdf reports	inside 100 years flood evaluation, + frequent Qdf reports	inside 100 year flood evaluation, + very frequent Qdf	floodway

[a]Dryweight in Lb/acre.
[b] In mm, hourly maximum probable precipitation.

gauge in the basin, it will be assumed that rainfall intensity there was 20–25 mm during the half-hour period.

According to the statistical analysis of NOAA rainfall data (Mears, 1977), the maximum 1-hour period rainfall to be expected in any given 100-year period at Glenwood Springs is 40.1 mm. The precipitation intensity observed during the 1977 storm appears representative of approximately a 50-year return period. Therefore, debris flows of this size probably have a return period of 50 years or less. Interviews with staff members of the Colorado Geological Survey and local residents revealed that flows occurred here 30 to 40 years ago (the exact date is in question). Thus the flows apparently have recurred in considerably less than a century. Subsequently, it will be assumed that the flows have a return period of 50 years (annual probability of 2%).

Debris-flow susceptibility algorithm.

The following algorithms were used to weight each of the primary factors controlling the susceptibility of a particular site to debris flows (Sdf):

$$\mathrm{Sdf} = \mathrm{slopedf}^*(\mathrm{aspect}^*7 + \mathrm{usc\text{-}casag}^*4 + \mathrm{sgmdf}^*9 + \mathrm{veg}^*8 + \mathrm{hgdf}^*5 + \mathrm{shrswell}^*2 + \mathrm{erosK}^*7 + \mathrm{lusess}^*3 + \mathrm{wsbuf}^*8 + \mathrm{femahist}^*10 + \mathrm{isohaa}^*4)/67 \quad (1)$$

$$\mathrm{trigger-ps} = (\mathrm{isoseismals}^*3 + \mathrm{PMP}^*5)/8 \quad (2)$$

$$\mathrm{Hdf} = \mathrm{Sdf}^*\mathrm{trigger-ps} \quad (3)$$

The results of these algorithms are normalized from 1 to 10. Eq. (1), which describes slope susceptibility to debris flows, is basically a weighted average, with the relative weighting for each physical factor indicated by the numerical suffix of that factor. For example, the "slopedf" factor receives a relative weighting of 6. These weighting factors are subjective and may be modified by the operator through a graphic user interface using the IPDS model (Mejía-Navarro et al., 1994). Each of the factors is internally normalized into 10 classes (Table 4), with a value of 10 assigned to the class representing conditions most susceptible to debris flows. The factors are:

slopedf: slope angles from 0°–89° (0–800%) are divided into 10 classes, with the highest rating (10) given to those angles most characterized by debris flows (here, 25°–45°, or 47–100%)

aspect: slope aspect, with 360° where 0° = north-facing, 90° = east-facing, 180° = south-facing, 270° = west-facing

usc-casag: this factor combines (i) regolith texture according to the Unified Soil Classification System (USC), reclassified by grain-size distribution in terms of susceptibility to infiltration and internal structural collapse, and (ii) regolith matrix proximity to the Casagrande A line (Casagrande classification: Lambe and Whitman, 1969) on a plot of liquid limit versus plastic index; a value of 10 is assigned to sediments most likely to fail according to these criteria

sgmdf: debris-flow-susceptible surficial geologic material; 34 exposed lithologic units — both bedrock and Quaternary deposits (Fig. 7) — were differentiated and reclassified into 10 categories according to their susceptibility to and historic influence on debris flows, with a higher weight of 10 to units such as debris accumulated in channels, very recent debris-flow deposits, colluvial deposits in high-elevation hollows, and the Maroon Formation, and a lower weight to units such as the oldest stable terraces, basalt flows, and the Sawatch Quartzite. (Because of the number of units in each category, this factor is not listed in Table 4.)

veg: vegetative cover of slope, with categories as calculated for the study area by the Soil Conservation Service (Harman and Murray, 1985)

hgdf, hgsub: hydrologic soil groups used to estimate runoff from precipitation (Harman and Murray, 1985); these are classified A (high infiltration rate, high water transmission and slow rise of internal pore pressure), B (soils with moderate infiltration rate when thoroughly wet), C (soils with slow infiltration rate when thoroughly wet), and D (soils with very slow infiltration rate when thoroughly wet, very slow water transmission and very fast rise of internal pore pressure); hydrologic soil groups were reclassified by their influence to accelerate landslides to trigger debris flows (hgdf) and to trigger subsidence (hgsub)

shrswell: clay content and mineralogy in relation to shrink–swell potential (Harman and Murray, 1985)

erosK: sheet erosion potential using Universal Soil Loss Equation factor K, a measure of the susceptibility of the soil to erosion by water; soils having the highest K values are the most erodible, facilitating debris flows

lusess: a zoning of land-use features based on their relative negative influence on slope stability; higher weighting (10) assigned to transportation because roads tend to truncate slopes, channelize runoff, and increase bank collapse; lowest weight assigned to government buildings, which generally have better engineering regulations than other structures in Glenwood Springs

wsbuf: the stream network was reclassified by order, and for single channels with a basin area greater than 1.5 km^2, the lower half of the stream was reclassified one order higher than the upper half; a 10 m wide buffer was built around all channels, ponds, and the Glenwood Ditch; an additional 10 m buffer was built around second-order channels, with a further 20 m buffer around third-order channels, and so on up to a buffer of 90–100 m around the Colorado and Roaring Fork Rivers. Weighting was assigned with a relative value of 10 for ponds and floodways of rivers; 9 for 10 m wide buffers around channels; and so on to 1 for a 90–100 m buffer around the largest channels.

femahist: the data are the result of a cross-tabulation and reclassification of surficial geology-debris flow maps, historical records from newspapers during this century, and FEMA Floodway maps (Federal Emergency Management Agency, 1986) (Fig. 9)

isohaa: isohyetals for average annual precipitation (1951–1980) from Colorado Climate Center (1984) and NOAA maps (1:500,000), extrapolated for the resolution of this study (1:24,000); weighting was assigned from 10 for areas with 600 mm to 2 for areas with 400 mm annual precipitation

Eq. (2), which describes seismic and climatic triggers for debris flows, is based on the factors isoseismals and PMP (hourly probable maximum precipitation), but it can be modified to independently evaluate hazards triggered by either seismicity or precipitation.

isoseismals: a reclass of a cross tabulation of; (a) buffers around two magnitude 5 and one magnitude 4 earthquake epicenters in the Glenwood Springs quadrangle, (b) buffers around faults, and (c) surficial geology and bedrock susceptibility to amplify seismic waves (conceptual analysis of texture, moisture, and geologic structure); the final reclass weights are 6 (areas with stronger potential (close to epicenters and faults), crossed with areas susceptible to amplify waves, such as the Eagle Valley Evaporite or debris flow deposits), 5 (areas with approximately 50 year interval of occurrence for earthquakes of intensity 5, and high susceptibility to earthquake damage), 4 (areas with lower susceptibility to earthquake damage), and so on to 1 (lowest susceptibility to potential seismic activity; e.g. Sawatch Quartzite)

PMP: Probable Maximum Precipitation (PMP) data were obtained from extrapolated isohyetal maps of average annual precipitation data from the Colorado Climate Center, and from the probable maximum precipitation evaluations of Mears (1977); weighting was based on the 1977 debris flow

Equation (3) indicates that debris-flow hazard at a site is the product of Eqs. (1) and (2), which is then normalized on a scale of 1 to 10 (lowest to highest hazard).

Vulnerability (vuln) is a result of a reclassification of the cross tabulation of population density, land use, transportation, and utility lines, where higher weighting values were assigned to areas of higher population density, assembly sites, and greater proximity to utility lines. The algorithm used to calculate the weighted average of the combination of (i) population density (higher weight to higher human concentration per hectare — U.S. Census of Population, 1990), (ii) landuse features reclassified by their vulnerability, from schools with higher weight (10) to farms with lowest weight (1), and (iii) lifelines (Highways 70 and 82, city roads, country roads, service lines such as phone and electricity) was:

$$\text{vuln} = (\text{density}*10 + \text{Lusevuln}*7 + \text{Lifelines}*2)/19.$$

5.3. Subsidence hazard

Surficial deposits in the Roaring Fork Valley can be considered metastable soils because they are subject to internal collapse and are very sensitive to changes in the soil moisture content. Usually soils from the debris-flow fans exhibit a collapse potential of 2–3% which could represent a metastable condition of moderate sensitivity. With piping processes, this condition could be moderately high to high. Debris-flow and colluvial deposits from the Maroon and Eagle Valley units are commonly characterized by moderate bearing capacity (below 1150 kPa) at natural low moisture contents (5–10%), but tend to collapse when subjected to wetting or underground water flow. The lower density and finer-grained soils would probably have higher subsidence potential than the coarser soils. For these type of conditions, it is recommended that runoff from slopes be carefully controlled to prevent erosion caused by landscaping practices, sheetwash, or seepage. It may be necessary to provide specific drainage ways and culverts to prevent excessive erosion along the steeper slopes.

Subsidence susceptibility algorithm.

The algorithms used to estimate site susceptibility to subsidence (Ssub) were:

$$\text{Ssub} = (\text{slopesub}*5 + \text{sgmsub}*9 + \text{lusess}*4 + \text{wsbuf}*3 + \text{veg}*2 + \text{hgsub}*7 + \text{erosK}*8 + \text{isohaa}*10)/48 \quad (4)$$

$$\text{trigger-sub} = (\text{PMP}*3 + \text{isoseismals}*5)/8 \quad (5)$$

$$\text{Hsub} = \text{Ssub}*\text{trigger-sub} \quad (6)$$

The results of these algorithms are normalized from 1 to 10. Eq. (4) uses the same factors as Eq. (1), as well as:

slopesub: slope angles at which collapse has been most prevalent, for the study area, 12° to 25° (21% to 46%)

sgmsub: as with the "sgmdf" factor in the debris-flow susceptibility algorithm, the 34 lithologic units (Fig. 7) were reclassified into 10 categories according to their susceptibility to be affected by subsurface solution and hydrocompaction, causing surface subsidence; a weight of 10 was assigned to units such as the Eagle Valley Evaporite, colluvial deposits, anthropogenically accumulated debris, and channel debris; the lowest weighting value was assigned to the most resistant units such as the Leadville Formation and Sawatch Quartzite; all other units are between weighting categories 2 and 9. (Because of the number of lithologic units, this factor does not appear in Table 4.)

Eq. (5), a weighting average of trigger factors, accounts for the relative importance of climatic and tectonic events in triggering subsidence, and Eq. (6) indicates that subsidence hazard is the product of Eqs. (4) and (5).

6. Hazard susceptibility map

Cartographic and statistical analyses of geological hazards require two different types of maps. The first uses a map of active geomorphic processes showing the location of actual geological hazards or of the most dynamic processes that have occurred during a specific time period. The other, a geological hazards probability

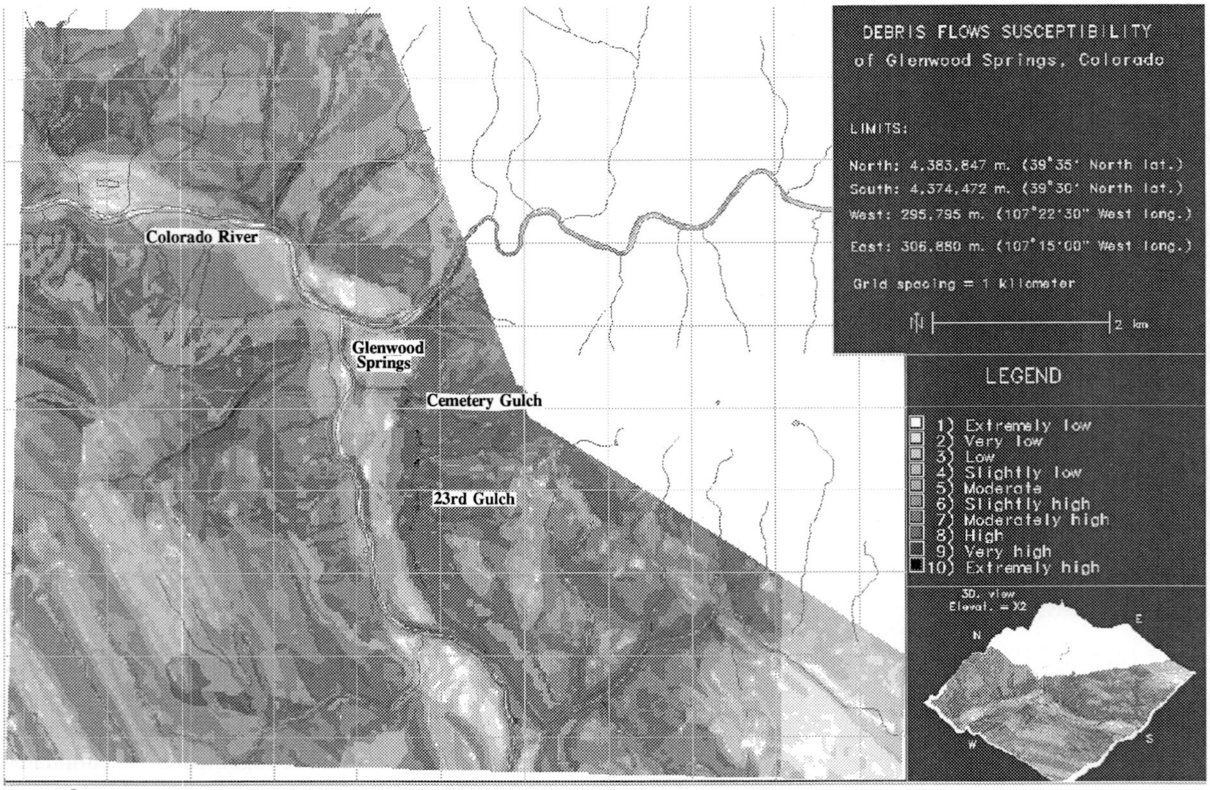

Fig. 10. Debris-flow hazard susceptibility map for the Glenwood Springs area.

map, illustrates the statistical occurrence of these types of hazards for a future time interval (Mark, 1992). Such a map does not give locations of actual geological hazards, but rather a statistical estimate of the likelihood that one or more of the studied hazards would occur in each pre-defined cell of the map grid.

The distribution of actual debris flows attributed to the July 24, 1977 storm (Morris and Weaver, 1978) is plotted in Fig. 9 by grouping one or more mapped debris flows into 10×10 m^2 cells that contain them. This map is reproducing the observed distribution as functions of the relevant physical factors of historical debris flows and floods, and on surface characteristics such as type, density and relative age of vegetative cover, and degree of weathering.

One of the first products of this project was a debris-flow hazard susceptibility map of the Glenwood Springs area at 1:24,000 scale (Fig. 10). The map provides an easily read analysis of selected areas related to debris flows. The debris-flow hazard susceptibility map was based on the factors of slope angle, aspect, surficial and bedrock geology, geotechnical characteristics, streamflow, precipitation, historical occurrence of debris flows, land use, vegetation, and seismic characteristics. Each of these factors was incorporated into the final map through the procedures described below.

The original surficial geology map, which is a match of Quaternary deposits and active geomorphic processes of the Glenwood Springs area, was compiled from existing sources (1:24,000 scale maps; Lincoln DeVore Testing Laboratory, 1976), and from 3 months of field work using 1:4000 scale maps and 1:10,000 scale air photographs. The slope map was developed from 1:24,000 Digital Elevation Map (DEM) data obtained from the U.S. Geological Survey. This DEM information is received on tapes and transformed into ARC/INFO for morphometry, aspect, and contour-lines coverage, and to GRASS 4.1 for detailed slope zoning using a grid cell of 10×10 m^2 resolution. Bedrock geology, soils, water bodies, land-use and utilities (highways, roads, lifelines, railroads) were digitized from existing maps.

Each individual map was then weighted as in these examples:

(1) The bedrock map was superimposed on the surficial geology and geomorphic processes maps in order to identify the rock units in which slope failure, erosion, or rilling processes have been occurring. The geologic formations were then listed in order of percentage of their outcrop areas that have failed by any of these

Table 5
Land use suitability indices for the Glenwood Springs study area

Class	Description
Suitable classes for urban development	
A	Very stable areas, almost free of any level of studied geological hazards. These areas can be used for any social purpose.
B_1	Relatively stable areas, minimum level of studied geological hazards. Their stability depends on social and infrastructure use. B_1 has the lowest requirements in the B category.
B_2	Relatively stable areas, minimum level of studied geological hazards. Their stability depends on social and infrastructure implementation. B_2 has more requirements than B_1.
B_3	Similar to B_2, but B_3 has more requirements in relatively stable areas to maintain the surface stability
C_1	Unstable areas with low to medium level of studied geological hazards. They can be transformed into relatively stable areas with low recuperation costs.
C_2	Unstable areas with medium level of studied geological hazards. Suitable recuperation and transformation into relatively stable areas require higher costs than C_1 areas.
C_3	Very unstable areas affected by high level of studied geological hazards, may be recovered with very high costs, but must be kept 'stable' because of social needs such as important roads or existing high cost urban infrastructure.
Unsuitable classes for urban development	
D_1	Very unstable areas affected by high level of studied geological hazards. They are low recoverable areas that require improvement with works to become restricted and relatively stable areas.
D_2	Extremely unstable areas, affected by very high level of studied geological hazards. They are not suitable areas for recuperation, and require improvement for protection of nearby areas.
E	Very stable areas, almost free of any level of studied geological hazards. These areas are prohibited for human settlement and are suggested for parks and recreational zones, available in contingency.
F_1	Relatively stable areas, with minimum to low level of studied geological hazards. Their stability depends on management quality. Forest reserves recommended for environmental protection, not fit for building
F_2	Relatively stable areas, with minimum to low level of studied geological hazards. Their stability depends on management quality. Stream corridors that should be prohibited for any human settlement, but that should be available in contingencies.
G_1	Unstable areas affected by moderately unstability to extremely unstability, need improvement to become F_1 category, recoverable but restricted for human settlement. They are recommended for forest and parks (environmental protection) that should be available in contingency.
G_2	Areas affected by high to extremely high level of studied geological hazards, which need works and removal of existing human settlement to become F_2 category.

processes. The result of this analysis, plus consideration of the response of each lithologic unit to land use, gave a category weighting that was reclassified for the whole area from 10 (lowest stability) to 1 (highest stability).

(2) Land use was mapped on the basis of seven categories; residential, commercial, public assembly sites (hospitals, schools, stadiums), government services (fire, police, administration), transportation networks, parks, and farmland.

(3) Following the method of Newman et al. (1978), the slope map was also superimposed on the surficial geology and geomorphic processes maps and systematically examined through a crosstab process to determine the slope intervals by percentage with maximum geomorphic process frequency. Those slope intervals having the greatest number of geomorphic processes such as landslides or rockfall were then labeled with the highest class number (10). Those slope intervals showing significantly fewer geomorphic processes were labeled with lower class numbers (1).

7. Land-use suitability indices (LUSI)

Land-use suitability indices represent the final product of the hazard assessment procedures described here. LUSI incorporate hazards, vulnerability, and risk concepts, and are determined from degree of hazard, planning restrictions, social conditions, and social needs combined through models (algorithms) that assign different values to types of land use. These algorithms define the relationship between landscape features and LUSI restrictions or demands.

There are 14 categories and subcategories of LUSI: A, B_1, B_2, B_3, C_1, C_2, C_3, D_1, D_2, E, F_1, F_2, G_1, and G_2. These LUS categories have been chosen based on needs and restrictions of land-use for urban planning considering different levels of social settlement, type of infrastructures, social services, environmental protection and risk mitigation, and are explained in Table 5. Categories A–C are or may become suitable for urban infrastructure, while categories D–G are reserved for environmental protection, contingency occasions, or avoided because of a high hazard level.

8. Conclusions

The procedure for geological hazard analysis presented here differs from previous work in that it incorporates the influences of multiple variables controlling the geomorphic processes that create hazards. These influences are analyzed using a series of algorithms that subjectively weight various factors. One of the strengths of this procedure is that the factor weightings and algorithms may be modified by each user through a graphical user interface, facilitating sensitivity analyses and providing the procedure with the flexibility to incorporate changes in climate or land use, for example, in the study area. The 14 land-use suitability classes that result from this procedure should be applicable to any region if the weighting factors that go into their delineation are changed appropriately. The use of GIS software in geological hazard analysis greatly facilitates the speed and ease of the calculations involved in these algorithms and, when combined with a menu-driven IPDS model, should make fairly sophisticated hazard analyses readily available to land managers and urban planners.

Acknowledgements

The Garfield County Planning Office, and particularly David Michaelson, provided much of the background data for this study. Special thanks are due to Ms. Jeanine Ford Artax, whose lovely hospitality made the entire field study much more enjoyable.

References

ATEC Associates, Inc., 1993. Final report for pretreatment facility and preliminary bearing capacities for associated structures, Glenwood Springs, Colorado. Unpubl. Tech. Rep., 33 pp.

Brabb, E.E., Pampeyan, E.H. and Bonilla, M.G., 1972. Landslide susceptibility in San Mateo County, California. U.S. Geol. Surv. Misc. Field Studies Map MF-360, scale 1:62,500.

Butler, D.R. and Walsh, S.J., 1990. Lithologic, structural, and topographic influences on snow-avalanche path location, Eastern Glacier National Park, Montana. Ann. Assoc. Am. Geogr., 80: 362–378.

Campbell, R.H., 1975. Soil slips, debris flows, and rainstorms in the Santa Monica Mountains and vicinity, southern California. U.S. Geol. Surv. Prof. Pap., 851: 51 pp.

Cardona, O.D., 1988. Estudios de Vulnerabilidad y Evaluación del Riesgo Sismico, Planificación Fisíca y Urbana en Areas Propensas. In: M. Mejía-Navarro (Editor), 2ª Conf. de Riesgos Geológicos del Valle de Aburrá, Agosto 2–6, 1988, 32 pp.

Carrara, A., 1983. Geomathematical assessment of regional landslide hazard. Fourth Int. Conf. on Applications of Statistics and Probability in Soil and Structural Engineering. Univ. di Firenze, Florence, pp. 3–27.

Carrara, P.E., 1979. The determination of snow avalanche frequency through tree-ring analysis and historical records at Ophir, Colorado. Geol. Soc. Am. Bull., 90: 773–780.

Chen and Associates, Inc., 1974. Preliminary engineering geology and subsoil investigation, proposed Van Rand Industrial Park State Highway 82, Garfield County Colorado. Prepared for Eldorado Engineering Company, 4 pp.

Chen and Northern, Inc., 1991. Supplemental subsoil study for foundation design addition to Wal-Mart Store, Glenwood Springs, Colorado. Prepared for Wal-Mart Stores, Inc., 8 pp.

Colorado Climate Center, 1984. Colorado average annual precipitation 1951–1980. Climatology Report 84-4, 1:500,000 scale map.

Ellen, S.D., 1988. Description and mechanics of soil slip/debris-flows in the storm, in the San Francisco Bay region, California. In: S.D. Ellen and G.F. Wieczorek (Editors), Landslides, Floods, and Marine Effects of the Storm of January 3–5, 1982, in the San Francisco Bay region, California. U.S. Geol. Surv. Prof. Pap., 1434: 63–112.

Ellen, S.D., Wieczorek, G.F., Brown, W.M. and Herd, D.G., 1988. Introduction. In: S.D. Ellen and G.F. Wieczorek (Editors), Landslides, Floods, and Marine Effects of the Storm of January 3–5, 1982, in the San Francisco Bay region, California. U.S. Geol. Surv. Prof. Pap., 1434: 1–5.

ESA Geotechnical Consultants, 1982. Final Report, Drainage and debris control plan for the City of Glenwood Springs, Colorado, vol. I. Report submitted to City of Glenwood Springs, 121 pp.

Federal Emergency Management Agency (FEMA), 1986. Floodway, Flood Boundary and Floodway Map, City of Glenwood Springs, Colorado, Garfield County. Panels 1043B, 1075, 1431B, 1432B, 1433B, 1434B, 1453B, 1465B, 1855B, 1856B, 1857B, 1858B, 1859B, 1880B. 14 maps.

Fox, F.M. and Associates, 1974. Roaring Fork and Crystal Valleys — an environmental and engineering geology study. Eagle, Garfield, Gunnison and Pitkin Counties, Colorado. Colorado Geological Survey, State of Colorado, 30 pp.

Gao, J., 1993. Identification of topographic settings conducive to landsliding from DEM in Nelson County, Virginia, U.S.A. Earth Surf. Process. Landforms, 18: 579–591.

Gingery Associates Inc., 1977. Floodplain information report, Roaring Fork River and Colorado River, Glenwood Springs, Colorado. Report prepared for the City of Glenwood Springs, Garfield County, and the Colorado Water Conservation Board, 27 pp.

The Glenwood Post. Daily newspaper of Glenwood Springs; archive at the Glenwood Springs Museum. (Also, The Free Weekly Newspaper.)

Harman, J.B. and Murray, D.J., 1985. Soil survey of Rifle area, Colorado, parts of Garfield and Mesa Counties, U.S.D.A. Soil Conservation Service, 149 pp.

Iverson, R.M. and Major, J.J., 1986. Groundwater seepage vectors and the potential for hillslope failure and debris-flow mobilization. Water Resour. Res., 22(11): 1543–1548.

Ives, J.D., Mears, A.I., Carrara, P.E. and Bovis, M.J., 1976. Natural hazards in mountain Colorado. Ann. Assoc. Am. Geogr., 66: 129–144.

Kirkham, R.M. and Rogers, W.P., 1981. Earthquake potential in Colorado. Colo. Geol. Surv. Bull., 43: 171 pp.

Kirkham, R.M. and Rogers, W.P., 1985. Colorado earthquake data and interpretations 1887 to 1985. Colo. Geol. Surv. Bull., 46: 105 pp.

Lambe, T.W. and Whitman, R.V., 1969. Soil Mechanics. Wiley, New York, 553 pp.

Lincoln DeVore Inc., 1991. Subsurface soils exploration proposed Wal-Mart addition Glenwood Springs, Colorado, 52 pp.

Lincoln DeVore Testing Laboratory, 1976. Bedrock geology (map sheet A) and surficial geology (map sheet B). Maps prepared under the provisions of House Bill 1041 for Garfield County, Colorado.

Mallory, W.W., 1971. The Eagle Valley evaporite formation northwest Colorado — a regional synthesis: U.S. Geol. Surv. Bull., 1311-E: 41 pp.

Mark, R., 1992, Map of debris-flow probability, San Mateo County, California, U.S. Geol. Surv. Map I-1257-M, 1:62,500 scale.

Mears, A.I., 1977. Debris-flow hazard analysis and mitigation, an example from Glenwood Springs, Colorado. Information Series 8, Colorado Geological Survey, 43 pp.

Mejía-Navarro, M., 1986. Geological risks: Analysis and suggestions about landslides in Antioquia. INGEOMINAS Report 2032, 34 pp.

Mejía-Navarro, M., 1992. Geological hazard and risk evaluation using GIS: Methodology and model applied to Medellín, Colombia. Unpublished MS thesis, Colorado State University, Ft. Collins, CO, 201 pp.

Mejía-Navarro, M., Oaks, S. and García, L.A., 1994. Model for integrated planning decision support (IPDS) in land-use incorporating geological hazard, vulnerability, and risk assessment. Conferencia interamericana sobre reduccíon de desastres naturales, Cartagena de Indias, Colombia, A-07, 22 pp.

Mora, S. and Vahrson, W.G., 1992. Determination a priori de la amenaza de deslizamientos utilizando indicadores morfodinamicos. In: J.B. Alzate (Editor), Memoria del Primer Simposio Internacional Sobre Sensores Remotos y Sistemas de Informacion Geografica (SIG) para el Estudio de Riesgos Naturales, Bogota, Colombia, pp. 258–273.

Morris, R.N., 1986. Geotechnical classification of debris-flow and debris floods in the Southern Rocky Mountains. Unpubl. M.S. Thesis, Colorado State University, 187 pp.

Morris, R.N. and Weaver, M.T., 1978. Geologic hazards of the glenwood springs metropolitan area, Garfield County, Colorado. Open-File Report 78-10, Colo. Geol. Survey, State of Colorado, 24 pp.

Murray, F.N., 1969. The stratigraphy and structural geology of the Grand Hogback Monocline, Colorado. Unpubl. PhD dissertation, University of Colorado, Boulder, 274 pp.

Newman, E.B., Paradis, A.R. and Brabb, E.E., 1978. Feasibility and cost of using a computer to prepare landslide susceptibility maps of the San Francisco Bay region, California. U.S. Geol. Surv. Bull., 1443: 27 pp.

Pack, R.T., 1985. Multivariate analysis of landslide-related variables in David County, Utah, in delineation of landslide, flash flood, and debris-flow hazard in Utah. Proc. Specialty Conf. Utah State University. General Series UWRL/G-85/03, 260–279, pp. 50–65.

Reid, M.E. and Iverson, R.M., 1992. Gravity-driven groundwater flow and slope failure potential 2. Effects of slope morphology, material properties, and hydraulic heterogeneity. Water Resour. Res., 28(3): 939–950.

Smith, M.I., 1986. Debris-flow costs and inventory of City of Glenwood Springs Area. The Mount Sopris Soil Conservation Board of Supervisors, 76 pp.

Smith, T.C., 1977. A method for mapping relative susceptibility to debris flows, with an example from San Mateo County. In: S.D. Ellen and G.F. Wieczorek (Editors), Landslides, Floods, and Marine Effects of the Storm of January 3–5, 1982, in the San Francisco Bay region, California. U.S. Geol. Surv. Prof. Pap., 1434: 185–194.

The Reinforced Earth Company, 1991. Reinforced earth wall design calculations for Wal-Mart, Glenwood Spring, Colorado, 46 pp.

US Census of Population, 1990. Tiger files, Topologically Integrated Geographic Encoding and Referencing System.

USGS National Earthquake Information Center (NEIC), 1993. Database of earthquakes in Colorado since 1880 until October 1993.

Walsh, S.J., Butler, D.R., Brown, D.G. and Bian, L., 1990. Cartographic modeling of snow avalanche path location within Glacier National Park, Montana. Photogramm. Eng. Remote Sensing, 56: 615–621.

Wieczorek, G.F., Harp, E.L., Mark, R.K. and Bhattacharyya, A.K., 1988. Debris-flows and other landslides in San Mateo, Santa Cruz, Contra Costa, Alameda, Napa Solano, Sonoma, Lake, and Yolo Counties, and factors influencing debris-flow distribution. In: S.D. Ellen and G.F. Wieczorek (Editors), Landslides, Floods, and Marine Effects of the Storm of January 3–5, 1982, in the San Francisco Bay region, California. U.S. Geol. Surv. Prof. Pap., 1434: 133–161.

Wieczorek, G.F., Lips, E.W. and Ellen, S.D., 1989. Debris-flows and hyperconcentrated floods along the Wasatch Front, Utah, 1983 and 1984. Bull. Assoc. Eng. Geol., XXVI: 191–208.

Woodward–Clevenger and Associates, Inc., 1972. Engineering geologic and ground water investigations for filings Nos. 3 and 4 of the GMCO Corporation property near Glenwood Springs, Colorado. Prepared for GMCO Corporation at Glenwood Springs, 12 pp.

Woodward–Clyde Consultants, 1983. Red Mountain Water Treatment Plant Glenwood Springs, Colorado. Prepared for James M. Montgomery Consulting Engineers, Inc.

The Binghamton Symposia in Geomorphology

List of Proceedings Volumes

Title	Editor(s)	ISBN
1. Environmental Geomorphology	D.R. Coates	Out of Print
2. Quantitative Geomorphology	M. Morisawa	Out of Print
3. Coastal Geomorphology	D.R. Coates	0-045-51038-5
4. Fluvial Geomorphology	M. Morisawa	0-045-51046-6
5. Glacial Geomorphology	D.R. Coates	0-045-51045-8
6. Theories of Landform Development	W.N. Melhorn and R.C. Flemal	0-686-10458-7
7. Geomorphology and Engineering	D.R. Coates	0-045-51040-7
8. Geomorphology in Arid Regions	D.O. Doehring	0-045-51041-5
9. Thresholds in Geomorphology	D.R. Coates and J.D. Vitek	0-045-51033-4
10. Adjustments of the Fluvial Systems	D.D. Rhodes and E.J. Williams	0-840-32108-2
11. Applied Geomorphology	R.G. Craig and J.L. Craft	0-045-51050-4
12. Space and Time in Geomorphology	C.E. Thorn	0-045-51056-3
13. Groundwater as a Geomorphic Agent	R.G. LaFleur	0-045-51069-5
14. Models in Geomorphology	M.J. Woldenberg	0-045-51075-X
15. Tectonic Geomorphology	M. Morisawa and J.T. Hack	0-045-51098-9
16. Hillslope Processes	A.D. Abrahams	0-045-51102-0
17. Aeolian Geomorphology	W.G. Nickling	0-045-51133-0
18. Catastrophic Flooding	L. Mayer and D. Nash	0-045-51142-X
19. History of Geomorphology	K.J. Tinker	0-045-51138-1
20. Appalachian Geomorphology	T.W. Gardner and W.D. Sevon	0-444-88326-6
21. Soils and Landscape Evolution	P.L.K. Knuepfer and L.D. McFadden	0-444-88692-3
22. Periglacial Geomorphology	J. Dixon and A. Abrahams	0-471-93342-2
23. Geomorphic Systems	J.D. Phillips and W.H. Renwick	0-444-89809-3
24. Geomorphology: The Research Frontier and Beyond	J.D. Vitek and J.R. Giardino	0-444-89971-5

SSDI 0169-555X(94)9999-Z